Studies on Lake Vechten and Tjeukemeer, The Netherlands

Developments in Hydrobiology 11

Series editor
H. J. Dumont

DR W. JUNK PUBLISHERS THE HAGUE-BOSTON-LONDON 1982

Studies on Lake Vechten and Tjeukemeer, The Netherlands

25th anniversary of the Limnological Institute of the Royal Netherlands Academy of Arts and Sciences

Edited by
R. D. Gulati and S. Parma

Reprinted from *Hydrobiologia*, vol. 95 (1982)

DR W. JUNK PUBLISHERS THE HAGUE-BOSTON-LONDON 1982

Distributors:

for the United States and Canada

Kluwer Boston, Inc.
190 Old Derby Street
Hingham, MA 02043
U.S.A.

for all other countries

Kluwer Academic Publishers Group
Distribution Center
P.O. Box 322
3300 AH Dordrecht
The Netherlands

Library of Congress Cataloging in Publication Data
Main entry under title:

Studies on Lake Vechten and Tjeukemeer, The Netherlands

(Developments in hydrobiology ; 11)
"Reprinted from Hydrobiologia, vol. 95 (1982)"
1. Limnology--Netherlands--Vechten, Lake--Congresses. 2. Limnology--Netherlands--Tjeukemeer --Congresses. 3. Vechten, Lake (Netherlands)--Congresses. 4. Tjeukemeer (Netherlands)--Congresses. 5. Limnologisch Instituut (Koninklijke Nederlandse Akademie van Wetenschappen) I. Gulati, R. D. II. Parma, S. III. Limnologisch Instituut (Koninklijke Nederlandse Akademie van Wetenschappen)

IV. Hydrobiologia. V. Series.
QH159.S78 1982 574.5′26322′094921 82-7791
AACR2

ISBN 90 6193 762 0 (this volume)
ISBN 90 6193 751 5 (series)

Cover design: Max Velthuijs

PRINTED IN THE NETHERLANDS

Contents

Preface VII

The twenty-fifth anniversary of the Limnological Institute, The Netherlands (1957–1982),
by S. Parma 1

PART ONE: LAKE VECHTEN

Lake Vechten: aspects of its morphometry, climate, hydrology and physico-chemical characteristics,
by C. L. M. Steenbergen & H. Verdouw 11

The phytoplankton species composition and the seasonal periodicity in Lake Vechten from 1956 to 1979,
by Marianne C. I. Blaauboer 25

The primary production of phytoplankton in Lake Vechten,
by W. A. de Kloet 37

Contribution of photosynthetic sulphur bacteria to primary production in Lake Vechten,
by C. L. M. Steenbergen 59

The aquatic macrophytes of Lake Vechten. Species composition, spatial distribution and production,
by Elly P. H. Best 65

Structural and functional characteristics of epiphyton and epipelon in relation to their distribution in Lake Vechten,
by Herman J. Gons 79

Macro-invertebrate communities associated with the macrophytes of Lake Vechten: structural and functional relationships,
by Jan Dvořák & Elly P. H. Best 115

The zooplankton: its community structure, food and feeding, and role in the ecosystem of Lake Vechten,
by R. D. Gulati, K. Siewertsen & G. Postema 127

Sedimentation and breakdown kinetics of organic matter in the anaerobic zone of Lake Vechten,
by Th. E. Cappenberg & H. Verdouw 165

Aspects of aerobic mineralization during spring in Lake Vechten with special reference to the ^{14}C-labelling technique,
by J. J. Olie & Th. E. Cappenberg 181

Nitrogen cycle of Lake Vechten: concentration patterns and internal mass-balance,
by H. Verdouw & E. M. J. Dekkers 191

PART TWO: TJEUKEMEER

Hydrology of Tjeukemeer,
by J. D. Leenen 199

Physico-chemical environment in Tjeukemeer with special reference to speciation of algal nutrients,
by H. de Haan 205

The algal periodicity in Tjeukemeer during 1968–1978,
by J. R. Moed & H. L. Hoogveld 223

Population dynamics and production of *Daphnia hyalina* Leydig and *Daphnia cucullata* Sars in Tjeukemeer,
by J. Vijverberg & A. F. Richter 235

Population dynamics and production of *Acanthocyclops robustus* (Sars) and *Mesocyclops leuckarti* (Claus) in Tjeukemeer,
by J. Vijverberg & A. F. Richter 261

INSTAR, a discrete event model for simulating zooplankton population dynamics,
by P. Hogeweg & A. F. Richter 275

Distribution and production of the larval chironomid populations in Tjeukemeer,
by D. M. Beattie 287

A note on the significance of larger bivalve molluscs (*Anodonta* spp. and *Dreissena* sp.) in the food of the eel (*Anguilla anguilla*) in Tjeukemeer,
by H. W. de Nie 307

Growth, condition and gonad development of bream (*Abramis brama* L.) in relation to its feeding conditions in Tjeukemeer,
by E. H. R. R. Lammens 311

The relations between 0+ fish density, zooplankton size and the vulnerability of pikeperch, *Stizostedion lucioperca,* to angling in the Frisian lakes,
by W. L. T. van Densen & J. Vijverberg 321

PART THREE: MISCELLANEOUS

Effects of thermal effluents from the Bergum Power Station on the zooplankton in the Bergumermeer,
by H. W. de Nie 337

Effects of entrapment and cooling water discharge by the Bergum Power Station on 0+ fish in the Bergumermeer,
by W. L. T. van Densen & R. H. Hadderingh 351

Temporal and spatial variations of dissolved oxygen concentrations in some Dutch polder ditches,
by R. Veeningen 369

Preface

The Netherlands, like the other industrialized countries in the northern hemisphere, is faced with increasing environmental problems. Being a country rich in water – at least in terms of surface area – and low-lying, problems related to water are as old as the nation itself. The problems in regard to water quality are becoming more acute and complicated due to the progressive rise in air and ground pollution. Glaring examples are the acid rains and contamination of ground-water resources by toxic, and often illegal, waste dumps.

Being a delta country, the Netherlands is also very vulnerable to pollution from the water sources outside the country's national boundaries, i.e. the rivers Rhine, Meuse and Scheldt.

Not only the national but also the international authorities are active in developing measures to both prevent and combat water pollution, and to restore as far as possible the natural balance or a 'reasonable' water quality. The pollution abatement programmes are based on ecological theories; accepting waterbodies as aquatic ecosystems, the authorities concerned try to influence and lead the progresses in the systems to reach or maintain the desired ecosystem structure. Suter II (1981) mentioned a number of reasons why the attempts to use ecosystem theories to assess the impacts on the environment have been disappointing to date. It is clear that one of the reasons for the failure is lack of proper knowledge.

The main mission of the Limnological Institute can be seen as work on fundamental problems, the solution of which can facilitate political decision making. Despite the economic recession, we as Dutch limnologists still have unique opportunities to do multidisciplinary, long-term, fundamental research on the structure and functioning of our aquatic ecosystems. That our scientific programmes should be subservient to the needs of society – though ambitious and challenging – is a realistic thought.

An important yardstick to measure the progress of work and the right of a scientific institute to exist is the quality and quantity of its research output. The slogan 'work, finish and publish' does not need christening. Thus the most festive way of celebrating our Silver Jubilee was to take stock of the situation, to pool our scientific efforts and compile the more recent works in the form of the present volume.

The main objects of study in the past decade have been:

a. Lake Vechten, a mesotrophic stratified sandpit in the middle of the Netherlands; and
b. Lake Tjeukemeer, a shallow, eutrophic lake in one of the northern provinces.

Most of the contributions deal with these two lakes, and reflect the research themes of the four working groups. In addition, recent work on Lake Bergum and on some Dutch polder ditches is included in a miscellaneous section.

Thanks are due to Dr W. Junk Publishers, The Hague, for providing us the opportunity to use an entire volume of Hydrobiologia, both as a scientific memoire and as a memento. Junk is international, of course. Its roots are, however, in the Netherlands and we are glad to use a 'Dutch' journal. The Institute is most grateful to Dr Henri Dumont, the editor-in-chief, and to Wil R. Peters and Cliff Morgan of the Publishers for the most pleasant cooperation. I am indebted to Miss Joke van Kleinwee and Miss Marion McLaughlin for the typing work and to Mr. E. M. Mariën, who took, the photographs. The Board of the Royal Netherlands

Academy of Arts and Sciences gave us the necessary financial support for this publication. The most important job was done by Dr Ramesh D. Gulati who spent several months correcting and improving the various typescripts. This loyal cooperation deserves the gratitude of the Institute.

Reference

Suter II, G. W., 1981. Ecosystem theory and Nepa assessment. Bull. Ecol. Soc. Am. 62: 186–192.

Nieuwersluis,
10 August 1982

Dr S. Parma,
Director

The twenty-fifth anniversary of the Limnological Institute, The Netherlands (1957–1982)

S. Parma
Limnological Institute, 'Vijverhof' Laboratory, Rijksstraatweg 6, 3631 AC Nieuwersluis, The Netherlands

Keywords: history, Limnological Institute, Netherlands, jubilee

Abstract

This paper marks the silver jubilee in 1982 of the Limnological Institute, The Netherlands. It describes the history, present research programme and organisation of the Institute, as well as its cooperative studies with other institutes and its training facilities.

Some remarks about its future plans of work are made.

History before 1957

As early as 1912 the 'Netherlands Society for the Prevention of Water, Ground and Air Pollution' tried to establish a Dutch Hydrobiological Station. Fischer (1912) stressed the need for a hydrobiological centre to study the biological aspects of water quality control, citing the existence of similar institutes abroad, such as those in Besse (France), Aneboda (Sweden), Plön (Germany), Lakeside Laboratory (Iowa, U.S.A.), etc. However, the society failed in getting the necessary funds.

After World War I the attempts were restarted. Romijn (1920) tried to give a new stimulus to the Netherlands Society, but due to the bad economic situation, without any success.

It was not until the early fifties that the 'Hydrobiological Society' formed a special 'Committee for the Foundation of a Hydrobiological Station'. In its opinion such a station must be intended to investigate fundamental and applied hydrobiological problems, and also play an active role in the training of university students. But again the government was not willing to give financial support. Nevertheless, the Hydrobiological Society continued its efforts and succeeded in mustering a few research projects, financed by different authorities. They were housed in a few rooms of the University of Utrecht. This situation lasted from 1954 until 1957 (Havinga 1957).

Not only the Hydrobiological Society, but also the Royal Netherlands Academy of Arts and Sciences, urged the need for a hydrobiological research centre. In 1952, a Committee on Ecology reported about the situation on ecological research in our country. It was stressed that in a country like the Netherlands, with its rich water resources, a biological institute studying fundamental fresh-water problems was urgently needed. Such an institute could function as a natural complement to the Netherlands Institute for Sea Research and a few other institutes dealing with applied fresh water research (Rapport 1952). Notwithstanding all this, the Committee on Ecology did not recommend the establishment of an institute. It feared a heavy strain on the governmental budgets and an undesirable competition with other claims for support of ecological research topics. The Committee did, however, ask for funds to support preliminary investigations, focussed on the significance of organisms in the cycle of elements in a freshwater ecosystem.

Due to the perseverance of the Hydrobiological Society, the Committee on Ecology of the Royal Academy installed in 1955 a subcommittee on hy-

Hydrobiologia 95, 1–9 (1982). 0018-8158/82/0951-0001/$01.80.

drobiology. Of the 5 members, Dr. Mrs. N. L. Wibaut-Isebree Moens and Dr. B. Havinga were deputed by the Hydrobiological Society; Prof. Dr. G. P. Baerends (chairman), Prof. Dr. A. W. H. van Herk and Dr. J. Verwey were members of the Committee on Ecology; and Dr. H. N. Kluyver was secretary. This subcommittee advised in 1956 the foundation of a Hydrobiological Institute in the lake area in central Netherlands. It was proposed that this institute study fundamental hydrobiological problems in their broadest sense, and provide facilities for research training of university students.

Simultaneously with these discussions in the subcommittee, the Committee on Ecology consulted several authorities about the desirability of founding a so-called Delta research laboratory in the estuarine SW part of the Netherlands. In this region the so-called Deltaplan (that is, the damming of several estuaries for safety reasons) was in progress. A study of the biological consequences of this 'macro' field experiment was strongly advocated. The recommendations of both the Committee on Ecology and the Subcommittee on Hydrobiology were combined, and the Royal Academy asked the Ministry of Education, Arts and Sciences for a budget to realise both proposals (Verwey 1977). In 1957 the Ministry put aside *f* 54 000 for the foundation of a Hydrobiological Institute with a Department for Delta Research. The Hydrobiological Institute was stationed at Nieuwersluis, near the Loosdrecht Lakes, and the Department for Delta Research at Yerseke in the middle of the Rhine-Meuse-Scheldt delta.

The 1957–1982 period

Introduction

From their inception the Hydrobiological Institute and the Department for Delta research functioned as independent units with their own directors, management committees and budgets. The first directors were Dr. Miss M. F. E. Nicolai and Dr. K. F. Vaas, respectively. In 1968 this actual situation was confirmed by a change of names of both departments. From this time, the Royal Netherlands Academy of Arts and Sciences has been supervising both the Limnological Institute and the Delta Institute for Hydrobiological Research. However, the Board of the Royal Academy has to re-evaluate every five years the relations between the two institutes.

In the next few paragraphs I will deal only with the Limnological Institute.

In the Subcommittee on Hydrobiology, two different approaches to the objectives of the Hydrobiological Institute already existed. The representatives of the Hydrobiological Society stressed the need to tackle applied limnological problems, particularly those concerning the management of aquatic ecosystems. On the other hand, the Royal Academy laid emphasis on 'free' and fundamental research. The current rules of the Institute show a synthesis of the two view points. The official objectives of the Institute are:

1) to undertake, perform and stimulate biological, chemical and hydrographical research of inland waters, all in the broadest sense,
2) to collaborate with other organisations in similar fields of interest,
3) to contribute to the training of limnologists and limnological students.

The Vijverhof Laboratory

The institute was accommodated in 1957 in an old country-house called 'Vijverhof' in the small village of Nieuwersluis, approximately 20 km north of Utrecht (Fig. 1). It selected Lake Vechten as object of research. This lake (Fig. 2) was created in 1941 by excavating superficial sand layers, needed for the construction of a highway. The lake, a sandpit (surface area 4.7 ha; max. depth, 11.9 m; mean depth, 6 m), is thermally stratified from about May to early November. It has a well developed littoral region. During the stagnation period the anaerobic zone extends maximally to the 6 m isobath. The phytoplankton composition points to a slightly eutrophic environment. The staff in the first years consisted of only a few members, so the research in Lake Vechten had mainly an inventory character. Gradually the staff increased and the initial stage was followed in the late sixties by a phase in which several processes in the lake were described more thoroughly. It was focussed at the species and population level and dealt with more or less isolated aspects of, for example, the energy flow, mineralisation and nutrient recycling, and population regu-

Fig. 1. The old country-house 'Vijverhof', now the main building of the Limnological Institute at Nieuwersluis.

lation. Around 25 papers have been published on studies in Lake Vechten.

Since about 1976, the studies in Lake Vechten have been concentrated in two work groups. The 'Primary and Secondary Production' work group is, as the name suggests, interested in the production and energy flow of the ecosystem. Its programme contains studies on primary production of phytoplankton, periphyton and macrophytes, and grazing of zooplankton and predation of *Chaoborus* larvae. The 'Mineralisation of Organic Matter' work group especially investigates the aerobic and anaerobic mineralisation of matter to its inorganic states, viz. carbon, nitrogen and phosphorus compounds. Also, ecological interrelations between micro-organisms observed in the field are being tested under defined laboratory conditions using chemostats. Their kinetic relationships using labelled compounds are being studied.

In the last decade there was a strong desire for both a coordinated and an integrated ecosystem approach to the limnological problems in Lake Vechten. This forms the basis of many of the studies in progress. In this regard, a three-pronged attack constitutes a simultaneous study of the processes controlling the production of organic matter by the autotrophs, its consumption by the herbivores, and its mineralisation by bacteria. The main research theme now is the study, on a daily basis, of carbon, nitrogen and phosphorus budgets.

Since 1968, a small part of the work of the Vijverhof Laboratory has been situated in the so-called 'Vechtplassen' (Fig. 2). Recently, both workgroups started a more intensified study in this area. The whole lake system was formed as a result of excavation for peat, followed by sand in some cases. The lakes range, depending on location, from nearly unpolluted, slightly eutrophic ones, to highly eutrophic and grossly contaminated ones. The latter ones have become polluted by the inlet of waste water and water from the River Vecht.

In the near future a gradual restoration of the

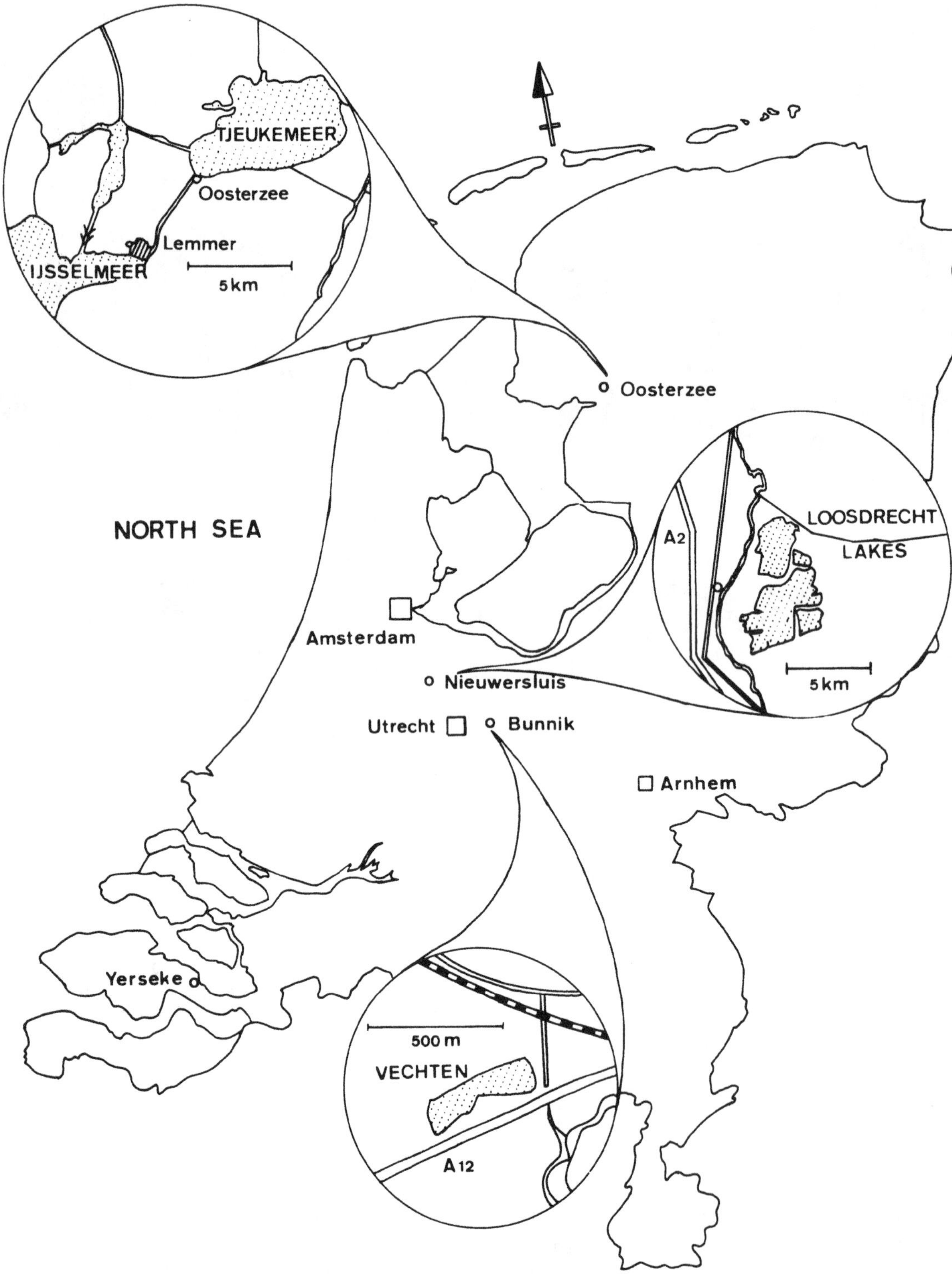

Fig. 2. Map of the Netherlands showing the main study areas of the Limnological Institute.

'original' biotic communities in a major part of the lake area (water surface, 1830 ha; depth 1–2 m) is to be expected. In this regard, the waste water treatment has recently been intensified, and from 1982 onwards supplementation will occur with dephosphatised water from the neighbouring less-polluted Amsterdam-Rhine canal. Because the decreased phosphorus input will in the course of time result in lower P-concentrations in the sediments and overlying waters, it is expected that this will be reflected in the species composition and production of algae and macrophytes and improvement in light climate. Coordinated studies have been started on the biotic and abiotic changes in this area accompanying the decrease in phosphorus loading. The research programme encompasses both fundamental and applied aspects.

Tjeukemeer Laboratory

From 1966 onwards the Institute participated in the International Biological Programme. It started an investigation of the productivity of Tjeukemeer. This shallow, freshwater lake (area, about 21 km^2; mean depth, about 1.5 m) is a part of the Frisian reservoir system and has a hydrology typical of this lake area (Fig. 2). In winter the lake receives nutrient and humus-rich water from the surrounding polders. In summer, however, water from the IJsselmeer is allowed to enter the lake for agricultural purposes. The humic-acid concentration of IJsselmeer water is much lower than that of the polder water but that of chloride much higher. The chloride concentrations increase by up to 200 $mg \cdot l^{-1}$, giving the lake an oligohaline character.

This IBP project, carried out in cooperation with the University of Liverpool, was concluded in 1971. However, the investigations in Tjeukemeer were continued by the Institute. Steadily, the number of researchers increased and the original field station grew into a well-equipped laboratory (Fig. 3). Now, two workgroups are studying several processes in Tjeukemeer and, for purposes of comparison, the other Frisian lakes as well.

The research of the 'Algology' work group is focussed on explaining the periodicity of abundant algal species in Tjeukemeer, monitoring the magnitude and periodicity of these algal populations, and obtaining information about the factors regulating their waxing and waning. These factors are also studied in laboratory experiments. Much attention is given to the speciation of different nutrients, and their availability to algae is tested in chemostats. Bio-assays are being performed to detect the factors co-regulating algal periodicity.

The 'Foodchains and Production Studies' work group is engaged in a study of the interrelations between the fish populations and their food organisms. Population structure and density, fecundity, mortality, growth, standing crop biomass and production are estimated for both the fish species and the most important fish-food species. The latter include zooplankton (copepods and cladocerans) and several macrofaunal benthic elements. Much work is directed at the ecology of fish-feeding in nature as well as under experimental conditions in aquaria. The results of these studies will be used in a simulation model in order to describe and to get more insight into dynamics of this subsystem.

Organisation of the Institute

The research work of the Institute is supervised by a Management Committee nominated by the Board of the Royal Academy, which, via the Ministry of Education and Sciences, provides the finances. An Annual Progress Report is published in the 'Verhandelingen der Koninklijke Nederlandse Akademie van Wetenschappen, afd. Natuurkunde, Tweede Reeks'.

Initially, the Institute started with only a few academic staff members and technicians. Gradually, their number increased and at present, of the 42 permanent positions, academic staff occupy 14. Due to part-time positions, contract jobs and students, the total working strength of the Institute is around 60 people, about 65% at Nieuwersluis, and the rest at Oosterzee.

Dr. Miss M. F. E. Nicolai, the first director, died suddenly in 1960, and was succeeded by Dr. H. L. Golterman. In 1972 the Institute obtained a second director in Dr. R. Soekarjo. Due to reorganisation in 1978, both directors received other assignments within the organisation of the Academy, and were succeeded by Dr. S. Parma. In the earlier years the research policy was mainly determined by the director. At present it is the outcome of discussions between director and staff, and finally authorised by the Management Committee.

As mentioned earlier, the Institute has two de-

Fig. 3. The Tjeukemeer Laboratory built in 1976 at the edge of the lake.

partments. The main building in Nieuwersluis (Fig. 1) has recently been reconstructed and restored. A second building adjacent to the main one was constructed in 1970 (Fig. 4). The Tjeukemeer Laboratory was built in 1976.

Both departments have good facilities for chemical, microbiological and experimental biological research. The Tjeukemeer Laboratory has a specialised room for fish experiments. A platform in Tjeukemeer, just in front of the laboratory, makes it possible to perform *in situ* bioassays and continuous registration of the physico-chemical environment of the lake. The Vijverhof Laboratory has environmental chambers and an isotope laboratory for experiments with ^{14}C, ^{3}H, ^{32}P and ^{35}S. In the middle of Lake Vechten is a raft for continuous sampling. The Institute has a well-equipped workshop for construction and modification of apparatusses and sampling devices. There is a terminal connected with the large computer centre of the University of Groningen. The library has all the well-known limnological journals and books. Moreover, in both departments there are research vessels and boats with outboard motors for biological and physico-chemical sampling and for net fishing. Finally, both departments have facilities for guest workers who can participate in the scientific programme of the Institute.

Collaboration

The scientific programme of the Limnological Institute is aimed at understanding the structure and functioning of freshwater ecosystems. This programme is congruent with that of the national 'Work Committee on Aquatic Ecology', a unit of

Fig. 4. The new building built in 1970 at Nieuwersluis.

the Foundation for Biological Research in the Netherlands. The Work Committee coordinates and stimulates fundamental aquatic ecosystem research in the Netherlands through interaction among the scientists. Several marine, brackish-water and freshwater institutes cooperate in this committee. Some of them, such as the Limnological Institute, have fundamental freshwater ecosystem research as their main goal.

The Department of Aquatic Ecology of the University of Amsterdam investigates the structure and function of the stratified Lake Maarsseveen. Moreover, this group is engaged in prominent research of microecosystems on a laboratory scale. The Laboratory for Aquatic Ecology of the Catholic University at Nijmegen studies aquatic ecosystems dominated by macrophytes. The Hydrobiology Division of the State Research Institute for Nature Management carries out research of several water-types (lakes, fens, ditches, etc.). It uses micro-ecosystems as a test object for environmental pollutants. The research at the Nature Conservation Department of the Agricultural University at Wageningen is mainly focussed on a study of the structure of running water. Lastly, some applied laboratories and institutes devote only a part of their research potential to aquatic ecological problems, or are interested only in special problems of the aquatic environment.

The Limnological Institute collaborates with a number of these institutes in the Netherlands. In this regard, some of these cooperative activities need to be specially mentioned. These are the studies on macrophytes and periphyton in cooperation with the Department of Aquatic Ecology (Amsterdam) in Lake Maarsseveen; the collaboration with the Laboratory of Microbiology, the University of Amsterdam, and the Research Institute for Nature Management in the Lake Loosdrecht project; and the collaboration of the 'Food Chains and Production Studies' work group with the 'General Fish Culture and Fisheries' section of the Agricultural University.

Also international contacts are numerous. In

1966, 1972 and 1976 Dr. H. L. Golterman organised international symposia entitled 'Chemical environment in the aquatic habitat', 'Humic substances' and 'Interactions between sediments and fresh water'. Recently (1981) the institute organised a workshop of the 'Plankton Ecology Group' and, in cooperation with the Hydrobiological Society, a meeting on 'Decomposition of aquatic macrophytes'.

As already mentioned, the Institute played an active role in the International Biological Programme. Worth mentioning is the participation in the Parakrama Samudra project of the University of Vienna in Sri Lanka.

Training

In the early sixties the Dutch universities did not have training facilities for limnological students. The Limnological Institute tried to fill this gap and organised training courses for undergraduate students from different universities. However, in the last decade, the universities of Amsterdam and Nijmegen have appointed professors in aquatic ecology, while other universities have also created training facilities for limnology students.

The Institute does, however, have a provision for postgraduate students to do a part of their Master's degree. The number of students working for 3–9 months at the Institute has steadily increased to about 25 per year. Thus, the students contribute reasonably to the scientific output of the Institute, and these contributions are included in the Annual Progress Reports.

A few places are reserved for Dutch scientists working for a Ph.D. degree.

The future of the Institute

The present strength of the staff, the reasonable equipment and the comparatively small teaching obligations together work favourably for long-term multidisciplinary research. Five-year programmes of the workgroups are intended to exploit these possibilities.

Our main objective will be to continue to do fundamental research of freshwater aquatic ecosystems. Up till now the main research objects of the two departments have been the shallow Tjeukemeer at the Tjeukemeer Laboratory and the stratified Lake Vechten at the Vijverhof Laboratory. In the near future the Vijverhof Laboratory will extend its scientific activity to shallow waters, the most common watertype in the Netherlands.

In the initial stage the Institute was interested mainly in pure research. However, applied problems were also studied, mostly on a contract basis for other Ministries. Examples are an intensive study on the effects of cooling water discharge of the electrical power plant at Lake Bergum and a study of the oxygen budget in ditches, both financed by the Ministry of Public Health and the Environment. The Netherlands is, however, increasingly confronted with problems concerning water quality and management of aquatic systems. The Ministries concerned need more fundamental information. In this respect, research at the ecosystem level is of great importance. Moreover, the national government is of the opinion that scientific research must be commensurate with the needs of society. This means that in future applied aspects shall take an important place in the choice of our fundamental research projects.

Summary

The joint efforts of the Royal Netherlands Academy of Arts and Sciences and the Hydrobiological Society resulted in the foundation in 1957 of the Limnological Institute.

The Institute is supervised by the Royal Academy, and has two departments. The Vijverhof Laboratory in Nieuwersluis has two work groups, studying production and mineralisation processes in the stratified Lake Vechten. The research programme in the near future will be extended to the shallow Loosdrecht Lakes, especially to investigate the effects of a decrease in phosphorus loading. The Tjeukemeer Laboratory in Oosterzee also has two work groups, studying algal periodicity and the interrelation between the fish populations and their food organisms.

The total permanent staff of both departments has a strength of 42. The Institute offers research and lodging facilities for guest workers and students.

The place of the Institute in the field of Dutch freshwater ecosystem research has been described.

Collaboration with other research establishments in the Netherlands has also been indicated.

Also, in the future, fundamental research will be the main objective. However, applied aspects, focussed on research into water quality and water management, will play an important role in the choice of fundamental research projects.

References

Fischer, J. C. H., 1912. Over het nut van hydrobiologisch onderzoek en hydrobiologische stations. Water, Bodem, Lucht 2: 14–18.

Havinga, B., 1957. De wordingsgeschiedenis van het Hydrobiologisch Station. Mimeographed report, dated 'Heemstede, Augustus 1957, 3 p. Archives of the Hydrobiologische Vereniging'.

Rapport van de Commissie voor Oecologie over het oecologisch onderzoek en onderwijs in Nederland, 1952. Koninkl. Nederl. Akad. Wetensch. Verslag 61: 105–116.

Romijn, G., 1920. Een internationaal hydrobiologisch instituut in Nederland. Vragen van den dag, febr. 1920.

Verwey, J., 1977. Iets uit de voorgeschiedenis van het Delta-Instituut voor Hydrobiologisch Onderzoek. Vakbl. Biol. 57: 210–215.

Lake Vechten: aspects of its morphometry, climate, hydrology and physico-chemical characteristics

C. L. M. Steenbergen & H. Verdouw
Limnological Institute, 'Vijverhof' Laboratory, Rijksstraatweg 6, 3631 AC Nieuwersluis, The Netherlands

Keywords: morphometry, climate, waterbalance, thermal regime, light penetration, ionic balance, Lake Vechten

Abstract

The physiography of Lake Vechten (The Netherlands) is described together with morphometric data. The lake (surface area 4.7 ha; mean depth 6.0 m) consists of two basins with maximum depths of 10.5 and 11.9 m. Meteorological conditions in the region and horizontal groundwater flow play an important role in the renewal time, which is about two years. The lake has in most years a circulation period from November till April but in some years, when the ice cover is prolonged, it is stratified in winter as well. The summer stratification extending from May to the end of October is very stable, with an anaerobic hypolimnion. Eddy conductivity in the stagnant water is calculated. The water transparency is strongly influenced by the presence of algal and bacterial populations and by resuspended particulate matter. Secchi disk depth ranges from 1.8 to 4.5 m. The 1% of the surface light reaches from 4 to 5 m depth in November and from 8 to 9.5 m depth in May. The ionic composition and nutrient status of the lake are given.

Introduction

Lake Vechten lies in the central part of the Netherlands in the municipality of Bunnik at 52°04′N and 5°05′E (Fig. 1). A first general description of the lake was given by Parma (1971). Lake Vechten was formed in 1941 by excavation of sand necessary for building a nearby highway. It has an area of 4.7 ha, a maximum depth of 11.9 m, and is stratified in summer, exhibiting hypolimnetic oxygen depletion. Since the foundation of the Limnological Institute in 1957, it has been the major research site of the Vijverhof Laboratory. For a summary of the development and progress of limnological studies on Lake Vechten see Best *et al.* (1978).

The majority of Dutch lakes are shallow (2–3 m depth) and were formed gradually by a combination of excavation of surface peat, used for fuel, and of wind and wave action (e.g. Redeke 1941; Beattie *et al.* 1978). But the sand digging has continued also in the shallow lakes, and nowadays artifically dug-out lakes are common in the Netherlands (see e.g. Leentvaar 1971). Some of them have been the object of limnological studies, such as Lake Wijde Blik (Gulati 1972) and the Maarsseveen lakes (Hallegraeff 1977; Ringelberg 1981). Generally these types of lake are large (20–265 ha) compared to Lake Vechten. Their depths range from 25 to 35 m and they are stratified in summer.

In the Netherlands, semi-natural small lakes (0.9–13.3 ha), which are relatively deep (6–15 m) and stratified in summer, are situated behind the dikes of the great rivers. These lakes, called 'Wielen', were created by breaches in dikes caused by high floods. New dikes were constructed around the pools isolating them from the river (Van Heusden 1945; Leentvaar 1958). Lake Vechten greatly resembles the 'Wielen' type in its morphometry, its stratification phenomena and in its isolated position, which protects the lake from external polluting substances.

In this paper, morphometric, climatic and hydro-

Hydrobiologia 95, 11–23 (1982). 0018-8158/82/0951-0011/$02.60.

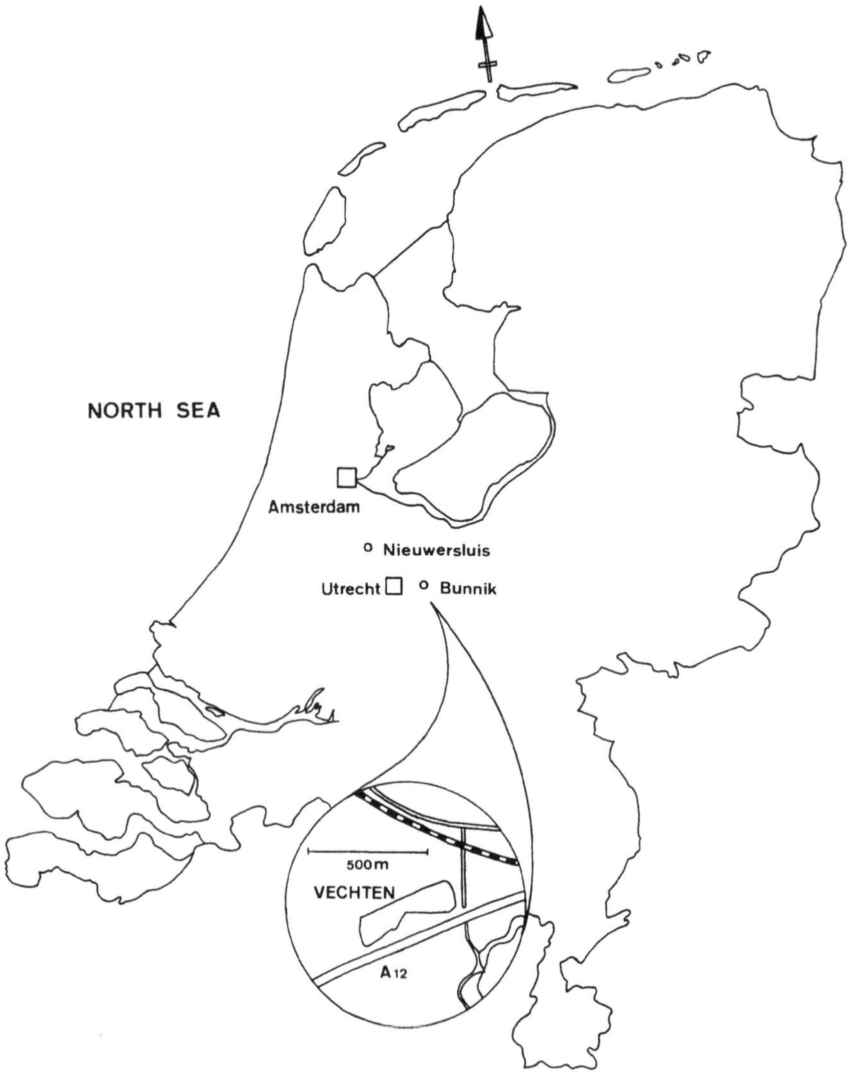

Fig. 1. Map of the Netherlands showing location of Lake Vechten.

logical data on Lake Vechten are presented and its thermal cycle, underwater light climate and chemistry described.

Morphometry

The morphometry of Lake Vechten was described by Parma (1971, 1980). Morphometric data (Table 1), the bathymetric map (Fig. 2) and hypsographic curves and depth–volume relations (Fig. 3) were all obtained from his studies.

The lake is divided into two basins by a shallow area (Fig. 2). The existence of two basins is of interest during the summer stratification, as the upper anaerobic strata of the two basins are separated by the ridge. Therefore the two basins can be treated individually for a study of relations between epilimnetic and hypolimnetic processes. The relative depths (z_r, see Hutchinson 1957) of the two basins are large, with values exceeding 4% (Table 1). Thus, the basins belong to the category of deep lakes with a relatively small surface area (Wetzel 1975). This also explains the great stability of summer stratification in the lake. Shore development (D_L) quantifies roughly the potential for development of littor-

Table 1. Morphometric data of Lake Vechten at a water level of 1 m above N.A.P. (Netherlands ordnance datum) (symbols from Hutchinson 1957; Wetzel 1975).

	East basin	West basin	Whole lake
Surface A ($\times 10^3$ m^2)	23.25	23.90	47.15
Maximum depth Z_m (m)	10.5	11.9	11.9
Mean depth $\bar{z}$ (m)	5.6	6.5	6.0
Depth of cryptodepression z_c (m)			10.9
Relative depth z_r	6.1%	6.8%	4.9%
$\bar{z} : Z_m$	0.53	0.55	0.50
Volume V ($\times 10^4$ m^3)	12.890	15.390	28.280
Length l (m)	225	245	470
Maximum breadth b (m)	125	130	130
Mean breadth $\bar{b}$ (m)	103	98	100
Shore line L (m)			1 130
Shore line development D_L			1.47

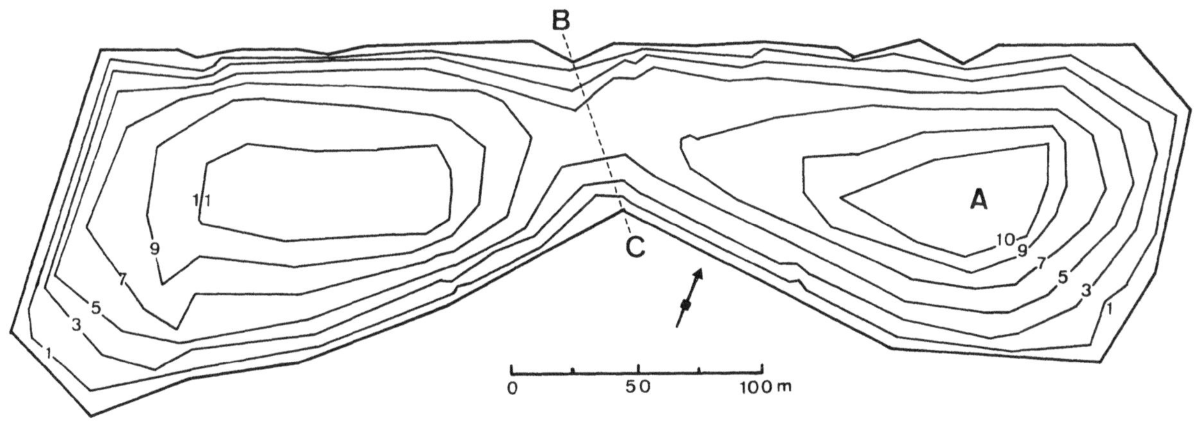

Fig. 2. Bathymetric map of Lake Vechten (contours in meters). A marks the position of a raft, which is the major sampling station. Broken line BC in the narrowest part of the lake represents the boundary between eastern and western basin.

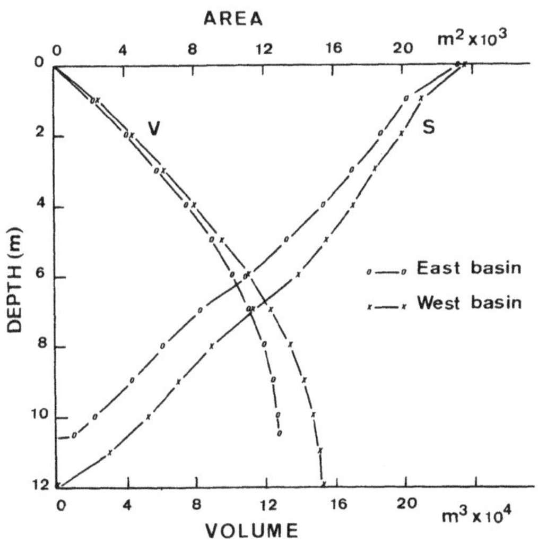

Fig. 3. Hypsographic curves (s) and depth–volume curves (v) of the two basins of Lake Vechten (data from Parma 1980).

al communities, which are to a great extent isolated from the main waterbody. The low value of D_L, namely 1.47 (Table 1), reflects the regular shore line of the lake, i.e. the absence of bays. The ratio $\bar{z}$:Z_m is 0.53 and 0.55 respectively for the eastern and western basins (Table 1). These values are close to the average for over hundred lakes of 0.47 (Neumann 1959), which suggests that the average shape of lake beds approximates an elliptic sinusoid.

The shore of Lake Vechten is bounded by pasture on which, except in wintertime, cattle graze. Extending from the northwest, north, and from northern to eastern sides, the lake is bordered by a narrow zone of trees, consisting of willows, alders and poplars. The trees, however, are of little significance as a barrier against wind, as the prevailing wind direction is SW (see Table 2), which is along the long axis of the lake (see Fig. 2). The importance

Table 2. Monthly averages for the years 1931–1960 of meteorological data measured at meteorological station KNMI, De Bilt (after Buisman 1978). For solar radiation values see Fig. 9.

	Jan.	Feb.	Mar.	Apr.	May	Jun.	Jul.	Aug.	Sep.	Oct.	Nov.	Dec.	Year
Air temperature (°C)	1.7	2.0	5.0	8.5	12.4	15.5	17.0	16.8	14.3	10.0	5.9	3.0	9.3
Air temperature, average maximum (°C)	4.3	5.2	9.5	13.3	17.7	20.7	21.9	21.8	19.2	14.1	8.8	5.4	13.5
Air temperature, average minimum (°C)	-1.0	-1.1	1.0	3.9	7.1	10.1	12.2	12.0	9.8	6.2	3.0	0.4	5.3
Numbers of days with maximum > 25 °C	0	0	0	0	2	6	6	6	2	0	0	0	22
Number of days with maximum < 0 °C	4	2	0	0	0	0	0	0	0	0	0	2	8
Number of days with minimum < 0 °C	16	15	13	4	1	0	0	0	0	2	6	13	70
Relative humidity (%)	87	84	79	75	74	74	78	80	82	86	89	90	82
Sunshine (h)	56	69	127	164	211	223	199	186	146	102	50	41	1 572
Wind force ($m \cdot s^{-1}$)	5.5	5.1	4.6	4.9	4.2	4.1	4.1	4.2	4.0	4.4	4.6	4.9	4.6
Prevailing wind direction	SSW	SW	SSW	WNW	NNW	W	WSW	WSW	SW	SW	SSW	SSW	SW
Precipitation (mm)	69	49	44	49	52	57	78	89	71	72	70	64	768

of allochthonous inputs of organic matter, mostly leaves and twigs from the surrounding trees, in Lake Vechten was evaluated using the relationship between the relative importance of airborne litterfall (K), lake area (A) and shoreline development (D_L) at different levels of phytoplankton productivity as presented by Gasith & Hasler (1976). Litterfall could potentially be important in Lake Vechten, as it is estimated to constitute 10% or less of annual phytoplankton production ($K \leqslant 0.1$) (cf. Gons 1981).

Climate

There is a meteorological station at De Bilt, KNMI (Royal Netherlands Meteorological Institute), located at a distance of 4.3 km to the north of Lake Vechten, from which the meteorological data presented in Table 2 were obtained (see Buisman 1978).

Lake Vechten lies on latitude 52°04'N, where the sun is visible for some 16 hours and 45 minutes at the summer solstice, and for 7 hours and 43 minutes at the winter solstice. May and June are the brightest months with an average of, respectively, 211 and 223 sunshine hours, and November and December are the darkest months with average values of 50 and 41 hours of sunshine (Table 2). Daily solar radiation varies greatly throughout the year (see Fig. 9). The normal value for December is 175 $J \cdot cm^{-2} day^{-1}$. Solar radiation increases gradually to its maximum in June, viz. 1 776 $J \cdot cm^{-2} day^{-1}$. Normal daily radiation values during July, August, September and October are respectively 1 583, 1 376, 1 011 and 582 $J \cdot cm^{-2} day^{-1}$.

Superimposed on the seasonal cycle of temperature determined by the sun, is a considerable and irregular variation depending on the origin of the air masses over the country. Those coming off the western and south-western Atlantic bring moist air which is relatively cool in summer and relatively warm in winter. Air from the opposite direction brings fine clear weather, warm in summer and cold in winter. Air coming from the north is cold at all times of year. Generally in winter, for a few days at least, parts of the lake are frozen. Occasionally (e.g. in the winter of 1979) the lake is completely frozen and observations can be made through the ice layer. As south-westerly winds prevail (Table 2), precipitation is spread over the whole year, with a relatively dry period from February to May (Table 2). August is usually the wettest month. A typical pattern is an approaching depression causing south-westerly winds at first and then westerly and north-westerly winds as it passes. The variation in average monthly air temperature may be seen in Table 2. June, July and August can be rather hot sometimes, with temperatures rising above 25 °C. January and February are the coldest months, with temperatures frequently below freezing point.

Hydrology

The data on soil composition (Fig. 4) of the region in which Lake Vechten is situated are based on drillings taken by the Rijks Geologische Dienst (State Geological Service) (see also Parma 1971; Bos 1981). The geological formation that makes up the region consists of the following main bands (see Fig. 4): Fluviatile deposits consisting of coarse sands (300–2 000 μm) mixed with gravel and intermittent thin bands of clay are found below a depth of −5 m N.A.P. (Netherlands ordnance datum). Above that, medium fine to medium coarse sands (150–300 μm) mixed with silt were laid down. The top layer generally consists of holocene clay.

In the absence of surface in- or effluents, precipitation on the lake surface, evaporation and seepage through the highly permeable sand layers surrounding the lake are the factors regulating the water level. The water level can fluctuate very quickly in dry periods during spring (cf. Table 2), when the lake's volume may decrease by about $1\% \cdot day^{-1}$ (Parma 1971).

Data on the lake's hydrology, particularly quantitative information regarding relations between water level, rainfall, evaporation and groundwater flow, are scarce. The lake's water balance was only studied in detail by Bos (1981) in 1980 from June to September. However, the period covered was atypical with regard to rainfall. So, the results are difficult to generalize from. For calculating groundwater transport, flow direction is assumed to be mainly horizontal, as the presence of clay bands greatly reduces the vertical permeability. Furthermore, inflow of groundwater through the lake bottom can probably be neglected as the bottom is covered for the greater part with clay originating from the top layer which was dumped into the lake after excavation. This is important since the interstitial water in the bottom deposits below the 8 m contour is nutrient rich. The flow of groundwater in the Lake Vechten region is generally from east to west. The pattern of isohypses, based on groundwater levels around the lake and constructed using an electric simulation model, is presented in Fig. 5. From this figure and from precipitation, evaporation and lake level data, in- and outflow was calculated to get an approximate water balance (Table 3). The vertical outflow through the lake bottom (leaking) appears to be very small. Total annual input corrected for evaporation was estimated at 13.2×10^4 m^3. The volume of the lake is 28×10^4 m^3 (Table 1), so the renewal time is approximately 2 years.

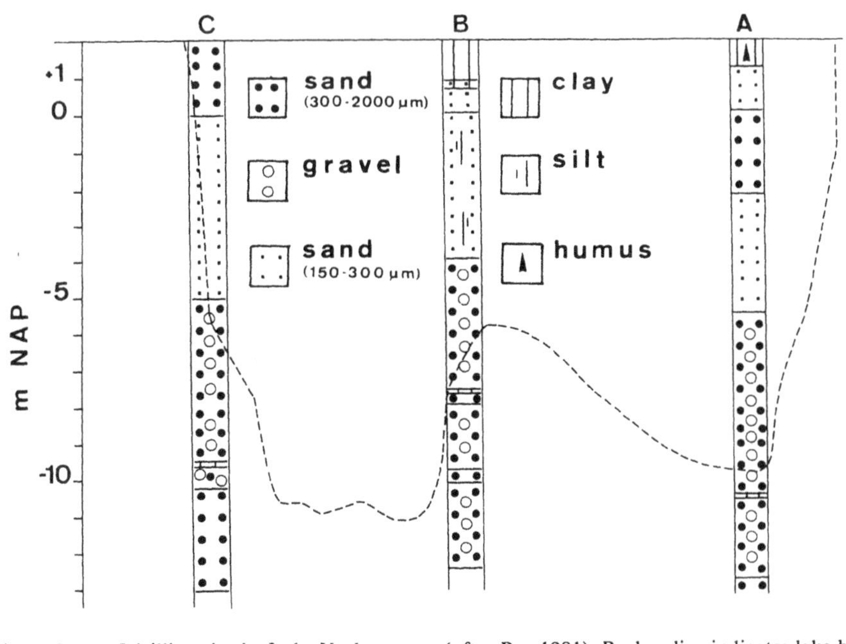

Fig. 4. Granulometric analyses of drillings in the Lake Vechten area (after Bos 1981). Broken line indicates lake bottom profile. For location of drilling sites see Fig. 5.

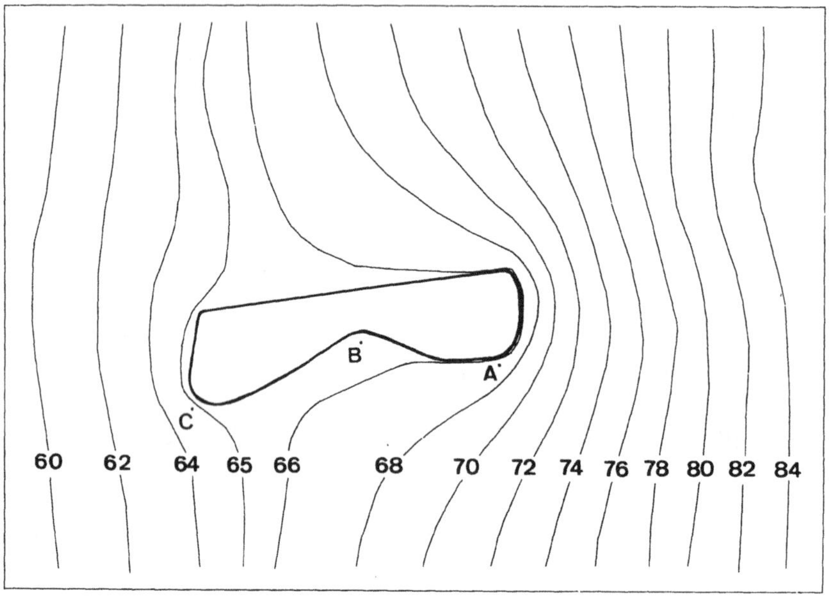

Fig. 5. Isohypses (cm + NAP) in the Lake Vechten area, 7 July 1980. Localities of drillings (Fig. 4; A, B and C) are indicated (after Bos 1981).

Table 3. Annual water balance of Lake Vechten (data from Bos 1981).

	Level (cm · yr^{-1})	Volume (× 10^4 m^3)	%
Input			
Horizontal groundwater recharge	257	12.1	77
Precipitation	75	3.5	23
Output			
Horizontal groundwater discharge	276	13.0	83
Evaporation	50	2.4	15
Vertical groundwater flow (leakage)	6	0.3	2

Thermal regime and stratification

According to the classification by Hutchinson (1957), Lake Vechten is in most years monomictic, but dimictic in years with a severe winter, i.e. with a long period of ice cover. The time-depth diagram of isotherms during 1977 in the eastern depression is given in Fig. 6. Thermal behaviour is identical for both basins. Vertical heterogeneity starts in late April leading to a stable stratification in May, which lasts to the end of October or to the beginning of November (cf. Parma 1971). The epilimnetic water temperature is around 20 °C, but, depending on daily weather conditions, values of 24 °C are not uncommon. Hypolimnetic temperatures range from 8 to 10 °C. The thermocline develops in the early stages of stratification at a depth of 4–5 m, and steadily descends until October (Fig. 6). The autumnal turnover is not drastic; however, the onset of complete circulation has a clear impact on suspended particulate matter in that the latter in the surface layers increases, as reflected in decreased light penetration (see later) and in higher sedimentation rates during that period (Verdouw & Dekkers, 1982).

An important parameter connected with stratification is the eddy conductivity coefficient (Hutchinson 1957), particularly in the stagnant water column. Eddy systems bring about turbulent transport of heat and dissolved matter and consequently play an important role in cycling of elements, e.g. in transport of nutrients (ammonia, phosphate) from the hypolimnion where concentrations are high to the epilimnion where they are low. This applies also

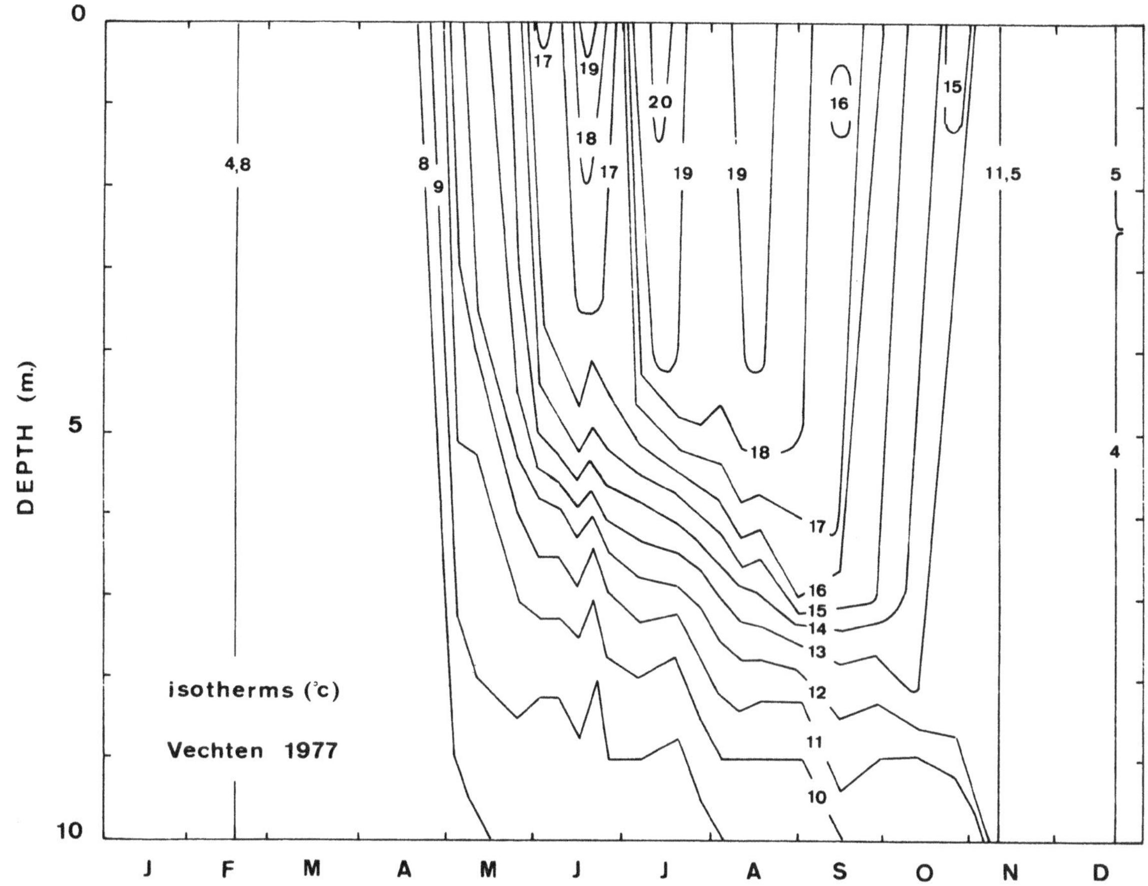

Fig. 6. Isotherms in eastern basin of Lake Vechten, 1977.

to the transport of oxygen and sulphate in the opposite direction. The eddy conductivity (calculated according to Hutchinson 1957) in Lake Vechten appears to be uniform in the water column from 5 to 9 m depth (Fig. 7, giving the required semilog plot of heating rate versus temperature increase). Average values for the years 1977–1980 are $(1.09 \pm 0.07) \times 10^{-2}$ cm^2 · s^{-1}. These values agree well with those found for other, sheltered lakes (see Hutchinson 1957; Cole 1975) and are approximately three times higher as the molecular thermal conductivity of water and three orders of magnitude higher than molecular diffusion coefficients. So, in Lake Vechten, the transport of heat and dissolved materials to and from the hypolimnion is apparently largely controlled by turbulent eddy flows.

Thermal stratification has a pronounced influence on distribution and concentration of several elements, primarily oxygen. As soon as thermal stratification starts, oxygen concentrations near the bottom start decreasing, resulting in a clinograde oxygen curve throughout the stratification period

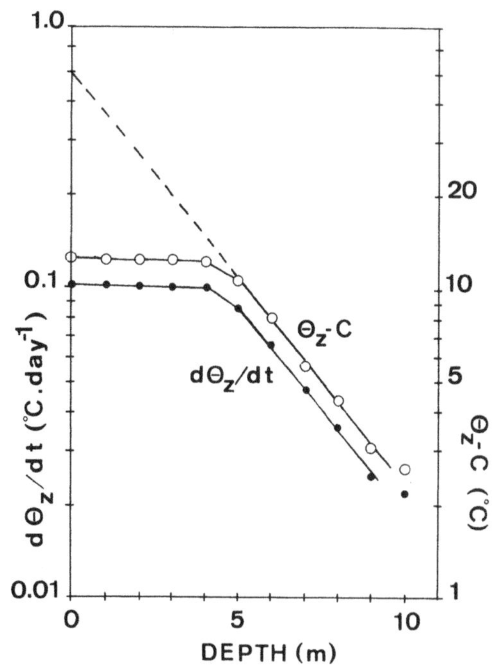

Fig. 7. Semilog plot of heating rate ($d\theta_z/dt$) and temperature increase (θ_z–C) in eastern basin of Lake Vechten for the period 19 March–7 July 1980 (symbols according to Hutchinson 1957).

(Parma 1971). The hypolimnion in Lake Vechten is anaerobic from May to October (Fig. 8); the upper boundary of the anoxic layer lies generally between 6 and 7 m depth. The thermal stratification pattern and accompanying oxygen regime in lake Vechten is very constant in successive years (cf. Parma 1971). As may be expected, climatic conditions influence the time of onset and end of thermal stratification. In some years, extremely strong south-westerly winds in October advanced the autumnal turnover process by several weeks.

Light penetration

Secchi disk depth

Secchi disk values (Fig. 9) are generally high (3.5 to 4.5 m) in May and in August–September, and low (2.0 to 2.3 m) from the end of October to November, coinciding with the period of autumnal turnover of the lake. During this period particulate matter, which has accumulated during summer in the deep strata, is resuspended throughout the entire water column, resulting in high turbidity. Another minimum in Secchi disk depth (1.8 to 2.5 m) occurs from February till the end of April. This minimum coincides with relatively high phytoplankton densities (Blaauboer 1982). At all other times Secchi disk depths vary from 2.5 to 3.5 m (Fig. 9). The Secchi disk depth correlates closely with percentage transmission and presents approximately 15% of subsurface light intensity (Beeton 1958; Vollenweider 1960). Hence, $\epsilon_{0-5} \times s = c$, where ϵ_{0-5} is the extinction coefficient for the mixed layer (0–5 m depth), s is the Secchi disk depth, and c is a constant which varies from 1.4 to 3.0, depending on local conditions (Vollenweider 1960). For Lake Vechten the constant c is 1.6 (range 1.2–2.2).

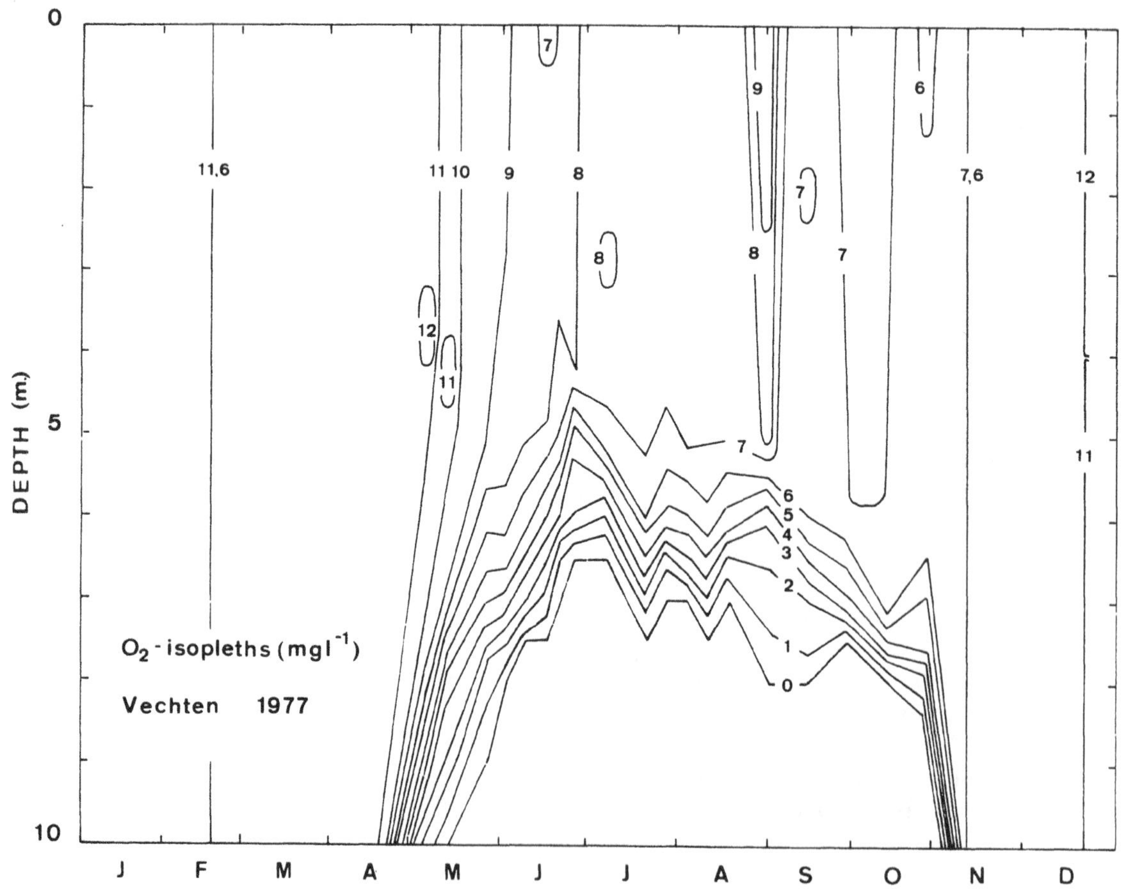

Fig. 8. Oxygen isopleths in eastern basin of Lake Vechten, 1977.

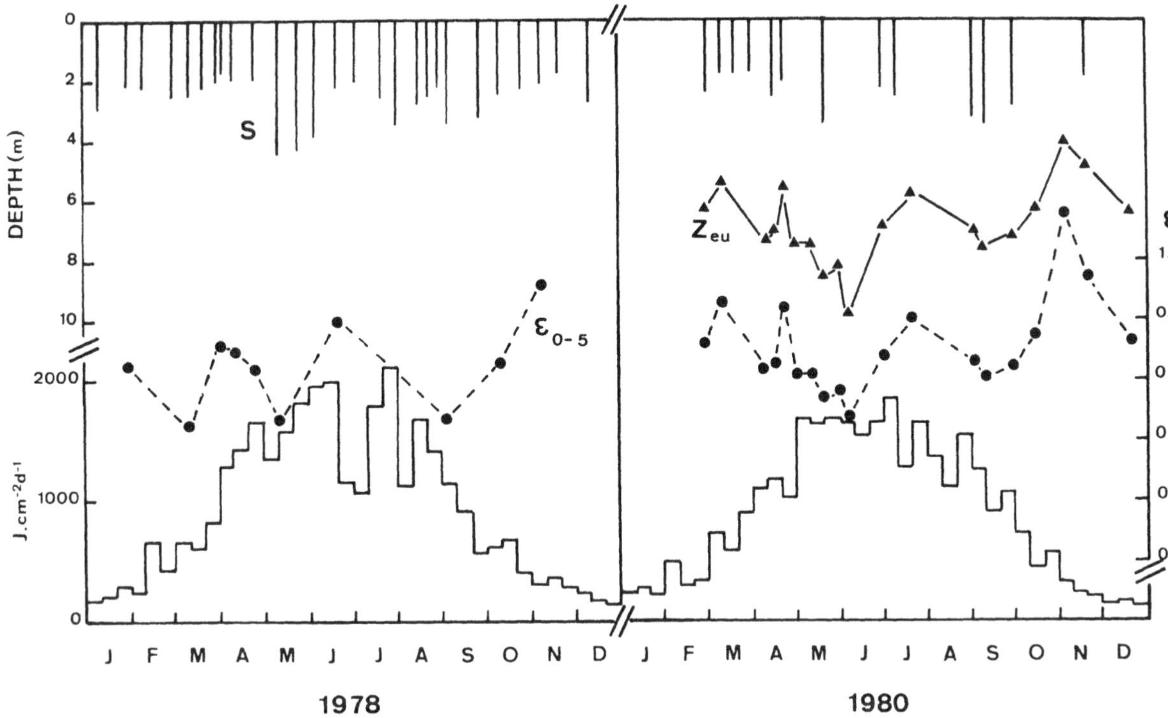

Fig. 9. Light penetration in Lake Vechten. Secchi disk depths (s) indicated by vertical lines. Vertical extinction coefficients (dotted line) are for the 0–5 m depth stratum only (ϵ_{0-5}). Euphotic depth (Z_{eu}, full line) calculated using ϵ_{0-5} values. Daily solar radiation values (vertical bars) are the average for ten consecutive days and are obtained at meteorological station KNMI, De Bilt.

Relative downwelling light penetrance

The vertical light attenuation of different wavelengths in Lake Vechten was measured (Fig. 10) using Standard Schott filters (Sauberer 1962) fitted to a photometer equipped with a silicon cell (B. Lange, Berlin). The most penetrating component of the spectrum is, as in many other lakes, in the green and yellow-orange range (e.g. Hutchinson 1957). In the period May–November the vertical extinction coefficient for green light (ϵ_g) varied from 0.38 to 0.85 m^{-1} and for yellow-orange light (ϵ_{yo}) from 0.38 to 0.75 m^{-1}. Red light penetrates less, ϵ_r varying from 0.73 to 1.02 m^{-1} and blue light the least, with ϵ_b between 0.69 and 1.42 m^{-1} (Steenbergen 1974).

The total attenuation coefficient in natural waters can be attributed to absorption by pure water, dissolved colour, and suspended particles (Hutchinson 1957;. Wetzel 1975). Light penetration in the deeper strata of Lake Vechten is affected mainly by the summer stratification. Dense populations of algae and bacteria develop in the metalimnion and in the upper part of hypolimnion (Blaauboer 1982; Steenbergen 1982). Besides, large amounts of detrital material accumulate in these strata. Consequently, light attenuates quickly and high extinction coefficients, i.e. up to 2 m^{-1}, are recorded in these strata (cf. Fig. 10). During autumnal turnover, light penetration is low, due to resuspension of particulate matter, and the extinction coefficient reaches values of more than 1 m^{-1} (Fig. 9).

A recurrent feature with respect to light penetration in Lake Vechten is the dramatic increase in transparency in May (Figs. 9 and 11). This so called 'clear phase' period coincides with the annual grazing maximum of herbivorous zooplankton and the concomitant decrease in phytoplankton density (Gulati *et al.* 1982). So, within approximately three weeks, ϵ_{0-5} decreases from 0.90 to 0.47 m^{-1} and Secchi disk depth increases from 2 to 4 m (Figs. 9 and 11).

At surface irradiance of less than 1%, net phytoplankton photosynthesis is conventionally assumed to be insignificant (Talling 1971; Harris 1978). The depth of the 1% light is, therefore, considered to be the lower limit of the euphotic zone (euphotic depth

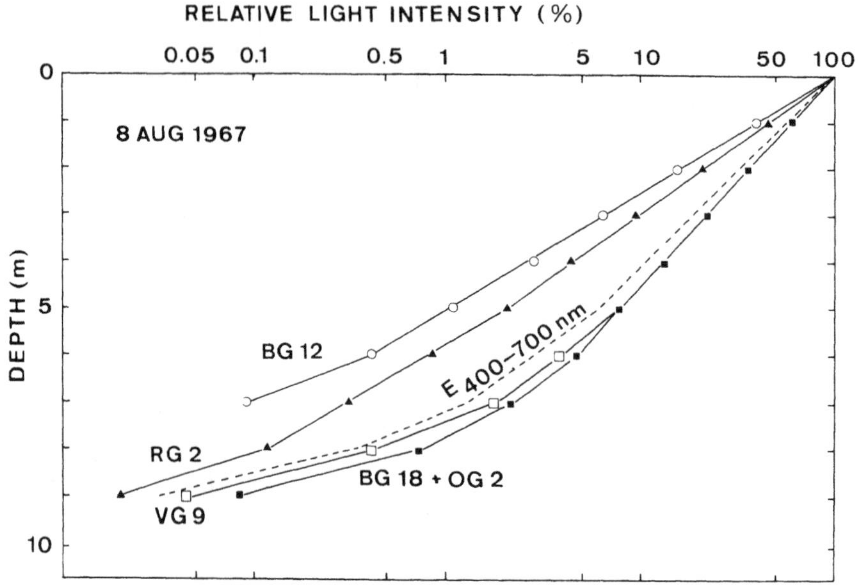

Fig. 10. Attenuation with depth of blue (BG 12), green (VG 9), yellow-orange (BG 18 + OG 2) and red (RG 2) light in Lake Vechten, 8 August 1967. Dotted line corresponds to PhAR light intensity (400–700 nm).

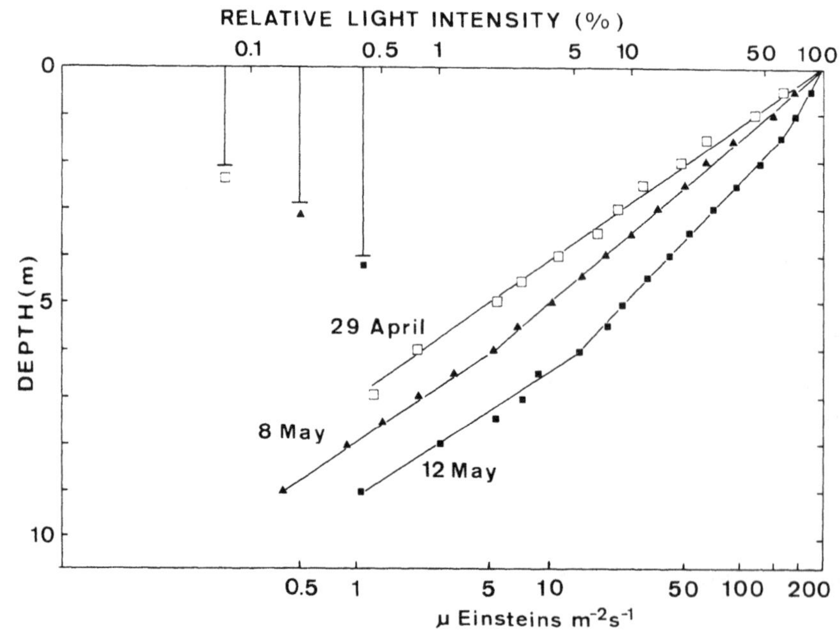

Fig. 11. Attenuation with depth of PhAR light intensity (400–700 nm), measured using quantum meter equipment (Lambda, U.S.A.), and corresponding Secchi disk depths (vertical lines) on three successive days during early 'clear phase' period in Lake Vechten, 1981. Units of light energy (μEinsteins · m^{-2} · s^{-1}) along the lower abscissa refer to the light profile of 8 May (moderately cloudy).

Z_{eu}). The calculated Z_{eu} in Lake Vechten, from the formula $Z_{eu} = \ln 0.01/\epsilon_{0-5}$, varies considerably (Fig. 9). Minimum Z_{eu} values (4–5 m) occur in November during the autumnal turnover; maximal values are recorded in May (8–9.5 m). During summer, Z_{eu} varies between 6 and 7.5 m. Apparently, during the last mentioned period, light reaching the metalimnion and upper hypolimnion is enough for algal and bacterial photosynthesis to occur (Steenbergen 1982). Besides, these deep populations may be adapted to very low light intensity and therefore show greater photosynthetic efficiency than algae from the mixed layer (Harris 1978).

Nutrient status and ionic composition

The nutrient status of Lake Vechten is summarized in Table 4. According to the criteria based on nutrient concentrations, the lake can be characterized as meso-eutrophic (Wetzel 1975). During the growing season nutrient concentrations in the epilimnion become very low, but they accumulate in the hypolimnion and form steep concentration gradients near the bottom; see for nitrogen Verdouw & Dekkers (1982) and for phosphorus Wisselo (1980).

The concentrations of the major ions and the ionic balance of the surface water of Lake Vechten are given in Table 5. The chemical composition closely resembles that of the groundwater in the neighbourhood of the lake (Bos 1981), which can be expected in view of the role of groundwater in the lake's water balance (Table 3). Calcium and bicarbonate clearly dominate the ionic composition. Rodhe (1949) has given approximate ion proportions for this type of lake: $Ca^{2+}:Mg^{2+}:Na^{+}:K^{+} = 64:17:16:3$ and $HCO_3^-:SO_4^{2-}:Cl^- = 74:16:10$. The cation concentrations in Lake Vechten are quite close to these proportions; of the anions, chloride is more abundant than sulphate. From 1960 onwards, i.e. the period in which the ionic concentrations have been regularly measured, no drastic changes occurred; only Cl^-, Na^+ and K^+ increased slightly. In 1966, concentrations of Na^+, K^+ and Cl^- averaged, respectively, 0.40, 0.04 and 0.46 meq · l^{-1}; in 1974 the values were 0.50, 0.06 and 0.60 meq · l^{-1}; for 1980, see Table 5.

The chemical composition of the water strata near the bottom is strongly influenced by the phenomena accompanying thermal stratification: CO_2 increase results in a pH decrease and enhanced solubility of carbonates; O_2 depletion leads to a reducing environment. Ionic concentrations can reach values of up to 6.5 meq · l^{-1}. The anion increase is caused by bicarbonate and that of cations by calcium, magnesium, iron, manganese and ammonium. Sulphate disappears in most years in the deepest strata during stratification and forms, like other ions, concentration gradients in the hypolimnion (Verdouw & Dekkers 1980). An example of the ionic balance at 10 m depth is given in Table 6.

Formation of carbonates is an important mechanism regulating the maximum concentrations of

Table 5. Ionic composition of surface water of Lake Vechten (eastern basin) on 9 September 1980; pH = 8.2.

Anions	meq · l^{-1}	%	Cations	meq · l^{-1}	%
HCO_3^-	2.62	68.8	Ca^{++}	2.45	64.6
Cl^-	0.79	20.7	Mg^{++}	0.74	19.5
SO_4^{2-}	0.40	10.5	Na^+	0.52	13.7
			K^+	0.08	2.1
Sum	3.81		Sum	3.79	

Table 4. Nitrogen, phosphorus and silica concentrations in Lake Vechten (eastern basin). Winter maxima apply to the whole (mixed) water column; epilimnetic minima to 0–6 m depth.

		Winter maxima ($\mu g \cdot l^{-1}$)	Epilimnetic nutrient minima ($\mu g \cdot l^{-1}$)
Inorganic-N	NH_3-N	200–300	<5
	NO_3-N	200–350	<5
Organic-N		300–500	
Total-P		20– 30	
PO_4-P		5– 10	<5
Si		~500	~100

Table 6. Ionic composition of water from 10 m depth of Lake Vechten (eastern basin) on 15 October 1974; pH = 7.0.

Anions	meq · l^{-1}	%	Cations	meq · l^{-1}	%
HCO_3^-	5.84	90.4	Ca^{2+}	3.75	58.9
Cl^-	0.58	9.0	Mg^{2+}	0.88	13.8
SO_4^{2-}	0.02	0.3	Fe^{2+}	0.69	10.8
PO_4-P	0.02	0.3	Na^+	0.50	7.8
			NH_4^+	0.31	4.9
			Mn^{2+}	0.18	2.8
			K^+	0.06	0.1
Sum	6.46		Sum	6.37	

several elements. During the circulation period, the water of Lake Vechten is saturated with calcium carbonate (Verdouw & Dekkers, unpublished results). During stratification, supersaturation caused by pH increase, is observed in the epilimnion, while the strata below 6 m depth are undersaturated due to pH decrease. On the other hand, during this period the solubility products of $FeCO_3$ and $MnCO_3$ are exceeded in the water strata close to the bottom (Verdouw & Dekkers 1980).

Summary

A description is given of Lake Vechten (The Netherlands) in terms of its morphometry, climate, hydrology, thermal and light regime, and chemistry.

The lake was formed in 1941 by sand digging and consists of two basins with maximum depths of 10.5 and 11.9 m, separated by a ridge at about 6 m depth. Surface area is 4.7 ha and mean depth 6 m (Table 1 and Figs. 2 and 3). The lake has no surface inlet or outlet, so its water balance (Table 3) is regulated by meteorological conditions (rainfall and evaporation, Table 2) and horizontal groundwater flow (Fig. 5). Vertical groundwater flow is restricted by the presence of clay layers in the highly permeable soil surrounding the lake (Fig. 4). The renewal time of the lake water is about two years.

Lake Vechten can be classified as warm monomictic. Only in years with a severe winter does it exhibit dimictic properties, due to prolonged ice-cover. The lake has a stable summer stratification from May till the end of October (Fig. 6). In the stagnant water below 5 m depth, eddy conductivity plays an important role in the transport of heat and dissolved substances (Fig. 7). Values over the years 1977–1980 averaged $(1.09 \pm 0.07) \times 10^{-2} \cdot cm^2 \cdot s^{-1}$. Light penetration is strongly influenced by the presence of particulate matter in the water (algal and bacterial populations and resuspended bottom material). Secchi disk depths are highest in May and August/September (3.5 to 4.5 m) and lowest at the autumnal turnover (2.0 to 2.3 m) and in February till April (1.8 to 2.5 m) (see Fig. 9). Extinction coefficients ranged from 0.38 to 0.85 m^{-1} for green, from 0.38 to 0.75 m^{-1} for yellow-orange, from 0.73 to 1.02 m^{-1} for red and from 0.69 to 1.42 m^{-1} for blue light (Fig. 10). A characteristic 'clear phase' period, coinciding with maximum zooplankton grazing, occurs in May with PhAR extinction coefficients about 0.5 m^{-1} (Fig. 9 and 11).

Lake Vechten can be characterized as meso-eutrophic with very low nutrient concentrations in the epilimnion during the growing season (Table 4). Ionic compositions of surface water and water near the bottom are given (Tables 5 and 6). Calcium and bicarbonate are the major ions, dominating the ionic balance of the lake.

Acknowledgements

The authors are indebted to R. D. Gulati and S. Parma for critically reading the manuscript, to W. A. de Kloet for generously providing data on light penetration, to M. J. B. Bär-Gilissen, E. M. J. Dekkers and E. M. Mariën for drawings and photography, and to J. van Kleinwee for typing.

References

Beattie, D. M., Golterman, H. L. & Vijverberg, J., 1978. An introduction to the limnology of the Friesian lakes. Hydrobiologia 58: 49–64.

Beeton, A. M., 1958. Relationship between Secchi disk readings and light penetration in lake Huron. Trans. Am. Fish. Soc. 87: 73–79.

Best, Elly P. H., Blaauboer, Marianne C. I., Cappenberg, Th. E., Gons, H. J., Gulati, R. D., De Kloet, W. A., Steenbergen, C. L. M. & Verdouw, H., 1978. Towards an integrated study of the ecosystem of Lake Vechten. Hydrobiol. Bull. 12: 107–118.

Blaauboer, Marianne C. I., 1982. The phytoplankton species composition and the seasonal periodicity in Lake Vechten from 1956 to 1979. Hydrobiologia 95: 25–36.

Bos, J. H., 1981. Hydrologisch onderzoek in 'Vechten' (with English summary). Student report No. 1981-14. Limnological Institute, Nieuwersluis/Oosterzee.

Buisman, J., 1978. Weer of geen weer. Bosch & Keunig, Baarn.

Cole, G. A., 1975. Textbook of Limnology. Mosby, St. Louis.

Gasith, A. & Hasler, A. D., 1976. Airborne litterfall as a source of organic matter in lakes. Limnol. Oceanogr. 21: 253–258.

Gons, H. J., 1981. Carbon cycle in Lake Vechten. In: Parma S. & Gulati R. D. (Eds.) Progress Report Limnological Institute 1980. Verh. Kon. Ned. Akad. Wet. afd. Nat., 2e Reeks 77.

Gulati, R. D., 1972. Limnological studies on some lakes in the Netherlands - I. A limnological reconnaissance and primary production of Wijde Blik, an artificially deepened lake. Freshwat. Biol. 2: 37–54.

Gulati, R. D., Siewertsen, K. & Postema, G., 1982. The zooplankton: its community structure, food and feeding, and role in the ecosystem of Lake Vechten. Hydrobiologia 95: 127–163.

Hallegraeff, G. M., 1977. Pigment diversity in freshwater phytoplankton - II. Summer-succession in three Dutch lakes with different trophic characteristics. Int. Revue ges. Hydrobiol. 62: 19–39.

Harris, G. P., 1978. Photosynthesis, productivity and growth: the physiological ecology of phytoplankton. Arch. Hydrobiol. Beih. 10: 1–171.

Hutchinson, G. E., 1957. A Treatise on Limnology, Vol. I: Geography, Physics and Chemistry. Wiley, New York.

Leentvaar, P., 1958. Observations on the plankton of some inland waters, especially 'Wielen', in the Netherlands. Beaufortia 6: 171–197.

Leentvaar, P., 1971. Biologische consequenties van zandwinning. Natuur Landschap 25: 116–124.

Neumann, J., 1959. Maximum depth and average depth of lakes. J. Fish. Res. Bd Can. 16: 923–927.

Parma, S., 1971. Chaoborus flavicans (Meigen) (Diptera, Chaoboridae): an autecological study. Ph.D. thesis, Univ. Groningen. 128 pp.

Parma, S., 1980. Historie en morfometrie van 'Vechten'. Internal report No. 1980-3, Limnological Institute, Nieuwersluis/Oosterzee.

Redeke, H. C., 1948. Hydrobiologie van Nederland; de zoete wateren. De Boer, Amsterdam.

Ringelberg, J., 1981. General introduction to the research in the Maarsseveen lakes area. Hydrobiol. Bull. 15: 3–4.

Rodhe, W., 1949. The ionic composition of lake waters. Verh. int. Verein. Limnol. 10: 377–386.

Sauberer, F., 1962. Empfehlungen für die Durchführung von Strahlungsmessungen an und in Gewässern. Mitt. int. Verein. Limnol. 11: 1–77.

Steenbergen, C. L. M., 1974. The role of light in the life cycle of Scenedesmus obliquus (Turp.) Kütz. Ph.D. thesis, Univ. Utrecht. 56 pp.

Steenbergen, C. L. M., 1982. Contribution of photosynthetic sulphur bacteria to primary production in Lake Vechten. Hydrobiologia 95: 59–64.

Talling, J. F., 1971. The underwater light climate as a controlling factor in the production ecology of freshwater phytoplankton. Mitt. int. Verein. Limnol. 19: 214–243.

Van Heusden, G. P. H., 1945. Waarnemingen in eenige 'Wielen' in de Betuwe. Tijdschr. Kon. Ned. Aardrijksk. Genootsch. 62: 118–141.

Verdouw, H. & Dekkers, E. M. J., 1980. Iron and manganese in Lake Vechten (The Netherlands); dynamics and role in the cycle of reducing power. Arch. Hydrobiol. 89: 509–532.

Verdouw, H. & Dekkers, E. M. J., 1982. Nitrogen cycle of Lake Vechten: concentration patterns and internal mass-balance. Hydrobiologia 95: 191–197.

Verdouw, H. & Dekkers, E. M. J., 1982. Nitrogen cycle of Lake Vechten (The Netherlands); role of sedimentation. Arch. Hydrobiol. 94: 251–263.

Vollenweider, R. A., 1960. Beiträge zur Kenntnis optischer Eigenschaften der Gewässer und Primärproduktion. Mem. Ist. Ital. Idrobiol. 12: 201–244.

Wetzel, R. G., 1975. Limnology. Saunders, Philadelphia.

Wisselo, A. G., 1980. Onderzoek naar de fosforkringloop in de plas Vechten in 1979 (with English summary). Internal report No. 1980-4, Limnological Institute, Nieuwersluis/Oosterzee.

The phytoplankton species composition and the seasonal periodicity in Lake Vechten from 1956 to 1979

Marianne C. I. Blaauboer
Limnological Institute, 'Vijverhof' Laboratory, Rijksstraatweg 6, 3631 AC Nieuwersluis, The Netherlands

Keywords: phytoplankton, species composition, seasonal periodicity, vertical distribution

Abstract

The species composition and seasonal periodicity of the phytoplankton in Lake Vechten (The Netherlands) have been studied over a 20 year period. The results show a more or less regular pattern of algal succession during the year. Phytoplankton growth starts early in the year leading to a maximum in early spring, dominated by Bacillariophyceae and some Chlorophyceae. In summer a second maximum develops mainly consisting of a chrysophycean and a dinophycean species, accompanied by several Chlorophyceae. In September the summer species are gradually succeeded by diatoms. By the end of the summer stratification in October algal numbers decrease rapidly. The winter community consists of several species of Cryptophyceae and Chrysophyceae.

When the data of 1975–1979 are compared with those from earlier records, distinct changes can be observed. These changes are discussed with reference to the trophic status of the lake. The vertical distribution of some species is described.

Introduction

Knowledge about the species composition and seasonal periodicity of the phytoplankton is essential when studying a lake ecosystem. Being the primary producers of the system, the algae represent a very important link in the food chain. Size and temporal as well as spatial distribution partly determine the grazing behaviour of primary consumers. Furthermore, a study of phytoplankton growth is in itself very interesting. There is a striking resemblance in the seasonal periodicity of algae year after year.

Therefore, phytoplankton studies have been carried out in Lake Vechten by the Limnological Institute, more or less continuously, since its foundation in 1957. The first records date from 1956 when the species composition of the phytoplankton was thoroughly investigated. In 1962 the investigations were continued. Over a period of 5 years the species present in the lake were carefully identified and their seasonal periodicity was studied. In 1974 the study of the phytoplankton was again resumed. Initially, special attention was given to the examination of several size fractions of the standing crop for use in a study of zooplankton grazing (see Gulati *et al.* 1982). Later on (1977–1979) the vertical distribution of phytoplankton species was emphasized as this was important for a study of the extracellular release by the phytoplankton (Blaauboer *et al.* 1982). In the meantime the study of the seasonal periodicity was continued.

This paper deals with different aspects of the phytoplankton. Firstly, the species composition and the seasonal periodicity of the phytoplankton during the period from 1977 to 1979 are surveyed. Secondly, the present situation is compared with that in 1956 and during the period from 1962 to 1967. Thirdly, attention is paid to the vertical distribution of some phytoplankton species during the

Hydrobiologia 95, 25–36 (1982). 0018-8158/82/0951-0025/$02.40.

summer stagnation. Finally some preliminary results are presented on the diurnal vertical migration of algae, particularly of *Ceratium hirundinella* (Dinophyceae).

The lake

Lake Vechten is a small, deep lake with a surface area of 4.7 ha, a maximal depth of 11.9 m and a mean depth of 6.0 m. The lake is stratified from May up to early November with a thermocline between 6.0 and 7.5 m. The oxygen in the deeper water layers is rapidly depleted, resulting in an anoxic hypolimnion (7.5–11.9 m) from July onwards. For a more detailed description of the lake see Parma (1971) and Steenbergen & Verdouw (1982).

Materials and methods

In the period from 1962 to 1967 surface and deep water samples were collected with a pump. Since 1974 a Ruttner or a 5 l Friedinger sampler has been used. Up to 1977 water samples, covering the whole water column from the surface to the sediment, were compounded. From 1977 onwards samples were collected at 1.2 m intervals and treated separately.

The samples were killed with Lugol's solution (10% final concentration) and concentrated by settling. The algal species were identified using the following works: Bacillariophyceae, Hustedt (1962), Pascher heft 11 (1976), and van der Werff & Huls (1957–1974), and for the remaining classes, Bourelly (1966, 1968, 1970), Prescott (1951) and Skuja (1948, 1957). The species were enumerated according to Utermöhl (1958), using a Zeiss inverted microscope.

Biovolumes were calculated using the data of Nauwerck (1963), Wetzel (1975) and Willén (1976). However, there are considerable differences in the biovolume reported by different authors, e.g. the biovolume reported for *Ceratium hirundinella* ranges from 40 000 (Wetzel 1975) to 75 000 μm^3 (Nauwerck 1963). In such cases biovolumes have been calculated from the measurement of the individual species in Lake Vechten.

Results and discussion

Species composition and seasonal periodicity; short-term cycles (1976–1979)

In temperate lakes phytoplankton growth starts in spring when solar radiation and daylength begin to increase. During this period nutrient levels are relatively high due to mineralization in winter. In Lake Vechten phytoplankton growth usually starts as early as January resulting in a spring maximum in February or March, consisting of diatoms (Bacillariophyceae) and two species of Chlorophyceae. At the onset of thermal stratification in May, the diatoms have disappeared and species like *Dinobryon divergens* (Chrysophyceae) start to develop. In summer the phytoplankton consists mainly of chlorophycean species accompanied by some members of the classes Chrysophyceae, Dinophyceae and Cyanophyceae. The dinophycean species *Ceratium hirundinella* persists through autumn, while the diatoms increase to a second maximum in September. At the end of October, when thermal stratification breaks down, algal numbers decrease rapidly. In winter, development of some cold water species of the classes Chrysophyceae and Cryptomonadaceae is observed.

Considering the species composition in more detail (Table 1) the *winter* (December) community consists of the following species in order of importance: *Rhodomonas minuta, Mallomonas akrokomos, Trachelomonas volvocina* and several *Cryptomonas* species. *R. minuta,* which may be present in high numbers in winter, and *T. volvocina* are perennial species showing some low peaks in spring and summer. *M. akrokomos* is a typical winter form (Hutchinson 1967). *Closteriopsis* sp. (Chlorophyceae) is present in low numbers (5 000 cells.l^{-1}).

In January, when these algae reach their maxima, several other algae start growing. In Lake Vechten an *early spring* increase of the phytoplankton can be observed. In many other lakes spring growth does not start until March or April, when the increasing light intensity exceeds a certain minimum value (Talling 1971; Fogg 1975). In Lake Vechten suitable growth conditions exist early in the year, presumably because of the optimal light conditions in winter (January and February). Extinction coefficients (ϵ) lower than 0.5 m^{-1} have

Table 1. Dominant phytoplankton species in Lake Vechten.

Species	Present in months	Average numbers during maximum ($\times 10^3.l^{-1}$)	Average biovol. during maximum ($\mu m^3 \times 10^6.l^{-1}$)
BACILLARIOPHYCEAE			
Stephanodiscus astraea (Ehr) Grun.	II–IV	500	10 000
St. hantzschii Grun.	I–IV	100	15
Asterionella formosa Hass.	III/IX	100	70
Centronella reichelti M. Voigt	III/IX	40	60
Cyclotella comta (Ehr) Kütz	III–V	?	–
Pennate diatoms (several species)	V/VIII–IX	60	600
CHLOROPHYCEAE			
Chlorella pyrenoidosa Chick.	I–III/VIII	20 000	40
Scenedesmus sp. (2-celled coen.)	III	400	48
Closteriopsis sp.	VI	100	35
Gemellicystis sp.	IV	30	–
Elakatothrix gelatinosa Wille	VII	20	1.5
Oocystis spp.	VII	80	32
Gloeocystis vesiculosa Näg.	VII	500	1 000
Botryococcus braunii Kütz.	VIII	20	–
Ankyra sp.	VI/VII	350	7
Selenastrum sp.	VII	<10	–
Sphaerocystis schroeteri Chod.	VI–VIII	50	32.5
Crucigenia quadrata Morren	VIII	<10	–
Tetraedron minimum (A.Br.)Hansg.	VIII	50	1.5
Staurastrum gracile Ralfs	VIII	<10	–
Phacotus lenticularis Stein	V–IX	<10	–
CHRYSOPHYCEAE			
Mallomonas akrokomos Ruttner	XII–I	75	37
M. caudata Iwanoff	VII–VIII	800	9 600
Dinobryon divergens Imhof	V	300	240
DINOPHYCEAE			
Gymnodinium helveticum Penard f. achroum Skuja	IV–V	10	150
Peridinium willei Huif.-Kaas	V–VIII	10	400
Ceratium hirundinella O.F. Müll.	VIII–X	350	17 500
CYANOPHYCEAE			
Filamentous blue green algae *Oscillatoria* and *Lyngbya* spp.)	VII–VIII	500	250
EUGLENOPHYCEAE			
Trachelomonas volvocina Ehr.	IX–III/VIII	10	18
Trachelomonas hispida (Perty) Stein	IX	<10	–
CRYPTOPHYCEAE			
Rhodomonas minuta Skuja	I/III–IV	1 000	200
Cryptomonas spp.	I/VII/IX	100	200

been observed against 0.7–0.8 m^{-1} in summer and even 1.15 m^{-1} just after the fall overturn. The low turbulence of the lake water is probably a consequence of the small surface area (4.7 ha) of the lake together with its sheltered position. Moreover, the lake rarely freezes. Apparently phytoplankton growth occurs in Lake Vechten in January and February at temperatures of 4 °C and below. Characteristic at this time of the year is an abundant growth of *Chlorella pyrenoidosa*, sometimes closely followed by *Stephanodiscus hantzschii*. They both show rapid increases in cell numbers even at a water temperature as low as 4 °C. Generation times of 11 days for *C. pyrenoidosa* and 5 days for *S. hantzschii* have been calculated from weekly counts (1978). At the end of January or in February *C. pyrenoidosa* reaches a peak of 20 to 40.10^6 cells.l^{-1}. Nauwerck (1963) described a comparable spring increase of *C. pyrenoidosa* and *S. hantzschii* var. pusillus in Lake Erken (Sweden). Pechlaner (1970), investigating the phytoplankton of the same lake in later years, calculated a doubling time for *S. hantzschii* of 3.3 days, i.e. even shorter than the doubling time of this species in Lake Vechten. After their maxima cell numbers of *C. pyrenoidosa* and *S. hantzschii* decrease gradually; *S. hantzschii* disappears completely by April, sometimes returning in autumn, although not as abundant as in spring. *C. pyrenoidosa* is present throughout the year.

In *spring* the next diatom species to increase in number is *Stephanodiscus astraea*. In some years, when numbers of *S. hantzschii* are low, growth of *S. astraea* starts earlier, almost taking the place of *S. hantzschii*. The population of *S. astraea* reaches an impressive maximum of $0.5–1.10^6$ cells.l^{-1}, usually in March or April. In the same period a *Scenedesmus* species (2-celled coenobia) and the diatoms *Cyclotella comta* and *Centronella reichelti* are present. The annual abundance of *Scenedesmus* sp. varies by an order of magnitude ($50–500.10^3$ coenobia.l^{-1}). *C. reichelti* usually shows a relatively low maximum. In some years cell densities are below the detection limit of the counting method. During or just after the *S. astraea* maximum, *Asterionella formosa* increases in number accompanied or closely followed by other pennate diatoms. The composition and abundance of the pennate diatoms vary greatly in the course of several years. This group is of limited quantitative importance in Lake Vechten and was not investigated on the species level.

A succession from *S. astraea* to *A. formosa* similar to that in Lake Vechten was found in other lakes. Nauwerck (1963), for example, reported successive growth of *S. hantzschii, S. astraea* and *A. formosa* in Lake Erken; this was confirmed by Pechlaner (1970). *S. astraea* seems able to grow at lower temperatures than does *A. formosa*, having its optimum temperature at 10–15 °C (Rodhe 1948). This is in accordance with the findings of Mackarewicz *et al.* (1979) in Lake Michigan near Chicago. In the earlier years of the study he observed a spring maximum in May-June (water temperature, 10 °C) dominated by *Asterionella* sp. and *Fragilaria* sp. In later years a shift in the dominant diatom species occurred resulting in a spring maximum in March, dominated by *Stephanodiscus* sp. and *Tabellaria* sp. (water temperature, 0–4 °C). However, temperature is not the only factor determining algal succession. Mackareth (1953) reported that *A. formosa* could take up phosphorus in concentrations down to 0.1 μg.l^{-1}. This may be advantageous when the phosphorus concentration of the water is low. Tilman & Kilham (1976) studied the phosphorus kinetics of *A. formosa* and *Cyclotella meneghiniana*. The former species appeared to be very efficient in using internal phosphate for growth. Competition experiments showed that *A. formosa* was competitively dominant when both species were P-limited (Tilman 1977). Thus, the diatom succession in Lake Vechten can be explained by the combined effects of temperature and nutrient status of the lake water. *S. astraea*, being able to grow vigorously at low temperatures, declines when the phosphorus content of the water falls to low levels (2–5 μg.l^{-1}). Under these circumstances *A. formosa* increases, being able to take up phosphorus at low concentrations. However, when the water circulation decreases, at the onset of thermal stratification, the population of *A. formosa* declines rapidly and the spring outburst comes to an end. Hereafter the algal populations develop mainly in the epilimnion (0–6 m), except for a few species, as will be discussed later.

In *early summer* (May) phytoplankton densities are relatively low, with only a limited number of species present in detectable quantities. *Gymnodinium helveticum*, present since early spring in low numbers, reaches a moderate maximum by the end of April or by early May when the diatoms have disappeared. Nauwerck (1963) also reported an increase of this species under similar circumstances in

Lake Erken, i.e. after the collapse of the spring maximum. However, the most important species in early summer is the flagellated Chrysophycea *Dinobryon divergens,* exhibiting a regular growth pattern, usually in May. The appearance of *D. divergens* after the decline of the diatom populations is a frequently observed phenomenon (Pearsall 1932; Hutchinson 1944; Nauwerck 1963). Hutchinson (1944) could not confirm all the explanations given by Pearsall (1932), although *D. divergens* seemed to be favoured by a fall in the phosphorus concentration. This was in accordance with Rodhe's (1948) later findings, that *D. divergens* has a low P-requirement and that a concentration of 5 $\mu g.l^{-1}$ inhibits growth in culture. Talling (1962), on the other hand, was able to culture *D. divergens* in a P-rich medium. During the spring outburst in Lake Vechten, PO_4-P concentrations fall from 10 to 15 $\mu g.l^{-1}$ in winter to less than 5 $\mu g.l^{-1}$ in April. So these observations correspond with those of Hutchinson, namely that growth of *D. divergens* occurs when PO_4-P has been largely removed by the spring phytoplankton. Its presence in the lake is restricted to 3 to 4 weeks only. In some years *D. divergens* has a second but comparatively low peak in August.

A third species characteristic of this season is the Chlorophycea *Closteriopsis* sp., already mentioned while describing the composition of the winter community. This alga shows a peak in June. Maximal numbers of *Closteriopsis* sp. can vary considerably from year to year ranging from 30.10^3 to 30.10^6 cells.l^{-1}.

In July algal biomass starts to increase rapidly to a large *summer* maximum in August, which extends well into September. Initially the algal community consists chiefly of chlorophycean species attaining their maxima in July. *Ankyra* sp. and *Sphaerocystis schroeteri* start to increase in June, within a few weeks followed by *Gloeocystis vesiculosa, Elakatothrix gelatinosa* and several *Oocystis* species. In the course of August their numbers decrease, while a large population of the Dinophycea *Ceratium hirundinella* develops. This alga is a very cosmopolitan summer species. In Lake Vechten it heavily dominates the phytoplankton standing crop in this period, at least on a biomass basis (see later). From May onward numbers of *C. hirundinella* increase gradually, resulting in a maximum of 0.3×10^6.cells.l^{-1} in September.

Another significant species growing abundantly in the summer months is *Mallomonas caudata* (Chrysophyceae). Maxima from 0.5 to 1.10^6 cells.l^{-1} have been recorded concentrated in the metalimnion of the lake (5–7 m depth). Other inhabitants of these water layers are the filamentous bluegreen algae (Cyanophyceae), represented by *Oscillatoria* and *Lyngbya* species, reaching densities of 1.10^6 filaments.l^{-1} in the same period. The aspect of stratification of algal populations will be discussed later in more detail.

Throughout the summer period, populations of *Chlorella pyrenoidosa* and of *Rhodomonas minuta* wax and wane at regular intervals. *Cryptomonas* spp. attain one or two relatively low maxima in summer and autumn respectively. By the end of August diatoms like *Centronella reichelti, Asterionella formosa,* some pennate diatoms and occasionally *Stephanodiscus astraea* start to increase, reaching a second maximum in *autumn.* The peak numbers of *A. formosa* in autumn always exceeded those in spring.

Algal numbers decrease in October when the turbidity increases due to the gradual autumnal turnover ($\epsilon = 1.15.m^{-1}$). As a consequence phytoplankton biomass is low after the overturn. By December transparency increases and *winter* species appear. Then the stage is set for renewed algal growth in early spring.

The description of the seasonal periodicity given above was based on cell counts. As the cell size of the individual species varies considerably, biomass data are also included. Total cell biovolume has been calculated for the results of two representative years, 1977 and 1979 (Fig. 1). The dominant phytoplankton species are indicated as well. It is evident that some of the larger species heavily dominate the total biomass at certain times of the year. In spring *S. astraea* accompanied by *Cyclotella comta* constitute at least 90% of the total algal volume, notwithstanding the very high numbers of *Chlorella pyrenoidosa* in this period (1977). In 1979 spring growth started late following a prolonged ice cover. In April *S. astraea* increased simultaneously with other diatom species, e.g. *A. formosa* and other pennates, although it usually preceded these species. In summer *Mallomonas caudata* accounted for about 50% of the total biovolume and *Ceratium hirundinella* for 40%; all other species accounted for the remaining 10%. In autumn the biovolume of *C. hirundinella* amounted to as much as 95% of the

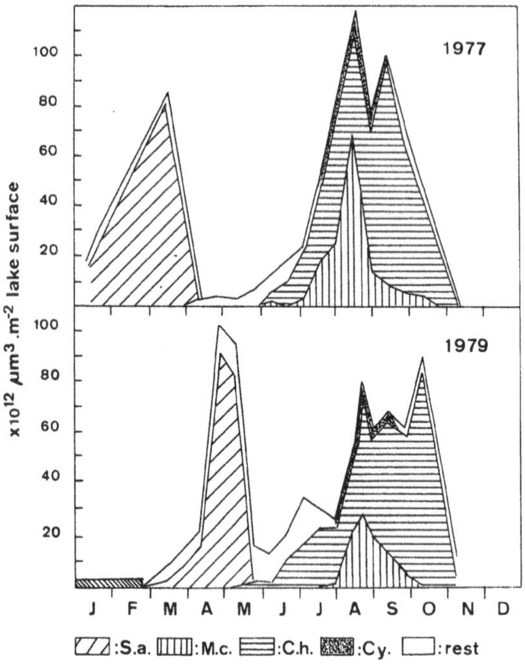

Fig. 1. Phytoplankton biomass in terms of cell volume during 1977 and 1979 in Lake Vechten. (S.a. = *Stephanodiscus astraea* accompanied by *Cyclotella comta;* M.c. = *Mallomonas caudata;* C.h. = *Ceratium hirundinella;* Cy = Cyanophyceae; rest = all other species); the hatched area in Jan–Feb. 1979 indicates ice cover.

total. It can be concluded that on a biomass basis the most important species in Lake Vechten are *S. astraea* in spring and *M. caudata* and *C. hirundinella* in late summer.

This record of the seasonal periodicity of the phytoplankton is generalized to some extent. Although there are many similarities in the succession of algal populations in the course of several years, the exact pattern differs from year to year, e.g. there is a variation of several weeks in the timing of the spring maximum (Fig. 1). In 1979 the spring increase was delayed until the ice cover disappeared. Also maximal cell numbers may vary considerably; some species, present in relatively low numbers in one year, may exhibit massive growth in another; moreover, species present in one year may be absent in the next year or vice versa. The species mentioned above were all regular inhabitants of Lake Vechten in the last 4 years (1976–1979) and are therefore considered as the principal species constituting the lake's phytoplankton.

Considering the trophic status of the lake it can be said to be slightly eutrophic. Species like *Dinobryon divergens, Sphaerocystis schroeteri, Gloeocystis vesiculosa, Oocystis* spp. and *Mallomonas caudata* are known to be common in less productive lakes (Hutchinson 1967). *Stephanodiscus astraea,* on the other hand, prefers eutrophic conditions, which exist in Lake Vechten in early spring. Nutrient concentrations decrease to a low level after the spring maximum and remain low throughout summer (for details see Steenbergen & Verdouw 1982). Consequently, after the onset of thermal stratification, species with low nutrient requirements can thrive well. The filamentous bluegreen algae, regarded as eutrophic species, are present in the metalimnion, apparently benefiting from the nutrient flux coming from the hypolimnion as well as from local regeneration of nutrients.

The species composition and periodicity from 1956 to 1979

The phytoplankton of Lake Vechten has been studied more or less continuously since 1956. The methods of sampling and enumeration have varied to some extent throughout the period of investigation. However, the greater part of the available data is sufficiently reliable to allow at least qualitative evaluation. In the periods from 1962 to 1967 and from 1974 to 1979, the species composition was studied in special detail.

Comparison of the earlier phytoplankton records with the present ones reveals definite changes in the species composition. To facilitate the discussion of similarities and differences between the two periods of more thorough investigation, the species were divided into three categories (Table 2).

1. Species that have been present continuously during the 23 years since 1956

Although the species belonging to this category show a relatively constant periodicity, some of them deserve special attention:

BACILLARIOPHYCEAE: *Stephanodiscus hantzschii* is found in early spring since 1962. From 1974 onward it has been frequently found in autumn as well, sometimes persisting throughout winter in low numbers. *Asterionella formosa* occurred regularly in autumn in most of the earlier years; since 1975 it has also been present in spring, although less abundant than in autumn. The pennate diatoms

Table 2. The phytoplankton species found in Lake Vechten from 1956 to 1979, divided into three categories

1. Species present throughout the investigation period

BACILLARIOPHYCEAE
Stephanodiscus hantzschii Grun.
St. astraca (Ehr) Grun.
Asterionella formosa Hass.
Cyclotella comta (Ehr) Kütz

CHLOROPHYCEAE
Elakatothrix gelatinosa Wille
Ankyra sp.
Oocystis spp.
Tetraedron minimum (A. Br.) Hansg.
Closteriopsis sp.
Scenedesmus sp. 2-celled coenobia
Selenastrum sp.

CHRYSOPHYCEAE
Dinobryon divergens Imhof
Mallomonas caudata Iwanoff

DINOPHYCEAE
Peridinium willei Huif.-Kaas
Gymnodinium helveticum Penard f. achroum skuja

CRYPTOPHYCEAE
Cryptomonas spp.
Rhodomonas minuta Skuja

EUGLENOPHYCEAE
Trachelomonas volvocina Ehr

2. Species, that have decreased in number or disappeared completely (marked*)

CHLOROPHYCEAE
Phacotus lenticularis Stein
Botryococcus braunii Kütz
Crucigenia quadrata Morren
Lagerheimia spp.*
Sphaerocystis schroeteri Chod

CHRYSOPHYCEAE
Mallomonas akrokomos Ruttner
Chrysococcus sp.*
Kephyrion rubri-claustri Conrad*

3. Species, that have increased in number or colonized (marked ″) the lake since 1956.

BACILLARIOPHYCEAE
Centronella reichelti M. Voigt″
CHLOROPHYCEAE
Gloeocystis vesiculosa Näg″
G. gigas Kütz.″
Gemellicystis sp.″
DINOPHYCEAE
Ceratium hirundinella O.F. Müll

EUGLENOPHYCEAE
Trachelomonas hispida (Perty) Stein″
CYANOPHYCEAE
Oscillatoria spp.
Lyngbya sp.

present in spring and again in autumn are *Synedra acus* and several *Tabellaria, Fragilaria* and *Nitzschia* species. Since their specific numbers are very low they are dealt with as a group of pennate diatoms.

CHLOROPHYCEAE: Most of the green algae belonging to category 1 (Table 2) show a very regular periodicity with a maximum in spring or in summer (July-August). Their maximum numbers vary considerably from year to year, especially for *Ankyra* sp., *Tetraedron minimum, Scenedesmus* sp. and *Closteriopsis* sp. In the past the last-named species has most likely been identified as *Closterium acutum* var. variable. It is not certain whether *Chlorella pyrenoidosa* has been present since the beginning of the phytoplankton study, as it was not identified as such before 1975. The 'small green cells' mentioned in earlier reports exhibited a growth pattern resembling that of *C. pyrenoidosa* in later years. Considering this, the 'small green cells' were assumed to be identical with *C. pyrenoidosa.* Since 1975 this green alga has evolved from an early spring species to a perennial one with additional maxima in spring and summer. Several desmid species were found occasionally in the lake throughout the investigated period in very low numbers. The most important 'species' is *Staurastrum gracile,* which occurs in the summer months.

CHRYSOPHYCEAE: *Dinobryon divergens* showed a long-term fluctuation in maximum numbers. In 1962 the maximum was very high, i.e. more than a million cells.l^{-1}. During 1965–1967 the numbers decreased to less than 50.10^3 cells.l^{-1}. From 1974 onward the maxima increased again to about one million cells.l^{-1}. *Mallomonas caudata* has been found in higher numbers since 1974. However, this 'rise' is most probably due to the method of sampling. In erlier years sampling was limited to the epilimnion. Since 1974 the water column down to 8.5 m has been sampled, including the *Mallomonas* population, which develops in the metalimnion (6.0–7.5 m). It was therefore missed in the past. This also applies to the appearance of filamentous blue-green algae in 1975. These algae show increased concentrations at 7–8 m depth (see later).

CRYPTOPHYCEAE: *Rhodomonas minuta* was initially restricted to the winter period. Since 1974 this alga has become increasingly perennial.

2. Species that decreased in numbers gradually in the course of time or completely disappeared

Many of the chlorophycean species listed in category 2 (Table 2) have reduced in number since 1967. All of them were dominant summer species in the sixties, reaching $0.5–1.10^6$ cells.l^{-1}. *Phacotus lenticularis* has been found only occasionally in the last few years. *Botryococcus braunii,* absent since 1974, showed a peak in 1977 of 30.10^3 colonies.l^{-1}. *Lagerheimia* sp. disappeared but for a small peak in 1974.

CHRYSOPHYCEAE: Maxima of *Mallomonas akrokomos* have been rather constant in time and magnitude throughout the investigation period. From 1963 to 1967 densities always exceeded 0.5. 10^6 cells.l^{-1}, but since 1974 they have been decreased by nearly an order of magnitude. *Kephyrion rubriclaustri,* dominant in spring, and *Chrysococcus* sp. occurred from 1963 to 1967, but have not been encountered since.

3. Species that have increased in numbers or colonized the lake since 1956

BACILLARIOPHYCEAE: *Centronella reichelti* was recorded for the first time in the spring of 1966. Since then it has occurred nearly every year in spring or in autumn or in both, sometimes persisting throughout winter in low numbers.

CHLOROPHYCEAE: *Gemellicystis* sp. was observed only in the springs of 1975 and 1977 when it reached a density of 350 and 45.10^3 colonies.l^{-1} respectively. *Gloeocystis vesiculosa* has been a regular summer species since 1974.

EUGLENOPHYCEAE: *Trachelomonas hispida* has been found in low numbers since 1975, mainly in autumn and winter, but was probably not recognized in earlier years.

DINOPHYCEAE: The most important species in category 3 is *Ceratium hirundinella.* In the first phytoplankton records of Lake Vechten (1956) *C. hirundinella* appeared in August and rose to a maximum of up to 50.10^3 cells.l^{-1}. In the course of 20 years the densities increased steadily, reaching 500.10^3 cells.l^{-1} in the last three years. It is present now for a largely extended period, from May up to November. As a consequence *C. hirundinella* strongly dominates the summer phytoplankton in Lake Vechten, as discussed earlier.

From the foregoing it can be concluded that some distinct changes have occurred in the species composition in Lake Vechten. The total number of chlorophycean species has decreased only slightly, i.e. from 17 to 15, but 6 species which were abundant in the past (1962–1967) declined sharply over a period of 10 years. Thus, at present, Chlorophyceae account for considerably less biomass in summer. This change is concomitant with an increase in the densities of *C. hirundinella.* Each year this Dinophycea starts growth earlier in the year with a steady increase in maximal numbers. Consequently a competition for light and nutrients may be expected. It is not improbable that this has led to the exclusion of species for which the requirements were not fulfilled.

Additional information about the earlier onset of growth of summer species in the seventies was deduced from an evaluation of the biomass data. The seasonal fluctuation in phytoplankton biovolumes in 1977 and 1979 (Fig. 1) shows a minimum in early summer, which lasts for one month. Comparable biovolume data are available for the period from March to 15 July 1956 (Fig. 2) showing a minimum

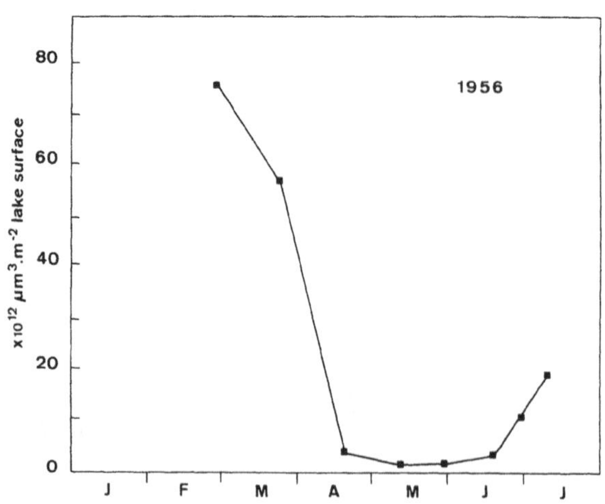

Fig. 2. Phytoplankton biomass in terms of cell volume during spring and early summer 1956 in Lake Vechten (Leussink, unpublished results).

after the spring maximum of at least two months. So in 1977 summer growth started about one month earlier (second half of May) than in 1956. For the period from 1956 to 1976 biovolumes could not be calculated, partly because of inadequate sampling methods. Therefore, the data of just one year (1956) are of limited value. However, the results do indicate that the summer peak now extends over a longer period, which suggests possible changes into a more eutrophic level of the lake in the course of the last 20 years. This would be in agreement with the findings of Davis (1964) about Lake Erie. The evaluation of chemical and physical parameters and of chlorophyll data will probably provide additional evidence.

Vertical distribution and migration of some phytoplankton species

In summer, when the lake is both thermally and chemically stratified, several phytoplankton species; i.e. *Mallomonas caudata* (Chrysophyceae) and *Oscillatoria* and *Lyngbya* sp. (Cyanophyceae), develop large population maxima in the metalimnion. *M. caudata* starts growth in July at 5–6 m depth. In the course of the following period the population sinks to 6–8 m depth, the densities increasing to one million cells.l^{-1} in some years (Fig. 3, Table 3). In the second half of September the stratification of the *M. caudata* population breaks down, resulting in an even distribution of the cells in the water column. Thereafter, cell numbers decrease rapidly and by the end of October the alga disappears. Cyst formation was not obvious, though Nygaard (1977) reports cyst formation in about 37% of the individuals present in populations in Lake Gribsø (Denmark). He also found increased O_2 and decreased P-concentrations in strata where *Mallomonas* was abundant. In Lake Vechten no such changes in O_2 and PO_4-concentrations associated with the *Mallomonas* populations were recorded. It should be noted that *Mallomonas* always occurred under mi-

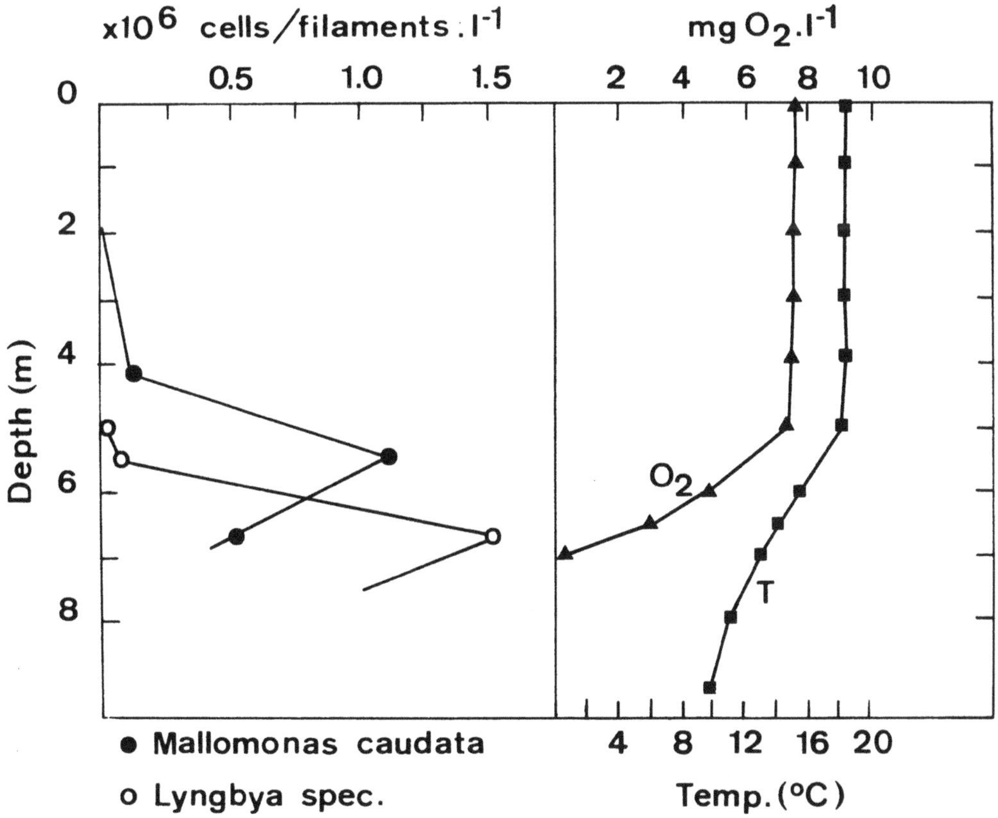

Fig. 3. Vertical distribution of *M. caudata* and *Lyngbya* sp. and profiles of oxygen concentration and temperature on 2 August 1977 in Lake Vechten.

Table 3. Vertical distribution of cell numbers ($\times 10^3.l^{-1}$) of *Mallomonas caudata* during 1977, 1978 and 1979.

	1977					1978					1979				
Depth interval (m)	19–7	2–8	16–8	31–8	13–9	17–7	1–8	15–8	29–8	12–9	17–7	30–7	14–8	28–8	11–9
3.6–4.8	52	69	–	–	44	–	30	11	–	–	–	–	939	51	+
4.8–6.0	183	1124	–	–	118	7	52	59	–	n.d.	–	86	411	68	7
6.0–7.2	741	500	2589	514	366	–	n.d.	140	138	n.d.	30	28	134	891	150
7.2–8.4	n.d.	n.d.	2313	427	n.d.	–	44	24	59	n.d.	–	+	–	351	722

n.d. = not determined.
\+ = present in very low numbers.
– = present in numbers below detection limit of counting procedure.

cro-aerophilic conditions. Light intensity at the depth of the *Mallomonas* maximum (6.5 m) was about 1% of the surface irradiance, which corresponds with Nygaard's results. From some other papers dealing with subsurface maxima of *Mallomonas* (Weimann 1933, 1942; Tilzer 1973) and from the observations in Lake Vechten, it can be concluded that *Mallomonas* prefers low light conditions. Moreover, diurnal vertical migration does not seem to occur. The depth of the *Mallomonas* maximum was rather constant, although maximal densities may vary considerably (Table 3).

The cyanophycean species *Oscillatoria* sp. and *Lyngbya* sp. exhibit a similar pattern of vertical distribution. These species develop dense populations, usually in the lower part of the metalimnion, from July to October (see Fig. 3). Numbers of filaments may rise to over 2 million.l^{-1}. This phenomenon has been observed frequently in many lakes in Europe and North America. These dense populations of bluegreens usually occur at a depth with low O_2 concentrations and a light intensity of about 1% of the surface light intensity (Zimmerman 1969; Fogg *et al.* 1973). In Lake Vechten the population of bluegreen algae developed at the boundary between the aerobic and anoxic water during stratification. Light intensities at these depths are around 1% of that in surface water and temperatures range from 9° to 14 °C. Zimmerman (1969) studied metalimnetic populations of *Oscillatoria rubescens* in the 'Vierwaldstätter See' (Lake Lucerne) and observed a vertical migration of the population in the course of summer. In Lake Vechten a vertical migration did not occur. Apparently bluegreen algae can grow vigorously in the deeper water layers at reduced light intensities. They are thought to benefit from the nutrient diffusion from the hypolimnion. However, there are many possible explanations for such subsurface maxima discussed extensively by Fogg *et al.* (1973). So far the conditions which favour the development of *Oscillatoria* sp. and *Lyngbya* sp. in the metalimnion of Lake Vechten have not been investigated in detail.

On 8 and 9 September 1980, the diurnal migration of the phytoplankton was investigated during a 24 hour experiment. The only species exhibiting a distinct vertical migration was *Ceratium hirundinella* (Dinophyceae), constituting about 90% of the total phytoplankton biomass. *C. hirundinella* cells tended to rise in the evening and at night (Fig. 4). In the morning they descended, which resulted in a maximum in the upper part of the metalimnion (5–6 m depth) at noon. The anoxic water of the hypolimnion formed the lower boundary for the migrating *Ceratium*. The results of counts of *Rhodomonas minuta* and *Cryptomonas* sp. (Cryptophyceae) point to a likely diurnal migration. They seem to move upward by day and downward at night. However, more data are needed before valid conclusions can be drawn.

Summary

In Lake Vechten phytoplankton growth starts in February or March with diatoms like *Stephanodiscus hantzschii* and *S. astraea,* and the Chlorophycea *Chlorella pyrenoidosa* resulting in a spring maximum in March or April. Thereafter, when thermal stratification sets in, a minimum in phytoplankton biomass occurs, which lasts about one month. From June onwards summer species be-

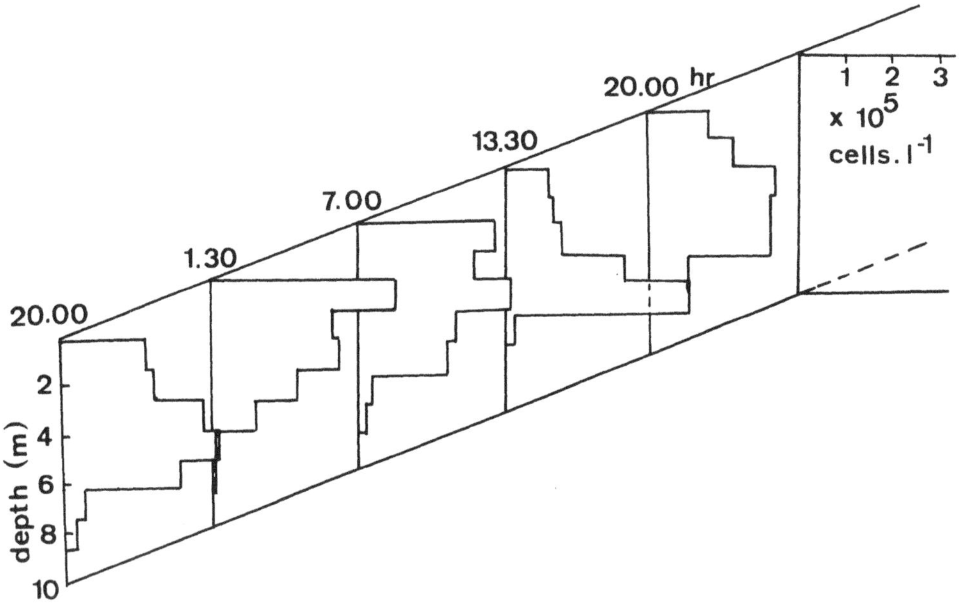

Fig. 4. Diurnal vertical migration pattern of *C. hirundinella* on 8 and 9 September 1980 in Lake Vechten.

longing to the Chlorophyceae, Chrysophyceae and Dinophyceae start to develop. The summer maximum in Lake Vechten is usually higher and extended over a longer period than the spring maximum. Dominant species in this period are *Mallomonas caudata* (Chrysophyceae) and *Ceratium hirundinella* (Dinophyceae), especially when biovolume data are taken into account. In August a distinct vertical distribution is found in numbers of *M. caudata* in the metalimnion and of *Oscillatoria* and *Lyngbya* species at the boundary of the meta- and the hypolimnion, where oxygen concentrations are very low. In September summer species are succeeded by diatoms like *Asterionella formosa* and several other pennate species. By the end of October phytoplankton biomass decreases rapidly with the gradual breakdown of thermal and chemical stratification. In December and January a typical winter community develops dominated by *Mallomonas akrokomos, Rhodomonas minuta* and several *Cryptomonas* species (Cryptophyceae).

In the course of 23 years (1956–1979) of phytoplankton study in Lake Vechten some changes in species composition and abundance have occurred. Numbers of several chlorophycean species have declined sharply. Consequently Chlorophyceae at present account for considerably less biomass in summer. Their place is taken by *Ceratium hirundinella,* which not only increased in number, but also extended the period in which it is present from 2 to 4 months (June-October). As a result of these changes the minimum in phytoplankton biomass in early summer shortened from 2 months in 1956 to one month only in 1977–1979. This might suggest a slight eutrophication of the lake in the course of the last 20 years.

Acknowledgements

During the past 25 years several workers in succession have contributed to the study of the phytoplankton in Lake Vechten. I would like to gratefully acknowledge their contributions to the results presented here, particularly B. Z. Salomé for his data from 1961 to 1967. Also many thanks are due to M. D. M. Trommel for careful identification and enumeration of the algal species from 1976 to 1979 and to P. J. Boesewinkel-de Bruyn for the enumeration in 1980.

References

Blaauboer, M. C. I., van Keulen, R. & Cappenberg, Th. E., 1982. Extracellular release of photosynthetic products by freshwater phytoplankton populations, with special reference to the algal species involved. Freshwat. Biol. (in press.).

Bourelly, P., 1966-1970. Les algués d'eau douce; initiation à la systematique. Boubée, Paris. 3 vols.

Davis, C. C., 1964. Evidence for the eutrophication of Lake Erie from phytoplankton records. Limnol. Oceanogr. 9: 275-284.

Fogg, G. E., 1975. Primary productivity. In: Riley, J. P. & Skirrow, G (Eds.) Chemical Oceanography, Vol. 2, pp. 386-445. Academic Press, New York.

Fogg, G. E., Stewart, W. D. P., Fay, P. & Walsby, A. E., 1973. The Blue-Green Algae. Academic Press, London.

Gulati, R. D., Siewertsen, K. & Postema, G., 1982. The zooplankton: its community structure, food and feeding, and role in the ecosystem of Lake Vechten. Hydrobiologia 95: 127-163.

Hustedt, F., 1962. Dr. L. Rabenhorsts Kryptogamen Flora von Deutschland, Österreich und der Schweiz. VII. Band: Die Kieselalgen. Johnson Reprint Corp., New York/J. Cramer, Weinheim.

Hutchinson, G. E., 1944. Limnological studies in Connecticut. 7. A critical examination of the supposed relationship between phytoplankton periodicity and chemical changes in lake waters. Ecology 25: 3-26.

Hutchinson, G. E., 1967. A Treatise on Limnology, Vol. 2. John Wiley, New York/J. Cramer, Weinheim.

Mackareth, F. J. H., 1953. Phosphorus utilization by Asterionella formosa Hass. J. exp. Bot. 4: 296-313.

Makarewicz, J. C., Baybutt, R. I. & Damann, K., 1979. Changes in the apparent temperature optima of the plankton of Lake Michigan at Chicago. J. Fish. Res. Bd Can. 36: 1169-1173.

Nauwerck, A., 1963. Die Beziehungen zwischen Zooplankton und Phytoplankton im See Erken. Symb. Bot. Upsal. 17: 1-163.

Nygaard, G., 1977. Vertical and seasonal distribution of some motile freshwater plankton algae in relation to some environmental factors. Arch. Hydrobiol. Suppl. 51: 67-76.

Parma, S., 1971. Chaoborus flavicans (Meigen) (Diptera, Chaoboridae): an autecological study. Ph. D. thesis, Groningen. 128 pp.

Pascher, A., 1976. Süsswasserflora Deutschlands, Österreichs und der Schweiz, Heft 10. Bacillariales by F. Hustedt. Reprinted by Otto Koeltz Sci. Publ., Koenigstein.

Pearsall, W. H., 1932. Phytoplankton in the English lakes. 2. The composition of the phytoplankton in relation to dissolved substances. J. Ecol. 20: 241-262.

Pechlaner, R., 1970. The phytoplankton spring outburst and its conditions in Lake Erken (Sweden). Limnol. Oceanogr. 15: 113-130.

Prescott, G. W., 1951. Algae of the Western Great Lakes Area. Cranbrook Inst. Sci. Bull. 31.

Rodhe, W., 1948. Environmental requirements of freshwater plankton algae. Experimental studies in the ecology of phytoplankton. Symbol. Bot. Upsal. 10: 1-149.

Skuja, H., 1948. Taxonomie des Phytoplanktons einiger Seen in Uppland, Sweden. Symb. Bot. Upsal. 9: 396 pp.

Skuja, H., 1957. Taxonomische und biologische Studien über das Phytoplankton Schwedischer Binnengewässer. Nova Acta Reg. Soc. Sci. Upsal. Ser. 4, 16: 1-404.

Steenbergen, C. L. M. & Verdouw, H., 1982. Lake Vechten: aspects of its morphometry, climate, hydrology and physicochemical characteristics. Hydrobiologia 95: 11-23.

Talling, J. G., 1962. Freshwater algae. In: Lewin, R. A. (Ed.) Physiology and Biochemistry of Algae, pp. 743-757. Academic Press, New York.

Talling, J. F., 1971. The underwater light climate as a controlling factor in the production ecology of freshwater phytoplankton. Mitt. int. Verein Limnol. 19: 214-243.

Tilman, D., 1977. Resource competition between planktonic algae: an experimental and theoretical approach. Ecology 58: 338-348.

Tilman, D. & Kilham, S. S., 1976. Phosphate and silicate growth and uptake kinetics of the diatoms Asterionella formosa and Cyclotella meneghiniana in batch and semicontinuous culture. J. Phycol. 12: 375-383.

Tilzer, M. M., 1973. Diurnal periodicity in the phytoplankton assemblage of a high mountain lake. Limnol. Oceanogr. 18: 15-31.

Utermöhl, H., 1958. Zur Vervollkommnung der quantitativen Phytoplankton-Methodik. Mitt. int. Verein Limnol. 9: 1-38.

Weimann, R., 1933. Hydrobiologische und hydrographische Untersuchungen an zwei teichartigen Gewässern. Beih. bot. Zbl. 51: 397-476.

Weimann, R., 1942. Zur Gliederung und Dynamik der Flachgewässer. Arch. Hydrobiol. 38: 481-524.

Werff, A. van der & Huls, H., 1957-1974. Diatomeeënflora van Nederland.

Wetzel, R. G., 1975. Limnology. W. B. Saunders, Philadelphia. 743 pp.

Willén, E., 1976. A simplified method of phytoplankton counting. Br. phycol. J. 11: 265-278.

Zimmerman, U., 1969. Ökologische und physiologische Untersuchungen an der planktischen Blaualge Oscillatoria rubescens D. C. unter besonderer Berücksichtigung von Licht und Temperatur. Schweiz. Z. Hydrol. 31: 1-59.

The primary production of phytoplankton in Lake Vechten

W. A. de Kloet
Limnological Institute, 'Vijverhof' Laboratory, Rijksstraatweg 6, 3631 AC Nieuwersluis, The Netherlands

Keywords: primary production, extracellular excretion, chlorophyll-*a*, phytoplankton, stratified lake, phototrophic bacteria, biomass, turnover rates

Abstract

The primary production of the phytoplankton of Lake Vechten (The Netherlands) (area, 4.7 ha; mean depth, 6 m), an unpolluted and stratified sandpit was investigated from 1969 to 1980 (except in 1971, 1975 and 1976) by the *in situ* ^{14}C-technique. Other data collected include: solar radiation, transparency, oxygen and thermal structure. In winter and spring diatoms, Cryptophyceae and Chlorococcales were important algal groups, while in summer Dinophyceae and Chrysophyceae were important. The chlorophyll-*a* concentration was compared to the cellular biovolumes (= fresh weight) of the most abundant phytoplankton species. The primary production maxima occurred in winter, spring and during the summer stratification. The vertical profiles of photosynthesis exhibit light inhibition at surface to a maximum of 4 m. The maximum of zooplankton grazing in May–June caused a sharp decrease in the phytoplankton biomass and seston concentration accompanied by the highest transparency ('clear water' phase).

The values for cellular C-fixation range from 10 to 1 307 mg C · M^{-2} · day^{-1} (annual mean of 280 mg C · m^{-2} · day^{-1}). High dark fixation (up to 100%) was encountered in the metalimnion and hypolimnion from July to October together with peaks of ^{14}C-fixation due to a crowding of phytoplankton and phototrophic anoxic bacteria. Extracellular excretion by phytoplankton, investigated in 1977 to 1979, was 15% of the annual mean of the total C-fixation. The photosynthetic efficiency, turnover rates, and activity coefficients were low, particularly in the summer months when *Ceratium hirundinella* was predominant. The seasonal variations were controlled mainly by solar radiation and probably phosphate, the former being more important in the non-stratification period and the latter during the stratification period.

Introduction

The primary production of the phytoplankton is one of the most important sources of the energy input in fresh waters (Golterman 1975; Wetzel 1975; Moss 1980). In order to obtain basic information on the energy flow in the ecosystem of Lake Vechten, the primary production rates of phytoplankton have been measured more or less regularly since 1969, at the same time as the other investigations on the lake. The chlorophyll-*a* has been employed as an index of living phytoplankton biomass. In this paper the results of the 10-year study of primary production on the lake are reported in terms of relationships with such biotic factors as chlorophyll-*a* and with the relevant abiotic factors.

Over these years several workers have studied the phytoplanktonic primary production (Gulati, 1969–1970; Liqui-Lung, 1972–1974; Blaauboer, 1976–1978; and Olie, 1979); the last two workers studied the extracellular primary products. The chlorophyll-*a* was measured by Boesewinkel (1969–1970), Liqui-Lung (1973–1974) and Steenbergen (1979–1980). The data of these authors are reported in the annual Progress Reports of the Limnological Institute.

Hydrobiologia 95, 37–57 (1982). 0018-8158/82/0951-0037/$04.20.

Material and methods

Description of the lake

Lake Vechten is situated north of the Utrecht-Arnhem highway (E 36) near the village of Bunnik (lat. 52°04′N, long. 5°05′E). The main morphometric characteristics of the lake are:

surface area:	4.71 ha
maximum depth:	11.9 m
mean depth:	6.0 m

The lake has two depressions, an eastern and a western one, more or less separated by a ridge. The present study was carried out mainly at a station in the pelagic part of the eastern basin of the lake (max. depth 10 m). The lake is stratified from May to October, the thermocline being found between 5 and 7 m. For more details see Steenbergen & Verdouw (1982).

Methods

Primary production rates were measured *in situ* using the ^{14}C-technique (Steemann-Nielsen 1952). The technique of sampling was similar to that of other phytoplankton studies in the lake (Blaauboer 1982). Before incubating, the water samples were filtered through a 125 μm mesh sieve to remove most zooplankton; in summer 170 μm mesh sieve was used to prevent the removal of the filamentous algae and large-sized dinophycean plankton, viz. *Ceratium hirundinella*. Light and dark bottles (approx. 125 ml) were filled with lake water to which 20 μCi $NaH^{14}CO_3 \cdot l^{-1}$ was added, and incubated at surface, 0.5, 1, 2, 3, 4, 5 and 6.5 m depth. From 1978 onwards, measurements were also done at 7.5, 8.5 and 9.5 m, at the last two depths during the summer months only. In general, the samples were incubated between 10.00 and 14.00 hr. After incubation the samples were placed in a lightproof box and transported to the laboratory in about half an hour.

Subsamples, mostly 100 ml, but less when filamentous algae were present, e.g. at metalimnion and hypolimnion, were filtered immediately through a 0.45 μm membrane filter (GmbH Sartorius SM 11306) under reduced pressure (>0.5 atm.). The filters were washed with 0.05 N HCl and 3 μm filtered lake water to remove inorganic ^{14}C. The radioactivity of the residue on the filters was measured in liquid scintillation counter (Packard Tri-Carb 2425) after transferring the filters to scintillation vials and adding 10 ml instagel and 5 ml distilled water. The external standard ratio method was used for the DPM conversion. Up to 1974 the samples were counted using a thin window gas-flow counter (Frieseke & Höpfner, F.R.G.). A comparison of the two counting techniques showed that the results with the scintillation counting were on the average 15% higher than those using the gas-flow technique (Liqui-Lung, pers. comm.).

The specific ^{14}C-activity in the incubation bottles was counted by using a 1 ml subsample in a scintillation vial to which 200 μl N NaOH and scintillation cocktail, as in the case of filters, were added. The use of NaOH is essential to retain inorganic ^{14}C in the cocktail, which is acidic and will result in loss of inorganic ^{14}C (see also Lean & Burnison 1979; Søndergaard 1980). Using the gas-flow counting technique, the $BaCO_3$ method (Steemann-Nielsen 1952) was applied to measure specific activity. The primary production rates were calculated according to Vollenweider (1969), subtracting the ^{14}C-uptake of the dark bottle from those of the replicates in light. An isotopic correction factor of 1.06 was employed.

Daily production rates were computed from the production measured during the incubation period and using the ratio between the total irradiance during the day and that during the incubation period, assuming that photosynthetic efficiency was not affected by changed in the quality of light. Global radiation data were obtained from the Royal Netherlands Meteorological Institute (K.N.M.I.) at De Bilt, about 5 km from the lake.

From 1977 to 1979 extracellular excretion rates of the phytoplankton were also determined using the same samples as for primary production. For details of this method see Blaauboer *et al.* (1982).

The primary production rates measured as C-fixation by phytoplankton are expressed as production (P), which together with extracellular production (E) constitutes the total C-fixation by phytoplankton (A). The other abbreviations used are:

ΣP_{hr}: areal production rates as mg $C \cdot m^{-2} \cdot hr^{-1}$
ΣP_d: areal daily production as mg $C \cdot m^{-2} \cdot day^{-1}$

For determination of chlorophyll-*a* concentration the lake water was sampled in a similar manner as for primary production. Subsamples, mostly 2 l, but 0.5 to 1 l if filamentous algae were present in metalimnion and hypolimnion, were filtered first through 125 or 170 μm plankton-gauze to remove zooplankton, as in the case of primary production samples. Thereafter, the samples were filtered through glassfibre filters (Gelman AE), but because of their relatively large mesh size (about 4 μm) compared with the cell size of some of the nanoplankters, from 1978 onwards they were replaced with Whatman filters (GF/F) which have a mesh size of 0.7 μm. Chlorophyll-*a* was extracted with 80% ethanol (Moed 1973; Moed & Hallegraeff 1978). Phaeopigments were measured after acidification with 0.4 N HCl (pH = 2.7); however, up to 1977 a relatively stronger acidification level was used following Golterman & Clymo (1969).

Underwater light penetration was determined by a quantum sensor (Li-192S) connected to the Li-185A meter (Lambda Instruments Co. Inc., U.S.A.), measuring quanta as photosynthetically active radiation (Ph. A. R.). The measurements are compared with those using the Secchi disc (black-and-white quartered disc 20 cm ∅). For further details see Steenbergen & Verdouw (1982).

For details of phytoplankton counting see Blaauboer (1982).

Results and discussion

Seasonal and spatial changes of primary production rates

The data of 1979 are described in detail to show the photosynthetic capacity at different depths and its seasonal changes together with the rates of extracellular excretion (see Fig. 1). In addition, the production of phytoplankton in the pelagic zone can be compared with those in the littoral region of periphyton (Gons 1982) and macrophytes (Best 1982), both of which were also measured in 1979.

The phytoplankton production of Lake Vechten shows seasonal cycles of great complexity. In other words, simple cycles with one or two peaks, such as often occur in dimictic lakes (cf. Hutchinson 1957), are lacking. Nevertheless, we will try to generalize the seasonal variation in production and extracellular excretion comparing their seasonal and spatial changes in 1979 with those of the other years (Fig. 1).

Due to the mild winter climate in the Netherlands and the lack or absence of long ice-cover in Lake Vechten – the winter of 1979 being an exception – the primary production starts increasing as early as January. Owing to a steady increase in solar radiation and water temperature from January to March, peaks of the ΣP_{hr} during spring were found mostly in March or in early April (44–73 mg C · m^{-2} · hr^{-1} in 1973, 1974 and 1977). In contrast, the ΣP_{hr} increased in 1978 to record relatively high values already in late January and early February, namely 66 and 79 mg C · m^{-2} · hr^{-1} respectively. These high rates during the early winter months are exceptional (Fig. 2), and probably due to an explosive increase of the diatom *Stephanodiscus hantschii* (Blaauboer 1982). For comparison, the maximum ΣP_{hr} in 1979, namely 46 mg C · m^{-2} · hr^{-1}, was observed on 14 August (Fig. 1). Besides, the ΣP_{hr} in 1979 soon after the ice period was unusually low in March, namely 14–17 mg C · m^{-2} · hr^{-1}, and the spring maximum of 30 mg C · m^{-2} · hr^{-1} was observed on 24 April (see Fig. 1).

The production rates increased during the first half of the stratification period, maximum ΣP_{hr} being encountered mostly in June. During May and early June the water usually became very clear with Secchi disc depths of up to 4.5 m, the annual maximum ('clear water phase'). Generally, the Secchi disc depth ranged between 2 and 3.5 m with an annual average of 2.7 m. The 'clear water phase' is most likely due to a sharp decrease in phytoplankton biomass (cf. Fig. 4) due to natural diminution, mainly of diatom species, and to the annual grazing maximum of herbivore zooplankton in May and June (Gulati *et al.* 1982).

The maximum production rate (P_{max}) in the epilimnion, found generally at a depth of 0.5 to 2 m, was observed at 4–5 m depth during the 'clear water phase'. During this phase a decrease in phytoplankton biomass was correlated with an increase of the activity coefficients (= Σ P per unit chlorophyll-*a*) which reached its maximum values, particularly, in May 1977 and 1978 (Fig. 6). During June to mid-July 1979 the ΣP_{hr} remained more or less constant. This is in contrast with the data of other years, except for 1977 (Table 1).

During the second half of summer stratification, i.e. from mid-July to mid-October, production profiles were characterized by high ^{14}C-fixation rates in the metalimnion and hypolimnion. P_{max} rates, mostly in the strata between 6.5 and 7.5 m, were in several cases even higher than those in strata in the epilimnion. In 1979 this phenomenon was observed first on 31 July and again on 11 September and 9 October; more than 60% of Σ P_{hr} was found below 6 m depth (see Fig. 1). Also, during the late summer of 1977 and 1978 ^{14}C-fixation rates below 6 m depth were high, although they comprised generally not more than 30% Σ P_{hr}, except on 2 August 1977 when it was 46% Σ P_{hr}. In 1973 and 1974 too the ^{14}C-fixation in metalimnion and hypolimnion and the values recorded were even higher than in the subsequent years. The maximum values encountered in 1973 and 1974 were 51.5 and 35.5 respectively compared to the range of maximum values in 1977–1979 of 8.5–14 mg C · m^{-3} · hr^{-1}.

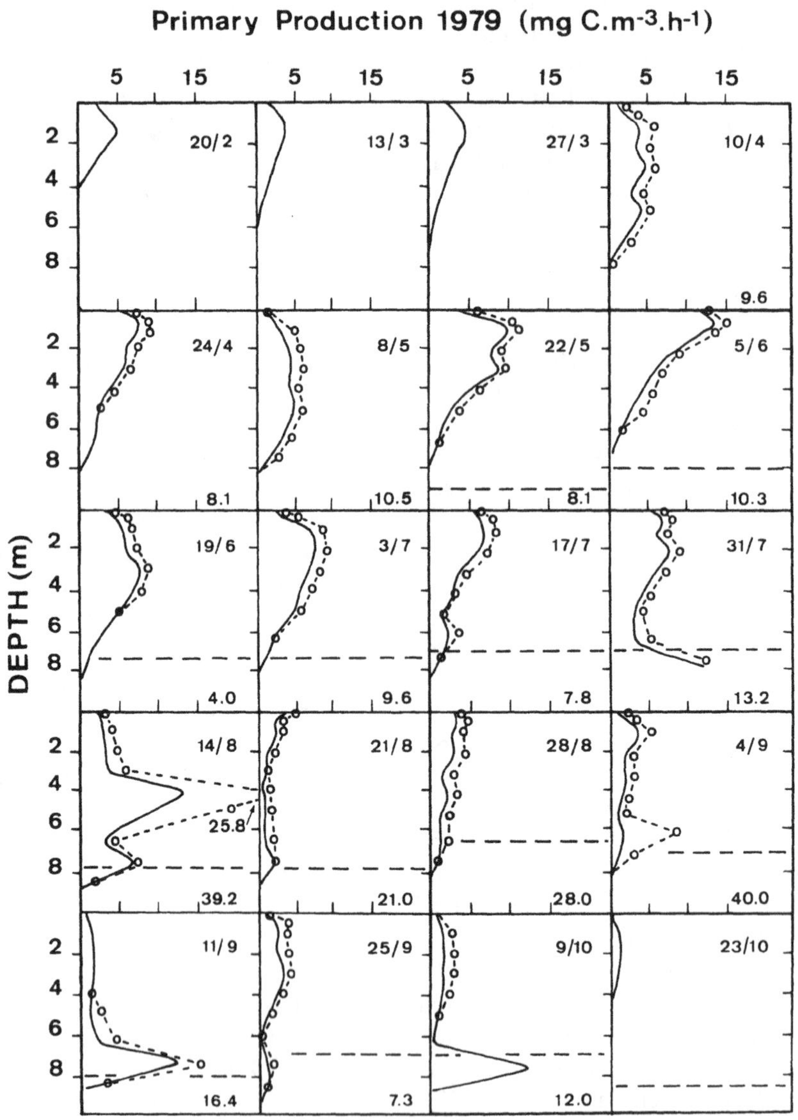

Fig. 1. Depth profiles of primary production rates of the pelagial phytoplankton in Lake Vechten during 1979. —: cellular C-fixation ---O---: total fixation, i.e. cellular + extracellular excretion. The dates of incubation are given in the right hand top corner of the boxes, and the excretion as percentage of total fixation in right hand bottom corner. The stripes indicate the depth below which oxygen is absent (Z_O-depth).

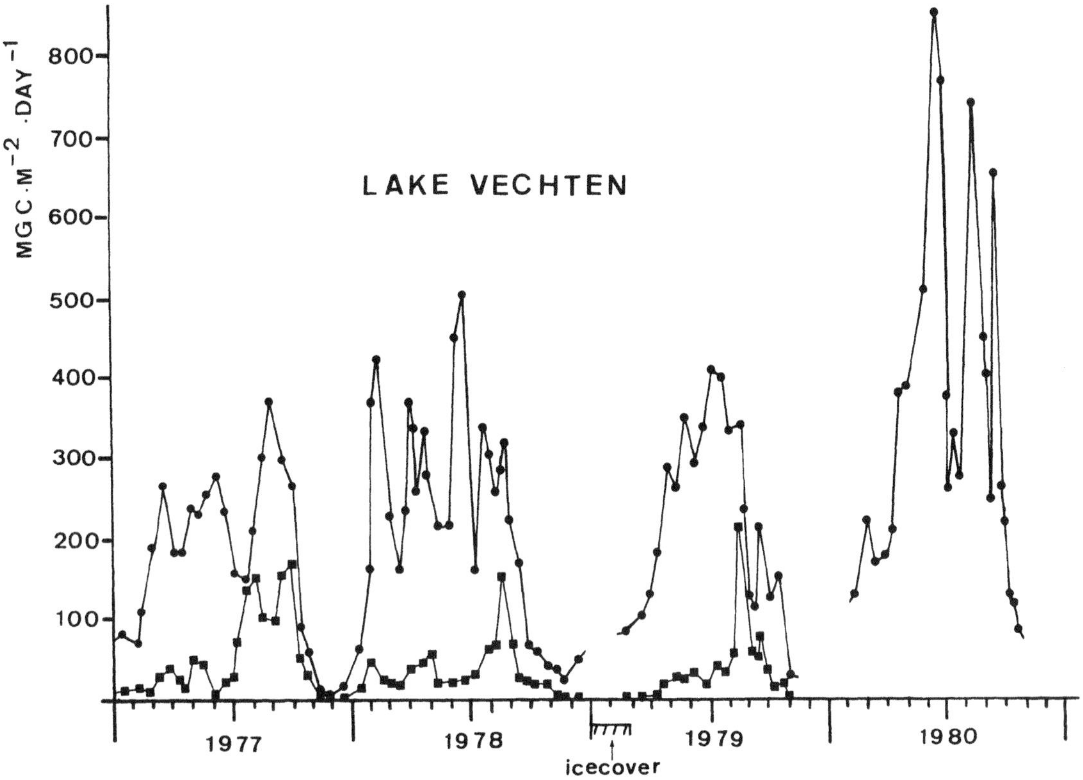

Fig. 2. Daily rates of production (ΣP_d) and extracellular excretion (ΣE_d) of pelagial phytoplankton in Lake Vechten during 1977–1980. —●—: ΣP_d-values; —■—: ΣE_d-values.

The peak values in metalimnion and hypolimnion were mostly recorded above the Z_O-depth (depth below which oxygen could not be detected, Fig. 1) but on several dates also in the anaerobic part, e.g. on 31 July, 25 September and 9 October in 1979. That the dark ^{14}C-fixation at these depths mostly comprised 25–50% of that in the light is striking. In deeper parts, i.e. 8 to 9 m, the ^{14}C-fixation in the light and dark bottles was comparable (Steenbergen 1982). Probably, the ^{14}C-uptake in the anaerobic hypolimnion was due to anoxygenic photosynthesis by bacteria (Chlorobiaceae and Chromatiaceae). The numbers of the filamentous blue-green algae *Oscillatoria* and *Lyngbya* species and the flagellate (Chrysophyceae) *Mallomonas caudata* were high in August and September. The last-named species became abundant in the 6–8 m stratum (Blaauboer 1982). The ^{14}C-fixation rate below 5 m was possibly too low due to diffusion of oxygen during the sampling. Increase in oxygen tension may inhibit the oxygenic photosynthesis and even lead to complete elimination of anoxygenic photosynthesis. Experiments recently carried out in Lake Vechten hint at this (Steenbergen 1982).

Table 1. Range of four fortnightly measurements of ΣP_{hr} in Lake Vechten during June–July.

Year	Period	Range ($mg C \cdot m^{-2} \cdot hr^{-1}$)
	June–July	
1974	11–23	46–116
1978	6–17	18– 65.5
1980	11–16	19– 74
1977	7–19	17– 27
1979	5–17	32– 42

The primary production by the epilimnetic phytoplankton (0–5 m) from mid-July to the end of October showed a normal pattern. The ΣP_{hr} in 1979 was low from 21 August until 4 September (9–14 mg C $\cdot$ m^{-2} $\cdot$ hr^{-1}) compared with that in spring and early summer but also with the rates in August and September in 1977 and in 1978 (20–33.5 mg C · m^{-2} · hr^{-1}). This was probably not due to changes in the abundance and the composition of phytoplankton. *Ceratium hirundinella, Chlorella* sp. and *Rhodomonas* sp. were present for all three years (1977–1979). Abnormally low solar radiation during the incubation period may have caused the low primary production rate on 21 August 1979 (45 J · cm^{-2} · hr^{-1} as against a normal of 200 J · cm^{-2} · hr^{-1}). The P_{max} value at 4 m depth on 14 August (Fig. 1) was probably due to *Mallomonas caudata,* being present in high numbers in the 3.6–4.8 m stratum (Blaauboer 1982).

After mid-October 1979 the production decreased sharply and, as was observed in 1977 and 1978 (Fig. 2), the annual minimum was reached in November.

Extracellular excretion

In the period 1977–1979, in addition to the ^{14}C-cellular fixation in phytoplankton, losses as extracellular excretion were also measured. In general these losses were no more than 20% of total ^{14}C-fixation (ΣA_{hr}); however, for short periods during summer stratification (August–September) excretion rates as high as 30–40% of ΣA_{hr} were recorded (cf. Figs. 1 and 2). During spring the extracellular release of assimilated carbon followed the increase in the epilimnion of dominant species such as *Chlorella* and *Stephanodiscus* spp. In 1977 the first E-value peak was recorded in early March, but in 1978 it was at the end of January followed by a second peak in April (Fig. 2). Due to an cover of ice in the first 2 months of 1979, spring data are missing, but from March to June excretion rates were similar to those of 1977 and 1978.

During the summer stratification of 1979, the highest excretion rates, as well as the highest total C-fixation, were observed in the water layer between 4 and 6 m, and on 14 August the E_{hr}-rate amounted from 40 to 50% of the total fixation in that layer (Fig. 1). In this stratum the abundant phytoplankton species, besides *Mallomonas caudata, Ceratium hirundinella* and *Chlorella* (see foregoing discussion), were *Rhodomonas minuta* and *Cryptomonas* sp. On 4 September even more than 60% of A_{hr} was observed between 5 and 7 m, probably due to *Mallomonas,* although the density of *Asterionella* in these layers was also high. In September and October the excretion rates decreased steadily and were observed only in the epilimnion. Similarly, in 1977 the extracellular release was also initially high in the metalimnion and later in the epilimnion; however, in 1979 the per cent excretion was relatively still high until the end of October and was 33% of ΣA_d (cf. Figs. 1 and 2). In contrast, during the summer stratification of 1978 the excretion rates were observed only in the epilimnion. For more details see Blaauboer *et al.* (1982).

Seasonal and spatial changes of chlorophyll-a

As an index of live phytoplankton biomass chlorophyll-*a* has been investigated in Lake Vechten for almost a decade. In relation to production studies during 1969–1976, the chlorophyll-*a* was determined mostly up to a depth of 7 m. However, due to the presence in bulk of algae and bacteria, which fixed carbon (Fig. 1) in strata below 7 m during stratification (July to October), the investigation was extended to the whole column water from 1977 onwards.

The annual course of chlorophyll-*a* concentration (Fig. 4) looks similar to that of phytoplankton production rates (cf. Fig. 2), although multiple regression analyses (using, amongst other data, that on chlorophyll and production in 1978; Gulati, pers. comm.) gives no significant relationship. This is most likely owing to the data of May when the lake starts to stratify and the chlorophyll-*a* concentration decreases to one of the lowest values during the year. In general, 2–3 mg · m^{-3} of chlorophyll was found in this period except in 1979 (Fig. 3) when the concentrations were still high and the minimum was delayed possibly due to the preceding thaw following the ice cover. This minimum level of chlorophyll-*a* and of seston $<105\ \mu m$ (Gulati *et al.* 1982) results in a 'clear water' phase, as mentioned earlier.

A second minimum of chlorophyll-*a* concentration (2–4 mg · m^{-3}) was – except for 1980 (cf. Fig. 4) – encountered in November–December just after breakdown of the thermal stratification. Corresponding to a decrease in the primary production rates, the chlorophyll-*a* concentration also decreased. This was caused by a sharp decrease of temperature and solar irradiance, thus limiting phytoplankton growth (Blaauboer 1982).

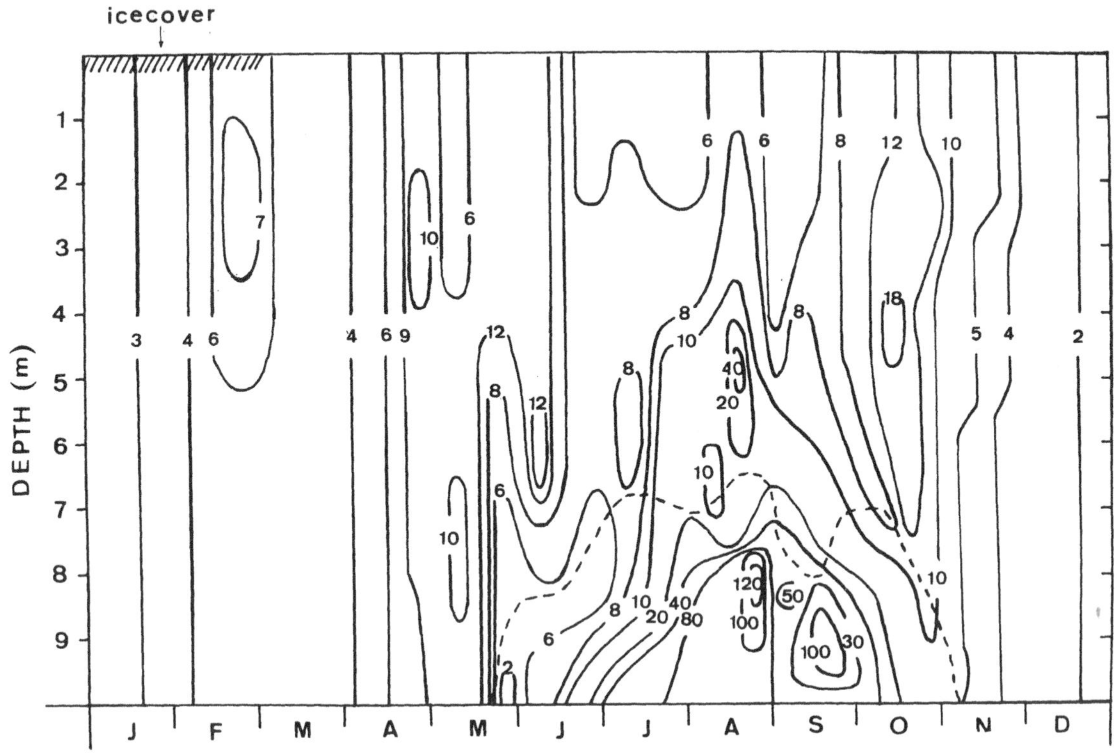

Fig. 3. Isopleths of chlorophyll-*a* concentration in the eastern, pelagial part of Lake Vechten during 1979. Broken line indicates the boundary between anaerobic layers below and aerobic zone above.

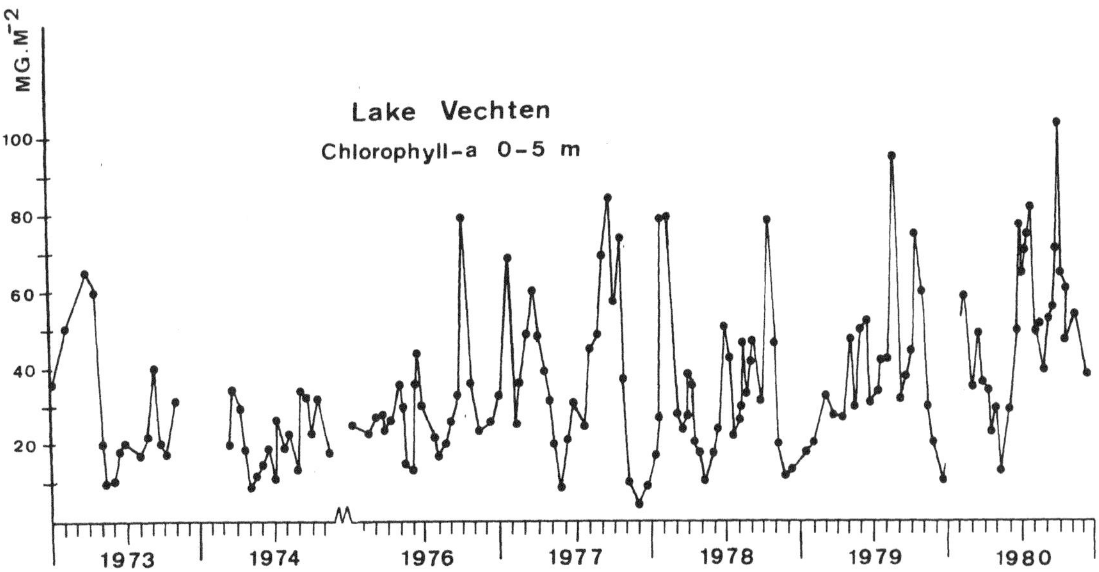

Fig. 4. Chlorophyll-*a* concentration in the upper 5 m stratum in the eastern, pelagial part of Lake Vechten.

From January to April, the chlorophyll-*a* concentration was high and main peaks were mostly found in January–February (12–16 mg · m^{-3}) and from mid-March to mid-April (7–12 mg · m^{-3}) (cf. Figs. 3 and 4).

The phytoplankton species abundant during this increase are *Chlorella* sp. and *Rhodomonas,* which possibly adapt to low temperature and light intensities. Cells adapted to lower light intensities during the winter months may have more chlorophyll-*a* content per unit biomass (e.g. Wetzel 1975; Desortová 1981). The chlorophyll-*a* maximum observed in end-January and in early February 1978 was due to high densities of the diatom *Stephanodiscus hantsschii.*

During the summer stratification the pattern of chlorophyll-*a* content in the epilimnion was irregular, with the main peaks mostly in September, but sometimes in October (e.g. 1978 and 1979, see Figs. 3 and 4). The chlorophyll-*a* concentration varied between 5 and 20 mg · m^{-3}; the maximum of 24 mg · m^{-3} was found on 17 September 1980 at 2 m depth. The chlorophyll-*a* maxima in September and October were due mainly to the abundance of the dinoflagellate *Ceratium hirundinella.* In September–October of 1977 and 1978, 70–80% of the chlorophyl-*a* concentration in the epilimnion belonged to the netplankton (Ø = 33–170 μm) (de Kloet, own observations), compared with 25–50% in June–August. In contrast, during the unstratified period chlorophyll-*a* belonged to the nanoplankton only.

During the summer stratification, especially from July to October, the chlorophyll-*a* concentrations in the metalimnion and hypolimnion were rather high (Fig. 3). In the metalimnion, generally between 5 and 7 m and just above the anaerobic zone, the mean concentration ranged between 10 and 40 mg · m^{-3} due to an abundance of *C. hirundinella, Mallomonas caudata* and *Chlorella* sp.; numbers of blue green algae were also high (Blaauboer 1982). The maximum chlorophyll-*a* concentration of 47 mg · m^{-3} on 14 August 1979 was due mainly to high density of *M. caudata,* giving a peak of primary production rates too (Fig. 1).

The highest chlorophyll-*a* concentrations were found in the hypolimnion, especially between 7.2 and 8.4 m depth (Fig. 3). Maxima observed in the different years were in second half of August:

1977: 99 mg · m^{-3}
1978: 81 mg · m^{-3}
1979: 123 mg · m^{-3}
1980: 188 mg · m^{-3}

An interesting point is not only the high chlorophyll concentrations but also as a consequence an increase below 5 m of the mean total chlorophyll-*a* content, especially in 1980 (Table 3). However, chlorophyll-*a* values in the anaerobic hypolimnion are questionable because of the methodological errors (Steenbergen, pers. comm.). On the other hand, the mean chlorophyll-*a* concentrations in the epilimnion (Table 2) also increased between 1970 and 1980. This increase may have been caused by eutrophication of the lake; this is indicated also by changes in the composition of phytoplankton species (Blaauboer 1982).

Table 2. Means and ranges of chlorophyll-*a* concentrations (mg · m^{-2}) in the upper 5 m stratum of the eastern, pelagial part of Lake Vechten (cf. Fig. 4). Period I, November–April (non-stratification); Period II, May–October (stratification). Number of measurements in parentheses.

Year	Period I	Period II	Annual	
	Mean	Mean	Mean	Range
1969	–	20 (11)	–	(4– 35)[a]
1970	11 (14)	26 (10)	20	4– 61
1973	–	19 (10)	(28)[b]	(9– 65)[b]
1974	–	22 (12)	–	(8– 34)[c]
1976	27 (11)	33 (13)	29	13– 79
1977	34 (11)	42 (12)	38	4– 84
1978	33 (14)	36 (15)	34	10– 78
1979	26 (9)	49 (13)	39	10– 95
1980	41 (9)	56 (18)	49	14–104

[a] Range from April to December.
[b] Mean and range from January to October.
[c] Range from mid-March to mid-November.

Table 3. Mean and ranges of chlorophyll-*a* concentrations (mg · m^{-2}) in the whole water column (A) and in the stratum below 5 m (B) of Lake Vechten. See further in text of Table 2.

	Year	Period I	Period II	Annual	
		Mean	Mean	Mean	Range
A	1977	64	171	120	7–352
	1978	65	145	109	21–284
	1979	48	181	126	19–392
	1980	80	324	230	26–538
B	1977	30	129	92	3–304
	1978	31	109	71	9–223
	1979	22	132	87	9–302
	1980	39	268	181	12–484

Comparison of the daily primary production rates in different years

The course of the daily integral production (ΣP_d) from 1977 to 1980 and of the extracellular excretion (ΣE_d) from 1977 to 1979 of the phytoplankton are shown in Fig. 2. The annual production means and ranges for the years 1969–1980 are given in Table 4. The average ΣP_d-values ranged from 162 to 427 mg C · m^{-2} · day^{-1}, giving a grand annual mean of 280 ± 100 mg C · m^{-2} · day^{-1}. The mean excretion rate during the 3 year study (1977–1979) was 32 mg C · m^{-2} · day^{-1} comprising about 15% of total carbon uptake by the phytoplankton (Table 5). Assuming that the excretion rates in the other years were also about 15% of the total carbon uptake, the mean daily total C-fixation for the whole 1969–1980 period would be about 330 mg C · m^{-2} · day^{-1}. Although the production was measured in the eastern basin of lake Vechten, a comparative study of phytoplankton production at stations in the western basin and in the littoral zone of the lake during 1979 and 1980 (de Kloet, unpublished data) confirms the validity for the whole lake of data collected in the eastern basin.

The daily values are computed using a light-ratio (see Methods); this calculation procedure has limitations since it accounts for the effects of the light quantity and not for the changes in spectral composition on the photosynthetic rates. The sum of carbon uptake of short-term incubations spread over the whole day have been shown to be nearly twice the production obtained by 12 and 24 hours exposures (e.g. Rodhe 1958; Kristiansen & Mathiesen 1964). This was confirmed by me for Lake Vechten in 1978 (unpublished data). A study of the diel patterns of the phytoplankton production in Lake Vechten during the summer of 1979, however, indicated that in general short-term ^{14}C incubations during midday, using a light factor, are a good indicator of the daily primary production (Opstelten 1981). Only on very sunny days may underestimates of up to 25% be obtained using the midday incubation data.

Table 4. Annual means and ranges of daily primary production rate (ΣP_d), AC-value and net photosynthetic efficiency (% of Ph.A.R., 400–700 nm) of the pelagial phytoplankton in Lake Vechten.

Year	Production (mg C · m^{-2} · day^{-1})		Activity coefficient (mg C · mg^{-1} chl · day^{-1})		Efficiency % (Ph.A.R.)	
	Mean	Range	Mean	Range	Mean	Range
1969	277	50– 630			0.29	
1970	354	50– 695			0.28	
1972	238	95– 645			0.20	0.12 –0.43
1973	414	110– 995			0.51	0.19 –2.2
1974	427	95–1 307			0.32	0.07 –0.67
1977	162	10– 375	2.3	0.6–15.2	0.16	0.05 –0.42
1978	202	25– 500	2.9	0.4– 9.9	0.22	0.055–0.81
1979	173	30– 415	2.4	0.3– 5.3	0.19	0.03 –0.41
1980	273	95– 850	2.55	0.4– 8.2	0.32	0.085–0.55

Table 5. Means and ranges of the daily rates of total C-fixation (ΣA_d) and the share of excretion (ΣE_d) in per cent for the pelagial phytoplankton of Lake Vechten.

Year	Excretion rates				Total C-fixation	
	(mg C · m^{-2} · day^{-1})		(% of ΣA_d-1)		(mg C · m^{-2} · day^{-1})	
	Mean	Range	Mean	Range	Mean	Range
1977	46	0–173	22.1	0–48	208	10–462
1978	27	0–143	11.8	0–34	228	25–527
1979	22	0–215	11.3	0–40	195	30–550

The daily rates of phytoplankton production of Lake Vechten and their seasonality are comparable with those of many other lakes in the temperature zone, e.g. Lawrence Lake in Michigan (Wetzel 1975). However, phytoplankton productivity data in lakes are voluminous and a good comparison is thus impracticable and perhaps also not very meaningful, because algae in each individual lake system have their own physiological characteristics and response to physico-chemical changes in their environments. Based on the productivity data (cf. Rodhe 1969; Wetzel 1975) the trophic status of Lake Vechten is in between an oligotrophic and mesotrophic one.

Most shallow Dutch lakes, which have originated from peat dredging, are naturally eutrophicated, but have become hypertrophic in the post-War period, owing to allochthonous input of polluted water. The average phytoplankton production in such lakes varies between 2 and 5 g C · m^{-2} · day^{-1} or more during the summer months (de Kloet 1971; Gulati 1972, 1976; Schmidt-van Dorp 1978; Golterman & Kouwe 1980). Lake Vechten is, however, relatively deep but comparable in depth with several lakes which have appeared recently in Holland, owing to sand digging, e.g. Wijde Blik, Vinkeveense Plassen, Lake Maarsseveen. Even though Lake Vechten resembles the other deeper lakes in its morphology and other physical phenomenon – it is small isolated and lacks a fluviatile input and has a well developed littoral zone – its annual phytoplankton productivity does not differ much from the other deep lakes mentioned (Gulati 1972, 1976; Flik & Keyser 1981).

The vertical extinction coefficient (ϵ_v) in Wijde Blik had a mean of 1.4 (m^{-1}) and ranged between 0.6 and 2.7 during 1969–1970, while I observed an average ϵ_v-value of 0.6 (range 0.45–0.95) in Lake Vechten during 1977–1978. Average Secchi disc value was 1.5 m in Wijde Blik (range 1.0–1.8 m) compared with 2.7 m (range 1.8–4.5 m) in Lake Vechten. Thus, comparing the light climate of, e.g., Lake Wijde Blik with that of Lake Vechten, the former was more turbid, especially during the period July–October due to a strong abundance of the blue-green alga *Microcystis aeruginosa* (Gulati 1972).

The primary production data can be evaluated better by comparing the shape of the production curve and by using the ratio $\Sigma P/P_{max}$. Vollenweider (1960) grouped the lakes on the basis of this ratio and their optical properties such as ϵ_v and Secchi depth. In Lake Vechten $\Sigma P/P_{max}$ during 1977–1979 had a mean value of 4.5 ± 1.8 (n = 77); the range did not differ much from year to year (2–12.5). The average $\Sigma P/P_{max}$ value of 5.1 ± 2.2 found during the stratification from May to October in Lake Vechten is comparable with those of Lake Erken (Ahlgren 1970) and Lake Varese (Vollenweider 1960). The high $\Sigma P/P_{max}$ ratio of Vechten compared with those of Wijde Blik (average 2; range 1–3.1; Gulati 1972) during the vegetation period from March to November is to be expected because Lake Vechten, being more clear, has a larger euphotic layer. The depth of P_{max}, Z_{max}, in Vechten was between 0.5 and 4 m (mean 1.6 m), while in Lake Wijde Blik it was between surface and 1 m.

About 70% of Σ P took place in the upper 2 m and 25% in the 2–4 m layer in Lake Wijde Blik, so the euphotic zone was hardly deeper than 4 m in this lake; in contrast, the euphotic zone extended generally to about 8 m in Lake Vechten (Fig. 1). On the average, about 40% of Σ P took place in the upper 2 m and 35% in the 2–4 m layer in Lake Vechten, not accounting for the production in metalimnion and hypolimnion during the summer months.

Nanoplankton versus netplankton production

The contribution of two size-fractions to the production rates of the pelagial phytoplankton, i.e. nanoplankton (∅ < 33 μm) and netplankton (∅ 33–125 or 170 μm), was investigated from March 1977 to March 1979. The two fractions were separated by a nylon filter of 33 μm mesh-size after the *in situ* incubations. This investigation is also intended to provide information on production rates of nanoplankton, which is a very important constituent in the food of the herbivore zooplankton in Lake Vechten (Gulati *et al.* 1982). The nanoplankton production varied between 7 and 440 mg C · m^{-2} · day^{-1} and contributed 82 and 85.5% to the annual phytoplankton production in 1977 and 1978, respectively. The netplankton production was of significance only during the summer period when it contributed on the average 24 and 25% of ΣP_d. These data correspond more or less with the percentages of chlorophyll-*a* in nanoplankton and netplankton described earlier. The production rates of the netplankton ranged then between 10 and 180 mg C · m^{-2} · day^{-1}., i.e., 6–40% of ΣP_d.

The nanoplankton was found to be very important in the primary production of not only the pelagial part of the lake, but also of the littoral zone (de Kloet, own observation). The larger contribution of nanoplankton to the primary production points to more oligotrophic conditions (cf. Rodhe *et al.* 1958; Kalff 1972) in Lake Vechten.

Gross or net production and respiration

The ^{14}C-technique has been a very useful method for measuring photosynthetic capacity of the phytoplankton in Lake Vechten – a low to moderately productive waterbody. However, it is still doubtful whether one measures gross or net production by this method (cf. Steemann Nielsen 1963; Golterman 1975; Dring & Jewson 1979; Harris 1980). Many investigatiors assumed that short-term ^{14}C-incubations, like the ones during the present study, measure net production rates or agree well with the net production of particulate organic matter available to consumers, e.g. herbivore zooplankton (e.g. Antia *et al.* 1963; Steemann Nielsen 1975; Wetzel 1975; Devol & Packard 1978).

Also, by using the ^{14}C-technique, dark respiration cannot, however, be estimated. Assuming that the algal respiration R is 10% of P_{max} (McAllister *et al.* 1964; Steemann Nielsen 1975) and that ^{14}C-uptake measures net production during the daylight (ΣP_d-values, Fig. 2), in Lake Vechten we found an average respiration at night of 28% of ΣP_d (range 8–59%; n = 45) using the data from March 1977 to April 1978 and of February–November 1979. Thus, 72% of the organic carbon assimilated every day by the phytoplankton is available for its further use in the ecosystem of Lake Vechten.

However, algal respiration and thus the P:R ratio will depend on many factors, e.g. temperature, not considered in the relation between R and P_{max}. Devol & Packard (1978) have accounted for this in the relation:

$$(R{:}Chl)_{epi} = 0.038\ e^{0.11T}$$

where R and Chl are, respectively, the respiration rate and the chlorophyll-*a* concentration in the epilimnion (epi), and T is the water temperature. An

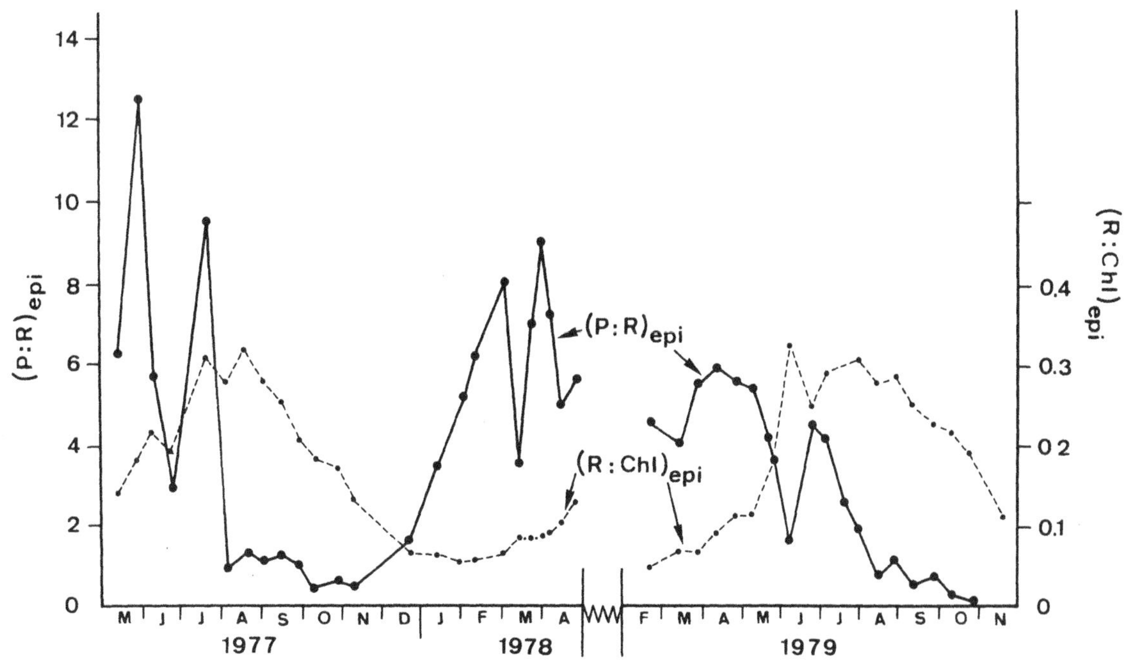

Fig. 5. Production:respiration (P:R) and respiration:chlorophyll (R:Chl) ratios of Lake Vechten calculated from integrated data of the upper 5 m (epilimnion). $(R{:}Chl)_{epi}$ is on an hourly basis. For further details, see text.

average R-value of 54% (range 8–230%) with respect to Σ P_d-values was computed using Devol & Packard's equation and Lake Vechten chlorophyll-*a* data for the March 1977–April 1978 and February–November 1979 periods (see Fig. 5). Thus, more than 50% of the net daylight production will be respired at night in the upper 5 m of lake Vechten. Low $(P:R)_{epi}$ values were found from August to November when more than 100% of the daily carbon production could be used up by respiration at night (Fig. 5). This corresponds with both maximal temperature (August) and maximum chlorophyll-*a* concentration in the euphotic zone (Fig. 4) due to predominance of netplankton such as *Ceratium hirundinella*. Since a sharp decrease in chlorophyll – due to high respiration calculated using Devol and Packard's equation – was lacking, the $(R:Chl)_{epi}$ ratio of these authors is not valid for Lake Vechten. From January to July, $(P:R)_{epi}$ ratios were generally lower than 0.25, i.e. $(P:R)_{epi}$'s were >4 (see Fig. 5).

Primary production per unit chlorophyll-a in relation to light

The photosynthetic capacity of the phytoplankton in Lake Vechten is illustrated in Fig. 6 (see also Table 4) by using the ratio between the daily primary production and the chlorophyll-*a* concentration, i.e. the activity coefficient (AC), or production rates per unit chlorophyll (mg C · mg^{-1} chl · day^{-1}; see e.g. Vollenweider 1969; Malone & Neale 1981). The means and ranges of the AC-values are based on the integral values of both the daily production rates and the chlorophyll concentrations for the whole column during 1977–1980 (Table 4). The AC-values of the years before 1977 are not calculated since the chlorophyll-*a* data of the deeper layers during the summer period are lacking. The ratio (%) of daily primary production of phytoplankton to daily photosynthetically active radiation (Ph.A.R.), expressed as the photosynthetic efficiency, was also calculated (Table 4).

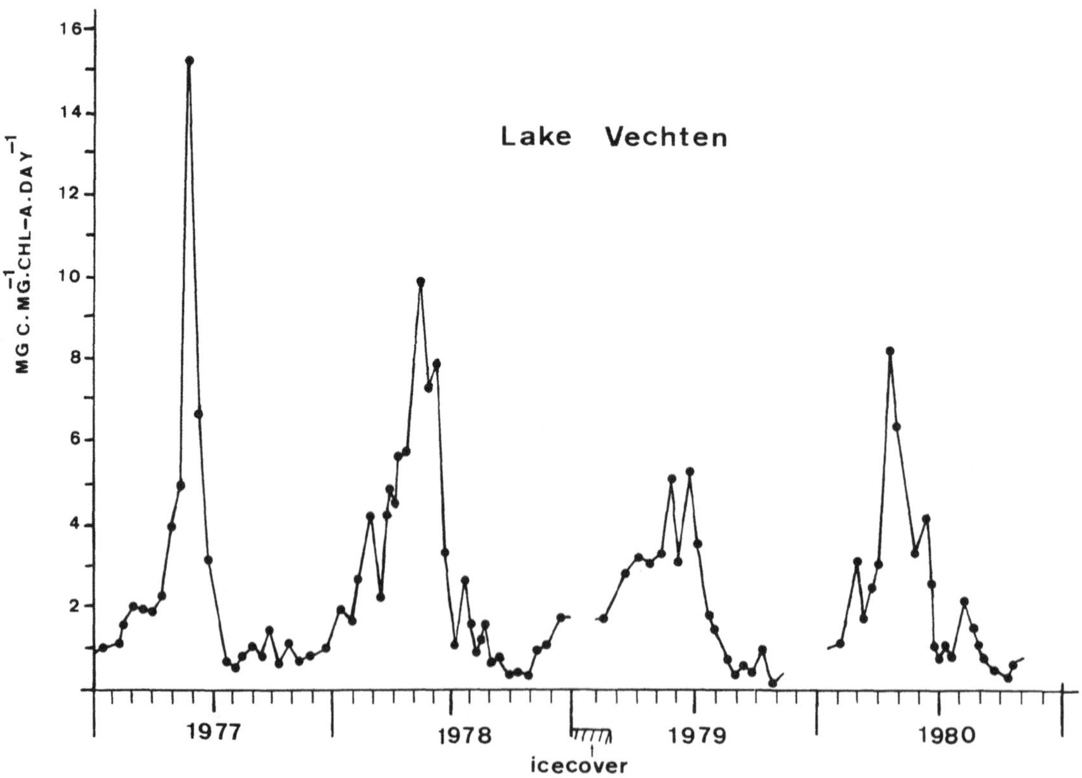

Fig. 6. Photosynthetic activity coefficients, i.e. cellular C-fixation per unit chlorophyll (AC-values), of pelagial phytoplankton in Lake Vechten.

In spring, when the lake was not yet stratified, the AC-values increased sharply after varying between 1 and 3 during the winter months (except for the end of February 1978; see Fig. 6), exclusively due to nanoplankton. This spring increase in AC-values coincided with that in the density of herbivore zooplankton and the maximum AC-values coincided with the grazing and excretion rates maxima of the zooplankton (Gulati *et al.* 1982).

The maximum AC-values were found mostly in May (but in April in 1980), correlated with minima for phytoplankton biomass, viz. 17 and 22 mg · m^{-2} chlorophyll-*a* on 24 May 1977 and 9 May 1978, respectively. Although the peaks of AC-value in 1979 were observed in May and June, the levels were much lower than in the other years. On the contrary, the chlorophyll-*a* content (63–67 mg · m^{-2}) was much higher than in the other years.

During the summer stratification, AC decreased sharply to values between 0.3 and 1.7 (except in July 1978 and August 1980, see Fig. 6). The lower AC-values are probably due to the predominance in summer of netplankton and its 1.7–2 times lower activity coefficient than those of the nanoplankton (de Kloet, own observation). This was especially the case during the abundance of the *Ceratium hirundinella* population in September – October when $(AC)_{max}$-values (= activity coefficients at depth of P_{max} in epilimnion; see Fig. 8) were at their minimum too. On the other hand, the lower AC-values during the summer months may be due to a self-shading effect, evidence of which in Lake Vechten is lacking.

A decrease of AC-values, when Lake Vechten is stratified, may be caused by changes in the ratio between the depth of the euphotic zone (Z_p) and of the mixed layer (Z_m). Ahlgren (1970, Fig. 21) found in Lake Norrviken that when the ratio Z_p:Z_m decreased, a large part of chlorophyll-*a* did not receive sufficient light for photosynthesis, so the AC-values decreased too. During the summer stratification the mixed layer (= epilimnion) of Lake Vechten was smaller than in the rest of the year (non-stratification). Thus, the ratio Z_p:Z_m increased without caus-

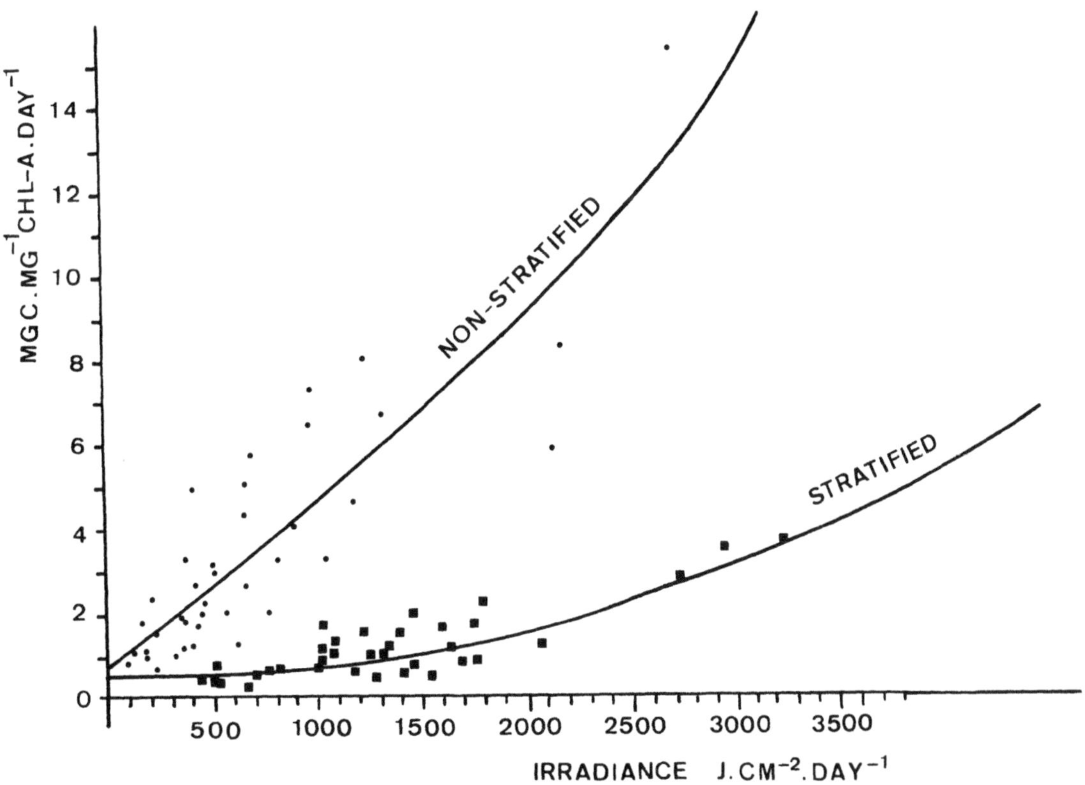

Fig. 7. The polynomial regression (quadratic) relation between the solar irradiance (x) and the AC-values (y) of the pelagial phytoplankton in Lake Vechten during the years 1977–1980 for two periods: i) non-stratified (Nov.–April): $y = 0.86 + 3.42 \cdot 10^{-3} x + 0.3 \cdot 10^{-6} x^2$; $r = 0.87$; $n = 35$; $P < 0.001$; and ii) stratified (July–Oct.): $y = 0.57 - 6.9 \cdot 10^{-5} x + 3.1 \cdot 10^{-7} x^2$; $r = 0.86$; $n = 35$; $P < 0.001$.

ing an increase in AC-values. This is in contrast with Ahlgren's (1970) conclusions.

In Lake Vechten, however, the situation is more complicated due to the high density of phytoplankton in the metalimnion adapted to low light, e.g. *Mallomonas caudata,* of blue-green algae, e.g. *Lyngbya* spp., and of phototrophic bacteria in deeper layers (6–8 m) from July to October. These organisms do contribute more significantly to chlorophyll than to primary production, thus causing a decrease in the AC-values in summer.

A correlation between AC-values and the corresponding daily energy input, i.e. the solar irradiance, was absent in Lake Vechten. However, treating the data of stratification and non-stratification periods separately revealed significant correlation (Fig. 7). This might be attributed to differences in the two periods of response to light conditions and to the availability of nutrients and changes in the composition of algae. In the winter period, solar radiation is an important governing factor (compare the slopes of the regression lines in Fig. 7). In summer, the irradiation has much less influence on the activity coefficients, and the algal productivity is possibly controlled mainly by nutrient concentrations.

The low AC-values – based on integral data – in the summer, especially from July to early October, were due to very low AC-values (0.1–0.6) of the bulk of the photosynthetic organisms in the metalimnion and hypolimnion. This is elucidated further by comparing the activity coefficients in summer with those in the upper few meters of the non-stratification period in the lake, particularly the activity coefficients $(AC)_{max}$ at depth of P_{max} during the period 1977–1979 (Fig. 8). The $(AC)_{max}$-values were 2.7–10.4 times higher than the AC-values during July to October, as against only 1.5–3 times during the rest of the year (cf. Figs. 6 and 8; except for data of November and December).

Nevertheless, daily solar irradiance and the $(AC)_{max}$ are also significantly correlated in both the stratified and non-stratified periods (Fig. 9). Again, this fact illustrates that the phytoplankton

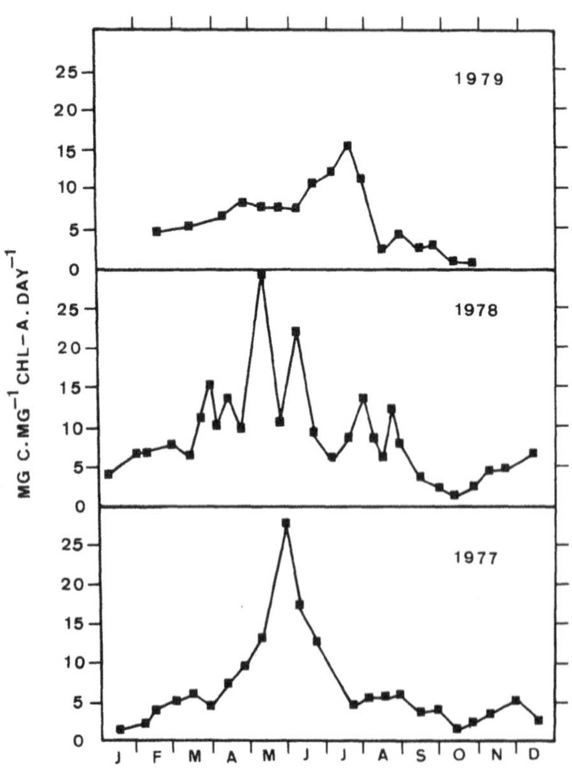

Fig. 8. The seasonal course of $(AC)_{max}$-values, i.e. P_{max} per unit chlorophyll, during 1977–1979.

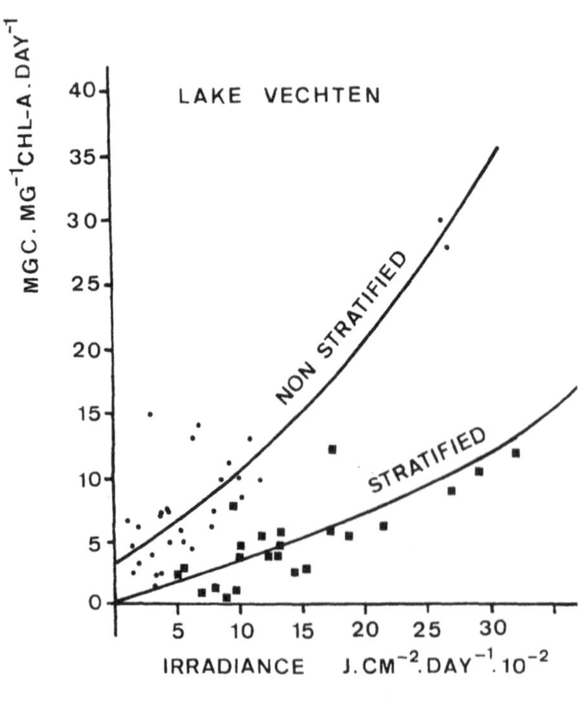

Fig. 9. The polynomial regression (quadratic) relation between the solar irradiance (x) and $(AC)_{max}$-values (y) during 1977–1979 for two periods: i) non-stratified (Nov.–April): $y = 3.47 + 5.2 \cdot 10^{-3} x + 1.6 \cdot 10^{-6} x^2$; $r = 0.90$; $n = 30$; $P < 0.001$; and ii) stratified (July–Oct.): $y = 0.093 + 3.1 \cdot 10^{-3} x + 2 \cdot 10^{-7} x^2$; $r = 0.79$; $n = 23$; $P < 0.001$.

production in Lake Vechten is governed by the solar irradiance. However, the lower slope of the regression line in the stratification period than in the non-stratification period does indicate that, along with light, other factors, e.g. nutrient limitation and change in algal composition, are important in lowering $(AC)_{max}$ in summer. In addition, in the non-stratified period, the water temperature is, along with light, an important controlling factor in Lake Vechten. The increase of AC in april to maximum values in May is most likely due to the increased availability of N and P regenerated by zooplankton as their excretion products related to high grazing activity during that period (Gulati *et al.* 1982).

The relationships between phytoplankton biomass, chlorophyll-a and primary production; and turnover rates

The AC-values so far discussed give information about the photosynthetic capacity of the phytoplankton in the lake. A comparison of chlorophyll-*a* concentration and production rates with the phytoplankton biomass may confirm these findings (Table 6).

Biomass concentrations, expressed as carbon content, were calculated from the biovolumes of the dominant phytoplankton species in 1977 and 1979 (Blaauboer 1982). The specific weight of the phytoplankton was assumed to be 1 and the carbon content to be 10% of the fresh weight (Vollenweider 1969; Winberg 1971). Of course, the estimates are rough owing to the variations in the conversion factors from volume to fresh weight to carbon. The factor would normally vary in nature because of the differences in phytoplankton species and their physiological stages (Vollenweider 1969; Hallegraeff 1976).

For comparing these parameters (Table 6), the euphotic zone was considered only to the Z_O-depth during the summer stratification. This was done to omit the bulk of organic matter due to phototrophic bacteria and thus confine ourselves to the upper layers which are relatively more important in regard to the photosynthetic process of phytoplankton.

The ratio (%) between chlorophyll-*a* and the carbon content of phytoplankton biomass varied between 0.8 and 6.5% with a mean of about 2% in 1977 and 3% in 1979 (see Table 6), excluding the abnormally high ratios of 11.2–17.1% from 12 April to 10 May 1977, due probably to estimation errors in biovolumes in this period. This became obvious comparing the carbon content derived from biovolumes (Table 6) with the corresponding carbon values of seston smaller than 105 μm (Gulati *et al.* 1982). Some disagreement in the relations between the chlorophyll content and biomass of phytoplankton species may occur just before and after the exponential phase of growth (Hallegraeff 1976).

A review of the relations between chlorophyll content and algal cell volume, in lakes of different trophic states, both with different groups of dominating algae and with unialgal cultures (Nicholls & Dillon 1978), shows that the chlorophyll-*a* content of the phytoplankton in Lake Vechten was rather low, but the range of 0.08–0.65% in Lake Vechten was comparable with those in a number of other temperate lakes (Nicholls & Dillon 1978; Jónasson & Adalsteinson 1979; Moss 1980; Desortová 1981). An interesting feature is that in the period when a single algal species comprised more than 80% of the total biovolumes in Lake Vechten, the chlorophyll-*a* concentration was relatively lower than when no single species was so dominant. This was observed in both years (Table 6), e.g. on 15 March 1977 and 24 April and 8 May 1979 when the diatom *Stephanodiscus astraea* formed 96%, 90% and 87%, respectively, of the total biovolume. This was also observed from 13 September to 11 October 1977, in which period *Ceratium hirundinella* formed 86–99% of the total cell volumes and also in the identical period in 1979 (Blaauboer 1982). The chlorophyll expressed as % carbon during the abundance of those algae was: *S. astraea*, 0.8–1.1%; and *C. hirundinella*, 1.3–2.7%). In both cases they are much lower compared with average values (calculated from literature using a conversion factor of 10 between algal cell volume and carbon; cf. Talling 1971; Tolstoy 1977; Nicholls & Dillon 1978), although there is also evidence of low cellular chlorophyll-*a* content on occasions when concentrations of *Ceratium* are high, as found in Esthwaite Water (George & Heaney 1978; Harris *et al.* 1979).

The turnover times calculated from the phytoplankton biomass and the primary production data varied between 3.4 and 73.5 days (Table 6); excluded are the data from 12 April to 10 May 1977 because of the abnormally high biovolumes, as mentioned earlier. The maxima for the turnover rates occurred in May and June, coinciding with maxima of zooplankton grazing.

The turnover rates of phytoplankton species in Lake Vechten are very slow, particularly during mid-July–October, compared with many other temperate lakes (e.g. Goldman *et al.* 1968; Ahlgren 1970; Happey-Wood 1978; Jónasson & Adalsteinson 1979; Golterman & Kouwe 1980). The relatively faster turnover rates during 1973 and 1974 in Lake Vechten (turnover time 1–15 days; Golterman & Kouwe 1980) may be due to, first, higher primary production rates found in 1973 and 1974 (cf. Table 2) and, secondly, to lower chlorophyll-*a* concentrations (Fig. 4). In addition, Golterman & Kouwe used a factor of 35 to convert chlorophyll into carbon, a factor which is much lower than the mean (± 50) found in 1977, but well comparable (33) to that in 1979 (Table 6).

The very slow turnover rates in Lake Vechten

Table 6. Comparison between biomass (via biovolumes) and chlorophyll-*a* concentrations and between biomass and daily primary production (turnover in days) of the phytoplankton in Lake Vechten. A stands for the water column below which oxygen is nil, and B for the upper 3.6 m stratum.

Date	A			B	
	Biomass ($g\,C \cdot m^{-2}$)	Chl-a/biom. (%)	Turnover (days)	Biomass ($g\,C \cdot m^{-2}$)	Turnover (days)
1977					
18- 1	2.7	2.8	32.5	–	–
15- 3	8.5	1.1	32.1	3.3	16.0
12- 4	0.4	17.1	2.0	–	–
26- 4	0.4	11.2	1.8	–	–
10- 5	0.3	14.7	1.4	–	–
24- 5	0.9	2.0	3.4	0.6	5.2
7- 6	1.3	2.8	4.5	0.5	4.0
21- 6	1.7	2.5	7.3	0.5	5.5
4- 7	2.1	–	16.6	1.1	18.4
19- 7	4.6	2.0	35.4	0.7	12.0
2- 8	7.1	2.8	39.2	3.2	45.1
16- 8	8.0	2.1	30.0	2.8	18.0
30- 8	7.2	2.0	20.6	3.7	18.5
13- 9	10.0	1.7	37.6	4.9	27.1
27- 9	6.9	1.5	26.5	3.1	21.8
11-10	4.8	2.0	54.0	2.9	69.0
1979					
13- 3	0.8	6.4	7.8	0.4	3.9
27- 3	1.5	–	11.6	–	–
10- 4	2.1	2.6	11.5	0.9	8.8
24- 4	10.4	0.9	36.1	4.5	20.1
8- 5	9.3	0.8	35.8	3.7	29.5
22- 5	1.6	4.1	4.7	0.4	1.6
5- 6	1.3	6.5	4.5	0.4	1.9
19- 6	1.8	2.5	5.8	0.9	4.3
3- 7	3.1	1.9	7.7	1.0	4.3
17- 7	2.7	3.3	7.2	0.5	2.0
31- 7	2.5	2.4	8.5	1.0	5.2
14- 8	5.4	4.3	16.8	2.7	28.1
21- 8	7.4	–	32.2	1.9	12.2
28- 8	3.5	2.9	28.0	1.0	12.1
11- 9	6.1	2.7	30.3	1.8	53.8
25- 9	3.7	1.7	34.3	2.4	41.4
9-10	7.7	1.3	73.5	4.5	113.0

from July to October were, particularly, correlated to the abundance of *Ceratium hirundinella.* This species, together with *Mallomonas caudata,* formed 90% or more of the total plankton volumes, of which the dinoflagellate contributed more than 60% (except on 16 August 1977 and 14 August 1979 when *M. caudata* formed 58% and 45% of the biovolume, respectively). Assuming the carbon to be 10% of fresh weight, as in the present study, is probably an underestimate of the carbon content of *Ceratium* cells, which may be 20% of fresh weight (Blaauboer 1982; cf. Lund 1964; Reynolds 1976). Thus, the present results may be even slower than computed using a conversion factor of 10.

The slow growth rate of the dinoflagellate *Ceratium hirundinella* found also in other lakes (e.g. Harris *et al.* 1979; Moore 1981) may explain the low turnover rates in Lake Vechten during the abundance of *C. hirundinella.* The ability of the dinoflagellate to migrate and the presence in high numbers of algae just above Z_0-depth result in slow turnover rates.

The turnover rates of the epilimnetic phytoplankton from 0 to 3.6 m layer (see Table 6) were generally faster than those in the column up to Z_0-depth. The turnover time in these layers during spring corresponds to the population doubling time of 5 and 11 days computed from increase of the cell numbers in 1978 by Blaauboer (1982) in Lake Vechten. Also, the range of the turnover time from June to August (4–45 days) in Lake Vechten was comparable to population doubling time (5–34 days) found in Esthwaite Water (Heaney 1976; Harris *et al.* 1979), where *C. hirundinella* also dominated during the summer months.

The turnover rates discussed so far appear to slow, particularly in the summer period when the phytoplankton community is dominated by netplankton. For the eight measurements during July–October 1977, when both the biomass of netplankton and nanoplankton and the share of the fractions in the total primary production were measured, the turnover rates were calculated separately (Table 7). It is striking to note that turnover times of nanoplankton were much faster (0.8–3.6 days) and those of netplankton were much slower (66–150 days) than those based on the mixed population as summarized in Table 6. The higher turnover rates of nanoplankton in summer are not reflected in its biomass, which remains rather constant under the strong selective grazing pressure of zooplankton. The much slower turnover rates of the netplankton thus found, though not clearly understood, may partly be due to senile algae in the deeper layers and partly due to the discrepancies in the production and biomass measurements related to the migration pattern of *Ceratium hirundinella.*

Fortnightly sampling at one station, as was done in Lake Vechten, will tend to give large errors in biomass data (Dorgelo & de Graaf Bierbrauwer 1981). Besides, increased variation in cell numbers of *C. hirundinella* may be caused by changes in vertical distribution owing to vertical migration (Blaauboer 1982). Such changes are induced by the diurnal change of the light climate (Heaney 1976; Harris *et al.* 1979).

Table 7. Biomass, p.p. rate and the share (%) of nanoplankton (Ø < 33 μm) in the total phytoplankton biomass and primary production in the watercolumn above Z_0 depth during July–Oct. 1977 in Lake Vechten. The turnover rates of nanoplankton (A) and netplankton (B) are compared.

Date	A					B
	Biomass (g C · m^{-2})	(%)	Production (mg C · m^{-2} · day^{-1})	(%)	Turnover (days)	Turnover (days)
1977						
4- 7	0.25	(11.0)	120	(90)	1.7	135
19- 7	0.10	(2.1)	100	(77)	1.0	150
2- 8	0.20	(2.8)	114	(63)	1.8	103
16- 8	0.40	(5.0)	202	(76)	2.0	117
30- 8	0.35	(4.4)	260	(75)	1.3	76
13- 9	0.15	(1.5)	195	(73)	0.8	141
27- 9	0.30	(4.3)	161	(62)	1.9	67
11-10	0.20	(4.2)	56	(63)	3.6	144

General discussion

The primary production study of the phytoplankton in Lake Vechten was aimed especially at providing information for the carbon balance study of the lake's ecosystem rather than at studying the effects of the environmental factors in detail. The relationship between the solar irradiance and the production rates per unit chlorophyll-*a* indicates that light may act as a controlling factor for the growth of phytoplankton (Figs. 7 and 9), particularly in the period of non-stratification.

The lower photosynthetic efficiency of the netplankton, e.g. *Ceratium hirundinella,* abundant in August and September may result in a lower response of AC as well as $(AC)_{max}$ values to the irradiance. However, the low nutrient concentration, of both $(NO_3 + NH_4)$-N and PO_4-P, from August to October (Table 8), in the epilimnion may indicate that these nutrients may limit production of algae, too. Nevertheless, vertical migration enables *Ceratium* sp. not only to utilize nutrients from the entire epilimnion, but also permits part of the population of this algal to utilize the reserve of the essential nutrients in the metalimnion and hypolimnion. The nutrient depletion in the epilimnion may cause a decrease in the cellular chlorophyll content (cf. Harris *et al.* 1979). In this regard, it may be added that the low PO_4-P concentrations and the high ratio between N ad P (Table 8) phosphate may serve also as a controlling factor for the phytoplankton during the rest of the year. However, further causal analysis is needed to get a better insight into the role of the abiotic factors.

Table 8. Average concentrations of dissolved inorganic phosphate (PO_4-P) and nitrogen in upper 5 m stratum and N/P ratios in the eastern part of Lake Vechten during 1979. Data of phosphate after Wisselo, of nitrogen after Verdouw & Dekkers (1982).

Date	PO_4-P ($mg \cdot m^{-3}$)	NO_3-N ($mg \cdot m^{-3}$)	NH_4-N ($mg \cdot m^{-3}$)	N/P
1979				
16- 1	5	140	245	77
13- 2	2	295	168	232
13- 3	2	333	97	215
10- 4	3	317	22	113
8- 5	<1	200	24	450
5- 6	3	128	21	50
31- 7	5	3	1	1
28- 8	4	21	25	11
25- 9	2.5	10	21	12
23-10	5	21	5	5
20-11	2	27	260	143
18-12	4	98	308	102

The correlation between N and P excretion of herbivore zooplankton and $(AC)_{max}$-values hints also at the role of nutrients (Gulati *et al.* 1982). A comparison of the N and P needs of the phytoplankton, calculated from primary production rates and the excretion of these nutrients by zooplankton in Lake Vechten, indicates that nutrient regeneration by zooplankton fulfilled substantially the nutrient requirements of the phytoplankton, and during the period of high grazing (May–June), supply via zooplankton even exceeded the phytoplankton demand (Gulati *et al.* 1981).

Among the autotrophs in the ecosystem of Lake Vechten, the phytoplankton was found to be the most important primary producer compared with the macrophytes and periphyton (Gons 1981). Unfortunately, owing to the moderate production level, only the ^{14}C-technique could be employed. Thus, the respiratory losses of algae are based on assumptions which rest on the theoretical relations with P_{max}, chlorophyll-*a* and temperature; a clear insight into the net production of organic matter by the phytoplankton, relevant to the carbon budget in the lake of this, is as yet not possible. Therefore, a study of the respiration rates of the phytoplankton in Lake Vechten is required.

Also relevant to the carbon budget of the lake's ecosystem is the decrease of about 40% in the primary production rates during the last few years compared with those in the period before 1975 (Table 2). An increase of netplankton, particularly the dinoflagellate *Ceratium hirundinella* (Blaauboer 1982), and of the larger size-fraction of the seston (33–105 μm, Gulati *et al.* 1982) in general may explain this. Nevertheless, it remains to be seen if the lake is passing through a change or if natural oscillations of phytoplankton populations can reverse the direction of this change.

Summary

The primary production rates of the phytoplankton in Lake Vechten (The Netherlands), a small (area 4.7 ha; mean depth: 6 m; max. depth: 12 m), stratified, unpolluted sand pit, was investigated with the ^{14}C-technique. This was done more or less regularly from 1969 to 1980 at one sampling station, which was assumed to be representative for the whole lake. As a part of the ecosystem research, other parameters measured simultaneously are: chlorophyll-*a*, phytoplankton biomass, and light climate.

The annual mean of cellular carbon fixation was 280 mg C · m^{-2} · day^{-1} (range 10 and 1 307 mg C · m^{-2} · day^{-1}). The production rates have decreased in the last few years (Table 2) and may have been caused by an increase of netplankton (∅ > 33 μm) during the summer, particularly of *Ceratium hirundinella*. The netplankton had a low photosynthetic efficiency and contributed on the average only 25% of total Σ P-values during the summer stratification. On the other hand, considering the data of 1980 which indicates increase in production rates, a natural oscillation of phytoplankton populations is not ruled out (Table 2). The minimum primary production rates were found during November to mid-January, i.e. from the start of circulation onwards, and the maxima occurred usually in early spring or summer. Maxima found sometimes already in winter were related to the early outburst of diatoms (e.g. *Stephanodiscus hantschii*). The P_{max}-values were observed between 0.5 and 4 m depth and during the spring maxima primary production was sometimes higher than that found during the summer. On the other hand, a second but high peak of ^{14}C-fixation rates occurred in the metalimnion and hypolimnion during July–October together with the bulk of chlorophyll-*a* due to a crowding of phytoplankton species such as blue-greens and dinoflagellates, and of phototrophic anoxic bacteria.

The extracellular excretion rates (Σ E_d) of the phytoplankton was on the average 15% of the total C-fixation (Σ A_d) during 1977–1979. Maximum E-rates found during August and September in the metalimnion were probably due to *Mallomonas* sp. The production rates in Lake Vechten are low to moderate (oligo-mesotrophic) compared with other Dutch lakes. This is possibly owing to morphometric and hydrological characteristics.

The production rate per unit chlorophyll-*a* (AC) and turnover rates indicated low photosynthetic capacity. The AC-values ranged between 0.3 and 15.2 mg C · mg^{-1} chlorophyll · day^{-1} (average 2.5), the maxima in May–June coincided with those of zooplankton grazing, minima of chlorophyll-*a* and seston (<105 μm) concentration an the highest transparency ('clear water' phase). The lowest AC-values were found during the summer months, when the chlorophyll-*a* concentration in the metalimnion and hypolimnion was high. But the AC-values in the epilimnion, e.g. $(AC)_{max}$, were also lower than those during the non-stratification. This low production capacity, particularly from July to October, was also shown by the turnover time of 20–40 days related to the abundance of *Ceratium hirundinella*. Maxima of turnover rates, found in May and June, coincided also with the zooplankton grazing maxima.

The relationship between activity coefficients and solar irradiance was significant, showing that light might be one of the factors controlling phytoplankton production, particularly during the period of non-stratification. In summer, the concentration of N + P in the epilimnion are low and may limit production, although the zooplankton excretion did partly fulfil the N and P demands of the phytoplankton in the lake.

Acknowledgements

I am much indebted to Dr. Sikko Parma and Dr. Ramesh D. Gulati for critically reading the manuscript and to Dr. Ramesh D. Gulati for correcting the English text. Mr. Klaas Siewertsen was very helpful with the drawings. Mrs. Marianne C. I. Blaauboer and Drs. Jaap Olie are acknowledged for supplying, respectively, their data on phytoplankton and extracellular excretion, and Mr. Ad G. Wisselo for his data on phosphorus. I thank Mr. Heb Roon for his assistance in the field.

References

Ahlgren, G., 1970. Limnological studies of lake Norrviken an eutrophicated Swedish lake. II. Phytoplankton and its production. Schweiz. Z. Hydrol. 32: 353–396.

Antia, N. J., McAllister, C. D., Parsons, T. R., Stephens, K. & Strickland, J. D. H., 1963. Further measurements of primary production in using a large volume plastic sphere. Limnol. Oceanogr. 8: 166–183.

Best, E. P. H., 1982. The aquatic macrophytes of Lake Vechten. Species composition, spatial distribution and production. Hydrobiologia 95: 65–77.

Blaauboer, M. C. I., 1982. The phytoplankton species composition and the seasonal periodicity in Lake Vechten from 1956 to 1979. Hydrobiologia 95: 25–36.

Blaauboer, M. C. I., Cappenberg, Th. E. & Van Keulen, R., 1982. Extracellular release of photosynthetic products by freshwater phytoplankton populations, with special reference to the algal species involved. Freshwat. Biol. 12 (6).

Desortová, B., 1981. Relationship between chlorophyll-*a* concentration and phytoplankton biomass in several reservoirs in Czechoslovakia. Int. Revue ges. Hydrobiol. 66: 153–169.

Devol, A. H. & Packard, T. T., 1978. Seasonal changes in respiratory enzyme activity and productivity in Lake Washington microplankton. Limnol. Oceanogr. 23: 104–112.

Dorgelo, J. & De Graaf Bierbrauwer, I., 1981. Phytoplankton in Lake Maarsseveen. I (1977–1979). Hydrobiol. Bull. 15: 29–40.

Dring, M. J. & Jewson, D. H., 1979. What does ^{14}C-uptake really measure? A fresh approach using a theoretical model. Br. phycol. J. 14: 122–123.

Flik, B. J. G. & Keyzer, A., 1981. Estimation of the primary production in Lake Maarsseveen I with an incubator. Hydrobiol. Bull. 15: 41–50.

George, D. G. & Heaney, S. I., 1978. Factors influencing the spatial distribution of phytoplankton in a small productive lake. J. Ecol. 66: 133–155.

Goldman, C. R., Gerletti, M., Javornicky, P., Melchiorri, U. & De Amezaga, E., 1968. Primary productivity, bacteria, phyto- and zooplankton in Lake Maggiore: correlations and relationships with ecological factors. Mem. Ist. ital. Idrobiol. 23: 49–127.

Golterman, H. L., 1975. Freshwater Limnology. Elsevier. Amsterdam.

Golterman, H. L. & Clymo, R. S., 1969. Methods for Chemical Analysis of Freshwater. I. B. P. Handbook, No. 8. Blackwell Scientific, Oxford. 172 pp.

Golterman, H. L. & Kouwe, F. A., 1980. Chemical budgets and nutrient pathways. In: LeCren, E. D. & Lowe-McConnell, R. H. (Eds.) The Functioning of Freshwater Ecosystems, pp. 85–141. I.B.P. No. 22. Cambridge Univ. Press.

Gons, H., 1981. Project 'Carbon cycle in Lake Vechten'. In: Progress Report Limnological Institute 1980. Verh. k. ned. Akad. Wetensch., Afd. Natuurk. 2de Rs. 77: 30–36.

Gons, H., 1982. Structural and functional characteristics of epiphyton and epipelon in relation to their distribution in Lake Vechten. Hydrobiologia 95: 79–114.

Gulati, R. D., 1972. Limnological studies on some lakes in the Netherlands I: Wijde Blik. Freshwat. Biol. 2: 37–54.

Gulati, R. D., 1976. Some limnological observations of the Oostelijke Vechtplassen. Hydrobiol. Bull. 10: 3–10.

Gulati, R. D., Siewertsen, K. & Postema, G., 1981. Phytoplankton-zooplankton interactions. In: Progress Report Limnological Institute 1980. Verh. k. ned. Akad. Wetensch., Afd. Natuurk., 2de Rs. 77: 12–16.

Gulati, R. D., Siewertsen, K. & Postema, G., 1982. The zooplankton: its community structure, food and feeding, and role in the ecosystem of Lake Vechten. Hydrobiologia 95: 127–163.

Hallegraef, G. M., 1976. Pigment diversity, biomass and species diversity of phytoplankton of three Dutch lakes. Ph.D. thesis, Univ. of Amsterdam. 177 pp.

Happey-Wood, C. M., 1975. Distinctions in algal ecology and production in two linked upland lakes, Gwynedd, N. Wales. Verh. int. Ver. Limnol. 19: 1045–1056.

Harris, G. P., 1980. The measurements of photosynthesis in natural population of phytoplankton. In: I. Morris, (Ed.). The Physiological Ecology of Phytoplankton. Studies in Ecology, Vol. 7. Blackwell Scientific, Oxford.

Harris, G. P., Heaney, S. I. & Talling, J. F., 1979. Physiological and environmental constraints in the ecology of the planktonic dinoflagellate Ceratium hirundinella. Freshwat. Biol. 9: 413–428.

Heaney, S. I., 1976. Temporal and spatial distribution of the dinoflagellate Ceratium hirundinella O. F. Müller within a small productive lake. Freshwat. Biol. 6: 531–542.

Jónasson, P. M. & Adalsteinsson, H., 1979. Phytoplankton production in shallow eutrophic lake Mývatn, Iceland. Oikos 32: 113–138.

Kalff, J., 1972. Netplankton and nanoplankton production and biomass in a northern temperate zone lake. Limnol. Oceanogr. 17: 712–720.

Kloet, W. A. de, 1971. Het eutrofiëringsproces van het IJsselmeer. Meded. Hydrobiol. Ver. 5: 23–38. (in Dutch)

Kristiansen, J. & Mathiesen, H., 1964. Phytoplankton of the Tystrup Barcelse Lakes, primary production and standing crop. Oikos 15: 1–43.

Lean, D. R. S. & Burnison, B. K., 1979. An evaluation of errors in the ^{14}C-method of primary production measurements. Limnol. Oceangr. 24: 917–928.

Malone, T. C. & Neale, P. J., 1981. Parameters of light dependent photosynthesis for phytoplankton size fractions in temperate estuarine and coastal environments. Mar. Biol. 61: 289–297.

McAllister, C. D., Shaw, N. & Strickland, J. D. H., 1964. Marine phytoplankton photosynthesis as a function of light intensity: a comparison of methods. J. Fish. Res. Bd Can. 21: 159–181.

Moed, J. R., 1973. Effect of combined action of light and silicon depletion on Asterionella formosa Hass. Verh. int. Verein. Limnol. 18: 1367–1374.

Moed, J. R. & Hallegraeff, G. M., 1978. Some problems in the estimation of chlorophyll-*a* and phaeopigments from pre- and post-acidification spectrophotometric measurements. Int. Revue ges. Hydrobiol. 63: 787–800.

Moore, J. W., 1981. Seasonal abundance of Ceratium hirundinella (O. F. Müller) Schrank in lakes of different trophy. Arch. Hydrobiol. 92: 535–548.

Moss, B., 1980. Phytoplankton biomass. In: LeCren, E. D. & Lowe-McConnell, R. H. (Eds.) The Functioning of Freshwater Ecosystems, pp. 146–149. I.B.P., No. 22. Cambridge Univ. Press.

Nicholls, K. H. & Dillon, P. J., 1978. An evaluation of phosphorus–chlorophyll–phytoplankton relationships for lakes. Int. Revue ges. Hydrobiol. 63: 141–154.

Opstelten, W., 1981. Diel patterns of phytoplankton productivity in Lake Vechten. Hydrobiol. Bull. 14: 219–221.

Reynolds, C. S., 1976. Succession and vertical distribution in response to thermal stratification in a lowland mere, with special reference to nutrient availability. J. Ecol. 64: 529–551.

Rodhe, W., 1958. The primary production in lakes: some results and restrictions of the ^{14}C-method. Rapp. Procès Verb. Réunions. J. Cons. perm. int. Explor. Mer 144: 122–128.

Rodhe, W., 1969. Crystallization of eutrophication concepts. In: Rohlich, G. A. (Ed.) Eutrophication, Causes, Consequences,

Corrections, pp. 50–65. Nat. Acad. Sci., Washington.

Rodhe, W., Vollenweider, A. & Nauwerck, A., 1958. The primary production and standing crop of phytoplankton. In: Buzzati-Traverso, A. A. (Ed.) Perspectives in Marine Biology, pp. 323–349. Univ. of California.

Schmidt-van Dorp, A. D., 1978. De eutrofiëring van ondiepe meren in Rijnland (Holland). Ph.D. thesis, State University, Utrecht. 254 pp. (in Dutch, with English summary).

Søndergaard, M., 1980. Adsorption of inorganic ^{14}C to poly-ethylene scintillation vials, a possible source of error of organic C. Arch. Hydrobiol. 90: 362–366.

Steemann-Nielsen, E., 1952. The use of radioactive carbon (^{14}C) for measuring organic production in the sea. J. Cons. perm. int. Explor. Mer 18: 117–140.

Steemann-Nielsen, E., 1963. Production, definition and measurements. In: Mill, M. H. (Ed.) The Sea, Vol. 2, pp. 129–164. Interscience, New York.

Steemann-Nielsen, E., 1975. Marine Photosynthesis. Elsevier Scientific, Amsterdam.

Steenbergen, C. L. M., 1982. Contribution of photosynthetic sulphur bacteria to primary production in Lake Vechten. Hydrobiologia 95: 59–64.

Steenbergen, C. L. M. & Verdouw, H., 1982. Lake Vechten: aspects of its morphometry, climate, hydrology and physico-chemical characteristics. Hydrobiologia 95: 11–23.

Talling, J. F., 1971. The underwater light climate as a controlling factor in the production ecology of freshwater phytoplankton. Mitt. int. Verein. Limnol. 19: 214–243.

Tolstoy, A., 1977. Chlorophyll-a as a measure of phytoplankton biomass. Acta Univ. upps. 416: 1–30.

Verdouw, H. & Dekkers, E. M. J., 1982. Nitrogen cycle of Lake Vechten concentration patterns and internal mass-balance. Hydrobiologia 95: 191–197.

Vollenweider, R., 1960. Beiträge zur Kenntnis optischer Eigenschaften der Gewässer und Primär Production. Mem. Ist. ital. Idrobiol. 12: 201–245.

Vollenweider, R., 1969. Primary Production in Aquatic Environment. I.B.P. Handbook No. 12, Blackwell Scientific, Oxford.

Wetzel, R. G., 1975. Limnology. Saunders, Philadelphia.

Winberg, G. P. (Ed.), 1971. Symbols, Units and Conversion Factors in Studies of Freshwater Productivity. I.B.P., London.

Contribution of photosynthetic sulphur bacteria to primary production in Lake Vechten

C. L. M. Steenbergen
Limnological Institute, 'Vijverhof' Laboratory, Rijksstraatweg 6, 3631 AC Nieuwersluis, The Netherlands

Keywords: primary production, photosynthetic bacteria, carbon cycle, secondary production

Abstract

The role of photosynthetic sulphur bacteria as primary producers in monomictic Lake Vechten (The Netherlands) is described. Lake Vechten has a surface area of 4.7 ha, a maximum depth of 11.9 m and a mean depth of 6 m.

Bacterial populations, appearing at the boundary layer of the oxidative and reductive zone from early June till late October, were composed of cyanobacteria, Chromatiaceae and green and brown coloured Chlorobiaceae. Predominating genera were *Synechococcus, Chloronema, Chromatium* and *Thiopedia*. The photosynthetic sulphur bacteria accounted for a primary production rate of 13.6–106.1 mg C.m^{-2} day^{-1}, which corresponded to 3.9–17.5% of total daily productivity in the pelagial zone. The percentage of photosynthetic bacterial production of total annual planktonic primary production calculated for the entire pelagial zone, taking into account compensation for decreasing volume of lower strata, was 3.6% (i.e. 127 against a total production of 3 510 kg C.lake^{-1}yr^{-1}).

Introduction

There are several reports on the carbon fixation by photosynthetic sulphur bacteria and the estimates of the importance of these organisms as primary producers in lakes vary widely (e.g. Wetzel 1975). Takahashi & Ichimura (1968) reported that in lakes having a permanent anoxic zone, photosynthetic bacteria were responsible for 9 to 25% of the total annual primary production. During summer stratification, they found values ranging from 3 to 5%. Other estimates greatly exceeding those reported by these authors come from Culver & Brunskill (1969) and Cohen *et al.* (1977). They found that photosynthetic sulphur bacteria accounted for 85 to 91% of the total primary production.

Recently, the occurrence of photosynthetic sulphur bacteria in meta- and hypolimnion of Lake Vechten was described and their productivity in relation to light intensity and sulphide concentration discussed (Steenbergen & Korthals, in press). Lake Vechten is a small (4.7 ha), relatively deep (maximum depth 11.9 m; mean depth 6 m) monomictic lake, located in the central part of The Netherlands (52° 04′ N; 5° 05′ E). Summer stratification lasts from May till November and is accompanied by hypolimnetic oxygen depletion resulting in clinograde oxygen curves throughout this period (Steenbergen & Verdouw, 1982). Only in severe winters when the ice cover is prolonged does the lake show dimictic properties. General limnological information on Lake Vechten is given by Parma (1971), Gulati (1975) and Steenbergen & Verdouw (1982).

This paper considers the contribution of bacterial primary production to total annual pelagial primary production. Organic matter produced by the meta- and hypolimnetic bacterial populations may be grazed by various zooplankton species dur-

Hydrobiologia 95, 59–64 (1982). 0018-8158/82/0951-0059/$01.20.

ing their diurnal vertical migrations and thus enter the food chain of the lake (e.g. Sorokin 1965). Thus, photosynthetic sulphur bacteria may play a role in the carbon cycle of Lake Vechten, particularly during summer stratification (see Best *et al.* 1978; Gons 1981).

Methods

Samples were collected at 0.5 m depth intervals with a Friedinger-type 5 l water bottle at a station in the centre of the eastern depression of the lake. To avoid contamination with atmospheric oxygen, the samples from meta- and hypolimnion were transferred via rubber tubing to the bottom of 100 ml serum bottles. These bottles were previously flushed with oxygen-free nitrogen, which had passed over BASF catalyser R3-11 at a temperature of 150 °C. Caps were quickly screwed, taking care to exclude air bubbles.

Photosynthetic CO_2-fixation rate was measured with the carbon-14 technique, essentially following Steemann Nielsen (1952). The rates are expressed as the difference in CO_2-fixation rates between the light and dark bottles. Into each 100 ml sample, 1 ml stock solution containing 2 μCi of sterile Na-$H^{14}CO_3$ (Radiochemical Centre, Amersham) was injected. Algal and bacterial photosynthetic rates were distinguished with DCMU (3-3, 4 dichlorophenyl-1, 1-dimethyl urea) at a final concentration of 2 μM. DCMU inhibits algal (oxygenic) photosynthesis by specifically blocking the photosynthetic electron flow at the reducing side of photosystem II (Trebst 1980), but does not affect bacterial (anoxygenic) photosynthesis that differs in many respects from plant photosynthesis (Gromet-Elhanan 1977). So, light bottles were incubated both with and without DCMU.

Light and dark bottles were suspended in the lake at the depths where the samples were collected for 3 hours from 08.00 to 11.00 a.m. Control experiments showed that the CO_2-fixation rate was linear with time for at least 2.5–3.0 hours levelling off gradually thereafter.

After incubation the bottles were covered with ice and transported to the laboratory in the dark within an hour and processed within 2 hours after arrival. Samples of 20–50 ml were filtered on membrane filters (0.45 μm porosity, Schleicher & Schüll), washed with diluted HCl, followed by filtered lake water. Filters were counted in Instagel, a xylene-based scintillation cocktail, with a Packard Tricarb liquid scintillation spectrometer (model 2425). Counting efficiency was determined by automatic external standardization.

Areal production rate was computed by integrating the rates at different depths and the daily rates by extrapolation using the ratio total irradiance during the day and irradiance during exposure time. Data on photosynthetically available radiation (PhAR in $W.m^{-2}$; 400–700 nm) were obtained from the meteorological station KNMI at De Bilt, located at a distance of 4.3 km from the lake. To determine as accurately as possible the importance of photosynthetic sulphur bacteria as primary producers in Lake Vechten production rates for the entire lake (kg $C.lake^{-1}.yr^{-1}$) have been corrected for the hypsometry of the lake (see Steenbergen & Verdouw 1982) by multiplying the rate at a given depth by the volume of water at that depth interval, and then summing these values. Total annual production values (kg $C.lake^{-1}.yr^{-1}$) were derived by planimetric integration.

Bicarbonate alkalinity was determined potentiometrically with a Quéré automatic titrator (Golterman *et al.* 1978). The water samples for sulphide detection were collected in screwcap vials containing highly alkaline anti-oxidant buffer, which converts HS^- and H_2S to S^{2-} (Baumann 1974). Concentrations of S^{2-} were determined with a silver/sulphide electrode (ORION 94–16) coupled to a specific ion meter (ORION 407A). Sulphide standards were titrated with lead perchlorate (Baumann 1974). Temperature and O_2 concentration were measured *in situ* with a combination temperature-oxygen probe.

Results and discussion

Bacterial populations

Dense populations of photoautotrophic bacteria are found in the metalimnion and in the upper zone of the hypolimnion during summer stratification in the lake (Steenbergen & Korthals, in press). The populations are composed of cyanobacteria, Chromatiaceae and green and brown coloured Chlorobiaceae with *Synechococcus, Chromatium, Thiopedia* and *Chloronema* as the predominant genera. In addition, both Chrysophyceae *(Mallomonas caudata)* and Cryptophyceae *(Rhodomonas* sp. and *Cryptomonas* sp.) were abundant in the metalimnion in summer compared with the epilimnetic phytoplankton dominated by Dinophyceae, namely *Ceratium hirundinella* (Blaauboer 1982).

Growth of photosynthetic sulphur bacteria started at the end of May in the deeper strata (8–9 m) of the lake at the oxygen-sulphide borderline. The phototrophic bacterial community developed maximal from late August to September. During this period, in particular, the gas-vacuole-containing *Chloronema*-type filaments dominated at 6–6.5 m depth, and *Chromatium* and *Thiopedia* at 7–7.5 m depth. Cells of the latter two genera contained densely packed sulphur granules, indicating that these cells had been using sulphide as an electron donor in the photosynthetic reaction (Pfenning 1978). The type of electron donor used by the *Chloronema*-type organism is less clear. Thiosulphate and simple organic molecules, e.g. acetate, may be used by this organism. Anoxygenic photosynthesis using sulphide as an electron donor by the cyanobacteria *(Synechococcus*-type cells) in Lake Vechten may be ruled out due to the extremely low sulphide concentration (<1 μM sulphide l^{-1}) at the depth of their abundance (cf. Table II in Padan 1979).

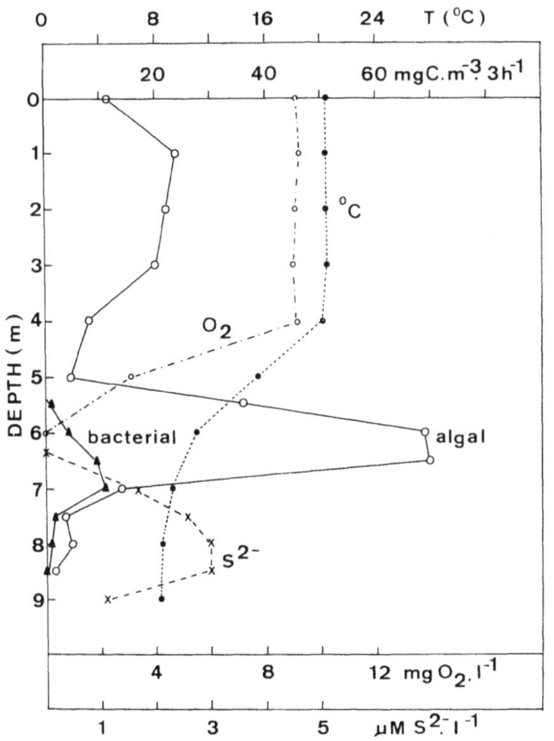

Fig. 1. Profiles of algal photosynthesis (DCMU-sensitive); bacterial photosynthesis (DCMU-resistant); sulphide (including S^{2-} + HS^{-} + H_2S) and oxygen concentrations and temperature in Lake Vechten (The Netherlands) on 27 August 1980.

From October onwards the bacterial community declined rapidly and disappeared at the onset of autumnal turnover in early November.

Bacterial photosynthesis

Typical vertical profiles during summer stratification in Lake Vechten of algal photosynthesis and bacterial photosynthesis and some environmental characteristics are depicted in Fig. 1.

Bacterial photosynthesis peaks coincided with the oxygen-sulphide borderline which during late summer occurs between 6.5 and 7.0 m depth. Maxima of algal photosynthesis were recorded in the epilimnion where *Ceratium hirundinella* dominates and in the lower strata of metalimnion (5.5–6.5 m depth; Fig. 1) where cells of *Mallomonas, Rhodomonas, Cryptomonas* and *Synechococcus* are abundant (Blaauboer 1982).

Integral algal primary production shows two annual maxima in early June and August with values of 30 and 33.5 kg C.lake^{-1}.day^{-1}, respectively (Fig. 2). The maximum in August could be attributed largely to the metalimnetic populations (cf. Fig. 1).

The anoxygenic photosynthetic activity of the sulphur bacteria became measurable in early June, with production rates of 4 mg C.m^{-2}.day^{-1} i.e. 0.07 kg C.lake^{-1}.day^{-1} (Fig. 2). The maximal bacterial production rate of 106 mg C.m^{-2}.day^{-1} recorded on 6 August (Table 1) is equivalent to 2.2 kg C.lake^{-1}.day^{-1} (Fig. 2).

The contribution of the primary production by photosynthetic sulphur bacteria to the total productivity has been determined in different kinds of lakes. Wetzel (1975), Cohen *et al.* (1977) and Pfenning (1978) have summarized the published literature on productivity of the photosynthetic bacteria. The daily rates range from 19 to 190 mg C.m^{-2} with exceptions, e.g. 2470 mg C.m^{-2}.day^{-1} in Fayetteville Green lake (Culver & Brunskill 1969) and 8 015 mg C.m^{-2}.day^{-1} in Solar lake (Cohen *et al.* 1977). It appears that in productive dimictic and meromictic lakes bacterial photosynthesis at certain periods easily can exceed that of algae and macrophytes (Wetzel 1975). Generally, the range of bacterial

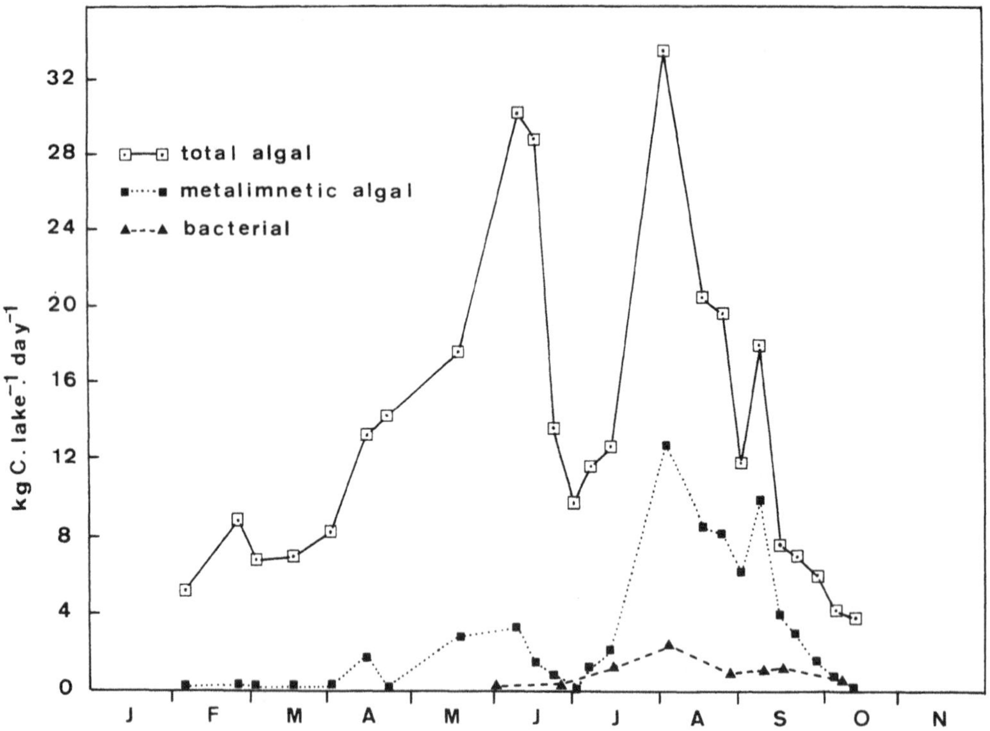

Fig. 2. Integral daily primary production of algae and photosynthetic sulphur bacteria of Lake Vechten (The Netherlands), 1980, computed for the entire pelagial zone of the lake, compensating for volumetric changes (kg C · lake^{-1} · day^{-1}).

photosynthesis in Lake Vechten is rather similar to that recently reported by Parkin & Brock (1980) for several lakes. They found during stratification a bacterial primary productivity in the range of 1.98–59.8 mg C.m^{-2}.day^{-1}, constituting 0.3–6% of the total daily primary production of these lakes (cf. Table 1).

The contribution of sulphur bacteria to the total annual pelagial primary production of Lake Vechten, calculated from the data presented in Fig. 2, is 3.6%, i.e. 127 kg C.lake^{-1}.yr^{-1} of bacterial against 3510 kg C.lake^{-1}.yr^{-1} of total production. This share of bacterial production will become still smaller when considering the macrophytic and periphytic primary production of Lake Vechten as well (Gons 1981). However, considering the summer stratification period only, the sulphur bacteria contribute on an areal basis between 3.9 and 17.5% of total primary production in the pelagial region of the lake (Table 1). Thus, their share in productivity and therefore their role in the carbon cycle of the lake during summer stratification cannot be ignored.

Table 1. Daily rates of integrated areal primary production (mg C.m^{-2}.day^{-1}) during summer stratification in 1980 of Lake Vechten (The Netherlands).

	Production (mg C.m^{-2}.day^{-1})			
Date	Algal	Bacterial	Total	% bacterial
25.06.80	336.5	13.6	350.1	3.9
16.07.80	286.0	52.7	338.7	15.6
06.08.80	1 012.6	106.1	1 118.7	9.5
27.08.80	596.6	38.0	634.6	6.0
09.09.80	593.1	42.1	635.2	6.6
17.09.80	217.8	46.1	263.9	17.5
08.10.80	106.8	21.9	128.7	17.0

In this respect the observations of Takahashi & Ichimura (1968) and Culver & Brunskill (1969) are of particular interest. They found that zooplankton just above the anoxic zone, where photosynthetic bacteria were abundant, were usually dark red, apparently a result of heavy grazing on red sulphur bacteria. Takahashi & Ichimura (1968) obtained direct evidence of grazing using ^{14}C-labelled photosynthetic bacteria. Also, Sorokin (1970) observed that *Daphnia* populations accumulate at the upper boundary of the H_2S-containing zone and consume large quantities of photosynthetic sulphur bacteria. Gophen *et al.* (1974) provided evidence that green sulphur bacteria may be a dietary source for zooplankton in Lake Kinneret. All the evidence suggests that photosynthetic bacteria may contribute significantly to the secondary production of a lake.

Finally, when considering aspects of the carbon cycle during summer stratification, one must bear in mind that photosynthetic sulphur bacteria form a link in the processes of mineralization of organic matter (Cappenberg 1981). They find their adequate growth conditions without competitors anaerobically in the light and metabolize available end products of anaerobic fermentative and respiratory processes, including H_2S, H_2, CO_2 and reduced simple organic molecules (Pfenning 1978).

Summary

Primary production measurements over an annual cycle were carried out in Lake Vechten, a small (4.7 ha) relatively deep (maximum depth 11.9 m; mean depth 6 m) monomictic lake located in the central part of the Netherlands, in order to assess the contribution of photosynthetic sulphur bacteria to total pelagial production.

The lake is stratified from May till November and exhibits hypolimnetic oxygen depletion. During summer stratification, dense populations of photoautotrophic bacteria develop in the metalimnion and upper strata of the hypolimnion. They consist of cyanobacteria *(Synechococcus*-type cells), Chromatiaceae *(Chromatium* sp. and *Thiopedia* sp.) and green and brown coloured Chlorobiaceae (particularly *Chloronema*-type filaments). Furthermore, algae are abundant at certain times in the metalimnion, viz. Chrysophyceae *(Mallomonas caudata)* and Cryptophyceae *(Rhodomonas* sp. and *Cryptomonas* sp.).

Algal and bacterial photosynthetic rates, measured by the ^{14}C method, were distinguished using the inhibitor DCMU, which specifically blocks the oxygenic algal without affecting the anoxygenic bacterial photosynthesis.

Anoxygenic photosynthetic activity of sulphur bacteria became measurable in early June at the oxygen-sulphide borderline with a rate of 4 mg $C.m^{-2}.day^{-1}$. Maximal bacterial production was recorded in August (106 mg $C.m^{-2}.day^{-1}$) (see Table 1). In October bacterial production decreased rapidly and the bacterial community disappeared in early November at the onset of autumnal turnover.

Considering the summer stratification period only, the sulphur bacteria contributed on an areal basis between 3.9 and 17.5% of total pelagial production (Table 1). So, it is suggested that in summer, sulphur bacteria play a significant role in the lake's carbon cycle.

Algal and bacterial production data were computed on a lake's basis taking the hypsometry of the lake into account. Integral algal production showed two maxima during the summer of 30 and 33.5 kg $C.lake^{-1}.day^{-1}$, respectively (Fig. 2). The latter maximum could be attributed largely to metalimnetic populations consisting mainly of *Synechococcus* and *Mallomonas caudata* (cf. Fig. 1). Integral bacterial production maximum recorded in August was 2.2. kg $C.lake^{-1}.day^{-1}$ (Fig. 2).

The contribution of sulphur bacteria to total annual pelagial production was 127 kg $C.lake^{-1}.yr^{-1}$ against an algal production of 3 383 kg $C.lake^{-1}$-yr^{-1}. So, sulphur bacteria accounted for 3.6% of total annual pelagial primary production.

Acknowledgements

I am greatly indebted to W. A. de Kloet for generously providing production data from epilimnion and lake's circulation period and for his help with tracer work and daybreak sampling; and to H. Roon for his adequate assistance in the field. I thank J. van Kleinwee, who typed the manuscript.

References

Baumann, E. W., 1974. Determination of parts per billion sulfide in water with the sulfide-selective electrode. Analyt. Chem. 46: 1345–1347.

Best, E. P. H., Blaauboer, M. C. I., Cappenberg, Th. E., Gons, H. J., Gulati, R. D., De Kloet, W. A., Steenbergen, C. L. M. & Verdouw, H., 1978. Towards an integrated study of the ecosystem of Lake Vechten. Hydrobiol. Bull. 12: 107–118.

Blaauboer, M. C. I., 1982. The phytoplankton species composition and the seasonal periodicity in Lake Vechten from 1956 to 1979. Hydrobiologia 95: 25–36.

Cappenberg, Th. E., 1981. Mineralization of organic matter. In: Parma, S. & Gulati, R. D. (Eds.) Progress Report, Limnological Institute 1980. Verh. Kon. Ned. Akad. Wet. afd. Nat., 2e Reeks 77: 17–19.

Cohen, Y., Krumbein, W. E. & Shilo, M., 1977. Solar lake (Sinai). 2. Distribution of photosynthetic microorganisms and primary production. Limnol. Oceanogr. 22: 609–620.

Culver, D. A. & Brunskill, G. J., 1969. Fayetteville Green Lake, New York. V. Studies of primary production and zooplankton in a meromictic marl lake. Limnol. Oceanogr. 14: 862–873.

Golterman, H. L., Clymo, R. S. & Ohnstad, M. A. M., 1978. Methods for physical & chemical analysis of fresh waters. IBP Handbook No. 8. Blackwell, Oxford.

Gons, H. J., 1981. Carbon cycle in Lake Vechten: In: Parma, S. & Gulati, R. D. (Eds.) Progress Report, Limnological Institute 1980. Verh. Kon. Ned. Akad. Wet. afd. Nat., 2e Reeks 77: 30–36.

Cophen, M., Cavari, B. Z. & Berman, T., 1974. Zooplankton feeding on differentially labelled algae and bacteria. Nature 247: 393–394.

Gromet-Elhanan, Z., 1977. Electron transport and photophosphorylation in photosynthetic bacteria. In: Trebst, A. & Avron, M. (Eds.) Encyclopedia of Plant Physiology, Vol. 5. Springer, Berlin.

Gulati, R. D., 1975. A study on the role of herbivorous zooplankton community as primary consumers of phytoplankton in Dutch lakes. Verh. int. Verein Limnol. 19: 1202–1210.

Padan, E., 1979. Impact of facultatively anaerobic photoautotrophic metabolism on ecology of Cyanobacteria (blue-green algae). In: Alexander, M. (Ed.) Advances in Microbial Ecology, Vol. 3. Plenum, New York.

Parkin, T. B. & Brock, T. D., 1980. Photosynthetic bacterial production in lakes: the effect of light intensity. Limnol. Oceanogr. 25: 711–718.

Parma, S., 1971. Chaoborus flavicans (Meigen) (Diptera, Chaoboridae) an autecological study. Ph. D. thesis, University of Groningen. 128 pp.

Pfennig, N., 1978. General physiology and ecology of photosynthetic bacteria. In: Clayton, R. K. & Sistrom, W. R. (Eds.) The Photosynthetic Bacteria. Plenum, New York.

Sorokin, Y. I., 1965. On the trophic role of chemosynthesis and bacterial biosynthesis in water bodies. Mem. Ist. Ital. Idrobiol. 18: 187–205.

Sorokin, Y. I., 1970. Interrelations between sulfur and carbon turnover in meromictic lakes. Arch. Hydrobiol. 66: 391–446.

Steemann-Nielsen, E., 1952. The use of radio active carbon (^{14}C) for measuring organic production in the sea. J. Cons. perm. int. Explor. Mer 18: 117–140.

Steenbergen, C. L. M. & Korthals, H. J., in press. Distribution of phototrophic microorganisms in the anaerobic and microaerophilic strata of Lake Vechten (The Netherlands). Pigment analysis and role in primary production. Limnol. Oceanogr.

Steenbergen, C. L. M. & Verdouw, H., 1982. Lake Vechten: aspects of its morphometry, climate, hydrology and physicochemical characteristics. Hydrobiologia 95: 11–23.

Takahashi, M. & Ichimura, S., 1968. Vertical distribution and organic matter production of photosynthetic sulfur bacteria in Japanese lakes. Limnol. Oceanogr. 13: 644–655.

Trebst, A., 1980. Inhibitors in electron flow: tools for the functional and structural localization of carriers and energy conservation sites. In: San Pietro, A. (Ed.) Methods in Enzymology, Vol. 69. Academic, New York.

Wetzel, R. G., 1975. Limnology. Saunders, Philadelphia.

The aquatic macrophytes of Lake Vechten. Species composition, spatial distribution and production

Elly P. H. Best
Limnological Institute 'Vijverhof' Laboratory, Rijksstraatweg 6, 3631 AC Nieuwersluis, The Netherlands

Keywords: macrophytes, species inventory, production rate, biomass, simulation models

Abstract

The macrophytic species compositions in Lake Vechten of 1963 and 1979–80 were compared and showed a considerable change. The main vegetation types were mapped in 1973, 1978, 1979 and 1980. All macrophytic communities, i.e. submerged, floating-leaved and emergent vegetation types, declined mainly due to increasing water turbidity, increasing tree-shading and, from 1978 onwards, grazing and trampling by cattle.

Production rates, derived from different combinations of measurement and calculation, were compared. The macrophytes contributed only about 7% to the total lake production in 1980.

Introduction

The role of the littoral region, viz. the zone where macrophytes occur, in the metabolism of shallow lakes has received increased attention lately. This region belongs to the most productive biotopes on earth (Wetzel 1975).

The littoral area with the various groups of aquatic macrophytes consists of an intricate habitat, which is affected by the lake's drainage basin as well as by its pelagial zone (Sculthorpe 1967; Bernatowicz 1967).

The aquatic macrophytes may contribute to this region by a) their primary production (e.g. Adams & McCracken 1974; Filbin 1980; Dykyova 1978); b) detritus formation (Wetzel *et al.* 1972); c) acting as nutrient source, by active or passive nutrient release, as well as sink, by nutrient uptake from the interstitial and ambient water (e.g. Carpenter 1980; Carignan & Kalff 1980); and by d) serving as substrate for other organisms such as periphytic algae, bacteria and macrofauna.

Considerable knowledge exists on primary production of a few macrophytic species. However, relatively little attention has been paid to the significance of the integral macrophytic unit for the carbon cycle of a lake, alongside the roles played by other primary producers such as periphytic and pelagic algae. Only a few studies can be mentioned in this respect (Wetzel 1964; Rich *et al.* 1971; Assis Esteves 1979; Jones 1980).

The studies on macrophytic production were part of an integrated ecosystem study in Lake Vechten. The aim of the latter was to get more insight into the interrelationships between the different biotic and abiotic components and the cycling of organic matter; most attention was given in this respect to the carbon cycle.

This paper surveys the changes in the species composition of macrophytes and the primary production of the main vegetation types during several years in Lake Vechten (The Netherlands). The production of macrophytes is compared with that of the other primary producers.

Site description

Lake Vechten (lat. 52° 04′N, long. 05′E) is a small lake, situated in the municipality of Bunnik,

Hydrobiologia 95, 65–77 (1982). 0018-8158/82/0951-0065/$02.60.

along the main road (E36) from Utrecht to Arnhem. It originates from sand digging in the early forties. The surface area is 4.7 ha; the average and maximum depth, respectively, are 6.0 and 11.9 m. Several historical, chemical and biological aspects of this lake have been illustrated by Steenbergen & Verdouw (1982), Verdouw & Dekkers (1982), Gons (1982), Gulati *et al.* (1982), Blaauboer (1982), Dvorak & Best (1982), de Kloet (1982) and Steenbergen (1982). The lake is surrounded by trees which have increased substantially in girth during the last decade, intercepting the solar irradiation in particular at the southern and eastern part of the lake. The lake is not open to the public; it is leased by a local association of anglers. It is situated on farmland, which is used most of the summer as pasture. During this period not only the terrestrial plants but also the aquatic ones are grazed as far as they are within the reach of the cattle. The land is fertilized by artificial and cattle manure.

Species composition and spatial distribution

Nowadays (1980) only a relatively small portion of Lake Vechten is reckoned to constitute the littoral region, i.e., the interface between the land of the drainage basin and the lake's pelagial zone. Based on the macrophytic vegetation, the littoral zone covers about 10% of the total lake area.

Within the littoral region three distinct zones can be demarcated (Fig. 1). The shallowest part, extending from the shoreline to about 1 m depth, is occupied by emergent species; here *Phragmites australis* predominates whereas *Sparganium erectum, Typha latifolia* and *Iris pseudacorus* occur locally in rather dense colonies. Bordering the emergent plants, *Polygonum amphibium,* a floating-leaved species, may occur rooting from 1 to about 3 m depth. Beyond this area, extending from the emergent vegetation to the limit of the photic zone, is the region of the submerged aquatics, predominated by, respectively, *Elodea nuttallii* together with *E. canadensis* and *Ceratophyllum demersum.*

The species composition has changed considerably during the last seven years.

The earliest known inventory of the littoral vegetation, mainly floating-leaved and emergent species, dates from 1963 (S. Parma, pers. comm.). In 1979 and 1980 the macrophytes were surveyed again and this time the submerged species were also recorded, the latter with the use of SCUBA. The methods used for surveys of submerged aquatics have been discussed by Best (1980). Identification of the species was done with several keys (Fassett 1975; Haslam *et al.* 1975; Heukels & van Oostrom 1970; van Raam 1977; Wood & Imahori 1965).

In Table 1 the macrophytic species present during 1963 and 1979–80 are listed. It is clear that both the species number and quantity have decreased,

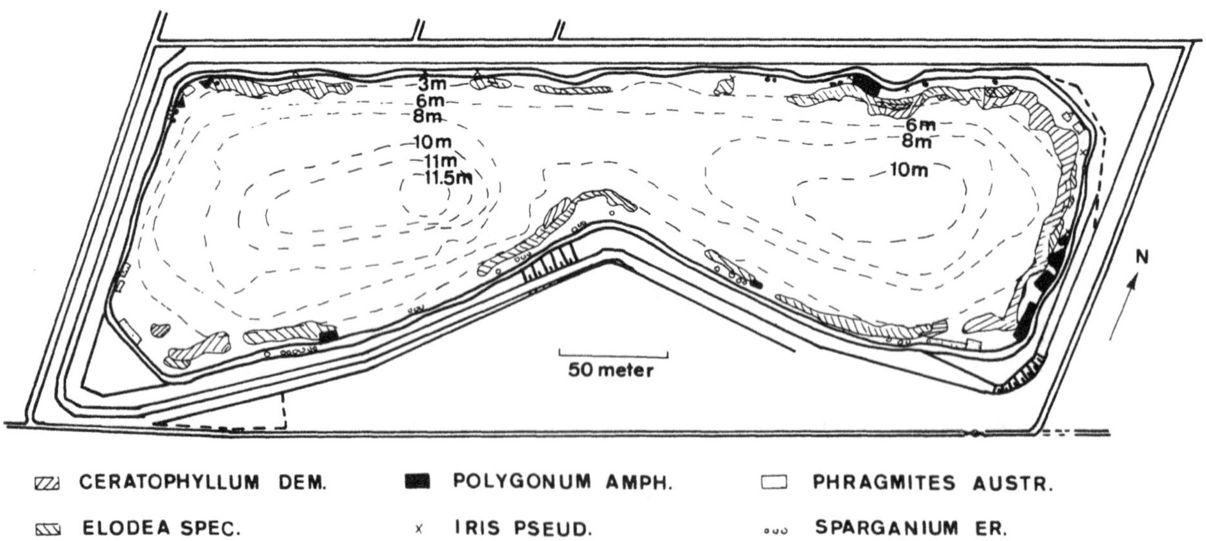

Fig. 1. Schematic representation of the main vegetation types of Lake Vechten in 1980. As criterion for a particular type of vegetation a cover percentage higher than 10% was used.

Table 1. Changes in the composition of macrophytic species in Lake Vechten in the course of time. p – predominant, a – abundant, i – incidental.

	Species	1963	1979–80
Emergent	*Alisma lanceolatum* With.	a	i
	Eleocharis palustris (L.) R. et Sch.	i	–
	Epilobium palustre L.	a	i
	Epilobium hirsutum L.	i	–
	Equisetum palustre L.	a	i
	Equisetum fluviatile L.	a	–
	Glyceria fluitans (L.) R. Br.	i	–
	Glyceria maxima (Hartm.) Holmb.	p	–
	Heloscadium nodiflorum Koch	i	–
	Iris pseudacorus L.	p	a
	Juncus articulatus L.	i	–
	Juncus compressus Jacq.	i	–
	Juncus inflexus L.	p	–
	Lycopus europaeus L.	p	i
	Lysimachia nummularia L.	a	i
	Lythrum salicaria L.	i	i
	Mentha aquatica L.	p	a
	Myosotis palustris Hill.	p	a
	Phragmites australis (Cav.) Trin. ex Steud.	p	a
	Sagittaria sagittifolia L.	a	i
	Scirpus maritimus L.	i	–
	Scutellaria galericulata L.	a	–
	Sium erectum Huds.	i	i
	Sparganium erectum L.	p	a
	Typha angustifolia L.	p	i
	Typha latifolia L.	p	i
Floating-leaved	*Polygonum amphibium* L.	p	p
	Potamogeton natans L.	p	–
Submerged	*Ceratophyllum demersum* L.		p
	Elodea canadensis Michx.	a	a
	Elodea nuttallii (Planch.) St. John.		p
	Fontinalis antipyretica L.		i
	Myriophyllum spicatum L.		i
	Potamogeton crispus L.		i
	Potamogeton lucens L.		i
	Potamogeton pusillus L.		i
	Ranunculus circinnatus Sibth.		i
	Chara globularis var. *virgata*		i
	Nitella mucronata (A. Br.) Miq. (f. *Leiopitys* (Wheld.) R.D.W.)		i

i.e. 42% of the emergent species have disappeared, and the species that were formerly predominant are now quantitatively reduced.

Of the floating-leaved species, only one of the two present in 1963 is still growing in the lake. Since the submerged plants were not recorded in 1963, a comparison with the survey in 1979–80 with respect to their presence or absence in the lake is not possible. From 1973 onwards the submerged macrophytes were surveyed using SCUBA diving. In that year the same submerged species as in 1979–80 were present but in higher abundance.

The main submerged vegetation types of the lake were also mapped (Table 2). In 1973 the predominant submerged species were *Ceratophyllum demersum* and *Elodea* sp. *Elodea* occupied mainly the shallow areas of the lake to about 2 to 3 m depth and *Ceratophyllum* grew at depths ranging from 2 to 5 m, both species showing a distinct zonation. The submerged littoral zone extended to 5 m depth, occupying about 37% of the total lake area.

In 1977, *Myriophyllum spicatum* started to thrive. In 1978 it already covered 4% of the total lake area, and caused a decline in the *Ceratophyllum* populations in the east and west of the lake. *Myriophyllum spicatum* deprives other submerged species of light by its canopy formation at the water surface (Titus 1977) and subsequently it starts growing on sites originally occupied by the other

Table 2. The main vegetation types of Lake Vechten. As criterion for a particular type of vegetation, a cover percentage higher than 10% was used. Exceptions: *1 cover percentage >25%; *2 plants irregularly distributed, number of plants counted.
†: extinct. ( ): not mapped. The stands were relatively pure; the contribution of other plant species than the principal one was neglected. No data were available for 1963.

Macrophytic species	Lake area occupied (%)			
	1973	1978	1979	1980
Submerged				
Ceratophyllum demersum	8.9*1	4.1	4.0	3.5
Elodea sp.	7.3*1	8.9	8.9	4.6
Myriophyllum spicatum	–	2.6	†	–
Nitella mucronata	–	0.2	0.2	†
	16.2	15.8	13.1	8.1
Floating-leaved				
Polygonum amphibium	(0.5)	0.5	0.5	0.5
Emergent				
Phragmites australis			1.6	0.2
Sparganium erectum			0.4	*2
Typha angustifolia			0.2	*2
	(2.2)	(2.2)	2.2	0.2
Total	18.9	18.5	15.8	8.8

species. In 1979 *Myriophyllum* vanished almost completely from the lake, with a few remains left in spring. In 1978 and 1979 two rather small meadows of *Nitella mucronata*, a characeous macroalga, were found along the southern border of the lake at 3 to 5 m depth.

All the areas covered by the submerged, floating-leaved and emergent vegetation types declined substantially in the course of time. The *Ceratophyllum* population decreased mainly between 1973 and 1978. The *Elodea* vegetation was stabilized for some time but showed a sharp decline from 1978 onwards, mainly at the south side of the lake and on both its shallowest and deepest sites. The other two species of submerged macrophytes thrived only briefly. Reasons for this may be a) diminution of radiation at the lake border due to substantial increase in tree girth, causing on several places a reduction of 50% of the light intensity, and b) increase in attenuation of incoming radiation in the lake water. The *Polygonum* biomass decreased mainly from the shoreside also, probably partly because of shading by trees and cattle grazing. Of all emergent species, mainly the plants growing nearest to the shore disappeared due to grazing and trampling by cattle. This was most conspicuous in *Phragmites australis* (see also Best *et al.* 1982; Dassen 1981).

Production

Problems arise in determining the annual production of freshwater macrophytes and therefore the estimates are mainly based on seasonal measurements of biomass. However, errors are unavoidable with the latter method. For instance, the estimates become less reliable when stands have stable or irregularly changing biomass (Rich *et al.* 1971; Thompson 1971). Although the importance of respiration data for the net carbon balance is recognized, such data are still scarce (Stanley & Naylor 1972; Hough 1974; Ondok & Gloser 1978; Titus *et al.* 1975; Søndergaard 1979; Filbin 1980). Losses of biomass due to grazing and the ensuing damage or mortality should be quantified (Matthews & Westlake 1969; Pieczyńska 1972; Kajak *et al.* 1972; Dawson 1973; Titus *et al.* 1975).

Since determination and, particularly, interpretation of biomass changes in terms of production is difficult, more research is nowadays directed towards the use of metabolic methods (Titus *et al.* 1975; Ondok & Gloser 1978, 1978a; Best 1981). In several cases a better understanding of the behaviour of the macrophytes *in situ* is gained (Ikusima 1970; Kvet 1971; Verhandelingen 1979). Much attention is nowadays paid to metabolism of macrophyte-periphytic complexes not enclosed in a cuvette (e.g. Edwards & Owens 1962; Goulder 1970; Best & Kersting 1981).

It is thus a precarious enterprise to estimate macrophyte production mainly from biomass determinations. Despite this, when these data are used in combination with relevant correction factors to make up for the shortcomings mentioned above, a production range thus obtained can be used, e.g., for the calculations of nutrient budgets.

In this section a summary is given of the data on primary production of the main macrophytic species collected during the last four years using various methods.

For the predominant submerged aquatics *Ceratophyllum demersum* and *Elodea* sp., the maximum relative growth rates, the seasonal pattern of photosynthetic activity (^{14}C method, terminal plant portions) and net production (biomass) were determined. A detailed study on the photosynthetic and growth characteristics of *Ceratophyllum*, is still in progress (Best 1981; Best & Dassen 1982). The floating-leaved and emergent plants have received more attention recently (Best *et al.* 1982; Dassen 1981). Their production estimates reported here are based on data collected by the multiple harvest method, which involves several samples taken with regular time intervals (2 weeks to 1 month). Hence, production rates calculated for the different macrophytic stands do not have the same level of accuracy.

Materials and methods

Measurements of photosynthetic activity in submerged macrophytes

Photosynthesis was determined with the carbon-14 technique (Steemann Nielsen 1952; Wetzel 1965; Adams & McCracken 1974). The modified method used in the present study is described elsewhere (Best & Meulemans 1979).

In 1977 the *in situ* photosynthetic activity of terminal portions of *Ceratophyllum* and *Elodea* species was compared.

In 1978 and 1979 data were collected on photosynthetic activity and capacity of *Ceratophyllum*. The photosynthetic activity was measured in plant portions incubated *in situ* enclosed in cuvettes; the photosynthetic capacity was determined in plant parts incubated in growth chambers in vessels filled with lake water, which were stirred, at a temperature of 20 °C and a light intensity of 14.4 W m^{-2} (an intensity often found at the plants' common rooting depth of 2.5 m in Lake Vechten). The plants were incubated for 2 h after a 1 h adaptation period. This time not only the terminal portions but also other plant parts were included. For *in situ* experimental work on submerged macrophytes and biomass sampling SCUBA was used.

Biomass sampling

Single plant marking. The growth curves and maximum relative growth rates of three submerged species, namely *Ceratophyllum demersum Elodea* sp. and *Myriophyllum spicatum*, were determined in 1978. For this, sites were chosen where the respective plant communities grew best. For *Ceratophyllum* depth classes at 1–2, 3 and 4–5 m were distinguished, whereas *Elodea* and *Myriophyllum* occurred only in narrow depth ranges. i.e. at 1–2 and 2–3 m, respectively. The plants were marked *in situ* (using SCUBA) with metal tags at the onset of the growing season and the biomass above the tags was recorded. At every subsequent harvest the biomass above the tags was determined. Three tagged plants were sampled every three weeks throughout the year. Thus confusion about age differences is avoided (Best 1980). The plants were freed from periphyton as described in Verhandelingen (1978). *Nitella mucronata* was not included in this study since the metal tags were too heavy for the plants and disturbed their growth pattern.

Biomass. The biomass of all submerged vegetation types was estimated by harvesting one square in the middle of a vegetation girdle every two months during 1978 and 1979, using SCUBA. For this a metal square of 0.25 × 0.25 m was used. In 1980 only the minimum and maximum biomass was recorded. This single sampling method with regular intervals was chosen since taking more replicates every sampling time damaged the vegetation significantly. Variation due to non-representative sampling inherent to this method was checked, being at the highest ± 17% (in dense vegetation types).

The biomass of the floating-leaved and emergent vegetation types was measured in 1979 and 1980. Rectangles of 0.2 × 0.5 m were harvested. In 1979 only minimum and maximum biomass was measured by duplicate sampling subsequently in February and July in *Polygonum, Iris, Sparganium* and *Typha.* This was repeated in 1980 for the latter 3 species. In 1980 the growth curves of *Polygonum* and *Phragmites* were recorded by monthly sampling of 3 plants; the biomass on an m^2 basis was calculated using these values and counting the number of plants per m^2. In addition at least 2 rectangles were harvested in the period when the plants had their highest biomass. In *Polygonum* two different vegetation types were distinguished on the basis of their growth characteristics growing in the sun (north in the lake) or in the shadow (south and east in the lake). In 1979 a detailed study on production of *Phragmites* was done which is described elsewhere (Best *et al.* 1982). The underground organs were sampled to a depth of 0.2 m in the sediment. No distinction was made between rhizomes and roots. In this way the largest part of the underground biomass was recovered since most emergent plants did not root much deeper in the bottom of Lake Vechten. The largest underestimate was certainly made in *Phragmites* and *Polygonum,* since these plant species were most difficult to sample due to a) their growth sites at relatively deep parts of the upper littoral and b) their irregular rootsystem. In *Polygonum* the error made in this way could not be quantified; in *Phragmites* it was 12% at the highest. The variation in biomass data inherent to the method of duplicate sampling only in the periods of lowest and highest biomass was 20% for aboveground and 25% for underground organs, respectively.

Since the aquatic plants – and submerged ones in particular – exhibit a patchy growth pattern, the whole aquatic vegetation was also mapped and the cover percentages were estimated in the period of maximum biomass.

All biomass samples were rinsed thoroughly and freeze-dried. Prior to drying the plant material of the emergent and floating-leaved species was divided into above- and belowground parts. The dry plant material was clipped into small pieces and several grams of this mixture were ground into a fine powder with liquid nitrogen in a JKA analysis mill (Janke & Kunkel KG, Stauffen, Germany). Several micrograms were used for ash determinations (in duplicate) by combustion at 450 °C. Biomass is expressed as ashfree dry weight.

The annual net production rates were calculated by subtracting the minimum biomass from the maximum one and dividing this by 365 (days). Similarly, seasonal net production rate was calculated, in this case by dividing by 230 (days) (the macrophytic growing season in Lake Vechten lasted from 1 April to 16 November).

Tissue analysis

In order to assess a possible nutrient limitation by both nitrogen and phosphorus, these two most probable limiting elements (Gerloff & Krombholz 1966) were determined in the dry matter. Also, carbon was measured for estimating the magnitude of the macrophyte compartment in the lake and for calculation of carbon flow through the system. Carbon and nitrogen were determined using an HCN analyzer (Perkin Elmer model 240) and phosphorus was by a modified molybdate method (Olsen 1967).

Results and discussion

Biomass and primary production of the main macrophytes

Production estimates based on biomass data. Maximum biomass of the principal macrophytes during 1980 is summarized in Table 3. The largest part of the littoral zone, about 8.1% of the total lake area, was occupied by submerged plants whereas the floating-leaved and emergent ones together covered only 0.4–0.5% of the lake area. The maximum biomass per m² stand area of the submerged species was much lower than those of the floating-leaved and emergent plants, the average values being, respectively, 36.9, 282.4 and 637.1 g ashfree dry wt.

Table 3. Cover and production of the aquatic macrophytes of Lake Vechten estimated from biomass harvests in 1980. Growing season 195 days (1 April–12 October). Biomass partitioning: A – above ground, U – underground.

Macrophytic species	Cover	Maximum biomass (ashfree dry. wt.)		Net production rate: Annual (g ashfree dry. wt. day^{-1})		Mean growing season (g ashfree dry. wt. day^{-1})		Mean growing season (g C day^{-1})		Average C content (% ashfree dry. wt.)
	stand area (m²)	m^{-2} stand area (g)	$lake^{-1}$ (kg)	m^{-2} stand area	$lake^{-1}$	m^{-2} stand area	$lake^{-1}$	m^{-2} stand area	$lake^{-1}$	
Submerged										
Ceratophyllum demersum	1 645	62.8	103.3	0.153	252	0.174	287	0.083	138	48.0
Elodea spp.	2 162	10.9	23.6	0.027	59	0.030	67	0.015	33	49.0
Floating-leaved										
Polygonum amphibium	247	216.3A	53.4	0.593	146	0.676	166	0.318	78	47.0
		66.1U	16.3	0.036	31	0.041	35	0.019	16	
Emergent										
Iris pseudacorus	5	185.6A	0.9	0.508	2	0.580	2	0.261	1	45.0
		178.2U	8.9	0.098	5	0.112	5	0.050	2	
Phragmites australis	84	1,422.6A	119.5	3.670	308	4.186	352	1.988	167	47.5
		291.9U	24.5	0.267	22	0.305	25	0.145	12	
Sparganium erectum	24	19.3A	0.5	0.053	1	0.060	1	0.027	1	45.0
		22.9U	0.6	0.013	0.3	0.015	1	0.007	0	
Typha angustifolia	70	240.3A	16.8	0.660	46	0.753	52	0.317	22	42.2
		188.6U	13.2	0.172	12	0.196	13	0.083	5	

The ratios of aboveground: belowground organs of the floating-leaved and emergent plants differed significantly. These values were 3.3:1 in *Polygonum,* 1.0:1 in *Iris,* 4.9:1 in *Phragmites,* 0.8:1 in *Sparganium* and 1.3:1 in *Typha.* Since a substantial part of the shoot biomass of *Sparganium* was grazed by cattle, the value for this species is much lower than under undisturbed conditions; in undisturbed plants this value would have been 2.5. From the data on maximum biomass per m^2 stand area, however, it might be concluded that only *Sparganium* and *Typha* are growing under suboptimal conditions (Assis Esteves 1979; Whigham *et al.* 1978; Westlake 1975), possibly due to shading by trees and untimely removal of photosynthetic tissues by cattle grazing.

Since the growth and production of *Ceratophyllum* and *Elodea* are depth-dependent, their vegetation types were divided into different depth-categories for the biomass estimates. Although it is known that growth and turnover properties of root systems are different from those of shoots (Dykyova 1978; Schierup 1970; Whigham *et al.* 1978) the available data are ambiguous and therefore not used here.

A comparison of the annual production rates of the different species (Table 3) shows that *Ceratophyllum* had a much higher rate than *Elodea,* namely 0.15 versus 0.03 g ashfree dry wt. m^{-2} stand area day^{-1}. Normally *Elodea* occupies shallower areas than *Ceratophyllum;* thus its production and maximum relative growth rate are higher. In the present case, however, *Ceratophyllum* started gradually to colonize shallower areas than it did earlier, possibly being forced by decreasing water transparency during its growing season (see later). Since *Elodea* received less and less radiation due to shading, decrease in water transparency and competition for radiation by *Ceratophyllum,* the growth and production of this species were poor. Of the floating-leaved and emergent plants, *Phragmites* showed the highest production rate of 3.94 g ashfree dry wt. m^{-2} stand area · day^{-1}. The annual macrophytic production rates recorded during 1979 and 1980 were compared. The production rate apparently increased in the submerged species from 0.06 to 0.08 g m^{-2} · day^{-1} of C, mainly due to the penetration of *Ceratophyllum* to shallower areas. It decreased in the floating-leaved *Polygonum* from 0.81 to 0.63 g m^{-2} · day^{-1} probably due to tree shading and, mainly, grazing by cattle. In emergent plants an apparent increase occurred in production rate from 0.94 to 1.18 g · m^{-2} · day^{-1}. In this case tree shading did not play an important role since the emergent plants grew mainly in unshaded areas. Since the cattle grazed particularly in the terrestrial part of the littoral zone a substantial part of the emergent vegetation types had disappeared by 1980 (Best *et al.* 1982). This apparent increase was due to the fact that in 1979 regions sparseley occupied by plants were included in the stand areas, whereas in 1980 the remaining plants were counted and only densely occupied areas were mapped.

On a unit lake area basis, the seasonal macrophytic production decreased significantly from 1979 to 1980, i.e., from 0.05 to 0.02 g · m^{-2} · day^{-1} or by about 66%. Since in both years the incoming radiation was very similar (2% difference over the growing season), this decrease can not be attributed to a decrease in incoming radiation.

Submerged production decreased by 33%, floating-leaved by 12% and emergent by 80%.

Production calculations based on interception of incoming radiation. As indicated earlier, objections can be made to production calculations from biomass data. On the other hand, if physiological properties of the plants concerned and their environmental conditions are sufficiently understood, including the growth limiting factors, production can be calculated using general simulation models with some adaptations.

Here, the model developed by Goudriaan & v. Laar (1978) for calculating gross photosynthesis of leaf canopies at different latitudes is used for the floating-leaved and emergent plants. All plants concerned are C_3-plants. Since no growth limitation due to nutrient scarcity (N, P) or water stress was noted, it is assumed not to exist. The lowest N-and P-concentrations in the plant material were higher than, respectively, 1.3% (N) and 0.13% (P) on a dry weight basis (Gerloff & Krombholz 1966). Therefore, according to Goudriaan & v. Laar (1978) an assumed A_{max} of 30 kg ha^{-1} h^{-1} of CO_2 is accepted, without accounting for photorespiration. A_{max} is the rate of leaf photosynthesis at light saturation. The model uses only the leaf area index (LAI), i.e. area per m^2 ground surface. The LAI of a closed canopy is 5. The relationship between cover percentage of a vegetation and the LAI can be formulated as: part covered (< 1) $= 1 - e^{-0.8 \times LAI}$.

The LAI depends on the physiological stage of the plants and therefore it changes in time (Dassen 1981). In the present case the plants are assumed to start unfolding their leaves on day 90 (1 April), reach half their maximum LAI on day 200 (mid July), their maximum on day 240 (late August) and have lost their leaves on day 320 (16 November). These assumptions are based on the growth curves observed for the different plant species. The maximum LAI values were calculated by the formula mentioned above using the cover percentages observed in the different vegetation types during the periods of maximum biomass. Thus, the maximum LAI was 5 for *Polygonum, Iris, Typha* and only 2 for *Phragmites* and *Sparganium*. For the calculation of gross CO_2 assimilation for stands in tree shade the attenuation of the incoming radiation by the trees is accounted for.

Fifty per cent of the solar radiation can be used by plants and in the Netherlands, at a latitude of 52°, the sky is overcast about 50% of the time (Heemst, H. v., pers. comm.). Therefore the available energy for plants here is 5.276 MJ $m^{-2} \cdot day^{-1}$.

For calculation of the gross photosynthesis of the submerged plants the preliminary growth model developed for *Ceratophyllum* (Best 1981) is used for both this plant and for *Elodea*. For the latter, adjustments for morphological differences are made.

The gross production rates of the different plant groups calculated via the models are shown in Fig. 2.

During the 'growing season', in this case defined as the period in which there is an increase in biomass, i.e. day 90 to 240, the simulated production values are higher than those derived from the harvest method (Table 4). The ratio S:H (simulated: harvest) is lower in the submerged plants than in the others. This might be due to the absence of a well-developed root system in the submerged plants, and consequently relatively lower respiration and turnover rates. On the other hand, the calculation of the LAI from the cover percentage of the vegetation type may give an overestimate in the second half of the growing season due to deterioration of photosynthetic tissues in senescing plant material which is included in the cover percentage. Assuming the average respiration rate to be about 33% of the production rate (growth: 30% of gross production; maintenance: 1% of biomass; Penning de Vries 1975; De Wit *et al.* 1978) the ratio S:H is still >1 in all plant groups, indicating extensive turnover and / or higher respiration rates. High respiration rates were reported at the end of the growing season for floating-leaved and submerged plants amounting to 50–60% of total primary production (Filbin 1980; Best & Kersting 1981).

The ratio S:H may be compared to the P:B ratios reported in the literature. The ratio of 1.5 for the submerged plants is lower than the value of 2.6 mentioned for *Myriophyllum spicatum* (although

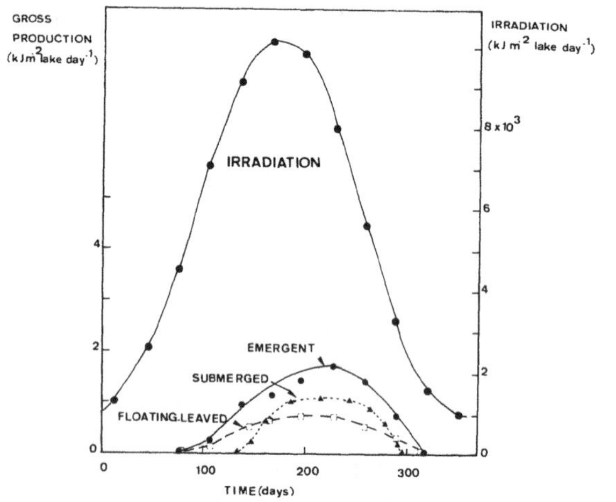

Fig. 2. Diagram in which total irradiance and the gross production rates (carbon assimilation minus photorespiration) of the submerged, floating-leaved and emergent plants are depicted. For converting the production values into grams carbon per m², divide by 41.85. For explanation see text.

Table 4. Comparison of daily production rates in g of carbon per m² calculated using, respectively, the calculation models of Goudriaan & van Laar (1978), Best (1981) and biomass data. Solar radiation input at a latitude of 52° is 5.276 MJ $m^{-2} \cdot day^{-1}$ (Goudriaan & van Laar 1978).

Macrophytic group	Production rate per unit lake area		
	Seasonal		
	Simulated (gross)	Harvest (net)	Ratio S:H
Submerged	249	171	1.46
Floating-leaved	228	94	2.43
Emergent	479	210	2.28

in the latter case the ^{14}C method is believed to measure net production, Adams & McCracken 1974).

The ratios found by me are higher than the 1.25 and 2.0 reported, respectively, by Winberg *et al.* (1972) and Sorokin (1972).

Reasons for the decrease in macrophytic growth 1973–1980

The decrease in biomass was conspicuous in the submerged and emergent macrophytes. The most likely factors causing this in the case of the emergent species have already been discussed.

Growth of submerged aquatics is largely limited by incoming radiation in eutrophic waters of the temperate zones. Possibly, therefore, a stronger light limitation has occurred since 1978, caused either by decreasing water transparency or decreasing solar irradiance or both. First, apparently due to increasing tree shading the incident radiation has been reduced in the shallowest areas. To test separately the effects of incident radiation and water transparency, the preliminary growth model for *Ceratophyllum* was used (Fig. 3). Substitution of the irradiance data of 1978 by those of 1979 resulted in even higher biomass than observed in 1978, since during the growing period of the macrophytes the irradiance was higher in 1979 than in 1978. Use of the extinction coefficients of 1980 instead of those of 1978 led to a decrease in biomass by about 50% (since extinction coefficients in 1979 were not measured the values of 1980 were used). Thus, the decrease in water transparency appears to be a major factor causing the disappearance of submerged plants.

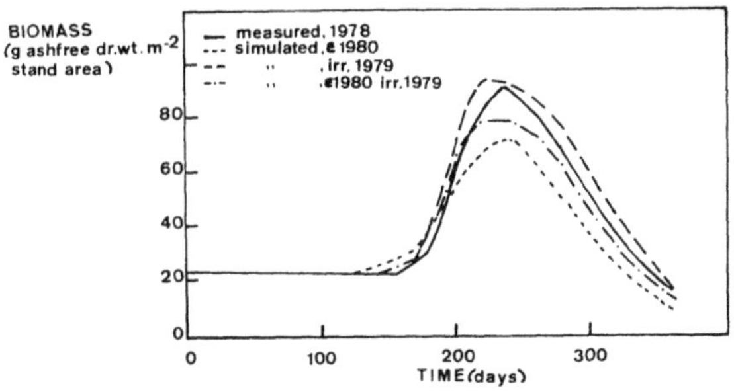

Fig. 3. Comparison of measured and simulated plant biomass of *Ceratophyllum* at a rooting depth of 2.5 m. For the simulation the effects of changes in, respectively, light extinction coefficient (ϵ) and total irradiance (irr.) on production are made visible.

Primary production in Lake Vechten

In Lake Vechten the contribution of macrophytes to the total primary production of the lake is small.

The annual net production of macrophytes in g C per unit lake area, derived from the annual gross production, was 7.4 g $\cdot$ m^{-2} yr^{-1}, i.e. seasonal (season = 195 days) daily production rate was 0.04 g $\cdot$ m^{-2} $\cdot$ day^{-1} or 1.60 kJ m^{-2} $\cdot$ day^{-1}. Since the daily rate of phytoplankton production amounts to 20 kJ m^{-2} $\cdot$ day^{-1} (Best *et al.* 1978) this would mean a macrophytic share of 7.4% in the total primary production. The macrophytic contribution to the total net primary production (being 21.2 kg C $lake^{-1}$ $\cdot$ day^{-1}) derived from the carbon budget in autumn of 1980 was 7.1%, whereas periphyton + phytobenthos and phytoplankton accounted for, respectively, 6.6 and 86.2%

Comparison of macrophytic production in different lakes

Only a few studies are known which quantify in the same lake the production of submerged as well as that of the floating-leaved and emergent macrophytes.

The production of the macrophytes per unit stand area in Lake Vechten lies in the range reported for the eutrophic Russian Lakes Batorin and Myastro (Table 5, Winberg *et al.* 1972). Of these waters Lake Batorin is highly eutrophic and Lake Myastro forms a transition between the former lake and the mesotrophic Lake Zaroch. In Lakes Batorin and Myastro both the submerged and emergent vegetation types are well developed, but in the former the submerged growth is dominated by angiosperms whereas in the latter characeans prevail.

The macrophytic production is higher in Lake Vechten than in the nearby eutrophic Schöhsee in Germany (Assis Esteves 1979); both lakes have many macrophytic species in common. However, the surface area and average depth of Schöhsee are, respectively, 17 and 1.8 times greater than in Lake Vechten. In both lakes the macrophytic production comprises a minor part of total lake primary production: 2% in Schöhsee and 7.3% in Lake Vechten. The differences in lake morphometry, and the ratio of littoral zone: total lake area account for the different relative contributions of phytoplankton and of macrophytes.

Table 5. Annual macrophytic production in different lakes.

Lake	Annual production rate ($g\ C\ m^{-2}$ lake y^{-1})	Relative contribution (%) submerged : floating : emergent
Lake Vechten, Netherlands (this study)	10.7 gross 7.4 net	26.7 : 23.3 : 50.0
Lake Batorin*, USSR (Winberg *et al* 1972)	9.4	54.3 : 45.7
Lake Myastro*, USSR (Winberg *et al* 1972)	19.6	35.7 : 64.3
Schöhsee, FRG (Assis Esteves 1979)	29.6	22.5 : 6.9 : 70.6
Marion Lake, USA (Davis 1970)	50.8	s + fl
Lake Zaroch*, USSR (Winberg *et al.* 1972)	86.5	98.8 : 1.1

* Original data recalculated with P/B = 2, instead of P/B = 1.25. 'Gross' production indicates gross production minus photorespiration (see text).

Summary

The changes in species composition and distribution of the macrophytes, and the production rates of the main vegetation types were surveyed in Lake Vechten, The Netherlands.

A comparison of the number of species present in 1963 and 1979–80 showed a conspicuous decline. About 40% of the emergent species disappeared and only one floating-leaved species remained. No comparison of the submerged species was made since their presence was not recorded in the 1963 survey.

With respect to the main vegetation types most attention was originally paid to the submerged ones, whereas later on the other two groups were included in the study. In 1973, the main submerged communities consisted of *Elodea* sp. and *Ceratophyllum demersum.* Both species formed distinct zones, the former from the lake border to 3 m depth, the latter from 3 to 5 m. *Myriophyllum spicatum* showed a short-term expansion during 1977–78 and affected the growth of the other two species during this period. The characean *Nitella mucronata* expanded considerably during 1978 and 1979.

Different methods were used to determine and calculate the production of macrophytes. For the submerged species maximum relative growth rate, biomass and seasonal photosynthetic rate were measured. Minimum and maximum biomass were determined in the floating-leaved and emergent plants. Nitrogen and phosphorus were assayed in the plant dry matter and proved to be above levels reported to be growth-limiting. The carbon content of the plant material was also measured and the average values varied between 42 and 49% ashfree dry wt.

In 1980 the submerged and floating-leaved plus emergent plants occupied, respectively, 8.1 and 0.5% of the total lake surface area. The maximum biomass per m^2 stand area was much lower in the submerged plants than in both floating-leaved and emergent plants, with average values of ashfree dry wt. of 36.9, 282.4 and 637.1 $g \cdot m^{-2}$. The production rates of the different plant species were calculated as g C on m^2 stand area basis. Of the submerged species, *Ceratophyllum* was more productive than *Elodea:* 0.15 vs. 0.03 $g \cdot m^{-2} day^{-1}$ of C. *Polygonum* had a production rate of 0.63 $g \cdot m^{-2} day^{-1}$ and *Iris, Phragmites, Sparganium* and *Typha,* respectively, had 0.61, 3.94, 0.07 and 0.83 $g \cdot m^{-2} day^{-1}$.

The gross production rates of the different plant groups were also calculated via a simulation model based on physiological properties and cover percentages of the plant species. These data were much higher than those based on biomass, indicating extensive respiration and/or turnover rates. The ratio Simulated: Harvested was lower in submerged plants than in the other two groups, possibly because of lack of a well-developed root system and therefore a relatively low respiration rate in the submerged plants.

All plant groups declined considerably in number and composition with time. An important reason for this was the decrease in solar irradiance at the lake fringe due to increasing shading by trees growing on the lake shore. The emergent and floating-leaved plants were substantially grazed and trampled by cattle. Submerged plant growth declined possibly also by increasing radiation attenuation by the lake water. This was suggested by the lower biomass values generated by the simulation model when light extinction coefficients of 1980 were used instead of those of 1978.

The macrophytes of Lake Vechten played a minor role in the total production of the lake. They contributed only 7.1% to the lake's total net primary production of 21.2 kg day $^{-1}$ of C (autumn, 1980), while periphyton + phytobenthos and phytoplankton accounted for, respectively, 6.6 and 86.2%.

The macrophytic production in Lake Vechten was in the same range as that reported for several Russian lakes, but higher than in the nearby German Schöhsee.

Acknowledgements

J. Goudriaan and H. H. v. Laar (Dept. of Theoretical Production Ecology, Agricultural Univ., Wageningen) advised on the handling of computer models. I feel indebted to R. Brouwer (Dept. of Botany, Univ., Utrecht) and J. Goudriaan for their critical reading of the manuscript. The technical assistance of J. H. A. Dassen is gratefully acknowledged. H. Roon assisted in the field work.

References

Adams, M. S. & McCracken, M. D., 1974. Seasonal production of the Myriophyllum component of the littoral of lake Wingra, Wisconsin. J. Ecol. 62: 457–467.

Assis Esteves, F. de, 1979. Die Bedeutung der Aquatischen Makrophyten für den Stoffhaushalt des Schöhsees. I. Die Produktion an Biomasse. Arch. Hydrobiol. Suppl. 57: 117–143.

Bernatowicz, S., 1969. Macrophytes in the lake Warniak and their chemical composition. Ekol. Pol. 17: 447–465.

Best, E. P. H., 1980. The submerged aquatic macrophytes in lake Maarsseveen – I. Species composition, spatial distribution and productivity. In: Ringelberg, J. (Ed.) Limnological Research in the Maarsseveen Lakes 1975–1980, pp. 205–219. Dept. of Aquatic Ecology, Univ. of Amsterdam.

Best, E. P. H., 1981. A preliminary model for growth of Ceratophyllum demersum L. Verh. int. Verein. Limnol. 21: 1484–1491.

Best, E. P. H., Blaauboer, M. C. I., Cappenberg, Th. E., Gons, H. J., Gulati, R. D., Kloet, W. A. de, Steenbergen, C. L. M. & Verdouw, H., 1978. Towards an integrated study of the ecosystem of Lake Vechten. Hydrobiol. Bull. 12: 107–118.

Best, E. P. H. & Dassen, J. H. A., 1982 (in prep.). Characteristics of growth and productivity of the submerged aquatic macrophyte Ceratophyllum demersum L.

Best, E. P. H. & Kersting, K., 1981. Comparison of different methods to measure the oxygen exchange in a Ceratophyllum community. Proc. 1st European Workshop on aquatic macrophytes. Illmitz, 3–10 may 1981 (in press).

Best, E. P. H. & Meulemans, J. T., 1979. Photosynthesis in relation to growth and dormancy in Ceratophyllum demersum. Aquat. Bot. 6: 53–67.

Best, E. P. H., Zippin, M. & Dassen, J. H. A., 1982. Growth and production of Phragmites australis in lake Vechten (The Netherlands). Hydrobiol. Bull. 15: 165–173.

Blaauboer, M. C. I., 1982. The phytoplankton species composition and the seasonal periodicity in Lake Vechten from 1956 to 1975. Hydrobiologia 95: 25–36.

Carignan, R. & Kalff, J., 1980. Phosphorus sources for aquatic weeds: water or sediments? Science 207: 987–989.

Carpenter, S. R., 1979. Submersed aquatic vegetation and the process of eutrophication. Ph. D. thesis, Univ. Wisconsin. 158 pp.

Dassen, J. H. A., 1981. Comparison of the productivities in emergent, floating-leaved and submerged aquatic macrophytes. Internal Rep. 1981–15, Limnological Institute, Nieuwersluis, The Netherlands (in Dutch, English summary).

Davies, G. S., 1970. Productivity of macrophytes in Marion Lake, British Columbia. J. Fish. Res. Bd. Can. 27: 71–81.

Dawson, F. H., 1973. The production ecology of Ranunculus penicillatus var. calcareus in relation to the organic input into a chalk stream. Ph. D. thesis, Birmingham, Univ. of Aston. 356 pp.

Dvorak, J. & Best, E. P. H., 1982. Macro-invertebrate communities associated with water vegetation in Lake Vechten. Netherlands) Hydrobiologia 95: 115–126.

Dykyjova, D., 1978. Primary production and production processes in littoral plant communities. In: Dykyjová, D. & Květ, (Eds.) Springer Verlag, Berlin.

Edwards, R. W. & Owens, M., 1962. The effects of plants on river conditions – IV. The oxygen balance of a chalk stream. J. Ecol. 50: 207–220.

Fassett, N. C., 1975. A Manual of Aquatic Plants. The University of Wisconsin Press, Madison. 405 pp (1st publ. 1940).

Fiala, K., 1973. Growth and production of underground organs of Typha angustifolia L., Typha latifolia L. and Phragmites communis Trin. Pols. Arch. Hydrobiol. 20: 59–66.

Filbin, G. E., 1980. Photosynthesis, photorespiration and primary productivity in floating, floating-leaved and emergent plants. Ph. D. thesis, Wayne State Univ., Detroit. 252 pp.

Gerloff, G. C. & Krombholz, P. H., 1966. Tissue analysis as a measure of nutrient availability for the growth of angiosperm aquatic plants. Limnol. Oceanogr. 11: 529–537.

Gons, H. J., 1982. Structural and functional characteristics of epiphyton and epipelon in relation to their distribution in Lake Vechten. Hydrobiologia 95: 79–114.

Goudriaan, J. & Laar, H. H. v., 1978. Calculation of daily totals of the gross CO_2 assimilation of leaf canopies. Neth. J. agric. Sci. 26: 373–382.

Goulder, R., 1970. Day-time variations in the rates of production by two natural communities of submerged freshwater macrophytes. J. Ecol. 58: 521–528.

Gulati, R. D., Siewertsen, K. & Postema, G., 1982. Zooplankton: its structure, food and feeding, and role in the ecosystem of Lake Vechten. Hydrobiologia 95: 127–163.

Haslam, S. M., Sinker, Ch. & Wolseley, P., 1975. British water plants. Field Stud. 4: 243–351.

Heukels, H. & v. Ooststroom, S. J., 1970. Flora van Nederland; 16e dr. Wolters-Noordhoff, Groningen. 908 pp.

Hough, R. A., 1974. Photorespiration and productivity in submersed aquatic vascular plants. Limnol. Oceanogr. 19: 912–927.

Hough, R. A. & Wetzel, R. G., 1977. Photosynthetic pathways of some aquatic plants. Aquat. Bot. 3: 297–313.

Ikusima, I., 1970. Ecological studies on the productivity of aquatic plant communities. IV: Light condition and community photosynthetic production. Bot. Mag. Tokyo 83: 330–341.

Jones, R. C., 1980. Primary production, biomass, nutrient, limitation, and taxonomic composition of algal communities epiphytic on the submersed macrophyte Myriophyllum spicatum L. in a hardwater, eutrophic lake. Ph. D. thesis, Univ. of Wisconsin. 200 pp.

Kajak, Z., Hillbricht-Ilkowska, A. & Pieczynska, E., 1972. The production processes in several Polish lakes. In: Kajak, Z. & Hillbricht-Ilkowska, A. (Eds.) Productivity Problems of Freshwaters. Warsaw & Krakow, Polish Sci. Publ. 129–147.

Kloet, W. A. de, 1982. The primary production of phytoplankton in Lake Vechten (The Netherlands). Hydrobiologia 95: 37–57.

Kvet, J., 1971. Growth analysis approach to the production ecology of reed swamp plant communities. Hydrobiologia 12: 15–40.

Mathews, C. P. &'Westlake, D. F., 1969. Estimation of production by populations of higher plants subject to high mortality. Oikos 20: 156–160.

Olsen, S., 1967. Recent trends in the determination of phosphate in the aquatic environment. In: Golterman, H. L. & Clymo, R. S., (Eds.), pp. 71–78. Amsterdam.

Ondok, J. P. & Gloser, J., 1978. Net photosynthesis and dark respiration in a stand of Phragmites communis. Trin. calculated by means of a model. I. Description by means of the model II. Results. Photosynthetica 12: 328–336; 337–343.

Penning de Vries, F. W. T., 1975. The cost of maintenance processes in plant cells. Ann. Bot. 39: 77–92.

Raam, J. v., 1977. Velddeterminatietabel voor de Nederlandse kranswieren. Gewest Gooi en Vechtstreek, Hilversum. 10 pp.

Rich, P. H., Wetzel, R. G. & Thuy, N. V., 1971. Distribution, production and role of aquatic macrophytes in a southern Michigan marl lake. Freshwat. Biol. 1: 3–21.

Schierup, H. H., 1970. Preliminary investigations on variation in biomass of Phragmites communis Trin., M. S. thesis, Univ. of Aarhus.

Sculthorpe, C. D., 1967. The biology of Aquatic Vascular Plants. Edward Arnold, London. 610 pp.

Søndergaard, M., 1979. Light and dark respiration and the effect of the lacunal system on respiration of CO_2 in submersed aquatic plants. Aquat. Bot. 6: 269–283.

Sorokin, Y., 1972. Biological productivity of the Rybinsk reservoir. In: Kajak, A. & Hillbricht-Ilkowska, A. (Eds.). Productivity Problems of Freshwaters, pp. 493–503. P. W. N. Warszawa-Krakow.

Stanley, R. A. & Naylor, A. W., 1972. Photosynthesis in Eurasian watermilfoil (Myriophyllum spicatum L.). Plant Physiol. 50: 149–151.

Steemann-Nielsen, E., 1952. The use of radio-active carbon for measuring organic production in the sea. J. Cons. 18: 117–1140.

Steenbergen, C. L. M., 1982. Contribution of photosynthetic sulphur bacteria to primary production in Lake Vechten. Hydrobiologia 95: 59–64.

Steenbergen, C. L. M. & Verdouw, H., 1982. Lake Vechten: aspects of its morphometry, climate, hydrology and physico-chemical characteristics. Hydrobiologia 95: 11–23.

Thompson, K., 1975. Production of Cyperus papyrus cited by D. F. Westlake. In: Cooper, J. P. (Ed.) Photosynthesis and Productivity in Different Environments. Cambridge University Press, Cambridge. 190 pp.

Titus, J. E., 1977. The comparative physiological ecology of three submersed macrophytes. Ph. D. thesis, the Univ. of Wisconsin, Madison.

Titus, J. E., Goldstein, R. A., Adam, M. S., Mankin, J. B., O'Neill, R. V., Weiler, Jr. P. R., Shugart, H. H. & Booth, R. S., 1975. A production model for Myriophyllum spicatum L. Ecology 56: 1129–1138.

Throughton, J. H. & Slatyer, R. O., 1969. Plant water status leaf temperature, and the calculated mesophyll resistance to carbon dioxide of cotton leaves. Aust. J. biol. Sci. 22: 815–827.

Velde, G. v. d., 1980. Studies in nymphaeid-dominated systems with special emphasis on those dominated by Nymphoides peltata (Gruel.) O. Kuntze (Menyanthaceae). Ph. D. thesis, Nijmegen, The Netherlands.

Verdouw, H. & Dekkers, E. M. J., 1982. Nitrogen cycle of Lake Vechten: concentration patterns and internal mass-balance. Hydrobiologia 95: 191–197.

Verhandelingen der Koninklijke Nederlandse Akademie van Wetenschappen, Afdeling Natuurkunde, Tweede Reeks, 1978. Steenbergen, C. L. M. & Mei, D. v. d. (Eds.): 71: 24–26.

Verhandelingen der Koninklijke Nederlandse Akademie van Wetenschappen, Afdeling Natuurkunde, Tweede Reeks, 1979. Parma, S. (Ed.) 73: 38–40.

Westlake, D. F., 1975. Primary production of freshwater macrophytes. In: Photosynthesis and Productivity in Different Environments. Cooper, J. P. (Ed.), pp. 189–207. Cambridge University Press, London.

Wetzel, R. G., 1964. A comparative study of the primary productivity of higher aquatic plants, periphyton and phytoplankton in a large, shallow lake. Int. Revue ges. Hydrobiol. 49: 1–64.

Wetzel, R. G., 1965. Techniques and problems of primary productivity measurements in higher aquatic plants and periphyton. Proc. I.B.P. Symp. on Primary Productivity in Aquatic Environments, Pallanza, Italy, pp. 251–267.

Wetzel, R. G., 1975. Limnology, pp. 355–418. W. B. Saunders, Philadelphia.

Wetzel, R. G., Rich, P. H., Miller, M. C., Allen, H. L., 1972. Metabolism of dissolved and particulate detrital carbon in a temperate hard-water lake. Proc. I.B.P.-Unesco Symp. on Detritus and its Role in Aquatic Ecosystems, Pallanza, Italy, pp. 187–243.

Winberg, G., Babitsky, V., Gabrilov, S., Gladley, G., Zakharenkov, I., Kovalevskaya, R., Mikheeva, T., Nevyadomskaya, P., Ostpenya, A., Petrovich, P., Potaenko, J. & Yakushko, O., 1972. Biological productivity of different types of lakes. Proc. I.B.P.-Unesco Symp. on Detritus and its Role. in Aquatic Ecosystems, Pallanza, Italy, pp. 383–404.

Whigham, D. F., Mc Cormick, J., Good, R. E. & Simpson, R. L., 1978. Biomass and primary production in freshwater tidal wetlands of the middle Atlantic coast. In: Good, R. E., Whigham, D. F. & Simpson, R. L. (Eds.) Freshwater Wetlands. Ecological Processes and Management Potential, pp. 3–21.

Wit, C. T. de, 1978. Simulation of assimilation, respiration and transpiration of crops. Simulation Monographs, PUDOC, Wageningen, The Netherlands.

Wood, R. D. & Imahori, K., 1965. A Revision of the Characeae, Monograph and Iconograph. Verlag von J. Cramer, Weinheim. 904 pp.

Structural and functional characteristics of epiphyton and epipelon in relation to their distribution in Lake Vechten

Herman J. Gons
Limnological Institute, 'Vijverhof' Laboratory, Rijksstraatweg 6, 3631 AC Nieuwersluis, The Netherlands

Keywords: periphyton, submerged macrophytes, sediment, artificial substrates, oxygen exchange, production, decomposition, sedimentation

Abstract

Epiphyton and epipelon were quantitatively collected, respectively, from the submerged macrophytes and the sandy lake bottom of Lake Vechten (The Netherlands). On a weight basis, epiphyton was maximal in autumn and epipelon in summer. In winter the chemical composition of epiphyton and epipelon was similar. In summer the epiphyton had on a unit weight basis more organic matter and carbonate, and had per unit organic matter a higher algal number, nitrogen and energy content than the epipelon. Algae predominating the epiphyton were filamentous greens and pennate diatoms; those in the epipelon were pennate diatoms and blue-green algae. In both cases, species known to frequent the phytoplankton were abundant. The diatoms were quantified using paper chromatographic pigment analyses. Both the epiphyton and the epipelon exhibited maximal photosynthesis in mid summer. That light was generally the limiting factor was evident from periphyton developed on artificial substrates. This periphyton differed widely in its composition from that on the natural substrates, mainly because the latter collected much more sedimenting matter.

In dense *Ceratophyllum* stands light was severely attenuated and the significant gradients in oxygen and pH were caused by the differences with depth in the proportions of photosynthesis and respiration. The oxygen content and pH at the bottom decreased owing to epipelic respiration. The epiphytic composition depended greatly on the degree of light attenuation. The epiphytic and epipelic respiration, except during part of the early summer, exceeded photosynthesis on a 24 h basis; this included the macrophytic photosynthesis during the time the vegetation was maximally developed. During the growing season import of organic matter, i.e. deposited seston, greatly exceeded that due to the photosynthetic production. After the summer maximum, the epipelon decreased faster than predicted from its oxygen exchange. It was concluded that sedimentation and resuspension determined mainly the changes in epiphyton and epipelon. Especially when covered with vegetation, the lower littoral of Lake Vechten plays a large part in the aerobic decomposition of sestonic organic matter.

Introduction

The present study deals with particulate matter collected from defined substrates in Lake Vechten, which are: submerged macrophytes; the sandy lake bottom; and perspex plates. The matter is composed of live organisms, detritus, and includes compounds generated by biological activity, such as calcium carbonate, but also clay particles. For practical purposes the size limits were 125 and 0.2 μm (see Methods). The above collection of materials will be referred to as periphyton. Thus the definition of periphyton by Wetzel (1975), i.e. microfloral growth upon substrates, was broadened into an analogue of the seston of the open water.

Periphyton may be subdivided referring to the substrate; Round (1973) used the terms epipelic and epiphytic for the algal communities associated

Hydrobiologia 95, 79–114 (1982). 0018-8158/82/0951-0079/$07.20.

with, respectively, sediments and macrophytes. As the species composition of the algal communities in the periphyton collected from the different substrates in the lake are comparable with Round's classification, the two kinds of periphyton are denoted as epipelon and epiphyton.

Traditionally, algae, being the more conspicuous component of periphyton, received more attention from limnologists. This has changed since studies revealed that functionally heterotrophic processes may often be important, and even predominant (Pieczyńska 1975; Mann 1975). A number of structural (Allanson 1973) and functional (Allen 1971; Wetzel & Allen 1972; Wetzel 1975) interrelationships within periphyton, and between periphyton and its substrate, have become apparent. However, much evidence is still circumstantial and scattered, especially if compared with that on planktonic communities.

For periphyton in lakes, as a rule, algae were regarded as the most important viable component. Consequently, attached organisms, considered to be the true periphyton, are distinguished from 'dependent' free-living organisms, or pseudo-periphyton (Sládečková 1962). The periphyton was, from the viewpoint of energetics, considered largely to be self-supporting, for which, at least during the growing season, algal photosynthesis provides the basis for the trophic dynamic structure. This was also the viewpoint at the start of the periphyton study (1977) on Lake Vechten, being aimed at periodicity and productivity of the epiphytic algae associated with the submerged macrophytes. The macrophytes covered most of the littoral zone down to 5 m depth, which comprises more than a third of the lake area. Floating-leaved and emergent macrophytes, in contrast, are less important substrates occupying about 5% of the lake area. Predominating submerged species were *Elodea* sp. and *Ceratophyllum demersum*, with *Myriophyllum spicatum* as a good third.

During the last few years, the submerged macrophytes declined; in 1980 there was nearly no growth below 3 m; *Ceratophyllum* predominated, while *Myriophyllum*, but for some isolated poorly developed individuals, had disappeared (for details, see Best 1982). These events led to an extension of the investigations on the epipelon. Seasonal changes could be studied quantitatively, which was allowed by the nature of the bottom: this is sandy, and has steep slopes, which prevents the accumulation of refractory matter over the years.

The periphyton research is a part of ecosystem studies on Lake Vechten with the main aim to quantify the pool sizes and fluxes of carbon. As a consequence the approach has been essentially non-taxonomic, but was directed towards characterization of the periphyton as an entity in the carbon cycle.

The following aspects have been investigated:

1. *Chemical composition:* The distinction between organic matter and other constituents was based on ignition losses. More information was obtained by elementary analyses. In this way, carbonate and ash content could be differentiated giving information on photosynthetic activity and degree of advance of mineralization. The latter may be further qualified by the elementary composition of the organic matter, from which the caloric equivalent and proximate molecular composition can also be derived (Spoehr & Milner 1949; Milner 1953); Hallegraeff (1976) applied this to seston of Dutch lakes. For other examples see Wetzel (1975) and Parsons *et al.* (1977).

2. *Composition of algal species and photosynthetic pigments:* The identification and counts of organisms were restricted to the algae, since their recognition is relatively simple, and their species composition provides information on functional aspects of the relevant biota (Cholnoky 1968; Hutchinson 1975; Round 1973). Also, this gave insight into interactions between the littoral and limnetic region. Chlorophyll *a*, of course, was measured to provide information on the autotrophic component of the periphyton. Paperchromatographic pigment analyses were carried out to supplement information on the importance of the taxa and the viability of the algae as demonstrated by Hallegraeff (1976).

3. *Oxygen exchange:* Since the environment of both epiphyton and epipelon is aerobic, oxygen measurements were used to establish material balances (Odum 1971; Parsons *et al.* 1977). The periphyton was characterized by the dark oxygen consumption, i.e. the potential community respiration (r), and the maximum gross oxygen production (p_g^{max}), which expresses the photosynthetic capacity. The ratio p_g^{max}/r is a measure of the relationship between the autotrophic and heterotrophic components. The *in situ* changes depended, of course, on the conditions of irradiation and on the depth distribution of light and periphyton.

During the first two years the studies involved average properties of epiphyton collected from plant shoots, or large sections, over which great differences might exist, in particular with respect to light. The subsequent research was focussed on the depth distribution of epiphyton within dense stands of *Ceratophyllum* and of epipelon. This included measurements on physico-chemical gradients in the field. From the summer of 1980 onwards, epiphyton and epipelon were studied along a fixed transect through the SE part of the lake (Fig. 1).

Materials and methods

Sampling procedures

For epiphyton plants, or their parts, were collected by SCUBA-diving. The epiphyton and macrophyte were separated using a Vibromixer (Chemap, Zürich), if necessary after cutting the macrophyte into sections. During the procedure the pH was lowered, using automatic titration with 0.1 N HCl, to 5 in order to dissociate carbonate encrustations. After this treatment (Fig. 2A) had been applied for 15–30 min, the epiphyton was quantitatively removed from its substrate, and the lake water's pH quickly returned to its pretreatment value. The method did not damage algal cells, filaments, or photosynthetic pigments. When applied with care, the macrophytic parts also remained intact. Possible interference of the lowered pH with respiratory and photosynthetic capacity was minimized by first removing the more loosely associated part by using the Vibromixer only: usually, this loose material comprised about three quarters or more of the total of epiphyton, also on the basis of chlorophyll.

Epipelon was collected using a specially designed stainless steel sampler (Fig. 2B). A known area of sediment can be isolated within the sharp-edged cylindrical chamber of this apparatus, which is forced into the sandy bottom by a weight, but prevented from penetrating too deeply by a wide rim. By applying suction through the butyl rubber tubing, water spurts through the narrow holes just above the rim into the chamber, thus eddying the

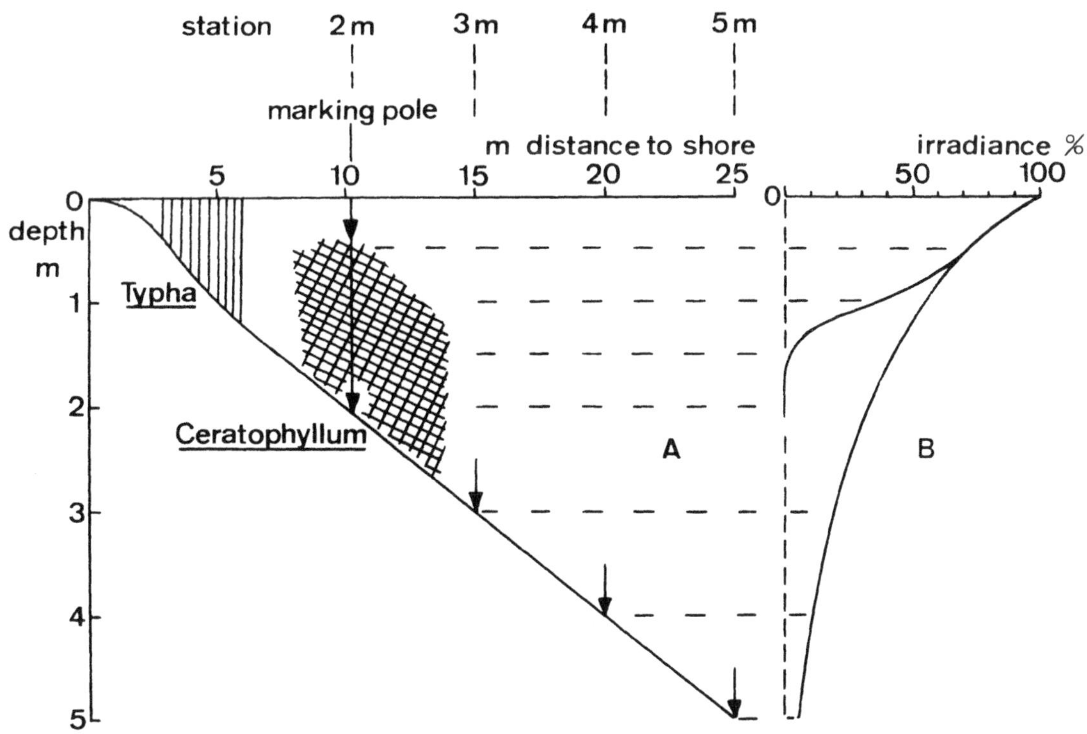

Fig. 1. Transect through the littoral zone of the SE part of the lake during 1980 and 1981. Conditions of light irradiance (B) as determined on 9 September 1980. The line on the left gives the light within the vegetation.

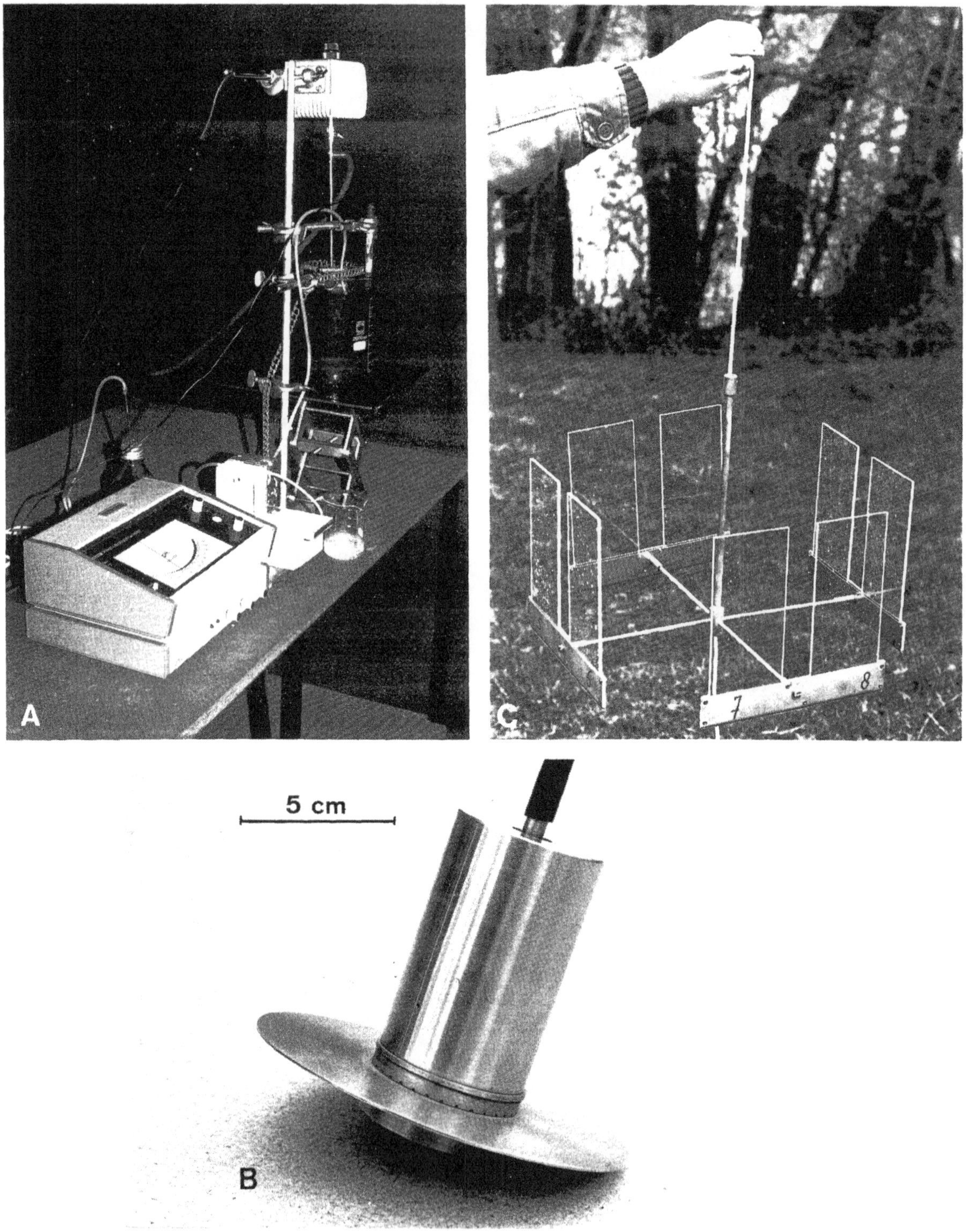

Fig. 2. The equipment used for the periphyton studies. A: Vibromixer combined with pH-stat for separating epiphyton from the macrophytes. B: Epipelon sampler. C: Assemblage for artificial substrates.

isolated volume of about 8 ml above the sand. The pumping was continued until all epipelic matter was removed from the sand, which involved a volume of about 0.7 l. The method is sufficiently powerful to bring up large sand grains. At each station (Fig. 1) five such samples were collected in one polyvinyl vessel.

Perspex plates used as artificial substrates (Fig. 2C) were exposed vertically so that 10 × 20 cm on either side could be colonized. At one particular depth eight such plates were hung. The first plate was removed and replaced by another after each interval. The others were removed one by one after subsequent intervals, in order to pursue the developments continuously. In the laboratory the periphyton was scraped off the sides and the edges.

All samples were treated with the Vibromixer, but the epipelon and 'perspex' periphyton were not exposed to the lowered pH. Thereafter, the suspensions were filtered over 125 μm mesh, to remove larger animals, leaf debris, etc. The 'perspex' periphyton sometimes clogged the filter due to abundance of filamentous green algae. The larger parts, notably snails and *Dreissena,* were then removed manually. Sand grains, present mainly in the epipelic samples, did not interfere with the subsequent analyses.

In general the suspensions were processed for analysis immediately. The samples for cell counts and pigment composition were conserved by, respectively, adding Lugol's solution and storage at -30 °C.

Water samples for physico-chemical analysis were collected using the sampler described or a similar apparatus (without wide rim) that could be lowered in dense *Ceratophyllum* stands without displacing plant shoots.

Gravimetric and elementary analyses

Aliquots of the suspensions were centrifuged and the pellets transferred into 25 ml glass vials. Dry weights were determined after freeze-drying and overnight storage in a desiccator.

After homogenizing, part of the dry matter was weighed into aluminium combustion boats (Perkin-Elmer, Oak Brook, Ill.) and ashed at 450 °C to determine the ignition loss. With this temperature it was safe to operate, since no decomposition of $CaCO_3$ occurs. Another part was used for elementary analysis using a Perkin-Elmer model 240 C-N-H analyzer.

For determining the contribution of $CaCO_3$ and the composition of the organic matter, elementary analysis was also carried out on the residue after ignition at 450 °C. The $CaCO_3$ as a percentage of the total was then computed according to:

$$\%\ CaCO_3 = (\%C_r \times \text{residue/total}) \times 100/12. \quad (1)$$

The carbon content of the organic matter was determined as:

$$\%\ C = \frac{\%\ C_t - (\%C_r \times \text{residue/total})}{\text{loss on ignition/total}} \quad (2)$$

where C_r = carbon contained in the residue, and
C_t = carbon contained in the total matter.

The hydrogen and nitrogen contents of the organic matter were computed in the same way as carbon (eq. 2).

All data on the contents and ratios of C, H and N are given on a weight basis.

Cell counts

The technique applied was the same as for the phytoplankton (see Blaauboer 1982). The data given in the Results are expressed as cell numbers, except for the blue-green algae, where the numbers refer to filaments.

Intactness of chloroplasts was used as a measure of viability of pennate diatoms.

Pigment analyses

Chlorophyll was extracted in methanol and computed according to Iwamura *et al.* (1970). Unless otherwise stated, the chl *a* values refer to 'total chl *a*', and may include considerable proportions of breakdown products.

The 2-dimensional paperchromatographical separation of the pigments (Fig. 3) essentially followed Hallegraeff (1976), but the extraction was in methanol. The pigments were concentrated by evaporation under nitrogen, and transferred in acetone prior to the separation. The yield of the procedure was fairly good: the integrated light absorption of the separated pigments was approximately 90% of that of the methanol extract. Not all pigments were

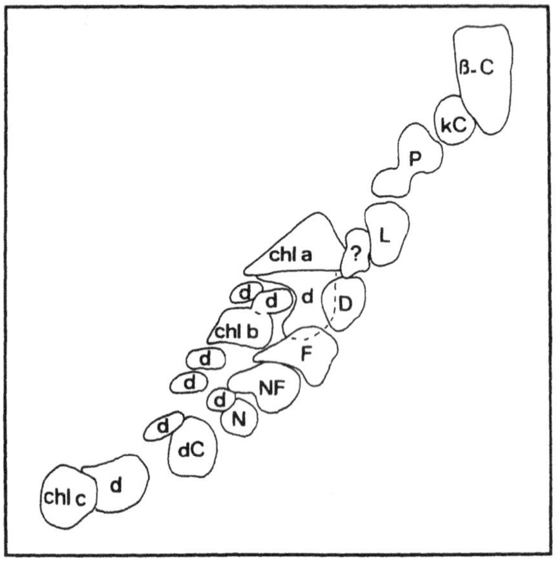

Fig. 3. 2-Dimensional paper chromatogram of epiphyton. chl *a*, *b*, *c* = chlorophyll *a*, *b*, *c*; β-C ≐ beta-carotene; dC = carotenoid derivative; kC = ketocarotenoid; ? = unidentified carotenoid; D = diadinoxanthin; F = fucoxanthin; NF = neofucoxanthin; N = neoxanthin; L = lutein; P = phaeophytin; d = chlorophyll derivatives.

completely isolated (Fig. 3), but in most cases the relevant pigments could be estimated accurately by applying spectrometric corrections. The results are expressed as percentage of total light absorption by the separated pigments.

Oxygen exchange

Two oblong glass containers, filled with water and maintained at *in situ* temperature under the control of a kryo-thermostat (Colora, Lorch/Württ.), served as the incubator. The containers received illumination (TL F 20 W/33 fluorescent lamps, Philips, Eindhoven) ranging from 3 to 5 $W.m^{-2}$ to saturating and often inhibiting values of 120 $W.m^{-2}$. For dark incubations the samples were wrapped in aluminium foil.

Oxygen consumption in the dark, i.e. the potential community respiration (r), was measured in duplicate in BOD-bottles containing approximately 260 ml of the suspension. During the incubation the suspension was not stirred, so the periphyton settled down on the bottom. The start and end concentrations of oxygen were measured using a YSI meter and self-stirring BOD-bottle probe (Yellow Springs, Ohio). In general the incubation time was about 4 h, excluding pre-incubation of 1–2 h, but the epipelic winter samples especially required overnight incubation for reliable results. Periodically, chemical oxygen consumption was checked. In general all oxygen consumption stopped on adding 10 ml of 37% formaldehyde.

For measuring the oxygen exchange vs. irradiance, the periphyton was incubated on filters. Depending on the concentration, 10–100 ml was filtered over 0.2 μm filter (Sartorius, Göttingen) with 5 cm diameter. Subsequently, the filter was cut into two equal halves, inserted into a glass vial containing on the average 26.4 ml of prefiltered (0.2 μm) lake water. The vials were tightly closed with screw caps, and inspected to ensure exclusion of air. The oxygen was measured using the YSI meter and probe, but the stirrer's boot was removed. After the incubation the filter half was first removed. Insertion of the probe isolated the content of the vial from exchange with air. The incubation time was similar to that of the BOD-bottles. For calculating net production the filtered lake water was incubated as blanks.

The results of the dark incubations in the BOD-bottles and of the filters were almost always comparable. The former, however, were used more often and the respiration values in the Results refer to these.

On one occasion the oxygen exchange vs. irradiance curve was determined for *Ceratophyllum*, when 4 plant tips of 15 cm length made free of the epiphyton were used. The plant tips were incubated in the above BOD-bottles, filled with lake water well below oxygen saturation. Each section was used for incubation at one irradiance and in the dark. Following a pre-incubation of 30 min, the sections were incubated for 1 h. Thereafter the four were transferred into the dark and, after 10 min pre-incubation, the respiration was measured over 90 min.

The irradiances were measured using R 3001 photometer equipped with an SD 101 photodiode detector with spectral range of 400–700 nm (Macam, Livingston, Scotland). The results were corrected for reflections by the glass of BOD-bottles and vials.

Physico-chemical parameters in situ

Light in stands of *Ceratophyllum* was measured

with the laboratory equipment already mentioned. The sensor could be moved vertically without causing significant changes in the position of the macrophytes shoots.

Oxygen and temperature were measured in the field, using both the BOD-bottles and the YSI meter and probe.

For the remaining analyses the samples were surrounded with ice and transported to the laboratory. The pH and HCO_3^- were determined on arrival, the latter potentiometrically with a Quéré automatic titrator (Jouan, Paris).

Results and discussion

Epiphyton

Distribution of epiphytic matter. Due to the loose association of epiphyton to macrophytes, stirring resulted in loss of particles *in situ.* Consequently, the total epiphytic matter was underestimated, and the results tend to be biased in favour of the matter more firmly attached to the substrate.

By far the greater part of the epiphyton consisted of detrital matter (Fig. 4A). In the epiphyton associated with *Chara* studied by Allanson (1973), algae were far more important as a structural component. Also, relative to the seston in Lake Vechten, algae were scarce, but bacteria and microfauna, namely Protozoa, were abundant. The significance of mucoid substances in keeping in place the epiphytic matter is shown in Fig. 4B.

The amounts of epiphyton were surprisingly high, considering the appearance, and the epiphyton dry weight often exceeded that of its substrate. The ash content was high; typically, organic matter amounted only to about 20%. This was for a great part due to carbonates, probably precipitated during photosynthesis of macrophyte and epiphytic algae (Fig. 4C). Despite this, the organic matter of epiphyton was also at times more than that of the macrophyte.

In 1978, epiphyton was compared from five populations of submerged macrophytes: *Elodea* sp., *Myriophyllum spicatum,* and *Ceratophyllum demersum* at three different depths (Fig. 5). Early in the season the plants had been tagged (see Best 1982). The five associations differed considerably on various dates, but the proportions of epiphyton to macrophyte on a weight basis did not differ significantly depending either on macrophyte species or depth when the seasonal averages were compared.

Speaking in general, the proportions of epiphyton were low from May to July, but high in early spring and autumn. Per area unit of littoral, the greatest quantities of epiphyton occurred in September and October (Fig. 30), with high proportions (Fig. 5) at the time of the maximum macrophyte standing crop.

Chemical composition. For the same categories of samples, average chemical characteristics (April-September) are summarized in Table 1.

The percentage organic matter was low, close to 20% for all groups. The carbon content varied more widely. These data include the carbon contained in carbonates. Thus, the variations in both % carbon (Table 1) and the high values of C:N and C:H ratios are explained by the high contribution of carbonate. The latter differed from category to category, assuming the carbon content of the organic matter to be 50%.

The % organic matter (Fig. 6) fluctuated seasonally, particularly when the macrophyte growth was fast; then the organic matter and carbon were maximal up to, respectively, 50 and 25%. This will have been due to comparatively low accumulation rates of both sediment and carbonates on the surfaces of the rapidly extending macrophytes, and to the development of epiphytic algae.

The decrease in organic matter and carbon content with depth of the epiphyton from the three *Ceratophyllum* stands (Table 1) is noteworthy, as are the great seasonal changes occurring with the epiphyton from the lesser depth (Fig. 6).

Large but unknown quantities of carbonate obscured the composition of the epiphytic organic

Table 1. Chemical characteristics of epiphyton during April-September 1978.

Substrate	Stratum (m)	% organic matter	% C	% H	% N
Ceratophyllum demersum	1-2	22.2	13.4	1.8	1.2
Ceratophyllum demersum	2-3	21.2	12.5	1.7	1.3
Ceratophyllum demersum	3-4	19.7	10.4	1.7	1.1
Elodea sp.		19.6	10.3	1.5	1.0
Myriophyllum spicatum		19.3	12.6	1.6	1.1

A 20
B 10

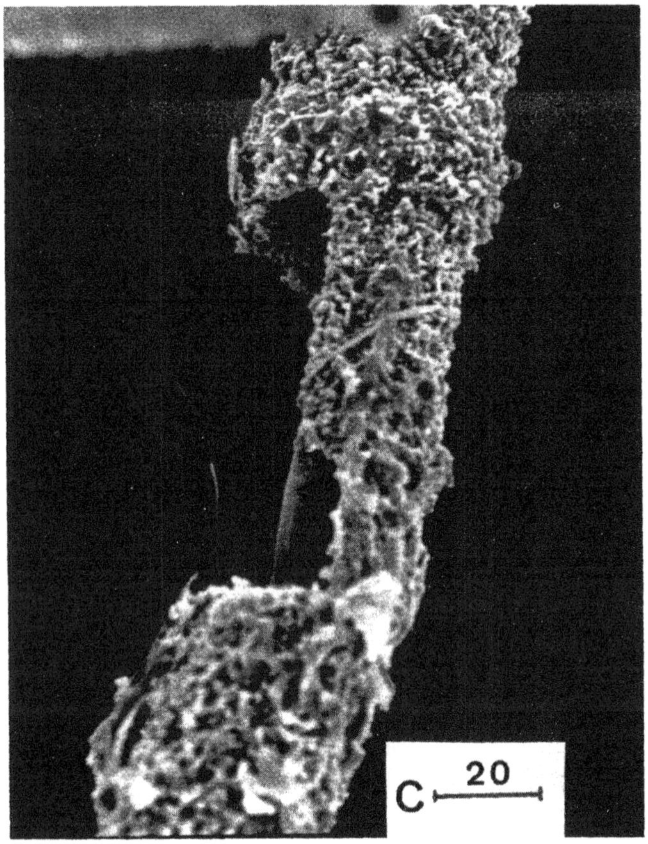

Fig. 4. Scanning electron photomicrographs of epiphyton. Scale in μm. A: Detrital aggregates with diatoms. B: Carbonate deposits and mucoid strands. C: Carbonate encrustation on filamentous green alga.

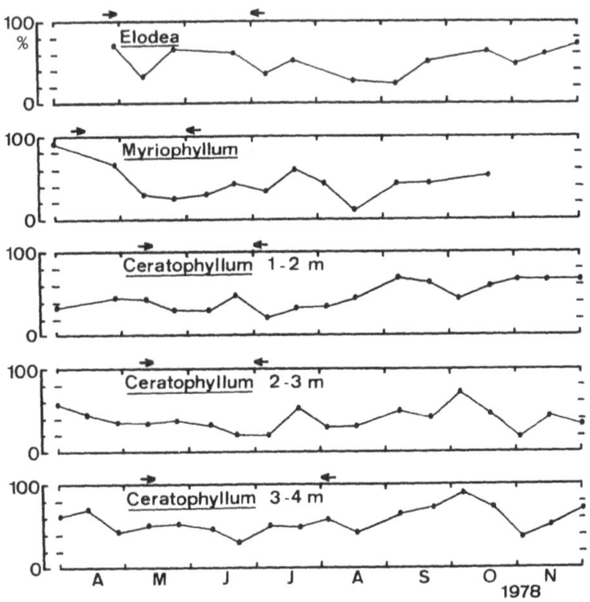

Fig. 5: Contribution of epiphyton on total dry weight basis to submerged vegetation. Arrows indicate the periods of main growth of the macrophytes.

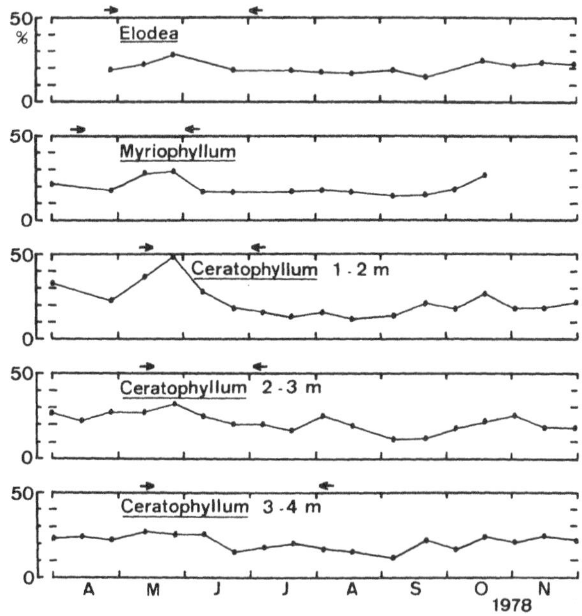

Fig. 6. Percentage organic matter of epiphyton from submerged vegetation. See also Fig. 5.

matter, until both total matter and the residue on ignition were analysed (see Methods) for the epiphyton collected from *Ceratophyllum* during 1980 and 1981 (Fig. 7A and Table 2).

Assuming all carbon in the residue came from $CaCO_3$, contributions of organic matter (i.e. the loss on ignition), $CaCO_3$ and ash (i.e. the residue minus $CaCO_3$), are given in Fig. 7. As in 1978 (Fig. 6), the organic matter was about 20%, but seasonal changes were less, notably during early summer. The $CaCO_3$ contributed the most in June and July and least in spring; this suggests a direct relation with photosynthetic activity. The decrease in $CaCO_3$ already in the second half of July is explained later.

The ash percentage indicated prevalence of the more viable epiphytic communities during summer, and increase in the detritus content during autumn and winter until spring next year. These changes were not substantiated by clear seasonal patterns in the carbon content of the organic matter, nor in organic C:N ratio.

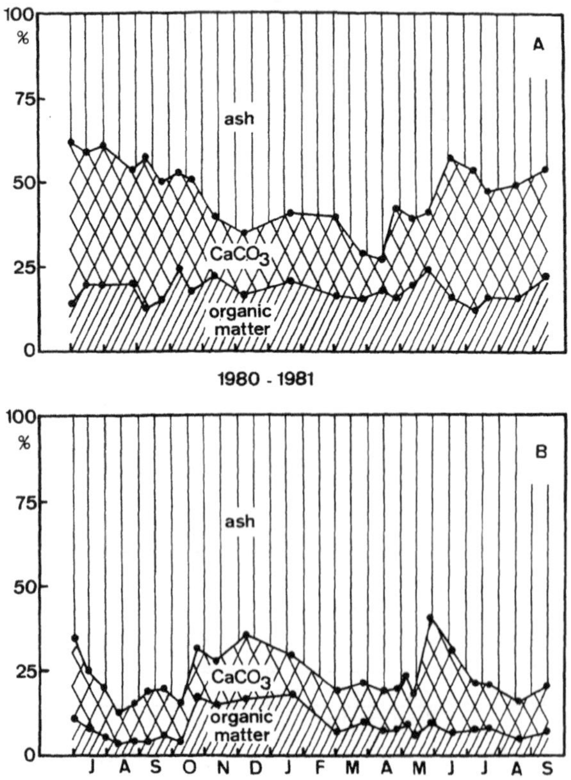

Fig. 7. Contribution of organic matter, calcium carbonate and ash to epiphyton (A) and epipelon at 4 m (B).

Table 2. Chemical characteristics of epiphyton and epipelon during July 1980–June 1981.

A. Composition of total matter	% organic matter	% $CaCO_3$	% ash	% C	% H	% N
Epiphyton	18.4	28.3	53.3	12.0	1.6	1.0
Epipelon, 3 m	10.4	15.2	74.4	5.8	1.0	0.5
Epipelon, 4 m	8.9	15.2	75.9	5.4	0.9	0.4
Epipelon, 5 m	10.1	15.2	74.7	6.4	1.0	0.5

B. Composition and R-value of organic matter	% C	% H	% N	R-value
Epiphyton	47.0	7.9	5.2	37.1
Epipelon, 3 m	42.0	7.7	4.2	31.8
Epipelon, 4 m	42.8	7.7	4.2	32.7
Epipelon, 5 m	45.2	8.1	4.6	35.7

The annual averages of the elementary composition of the epiphytic organic matter are given in Table 2B. The nitrogen percentages were fairly high, indicating a fair contribution of proteinaceous matter. For degree of reduction, the concept of the R-value (Spoehr & Milner 1949) was used. The R-value of carbon compounds ranges from zero, for carbon dioxide, to 100, for methane, and may be applied equally well to mixtures of organic compounds such as we find in plant materials (Milner 1953). Hallegraeff (1976) established the following equation for sestonic matter:

$$\text{cal. mg ash-free dry weight}^{-1} = -0.245 + 0.144 \cdot R.$$

Epiphytic algae. Detritus and carbonates were the major fractions in the epiphyton. Live algae were only a small part, smaller than the phytoplankton in the seston of Lake Vechten. This was evident from microscopical observations and pigment analyses.

During 1977 cell counts were made of the algae collected from 75 cm shoot lengths of *Ceratophyllum*. On the basis of numbers three categories of algae were most important (Fig. 8), and showed a marked periodicity. In spring small green algae predominated, mainly *Chlorella* and *Scenedesmus* species. As the growing season advanced, the frequency of these species decreased; abundant development of filamentous green algae was followed by predominance of pennate diatoms later on.

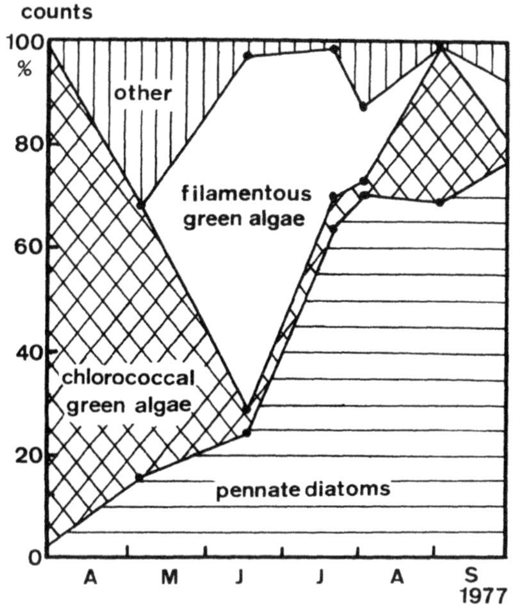

Fig. 8. Succession in the percentage composition of main algal groups of epiphyton from *Ceratophyllum*.

Table 3. Pennate diatom species in the epiphyton collected from submerged macrophytes.

Araphidineae	**Biraphidineae**
Asterionella formosa	*Amphipleura pellucida*
Centronella reichelti	*Amphora ovalis*
Diatoma elongatum	*A.* sp.
D. sp.	*Cymatopleura elliptica*
Fragilaria brevisstriata	*C. solea*
F. capucina	*Cymbella affinis*
F. construens var. *venter*	*C.* spp.
F. intermedia	*Epithemia zebra*
F. sp.	*E.* sp.
Synedra acus	*Gomphonema acuminatum*
S. acus var. *angustissima*	*G. constrictum*
S. pulchella	*G. olivaceum*
S. rumpens var. *familiaris*	*G. parvulum*
S. ulna	*G.* sp.
S. vaucheriae	*Gyrosigma attenuatum*
	G. sp.
Raphidioidineae	*Navicula hungarica* var. *capitata*
Eunotia lunaris	*N. radiosa*
E. spp.	*N. viridula*
	N. spp.
Monoraphidineae	*Nitzschia acicularis*
Achnanthes lanceolata	*N. amphibia*
A. lanceolata var. *rostrata*	*N. dissipata*
A. spp.	*N. palea*
Cocconeis pediculus	*N. sigma*
C. placentula	*N. sigmoidea*
C. sp.	*N. subrostrata*
Rhoicosphenia curvata	*N.* spp.
	Pinnularia microstauron
	P. viridis
	Rhopalodia gibba
	R. sp.

Important species of filamentous green algae belonged to the genera *Mougeotia, Spirogyra, Oedogonium* and *Bulbochaete*. Within this group *Oedogonium* filaments were most abundant all through summer and autumn, sometimes up to 100%.

Pennate diatom species are listed in Table 3. Most of the araphid diatoms have been encountered frequently in the phytoplankton (Blaauboer 1982), but the others were either absent or present in very low numbers. The biraphid species were especially abundant, notably of the genus *Nitzschia*. This makes the diatom association not typically epiphytic but rather epipelic; see Round (1973) and Hutchinson (1975).

The other algae (Fig. 8) are mainly species known to occur planktonically. In May high numbers of small unidentified flagellates occurred. During the months thereafter, blue-greens of the genera *Oscillatoria, Lyngbya* and *Anabaena* dominated the other algae.

In the following years similar patterns were found (see also Fig. 17). Also, in 1977 and 1978, the species composition of the epiphyton from *Ceratophyllum* was comparable with that from *Elodea* and *Myriophyllum*.

Relationships of species composition between epiphyton and phytoplankton are shown in Fig. 9, for three characteristic times in the year: A. late March, when *Ceratophyllum* was still stretched over the bottom; B. June, between the two periods of main phytoplankton development and with exponential growth of the macrophytes; C. late August, when the littoral development had reached its maximum.

In March (Fig. 9A) the first three taxa in the epiphyton had similar frequencies in the phytoplankton; except for the not-determined pennate diatoms, all taxa in the epiphyton were also present in the plankton. The phytoplankton was more diverse: several species with fairly high frequencies did not occur in the epiphyton.

In June (Fig. 9B) a completely different situation existed. The epiphyton was richer in species, with

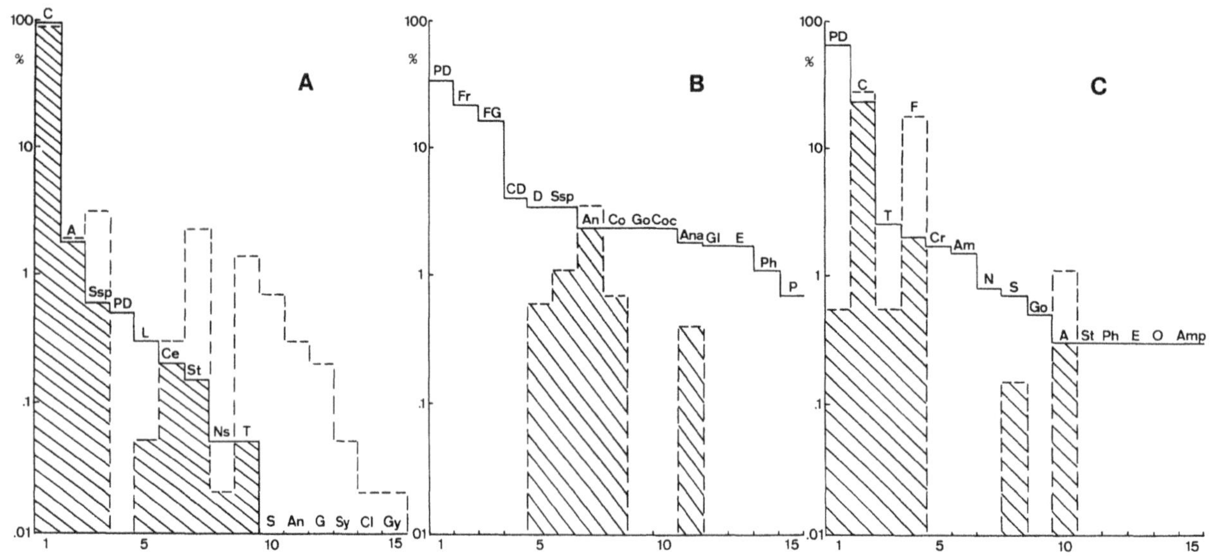

Fig. 9. Comparison of species composition of epiphyton with that of the phytoplankton. Hatched areas represent frequencies common to both groups. A: March. B: June. C: August 1977. See text for further explanation. A = *Asterionella formosa;* Am = *Amphipleura pellucida;* Amp = *Amphora* sp.; An = *Ankyra* sp.; Ana = *Anabaena* sp.; C = *Chlorella* sp.; CD = centric diatoms spp.; Ce = *Centronella reichelti;* *Cl* = *Closteriopsis* sp.; Co = *Coelastrum microporum;* Coc = *Cocconeis* sp.; Cr = *Crucigenia quadrata;* D = *Dinobryon divergens;* E = *Elakatothrix gelatinosa;* F = small flagellate spp.; FG = filamentous green algae spp.; Fr = *Fragilaria* sp.; G. = *Gloeocystis gigas;* Gl = *Gloeotrichia* sp.; Go = *Gomphonema* sp.; Gy = *Gyrosigma attenuatum;* L = *Lyngbya* sp.; N = *Navicula radiosa;* Ns = *Nitzschia sigma;* O = *Oocystis* sp.; P = *Peridinium willei;* PD = *pennate diatom* spp.; Ph = *Phacotus lenticularis;* S = *Scenedesmus* cf. *quadricauda;* Ssp = *Scenedesmus* spp.; St = *Stephanodiscus astraea;* Sy = *Synedra ulna;* T = *Tetraedron minimum.*

predominance of pennate diatoms and filamentous green algae, which groups were absent in the phytoplankton. Of the 15 epiphytic taxa listed, only five were also planktonic.

The situation found in August (Fig. 9C) was intermediate. The first four taxa were present in both epiphyton and phytoplankton. Of the remaining 11 taxa, only two were common in both habitats.

Principles underlying the above described differences and similarities in species composition will be discussed in the final section.

Photosynthetic pigments. The chlorophyll *a* in the epiphytic organic matter (Figs. 10 and 11) had about 5 mg . g^{-1} as the upper limit. Generally, two thirds of the chlorophyll was chromatographically pure (Fig. 12). Consequently, epiphytic algae cannot be expected to contribute much more than 50% to the epiphyton on an organic weight basis; see Sládeček & Sládečková (1964) and Weber (1973).

During the summer of 1978, the chlorophyll content was maximal in June, when the filamentous green algae predominated. The chlorophyll contents of the epiphyton from *Ceratophyllum* and *Myriophyllum* were similar, but that from *Elodea* tended to be lower than with the first two categories (Fig. 10).

Results of one year, 1980–1981, are given in Fig. 11. Between the summer maxima and spring minimum in these years, the chlorophyll content changed markedly. There is evidence that these were caused by variations in the contribution of species from the limnetic region rather than by specifically epiphytic forms. For example, in January, very high numbers of the centric diatom *Stephanodiscus* coincided with its massive growth in the open water.

It has been shown (Fig. 8) that three algal groups were most important numerically; with respect to pigmentation this involved only two types, i.e. diatoms and green algae. As these differ markedly in their auxiliary pigments (e.g. Round 1973 and Van den Hoek 1978), paper chromatography gave quantitative information on the contribution of the two types to the pigments in the epiphyton.

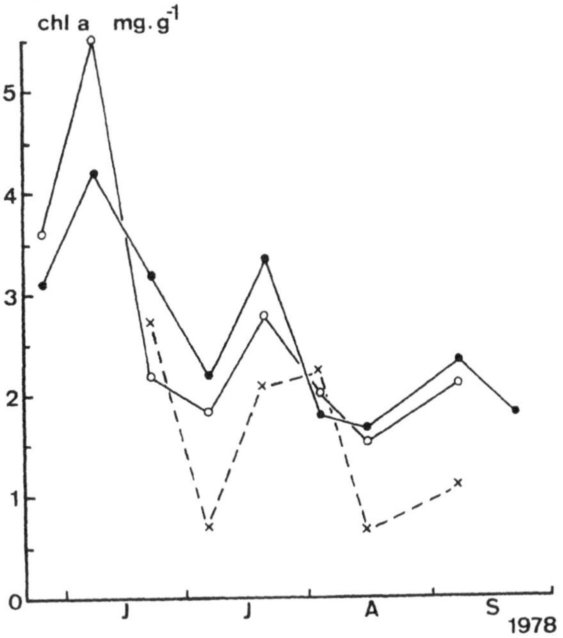

Fig. 10. Summer time changes in chlorophyll *a* content of epiphytic organic matter from *Ceratophyllum* at 1-2 m depth (dots), *Myriophyllum* (circles) and *Elodea* (crosses).

Neither pigment diversity nor pigment species composition changed substantially during the study, contrary to the results of seston (Steenbergen & Steketee, unpublished).

In Fig. 12 the course of six groups of pigments of the epiphyton collected from *Ceratophyllum* growing at 1-2 m depth is given.

About 20% of the total light absorption was by the chlorophyll degradation products. Similar percentages occurred also with seston, but in that case there was more fluctuation (Steenbergen & Steketee, unpublished; see also Hallegraeff 1976). Chlorophyll *a* varied between 35%, in August, and 45%, in June and autumn. The highest contribution of chlorophyll *b* + lutein was in June and autumn, while the diatom pigments were most important in the intervening period.

With the auxiliary pigments, chlorophyll *b* and fucoxanthin determinations allow the further quantification of the importance of the two groups of algae. In Fig. 13 the chlorophyll *b*: chlorophyll *a* ratio, which indicates the contribution of the green algae, has been plotted against the fucoxanthin: chlorophyll *a* ratio as an index for the diatoms.

For the epiphyton of the above depth category of *Ceratophyllum*, least-squares linear regression gave:

$$\text{chl } b/\text{chl } a = -0.47 \cdot \text{fucoxanthin}/\text{chl } a + 0.44$$
$(r = 0.94)$.

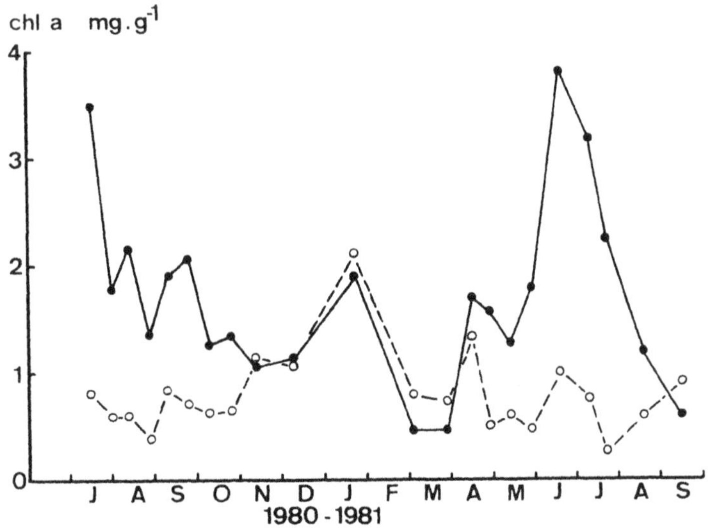

Fig. 11. Seasonal changes in chlorophyll *a* content of epiphytic organic matter from *Ceratophyllum* and epipelic organic matter at 4 m depth (broken lines).

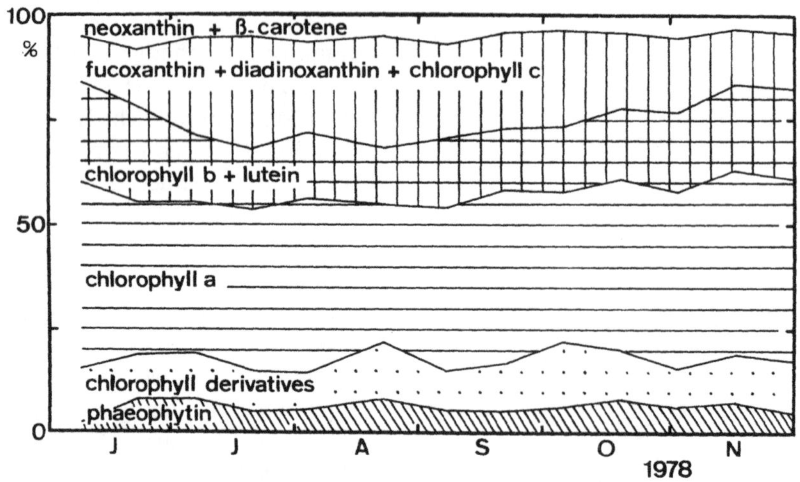

Fig. 12. Seasonal changes in pigment composition of epiphyton from *Ceratophyllum*. Chlorophyll *b* and lutein occur in green algae; fucoxanthin (sometimes including small amounts of neofucoxanthin), diadinoxanthin and chlorophyll *c* in diatoms; the remaining pigments in both groups of algae.

The intercept on the abscissa expresses the fucoxanthin: chl *a* ratio of a pure diatom chromatogram, which was 0.95 in this case. This is within the range of values for mono-algal diatom cultures and natural populations (Jeffrey 1968; Hager & Stransky 1970b; Jensen & Sakshaug 1973). We found that with light-limited cultures of *Nitzschia palea,* values ranged from 0.94 to 1.10 (Gons & Steketee, unpublished).

On the other hand the intercept, 0.44, on the ordinate represents exclusively green algae. Also, this chl *b*: chl *a* ratio is comparable to that for cultured green algae, both filamentous and chlorococcal species (e.g. Jeffrey 1961; Hager & Stransky 1970a).

The linear fit indicates that the epiphytic algae were not subject to marked changes owing to limiting factors; Jensen & Sakshaug (1973) reported that the fucoxanthin: chl *a* ratio of natural diatom communities may vary by a factor of 5.

If the ratios for epiphyton collected from *Elodea* and *Myriophyllum* are included in the regression, the relationship remains much the same. Yet, the values for the epiphyton from *Elodea* deviated in November (Fig. 13), while for epiphyton from *Myriophyllum* exceptional ratios occurred on one date in September, but those were in line with the given relationship.

The value of 0.95 was used to calculate the chlorophyll *a* associated with fucoxanthin. Further it was assumed that neoxanthin and β-carotene were distributed over the diatoms and green algae in the same proportion as chlorophyll *a*. Thus the contribution of the diatoms to the light absorption of the intact photosynthetic pigments was obtained for the epiphyton from *Ceratophyllum* (Fig. 14). This varied from about 25% in June and November to 62% in August. Similar high summer values for the diatoms also applied to the epiphyton from *Elodea* and *Myriophyllum;* for the latter the values were even higher, more than 80%, in September and October.

Epipelon

Distribution of epipelic matter. The study on the littoral sediments started after the submerged macrophytes stand area had become markedly reduced over recent years. The sediments are permanently aerobic down to about 5 m depth (Fig. 19). The material was collected along a fixed transect in the SE part of the lake (see Fig. 1 and Methods); here the bottom was sandy, and the thin layer of epipelon could be removed quantitatively by applying suction (see Methods).

The amounts of epipelon showed quite spectacular temporal changes (Fig. 15); the values for 5 m changed almost two orders of magnitude from a winter minimum of about 30 to a summer maximum of nearly 1 900 g. m^{-2}. The epipelon generally increased markedly with depth, especially in the

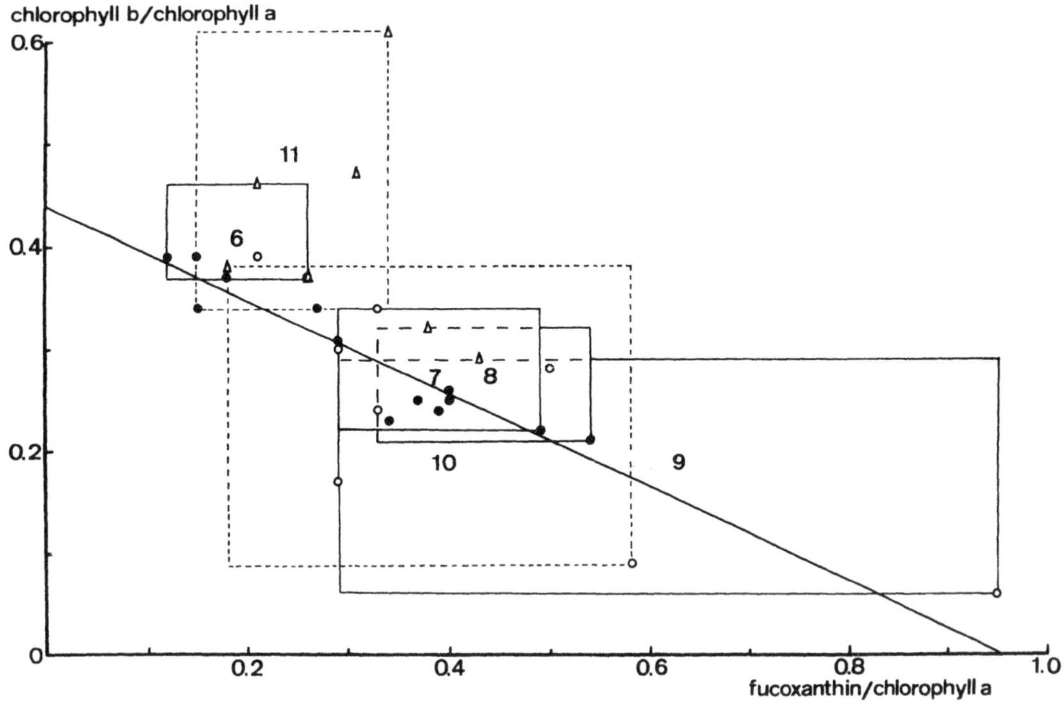

Fig. 13. Relationship between chlorophyll *b*/ chlorophyll *a* (index for green algae) and fucoxanthin/ chlorophyll *a* (index for diatoms) for epiphyton from *Ceratophyllum* (dots), *Myriophyllum* (circles) and *Elodea* (triangles) in 1978. The numbers refer to the months.

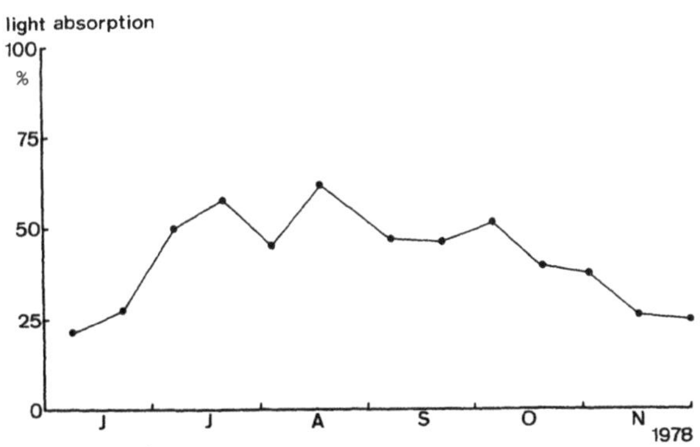

Fig. 14. Contribution of diatom pigments to the light absorption by the photosynthetic pigments in epiphyton from *Ceratophyllum*.

summer. Yet, the values were occasionally very close, notably following a rapid decrease, as in September both years.

As most changes in the epipelic densities cannot be explained by biological processes (see section on oxygen exchange), the dynamics are largely due to redistribution of matter within the lake.

So far, little is known about the densities and

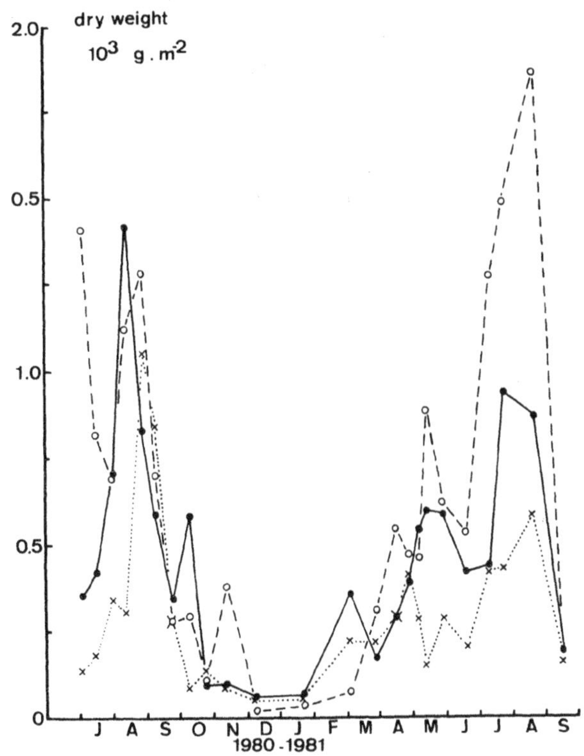

Fig. 15. Seasonal changes in epipelic dry weight at depths of 3 m (crosses), 4 m (dots) and 5 m (circles).

depth distribution of epipelon in the other lake parts. Some indications were obtained when the sampling was extended to the other side of the lake, and measurements were carried out at weekly intervals (Fig. 16).

The relative depth distributions showed a high degree of correspondence, indicating that redistribution patterns in both sides of the lake were similar. From 28 April to 12 May there was a general decrease at 2 and 3 m depths; excluding the high value at 5 m depth of 28 April along the North transect, the values increased especially at 5 m depth but also at 4 m. This suggests that the material lost from the lower depths was not resuspended, and thus mixed over the entire water column, but transported and redeposited along the bottom. It must be mentioned here that the bottom of Lake Vechten has steep slopes; along the South transect the slope is 11° and gradual; on the North side this is on the average 14°, and varies from 24° (between 1 and 2 m depth) to 10° (between 2 and 4 m).

For organic matter, the averages of all depths were not significantly different (Fig. 16), but the percentage organic matter was much higher on the North than on the South side (Table 4). From this it may be conceived that detritus in an advanced state of mineralization, having a small particle size and high ash content, tend to accumulate proportionally more on the South than on the North side of the lake.

The significance of the redistribution with respect to the development in the littoral and the lake's material cycles will be discussed in the final section.

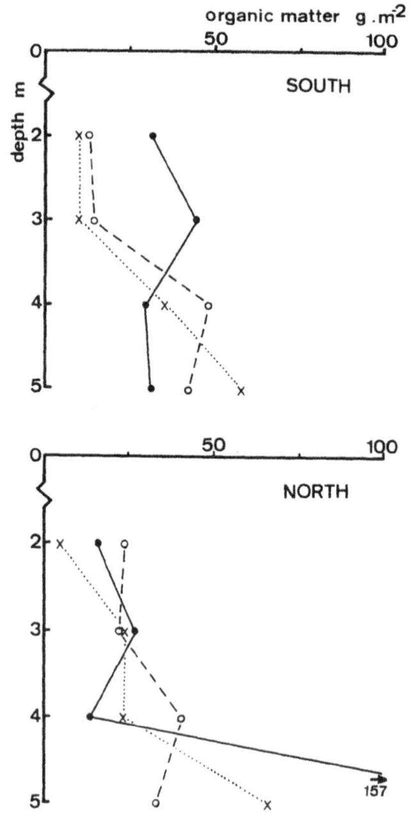

Fig. 16. Comparison of the distribution of epipelon at southern and northern sides of transect across the lake on 28 April (dots), 6 May (circles) and 12 May (crosses) in 1981.

Chemical composition. The study in 1980 and 1981 included that on epiphyton from *Ceratophyllum*, so the chemical compositions of epipelon and epiphyton can be compared (Table 2).

Table 4. Characteristics of epipelon on the southern and northern lake side. See text for further explanation.

	South transect			North transect		
Date	28 April	6 May	12 May	28 April	6 May	12 May
A. Total matter						
% organic matter	11.0	8.2	8.2	13.9	18.3	12.4
% $CaCO_3$	12.5	13.0	11.6	9.9	11.0	11.5
% ash	76.5	78.9	80.3	76.2	70.7	76.1
B. Organic matter						
% C	42.5	41.9	47.7	48.8	46.0	50.4
% N	4.1	4.3	5.6	4.3	4.1	4.7
C:N	10.3	9.6	8.5	11.4	11.2	10.8
chl *a* $mg.g^{-1}$	0.7	0.9	0.9	1.1	1.1	1.0

The proportion of organic matter in the dry weight of epipelon was on the average much lower than that of the epiphyton, resulting, of course, in the lower values for carbon, hydrogen and nitrogen (Table 2A). Elementary analysis of the residue on ignition demonstrated that the epipelon contained much less carbonate, but more ash than epiphyton; evidently the former included even more detritus than the epiphyton. This was demonstrated also by the elementary composition and R-value of the organic matter (Table 2B). The epipelon also differed qualitatively depending on the depth it came from, i.e. the energy content increased with depth. Nevertheless, for comparison with the epiphyton it is sufficient to deal with the data of the epipelon from 4 m (Fig. 7B).

The course of the percentages of the main constituents differed widely from that of the epiphyton. The minimum of organic matter, and the maximum of ash, occurred in August. After the fall recirculation, the increase in percentage organic matter was striking, which might be attributed to either input from the hypolimnion and deeper sediments, or differentially greater resuspension of detritus in an advanced state of mineralization. This change may also reflect input following decline of the macrophytes or leaf fall, but the former possibility was excluded by the low macrophytic biomass, and the latter by the exclusion of large fragments from the analyses. During the subsequent months the composition of the epiphyton and epipelon were similar, as were the chlorophyll contents (see below).

The carbonate content of epipelon changed less than that of epiphyton; only in May and June did the values increase, which may have been associated with photosynthetic activity in the epipelon itself: at 3 m depth calcium carbonate reached a maximum of 41%, but at 5 m it remained low (about 15%).

As mentioned above, the epipelon from opposite sides of the lake differed in organic matter and ash content. Depth averaged composition of the matter on the three dates are given in Table 4. Most characteristics support the conclusion that the epipelon from the North transect was in a less advanced state of decomposition, as was explained by differential redistribution over the lake of particulate matter, including $CaCO_3$. This was not apparent, however, from the percentage organic nitrogen: at the North side of the lake the organic C:N ratios were consistently higher.

Epipelic algae. For the greater part of the year the epipelon consisted almost exclusively of detrital associations, of which the components ranged from recently deposited plankton to fine-grained particles of undefined origin. In periods of clear water, however, conspicuous patches of algae developed upon the sediment surface, and a diver was able to distinguish between the various algal types, viz. filamentous green algae, diatoms and blue-green algae. Especially during the clear water phase of early summer, abundant, but temporary, growth of filamentous greens invariably occurred, sometimes down to depths over 4 m.

In relation to the total epipelic matter, the algae were considerably lower in number than in the epiphyton. Per unit organic matter the epipelon contained only a fifth, or less, of the numbers encountered in the epiphyton (Fig. 17).

Pennate diatoms and blue-green algae, as in case of epiphyton, occurred significantly more in the epipelon than in the epilimnetic phytoplankton. All pennate diatom genera listed in Table 3 also occurred in the epipelon. But in contrast with the epiphyton, filamentous green algae had very low frequencies or were absent. Proportionately, the epipelon contained less diatoms, but more blue-greens, than did the epiphyton. The greatest difference was in the contribution of the remaining taxa, i.e. mainly species which frequent the phytoplankton (see Blaauboer 1982). These, though similar in both epiphyton and epipelon, were relatively far more abundant in the latter.

The composition of epipelic algae did not differ consistently with depth.

Photosynthetic pigments. Year-round values for the chlorophyll *a* content of epipelic organic matter from 4 m depth are given in Fig. 11. From November to April the values were as good as equal to those of the epiphyton. This suggests that during these months the contribution of pigments of deposited seston was predominant in both associations, as was also apparent from microscopical observations. In summer, when the epiphytic communities thrived, the epipelic chlorophyll content remained low, although some increase did occur when the transparancy was high. Broadly speaking, the differences between epiphytic and epipelic chlorophyll content confirmed those in the algal numbers of the two associations during July-September 1980 (compare Fig. 17). The epipelic

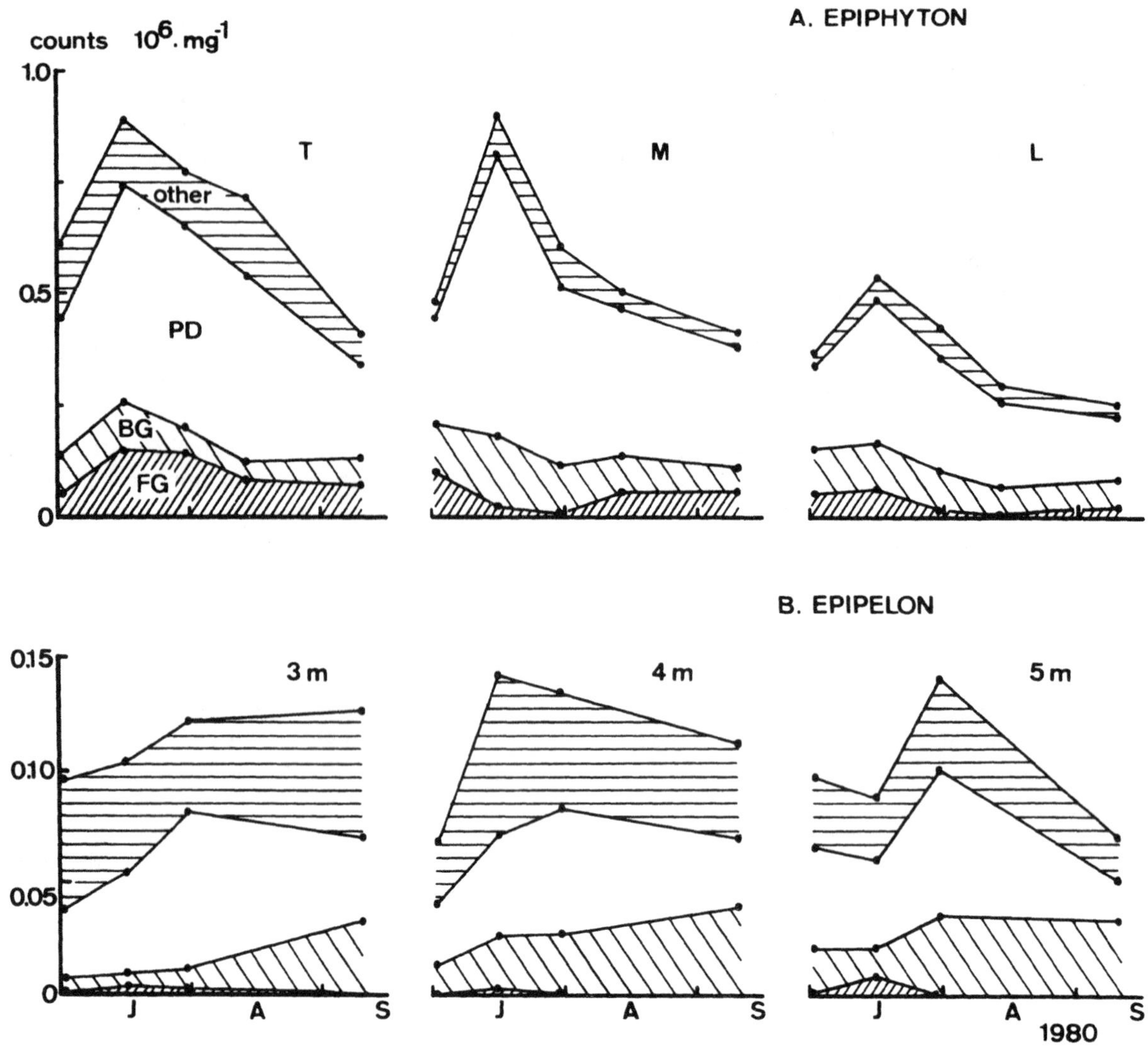

Fig. 17. Main algal groups in the epiphyton and epipelon during the summer of 1980. Abbreviations: T, M and L = top, middle and lower part of vegetation, see text; PD = pennate diatoms; BG = blue-green algae; FG = filamentous green algae.

chlorophyll content at the three depths did not differ consistently from one another.

With epipelon, paper chromatographical pigment analyses were carried out on the samples of 9 September 1980 only (Fig. 18). Generally, the percentage light absorption by chromatographically pure chlorophyll *a* of epiphytic and epipelic pigments was similar. The greatest difference was in chlorophyll *b*, reflecting the small proportion of filamentous green algae in the epipelon (Fig. 17). Other aspects of the distribution (Fig. 18) will be dealt with below.

Structure and function in relation to environmental conditions, with particular reference to light

Seasonal changes in physico-chemical characteristics of the lower littoral zone. For the growth of submerged macrophytes and the associated epiphytic algae and epipelic algae, the depth distribution of light is of foremost importance. Excluding those clear lakes where macrophytes may occur at such depths that hydrostatic pressure may inhibit their growth (Hutchinson 1975), the lower boundary up to which the macrophytes may occur un-

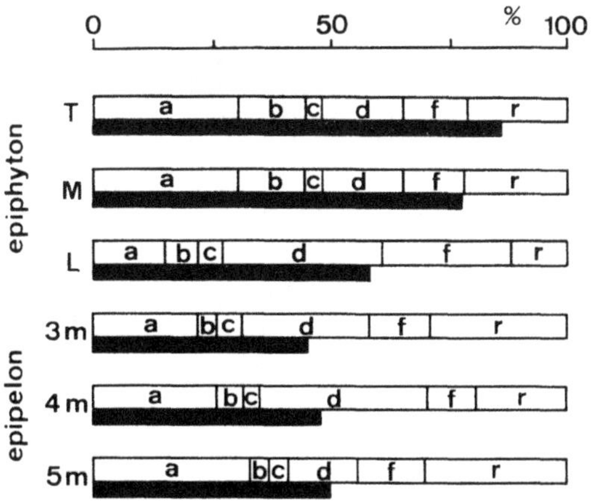

Fig. 18. Pigment composition and viability of pennate diatoms (black bars) on 9 September 1980. Abbreviations: T, M and L = top, middle and lower part of vegetation, see text; a, b, c, d = chlorophyll *a, b, c,* chlorophyllous breakdown products; f = fucoxanthin (including neofucoxanthin); r = remaining pigments.

doubtedly depends principally on the availability of light and water transparency. This, of course, also determines the distribution of the epiphytic and epipelic algae. Because of the generally poor conditions of irradiance in the lower littoral of lakes in the temperate regions, the growing season tends to be short (Kajak *et al* 1972) (see also the next sections).

The vertical extinction values in the littoral and the limnetic region (Steenbergen & Verdouw 1982) are generally similar. At the onset of the development of *Ceratophyllum* and of the associated epiphytic algae at 2 m depth, 15–30% of the surface light reached this depth. In early September 1980 the macrophytes apices reached up to 0.5 m below the surface, with about 75% of the surface irradiance, but a considerable part of the vegetation received no light owing to mutual shading (Fig. 1). The epipelon at 3, 4 and 5 m depth received, respectively, 5–20, 2–15 and 1–8% of the light at the surface. These values are of the incident light reaching the communities, but very little is known about the microdistribution of light within epiphyton and epipelon.

Oxygen content, temperature and pH were measured of the water above and within the *Ceratophyllum* stand and at the bottom along the transect. The isopleths are given in Figs. 19 and 20.

In 1980 the vegetation at station 2 m was well developed, but very poor in the next year: the decline of the macrophytes was continued and dramatic in 1981.

Since the thermocline was generally below 5 m (Steenbergen & Verdouw 1982), temperature variations with depth were slight, but from May to July a decrease of 3–5 °C might occur going from the surface (at station 2 m) to 5 m depth. In August and September the temperature somewhat increased with depth, possibly due to the initial cooling down. The gradients diappeared almost completely during the autumnal overturn. The icecover was insignificant and the mixing continued until April when the water was rapidly warmed and stratification started, but was terminated by a spell of cold weather. Early in May the stratification became stable, as is generally the case in the lake.

The oxygen gradients were more complicated (Fig. 19B). In July 1980 the oxygen concentration dropped by more than 9 mg. l^{-1} from station 2 m, above and within the upper part of the plant bed, down to 5 m. At 6 m the sediment was already anaerobic. The summer time concentrations at the bottom were lower than at the corresponding depths in the limnetic region (see Fig. 21B and Steenbergen & Verdouw 1982).

After July 1980, oxygen was minimal at the bottom between the plants. This was caused by the accumulation of sedimented matter, dissociated epiphyton and the macrophytes on the bottom where it was dark due to shading of the overlying vegetation. The oxygen concentrations at 3 m, i.e. outside the plant bed, were, however, considerably higher, although influenced by the water with low oxygen at 2 m: the highest values occurred between 3 and 4 m depth, where the oxygen concentration was even higher than that above the plants.

From mid September the macrophytes sank to the bottom and started to disintegrate in October; the oxygen gradients diminished, and the values became uniformly low due to the fall recirculation. On the bottom at station 2 m, however, the oxygen content remained noticeably lower until January. Oxygen stratification, like that of temperature, started in April and was firmly established from the beginning of May. That during the summer of 1981 a well developed vegetation at station 2 m was absent, was shown by the oxygen isopleths: the values were not as high as in July or as low as in August 1980.

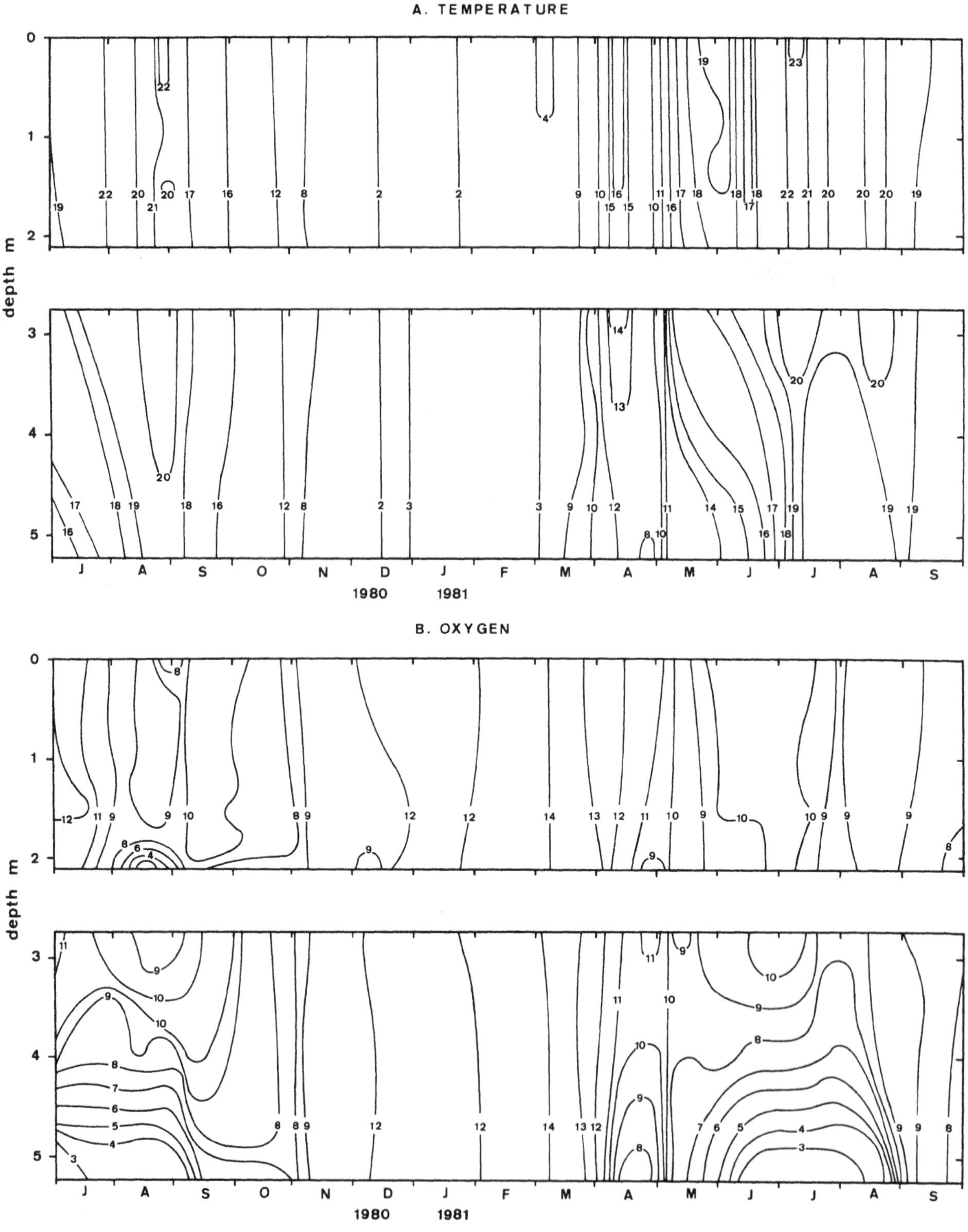

Fig. 19. Isopleths of temperature (A) and oxygen content (B) of littoral water collected along transect. See Fig. 1.

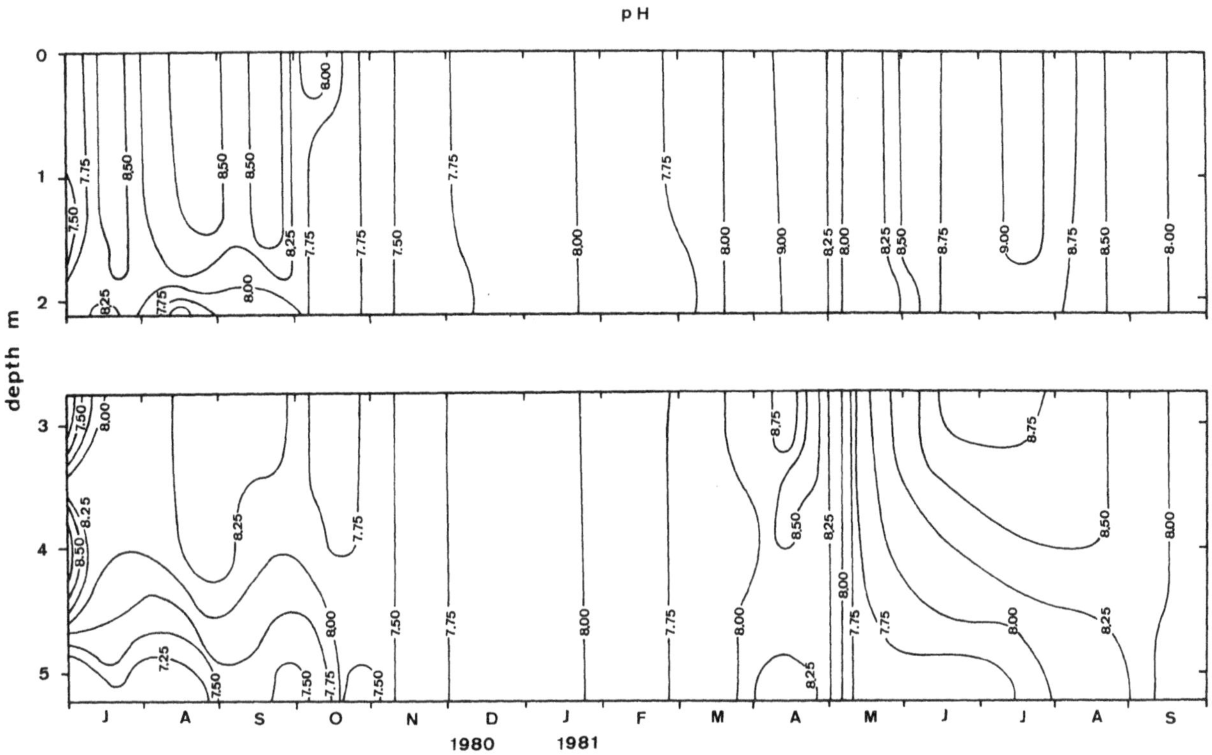

Fig. 20. Isopleths of pH. See also Fig. 19.

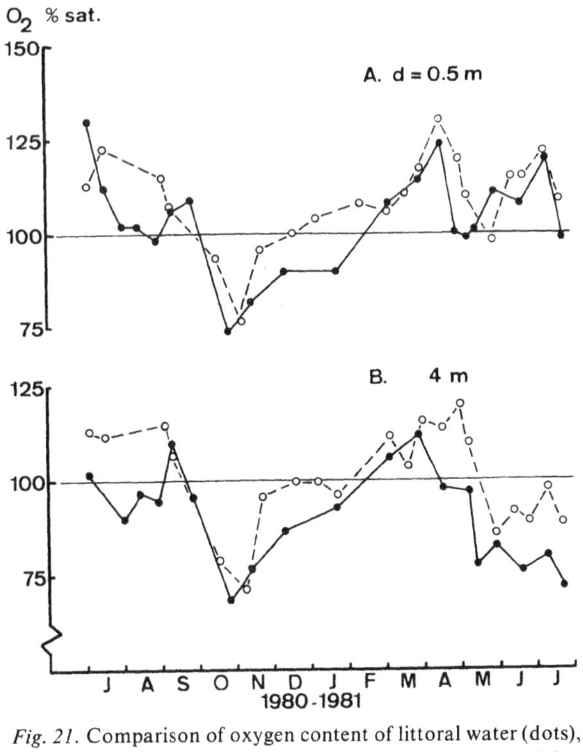

Fig. 21. Comparison of oxygen content of littoral water (dots), with that in the limnetic region (circles), at 0.5 and 4 m. A: 0.5 m depth, i.e. littoral water above vegetation at station 2 m. B: 4 m depth, i.e. littoral water at the bottom at station 4 m.

The changes for pH were similar to those of oxygen (Fig. 20); this is owing to the effects of photosynthesis and respiration on the bicarbonate equilibrium (Wetzel 1975). This also influences markedly the chemical composition of the epiphyton and epipelon (Fig. 7), since $CaCO_3$ precipitates at high pH during photosynthesis and may again dissolve when the pH is low due to the predominance of respiratory activity. Moreover, these processes have side-effects on the lake metabolism since organic substances and nutrients such as phosphorus may co-precipitate (Wetzel & Allen 1972).

Although the seasonal changes in the littoral depend greatly on the processes in the limnetic region, it is clear that biological activities may significantly influence oxygen and pH in the littoral water. The oxygen contents of the littoral and limnetic regions are compared in Fig. 21.

In summer 1980 the epilimnetic water was supersaturated with oxygen. After the autumnal overturn in November the saturation value dropped to 75%. In January equilibrium became re-established. The phytoplankton (Blaauboer 1982; de Kloet

1982) undoubtedly contributed to supersaturation during spring and summer.

The littoral water at 4 m had, except during the overturn, a lower oxygen concentration than the limnetic region at 4 m. This was particularly so in the summer when the epipelic matter was high (Fig. 15), but from November to January considerable differences existed as well.

One may anticipate that above a submerged macrophyte bed, such as at station 2 m in 1980, the oxygen concentration was higher than in the open water. In general, however, the reverse was found (Fig. 21A). That the plant bed may consume rather than produce oxygen will be dealt with in the section on the depth distribution with *Ceratophyllum* stands.

Growth upon artificial substrates. It has been documented extensively (e.g. Castenholz 1960; Sládečková 1962; Tippett 1970; Round 1973; Wetzel 1975) that artificial substrates may yield periphyton different from that associated with the natural ones. Nevertheless, artificial substrates may be useful tools in the study of the mechanisms underlying the distribution of periphytic organisms (Cattaneo 1978) or food chain aspects (Madsen 1972; McMahon *et al.* 1974), and several workers concluded that carbon fixation rates of algae growing on artificial substrates were similar to those of epiphytic algae (Allen 1971; Hooper & Robinson 1976). The application in the present study was to gain insight into the seasonal development of periphytic algae in relation to the depth.

The area of the substrates was sufficient to allow for weight analyses. The perspex plates (see Methods) were spaced on the basis of the average vertical extinction of the lake water. The plates were hung in a vertical position, to minimize the accumulation of sedimenting material, in the littoral of the SE part of the lake using a float and anchor. Around the anchor the bottom was cleared of macrophytes in order to prevent the shading of the substrates.

After exposure, the plates were covered rather uniformly by algae, but definite colonies also occurred, especially of blue-green and chlorococal green algae. Filamentous green algae and pennate diatoms predominated. Contrary to those in the epiphyton and epipelon, the diatoms adhered tightly to the substrate. In most cases large quantities of carbonate were deposited, and a considerable amount of detritus had been trapped, but the proportions of the latter were small compared with those in the epiphytic samples. After long exposure the periphyton contained conspicuous numbers of molluscs and insect larvae.

The growth per sample interval upon new plates (Fig. 22) was the highest in June, due mainly to filamentous green algae of the species observed also in the epiphyton. Subsequently the development was comparatively small; it increased again during August and September, and was insignificant thereafter.

During May and June most growth occurred at 0.5 m, and decreased with depth; in July little variation with depth was found, while later it was more at 1 and 2 m than at 0.5 m depth. Thus, the light inhibition of photosynthesis (see also de Kloet 1982) was expressed in the formation of biomass, but not in the case of the filamentous greens in May and June. Every time, the growth at 4 m depth was the lowest.

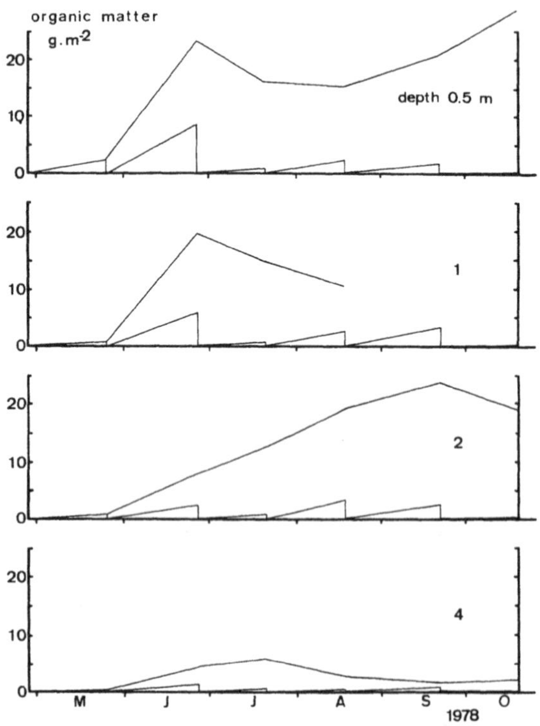

Fig. 22. Developments of periphyton on perspex plates. The lower lines represent growth on new plates per interval, the upper lines the developments on plates exposed from April.

The foregoing results suggest that during the greater part of the summer light was the key factor in the periphytic algal development. In May and June light limitation occurred already at 1 m depth or less, and this was always so at 4 m. In July, however, phosphorus was most likely the growth-limiting factor (Steenbergen & Verdouw 1982). Such a mid summer depression in periphytic growth has often been observed (Cattaneo & Kalff 1980). During the periods before and after summer the growth also decreased with depth. Buth then the lower temperatures may also have played an important role in affecting rates of colonization and of light saturated photosynthesis.

By extrapolation, the compensation depth for the growth of the periphytic algae, say diatoms, was more than 6 m during the greater part of the investigation period.

By serial sampling of the plates the wax and wane of the developed communities was followed. It may be remarked that the changes are not simply functions of algal photosynthesis and respiration, but might also reflect the history of colonization, degree of mutual shading, detachment from the substrate, mortality and mineralization, and grazing.

The relevance of the history of colonization (see Madsen 1972; Hutchinson 1975; Cattaneo *et al.* 1975; Siver 1977, and Perkins & Kaplan 1978) was evident from the following results. In June the periphyton upon the already-colonized substrates increased more than that upon the new ones. This can be explained by a time lag in the development of the latter, but was certainly also due to the filamentous green algae enlarging the substrate area for colonization by diatoms and for the trapping of detritus. In the September-October interval, there occurred little growth on the new substrate at 0.5 m depth, but apparently the conditions still allowed for substantial increase in an already existent community at the same depth.

Probably the detachment of filamentous greens after senescence was the main cause of the loss during July and August at 0.5 and 1 m depth: at 2 m there was still growth, and the development of the filamentous green algae had not been as abundant.

The community at 4 m declined during August and September, while at 2 m and on the new plates the growth was still considerable. It is likely that light was too poor to allow for compensation of the losses due to grazing. It is also possible that due to mutual shading the algae received insufficient light for balancing the requirements of energy for maintenance (Gons & Mur 1980).

Although consumption was not quantified, it can certainly have affected the development, in view of the presence of macro-invertebrates. Consumers of unknown trophic level caused the fall of two plates from 1 m down to the bottom.

The chemical characteristics of the periphyton collected from the plates during May to October are presented in Table 5. The contents carbonate and ash were calculated assuming the percentage carbon of the organic matter to be 50.

In the periphyton of the first category (Table 5A), the data on the depths from 0.5 to 2 m were similar, but differed markedly from those of 4 m, particularly in the lower percentage of carbonate in the latter case. This reflects the fact that light reached saturating values down to 2 m, but was growth-limiting at 4 m. The lower carbonate percentage at 0.5 m than at 1 and 2 m depth reflects light inhibition at 0.5 m.

The percentages of ash and nitrogen were higher at 4 m than at the depths above. This may be explained by differences in species composition. Relatively more diatoms occurred at 4 m, and filamentous greens have a lower percentage of ash and nitrogen (Strathmann 1967; McMahon *et al.* 1974) than diatoms.

The periphyton of the second category (Table 5B) differed considerably from the first one. Here, the filamentous green algae, which predominated after the first interval, played the foremost role. They remained an important component of the

Table 5. Chemical characteristics of periphyton collected from artificial substrates. See text for explanation.

Depth (m)	% organic matter	% $CaCO_3$	% ash	% C	% H	% N
A. Periphyton after one interval of exposure						
0.5	28.9	36	35	18.8	2.2	1.1
1	25.2	45	30	18.0	1.9	0.9
2	28.7	44	28	19.6	2.1	1.4
4	34.8	16	49	19.3	2.9	2.4
B. Periphyton after 2–5 intervals of exposure						
0.5	20.2	60	19	17.4	1.5	0.5
1*	22.3	47	31	16.8	1.3	0.5
2	21.4	54	24	17.2	1.4	0.6
4	20.1	60	20	17.3	1.6	1.1

* Excluding the data of September; see Fig. 22 and text.

substrates all over the investigation period, although less pronounced at 4 m. This was confirmed by the lower ash and nitrogen contents here than in the case of the first category, which contained proportionately more diatoms.

The lower percentage of organic matter was owing to a very high proportion of $CaCO_3$ – even at 4 m depth – indicating that carbonate precipitation was associated not only with photosynthesis per unit biomass, but also depended greatly on that per unit substrate area; with the new plates the specific photosynthetic rates may well have been higher, but this did not result in the same percentage of $CaCO_3$ as the second category, since the biomass was much lower.

The depth distribution of periphytic development on submerged macrophytes and on the artificial substrates used may differ due to several factors, important among which are:

1. *The material nature of the substrate.* While the artificial perspex substrate had fixed dimensions and was biologically inert, the macrophytes have a continuously varying area for colonization and may have influenced the epiphytic development, for example, by precipitation of carbonate and acting as a shelter for macro-invertebrates grazing the epiphyton (Mason & Bryant 1975; Dvořák & Best 1982).

2. *The morphology.* The macrophyte formed a geometrically more diverse substrate for microfloral growth, and collected more sedimenting matter than the perspex plates.

3. *The light regime.* The plates were orientated and spaced in such a way that only the vertical extinction of the lake water determined the percentage irradiation received by the community. In the case of macrophytes light attenuation due to mutual shading plays an important role as well, since the euphotic zone may become reduced dramatically.

It appeared also that the 'perspex' periphyton was both qualitatively and quantitatively different from the epipelon collected at the corresponding depths during 1980 and 1981. This was due to the dominant role of sedimentation and resuspension in changes in the epipelon.

Depth distribution with dense populations of Ceratophyllum. Epiphyton is not evenly distributed over the lengths of submerged marcrophytes because of the gradients both with respect to age, i.e. the time span after first colonization, and the physico-chemical environment, particularly the irradiance. Along these gradients within plant stands, the photosynthetic and respiratory activities may differ greatly; Buscemi (1958) showed that light modulated oxygen stratification occurred in vegetation of *Elodea;* the observations on dense stands of *Ceratophyllum* in Lake Vechten (Figs. 23 and 24), though not as marked, were similar.

During the summer of 1979 and 1980 the depth distribution of light, oxygen, temperature and pH were measured 1–2 h before noon at a fixed location in the SE part of the lake. From the end of July to mid September similar observations were made (Fig. 23).

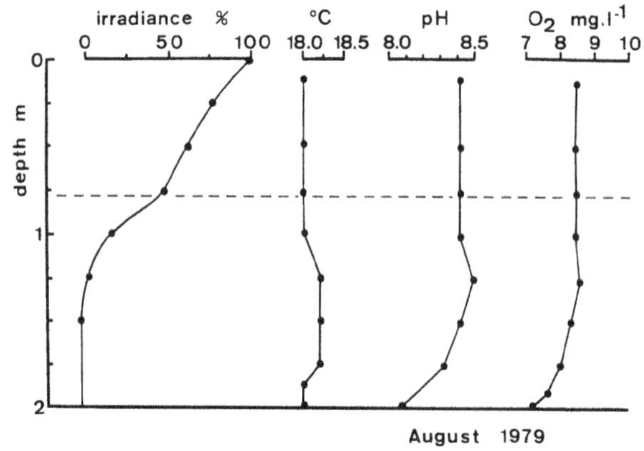

Fig. 23. Depth distribution of light, temperature, pH and oxygen in a dense stand of *Ceratophyllum.*

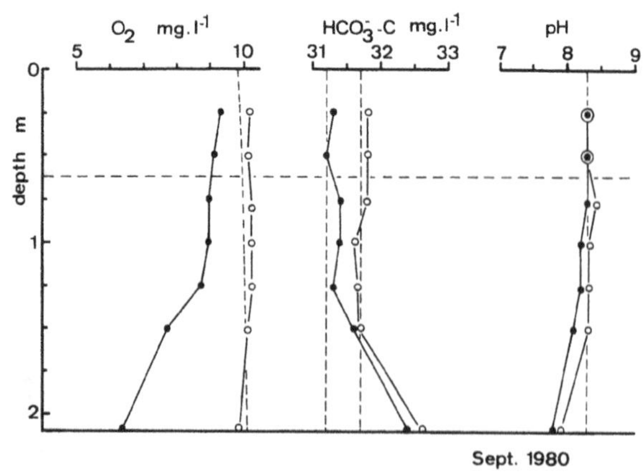

Fig. 24. Diurnal changes in the concentration of oxygen and bicarbonate and of pH in a dense stand of *Ceratophyllum.* Values at midnight given by dots, those at noon by circles. The dashed lines give the values at corresponding depths in the limnetic region.

Between the macrophytes the light decreased sharply: the lower half of the stand was virtually in the dark. The course of oxygen and pH changes was parallel, both exhibiting small peaks well below the plant apices, and decreasing towards the bottom, where both the oxygen concentration and pH were considerably lower than near the surface.

Diel variations in oxygen, pH and HCO_3^- are presented in Fig. 24, and compared with those in the open water. The latter exhibited no diel changes, but between the plants both the oxygen concentration and pH were lower at night than in the day, and at night well below the limnetic values. At noon, though the gradients were very weak because of strong wind, they were similar to those in Fig. 23. Reciprocal gradients were shown by the HCO_3^- concentrations.

These changes were related to the vertical distribution of light (Fig. 23). Within the lower half of the plant stand and upon the bottom only respiration occurred, and even with full sunlight this resulted in markedly lower oxygen concentrations than in the overlying water. The daytime changes within the upper half of the stand, caused by photosynthesis, were never pronounced. The main reasons for this are: 1) diffusion, which will smooth such changes, and 2) community respiration being high, a significant amount of oxygen produced will be consumed.

This was examined in relation to the depth distribution of mass and metabolism of macrophyte and epiphyton. In the same stand where the measurements on oxygen etc. were made (Fig. 23), three similar plant shoots (± 115 cm) were cut at the bottom. These were divided in the laboratory into three portions: the top 15 cm (T), and two equal lengths (± 50 cm) of middle (M) and lower (L) parts (Figs. 25 and 27). These portions were exposed *in situ* to irradiances of, respectively, 50 to 15, 15 to 0, and 0% (Fig. 27).

The concentration of macrophytic and epiphytic matter in the M-portion was the highest (Fig. 25A), and the chemical composition of epiphyton depended greatly on depth. Relative to the organic matter, the highest percentage ash was in the L-portion. The epiphyton of the M-portion had the highest carbonate content, but this was much lower below.

That the deeper part was aphotic was also evident with microscopical observation; very few live organisms were detected, and whereas the material from

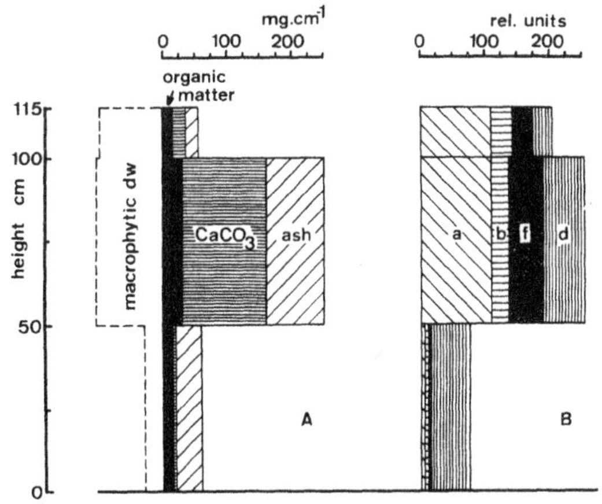

Fig. 25. Depth dependence of the distribution and constitution of epiphyton in a dense stand of *Ceratophyllum*. A: Chemical constituents. Values refer to three shoots. B: Pigments. Abbreviations: a, b, d = chlorophyll *a*, *b*, chlorophyllous breakdown products; f = fucoxanthin.

the upper parts contained large aggregates of carbonate and organic matter, in the lower small greyish fragments were mainly encountered.

The results of the pigment analyses were consistent with these findings (Fig. 25B). In the dark deeper parts the epiphyton contained almost exclusively chlorophyll derivatives, but in the upper parts considerably more intact pigments. The pigment per unit organic matter was the highest in the T-portion, with relatively more chlorophyll *b* and less fucoxanthin than in the middle. The microscopical observations confirmed this, because the epiphyton in the M-portion contained comparatively more diatoms and less filamentous green algae (see also Fig. 17).

Using the same samples as for the aforementioned observations, respiration rates in the dark of plant parts and epiphyton of all fractions were measured, and oxygen exchange vs. irradiance curves were established for the 15 cm plant top sections and the epiphyton of the M-portion (see Methods).

The respiration rate per unit organic matter of the macrophyte decreased from 1.7 in the top, and 1.0 in the middle, to 0.5 mg $O_2.g^{-1}.h^{-1}$. The epiphytic rates were about 5 times higher than those of the plant. The upper two fractions did not differ in their specific respiration, which was in both cases 8.0 as against 3.0 mg.$g^{-1}.h^{-1}$ in the L-portion.

The gross photosynthetic capacity of epiphyton was also markedly higher than that of *Ceratophyl-*

lum (Fig. 26). The value for the macrophyte was 6 mg $O_2.g^{-1}.h^{-1}$, which, converted to carbon fixation, was within the range reported by Best (1979), see also Westlake *et al.* (1980). Per mg chl *a*, the epiphytic value was 3.3 mg $O_2.h^{-1}$, a rate close to that reported for benthic diatoms in culture and field samples also incubated on filters (Admiraal 1977), and to that for *Nitzschia palea* from light-limited continuous cultures (Gons & Werkhoven, unpublished). Compared with the respiratory rate, however, the epiphytic photosynthesis was low (cf. Admiraal 1977). Apparently, the included heterotrophic organisms contributed significantly to the oxygen consumption (see e.g. Parsons *et al.* 1977; Harris 1978; and Westlake *et al.* 1980).

The respiration and photosynthesis were computed in relation to depth (Fig. 27). The depth distribution of macrophytic and epiphytic organic matter was smoothed, and the response to *in situ* light (Fig. 23) calculated for two 7.5 cm depth intervals, for the top 15 cm, and 10 cm intervals for the parts below. The specific oxygen exchange in the light of the middle 50 cm and top 15 cm of *Ceratophyllum* was assumed to be the same, and the epiphyton from these parts to have identical curves (Fig. 26). The sum of the net oxygen exchange in light and the respiration in dark of the relevant depth intervals was taken as gross photosynthesis.

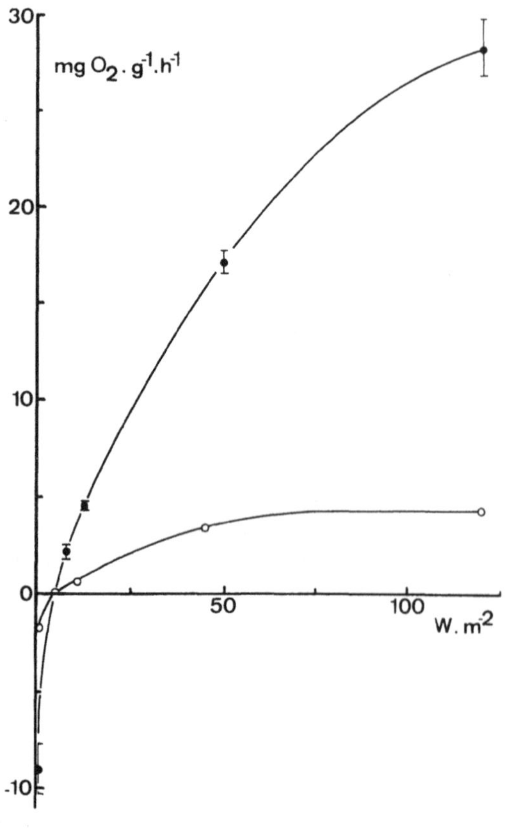

Fig. 26. Photosynthesis per unit organic matter vs. light curves of epiphyton from middle fraction (dots) (see text), and of *Ceratophyllum* apices (circles).

Both in the dark and in moderate to full sunlight epiphyton was the dominant functional component. Photosynthesis was restricted not only to a thin layer on the top of the plant bed but also extended to deeper parts, where the epiphytic matter became more important than the macrophytes.

The oxygen profile (Fig. 23) indicated that the community respiration might be high compared with its photosynthesis. The integrated values of the net productivity of the community (Fig. 27) were so low compared with the respiratory activity that in a 24 hour period the macrophyte stand was oxygen consuming, despite the high photosynthetic capacity. Note that this was even more so since the measurements did not include the macrophytic, epiphytic and epipelic respiration on the bottom in the vegetation. At that time (August) neither the macrophyte nor the total algal community showed signs of decay, but indeed still increased; it was therefore concluded that the heterotrophic utilization required imports of matter produced elsewhere. This aspect was given special attention in the subsequent work.

In the *Ceratophyllum* stand, the light attenuation markedly influenced the epiphytic characteristics, but not so much for the epipelon from 3 to 5 m depth (Figs. 17 and 18). The latter may be explained by the comparatively small change in irradiance: on 9 September 1980 this was no more than that over a 25 cm depth interval with the *Ceratophyllum* stand (Fig. 1).

Seasonal changes in epiphytic and epipelic photosynthesis and respiration. The maximum specific rates of gross oxygen production (p_g^{max}) for epiphyton, collected from the whole length of plant shoots, and epipelon are given in Fig. 28A.

The values were similarly low from October to April; the photosynthetic capacity was markedly higher only during a brief period in summer. In autumn and winter the chlorophyll content of epiphyton and epipelon was similar, but in the summer

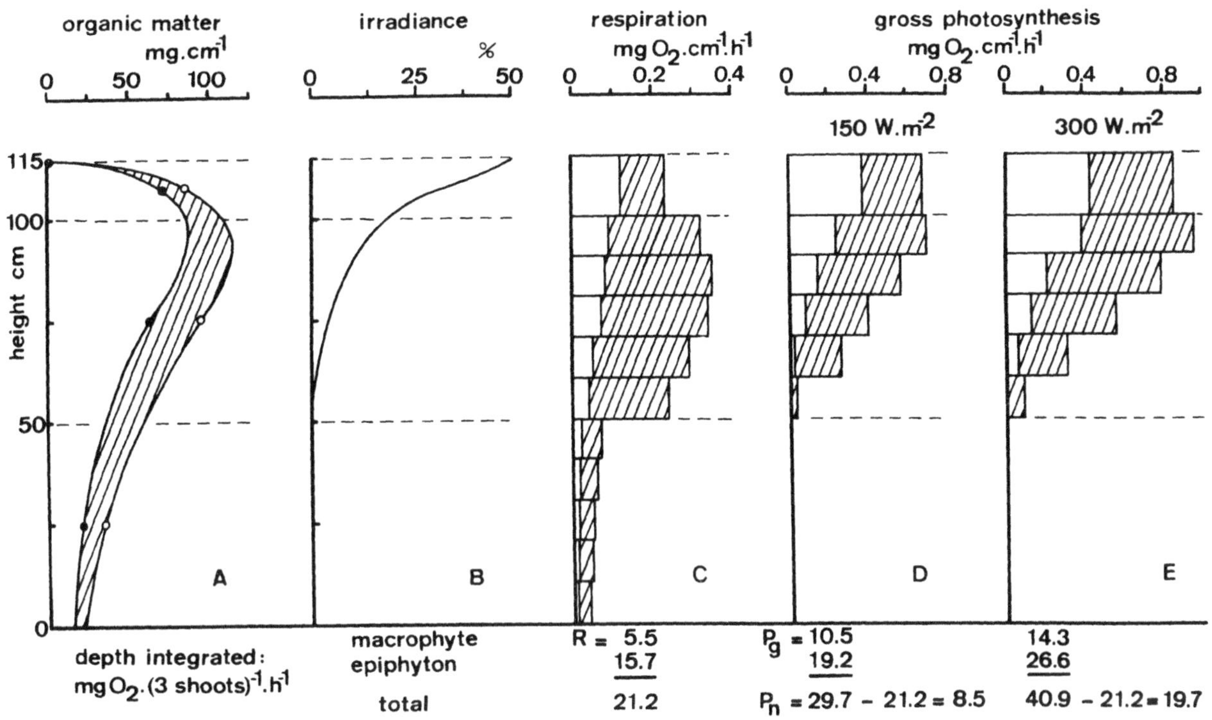

Fig. 27. Depth distribution of organic matter, irradiance, respiration and gross photosynthesis in conditions of moderate and full sunlight. Shaded parts represent the epiphytic contribution to mass and metabolic activity of the vegetation. All values refer to epiphyton and macrophyte of 3 shoots.

months it was quite different (Fig. 11); the latter was true also for the number of algae per unit organic matter (Fig. 17). That the epipelic oxygen production rate in summer was similar to that of the epiphyton was therefore due to a higher photosynthetic capacity of the epipelic algae. This may have reflected the differences in species composition, notably with respect to the contributions of 'planktonic algae' and filamentous greens, but may also be explained by a '*Chlorella*-type' adaptation to irradiance (Jørgensen 1969). Per unit chl *a*, the highest epiphytic value occurred in May 1981, namely 5.4 mg $O_2.mg^{-1}$ for the epiphyton from the plant tips during a bloom of filamentous green algae. In July 1981 a maximum of 21 mg O_2.mg chl a^{-1} was found for the epipelon from 4 m depth.

Although exchange of organic matter with the seston may also result to change in photosynthetic capacity, light and temperature were evidently of primary importance. The rise of water temperature (Fig. 17) during March to mid April 1981 from 3.5 to 14.5 °C, may alone have caused a doubling or more in p_g^{max} (e.g. Steemann Nielsen 1975). This was followed by a period of low light and temperatures: by the end of April the water temperature had decreased to 10 °C. However, in early May it was calm and sunny weather and a clear water phase started. These factors induced a rise of 30–40 cm in the upper parts of the macrophytes, which lay hibernating on the bottom, and dense growth of filamentous green algae upon the raised parts. During the rest of May the irradiation and transparency decreased again, and further development of the macrophytes was arrested until June.

The changes in light and temperature were reflected in the variations in the photosynthetic capacity of epiphyton and epipelon, and also on a chlorophyll basis; but the high values in mid summer and the decrease thereafter were also related to the contribution of detrital matter, which in turn depended on sedimentation and heterotrophic utilization rates (see the final section).

In Fig. 28B the values of the community respiration rates (r) are given. The epiphytic and epipelic values were below 2 mg $O_2.g^{-1}.h^{-1}$ during September-April, but while the epipelic rates remained low

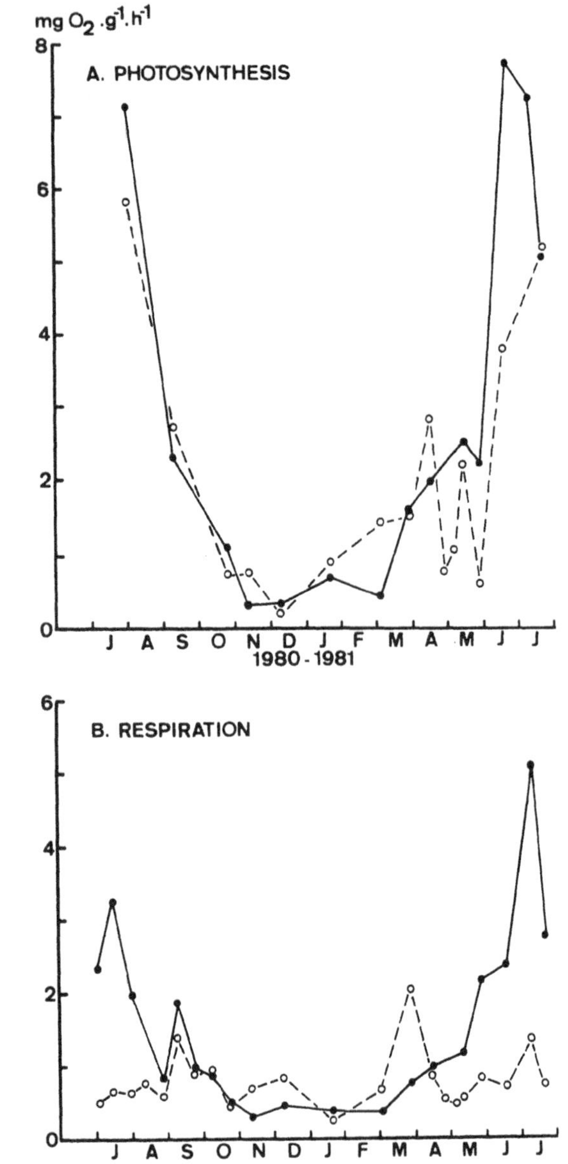

Fig. 28. Seasonal changes in maximum gross photosynthesis (A) and respiration in the dark (B) per unit organic matter of epiphyton (dots) and epipelon (circles).

in summer, those of the epiphyton increased markedly. The respiration of epipelon exhibited no correlation with the photosynthetic capacity, nor any influence of temperature. The seasonal changes in epiphytic respiration rate, however, tended to follow those in p_g^{max}.

The ratios of p_g^{max} and r (Fig. 29) indicated the extent to which the respiration rates can be attributed to either algae or heterotrophic organisms. Most literature data (Steemann Nielsen 1975; Parsons *et al.* 1977; Harris 1978; Westlake *et al.* 1980) indicate the ratio maximum photosynthesis to respiration in dark of a pure algal community to be 5–10. The ratio for epiphyton and epipelon fluctuated around 2. The latter value implies that for the greater part of the year the algae did not contribute more than 40%, but the heterotrophic organisms at least 60% to the oxygen consumption in the dark. Furthermore, since the average daily photosynthetic production rate *in situ* would be much lower than p_g^{max}, the heterotroph consumption would as a rule exceed the autotrophic production.

The relations between *in situ* photosynthesis, depth and p_g^{max} / r are illustrated using the data of 9 September 1980 (Table 6). On this date irradiance conditions were fairly good, the subsurface light reaching a maximum of 165 $W.m^{-2}$ (400–700 nm), and the vertical extinction being 0.56 m^{-1} (see also Fig. 1). The epiphyton was subdivided in the same way as described in the section on the depth distribution. In the stand along the transect (Fig. 1) the macrophytes had on the average 100 shoots per m^2, and this was used as the basis for computation of the integral values; at the central part of the stand the maximum standing crop was more than twice as high.

The epipelon received less light than most of the epiphyton. The epiphyton from the L-portion was photosynthetically inactive, and the response of epipelon from 5 m depth was weak. The highest photosynthetic capacity, both per unit organic matter and chlorophyll, was found in epipelon from 3 and 4 m. Both total and intact chl *a* was lowest in

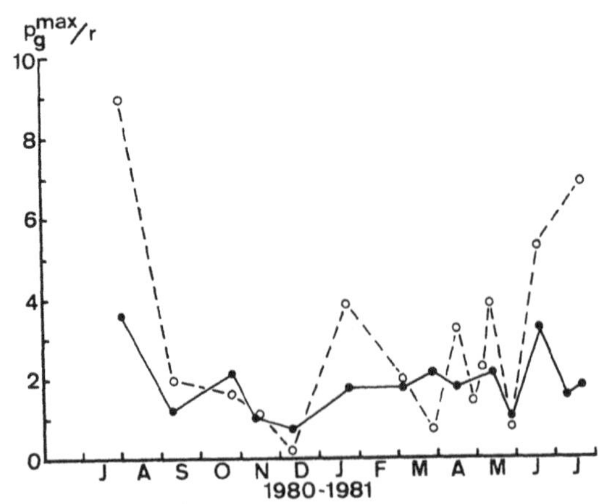

Fig. 29. Seasonal changes in p_g^{max}: r ratio of epiphyton (dots) and epipelon (circles).

Table 6. Epiphytic and epipelic production and respiration on 9 September 1980. See also Fig. 1.

		% surface irradiance	Organic matter ($g.m^{-2}$)	chl *a* ($mg.m^{-2}$)	p_g^{max}/r	P_g^{max}	P_g ($g\ O_2.m^{-2}$.	R day^{-1})	P_n
Epiphyton	T (15 cm)	62–58	4.0	6.7	1.4	0.28	0.12	0.21	-0.09
	M (62.5 cm)	58– 5	37.8	96.4	1.6	2.83	0.80	1.73	-0.93
	L (62.5 cm)	5– 0	15.1	2.6	0.0	0.00	0.00	0.63	-0.63
	Integrated	62– 0	56.9	105.6	1.2	3.11	0.92	2.56	-1.65
Epipelon	3 m	17	23.1	10.4	2.0	0.96	0.40	0.48	-0.09
	4 m	10	24.7	20.4	2.0	1.61	0.35	0.81	-0.47
	5 m	6	48.7	21.5	0.9	1.02	0.03	1.11	-1.08

Note: specific maximum gross photosynthesis and respiration symbolized p_g^{max} and r; daily integrated gross and net production, and respiration have the symbols P_g, P_n and R; P_g^{max} is the potential daily integrated gross production. See text.

the epiphytic L-portion, which also contained less viable diatoms than the remaining epiphyton (Figs. 17 and 18). The epipelon differed from the epiphyton in algal and pigment composition, and had a significantly lower percentage viability of the diatoms.

The daily integrated productions (P_g and P_n) were computed from the diel changes in irradiance and the oxygen exchange vs. irradiance curves over hourly intervals. For all categories P_n was negative. The P_g for epiphyton was greatly exceeded by R. The P_n for epipelon at 3 m was only slightly negative, but due to the decrease in irradiance R was more than twice the P_g at 4 m, and at 5 m depth photosynthesis was negligible.

The seasonal change in p_g^{max}/r (Fig. 29) is now considered in the following perspective. The P_g^{max} values in Table 6 express the gross production should the epiphyton and epipelon receive optimal light during the 24 hours. The epiphytic top fraction realized 43% of this potential, but with the integrated epiphyton this was only 30%. For the epipelon from 3, 4 and 5 m the percentages were, respectively, 41, 21 and 3. Gross photosynthesis of more than 43% of P_g^{max} will only be attained when conditions of light are more favourable than on 9 September 1980, which involves irradiation, daylength and lake water transparency. Such conditions may exist from part of May onwards in summer (Steenbergen & Verdouw 1982), but were rare during 1980 and 1981. Taking 50% of P_g^{max} as the maximum for daily gross photosynthetic production, this balances respiratory losses only when p_g^{max}/r is at least 2.

For the epiphyton (Fig. 29) considerably higher ratios occurred only in mid summer. The epipelon (also at 3 and 5 m depth, Gons, unpublished) more often had values of over 2. Because of low light, however, daily net production of epipelon would be negative, except for favourable conditions of irradiation and vertical extinction during the summer.

Thus, the heterotrophic processes dominated the oxygen exchange of epiphyton and epipelon during the major part of the year. The general increase in epiphyton and epipelon during January to, respectively, October and August, can be explained by imports, the sources and significance of which for the lake's nutrient cycles are discussed below.

Autotrophy, heterotrophy and redistribution processes. Two energy flows drive the processes in the lower littoral of Lake Vechten. In the first place solar energy is stored in the biomass of the submerged macrophytes, epiphytic and epipelic algae. Secondly, energy contained in organic matter, produced both autochthonously and allochthonously, is continuously redistributed and utilized in the various lake parts. Evidence was obtained that submerged vegetation in particular is effective in trapping this form of energy as well. Thus, the lower littoral may be regarded as an energy subsidized lake part (Odum 1971) and be compared with environments such as tidal mud flats and coral reefs.

Changes in epiphytic and epipelic organic carbon can be described as:

$$\Delta C = P_g + I - R - E$$

where: ΔC = net change in organic carbon,
P_g = gross photosynthetic carbon fixation,
I = import,
R = respiratory carbon loss,
and E = export.

The proportions of P_g and I depend on depth, since production by the autotrophs decreases with depth due to the light attenuation, while sedimentation rates tend to increase. This will be complicated, however, with the development of dense macrophyte stands, which changes the light climate and may influence sedimentation rates, and provide an environment for physico-chemical mechanisms for accumulation of organic matter.

From October through winter and early spring, the light conditions in the lower littoral were poor for algal growth. That at these depths light was indeed a key factor was evident from epiphytic and epipelic growth of filamentous green algae near the shore in winter (Gons, unpublished) and the growth upon artificial substrates (Fig. 22). The epiphyton and epipelon were similar in their chlorophyll *a* content and chemical composition, especially in winter (Figs. 7 and 11), and algae known to frequent in the phytoplankton were abundant (Fig. 9A).

Depending on meteorological conditions and lake water transparancy, epiphytic and epipelic algal growth was stimulated in spring and early summer. This first involved filamentous green algae (Fig. 8). During accelerated macrophytic growth these benefitted from improved light conditions because of the substrate reaching close to the water surface, while the accumulation rates of sedimenting matter per unit substrate area were low. As a consequence the 'loose' fraction of the epiphyton was relatively small, while carbonate (Fig. 7A), organic matter (Fig. 6), chl *a* (Figs. 10 and 11) and photosynthetic capacity per unit organic matter (Fig. 28) reached their maximum. In this period, similarity between the species composition of epiphyton and phytoplankton was the least (Fig. 9B).

Thereafter, the epiphytic heterotrophic component became increasingly important, since within the vegetation light was severely attenuated due to mutual shading, and due to the accumulation of deposited matter. Nutrients also possibly limited the growth, particularly of the filamentous greens. Later on, the amounts of loose material increased so that the vegetation might be regarded as sediment enlarged in area and volume. The sediment character was expressed by the abundance of freely motile biraphid diatoms, both numerically (Fig. 8 and 17) and on the basis of pigments (Fig. 14).

When the submerged macrophytes had reached their maximum standing crop, the epiphyton exhibited both great photosynthetic potential and high oxygen consumption rate (Fig. 27, Table 7). That the epiphytic production may be important on a lake basis – even surpassing that of submerged macrophytes – has been concluded by several investigations using different approaches (e.g. Wetzel *et al.* 1972; Kajak *et al.* 1972; Spodniewska *et al.* 1975; Cattaneo & Kalff 1980; Jones 1980). The share of epiphytic production to that of the lake provides information on the trophic state. Cattaneo & Kalff (1980) proposed that increased nutrient loading stimulates the epiphytic algal growth more than that of the macrophytes (see also Fitzgerald 1969). The epiphyton may in turn adversely affect the macrophytic development due to shading (Eminson & Phillips 1978; Phillips *et al.* 1978).

In the case of Lake Vechten, however, it was not

Table 7. Seasonal maxima of organic carbon, chlorophyll, net increase of organic carbon (Δ C), gross photosynthetic carbon fixation (P_g) and respiration (R). The figures in parentheses are the months.

	Organic C ($g.m^{-2}$)	chl *a* ($mg.m^{-2}$)	Δ C	P_g ($gC.m^{-2}.day^{-1}$)	R
A. Epiphyton of station 2 m and epipelon of station 4 m during 1980 and 1981					
Epiphyton	80 (10)	206 (10)	4.0 (9/10)	0.6 (8)	1.3 (10)
Epipelon	35 (7)	42 (5)	1.1 (4/ 5)	0.3 (7)	0.5 (7)
B. Periphyton of perspex plates during April-October 1978					
Depth 0.5 m	14 (10)	77 (10)	0.4 (6)	–	–
4 m	3 (7)	23 (10)	0.06 (6)	–	–

the algae but the detrital aggregates that mainly constituted the epiphyton. The data on the oxygen exchange indicated that, except during favourable light conditions such as occurred in the first half of the summer, heterotrophic utilization exceeded the algal production – the import of organic matter explains this. Such an import will consist largely of particulate matter, but may also involve the adsorption or co-precipitation of dissolved compounds (Wetzel & Allen 1972; Wetzel 1975). The imports were massive, thus concealing any specific interaction between substrate and epiphyton: similar chemical and algal species compositions were found with the epiphyton of *Ceratophyllum, Elodea* and *Myriophyllum,* the differences found being rather quantitative and due to the location of the stands.

While newly deposited matter was only loosely associated, the carbonate and mucoid substances secreted by epiphytic bacteria and algae (see Fig. 4 and Allanson 1973) cemented particles to more firmly associated larger aggregates on the macrophyte. This allowed for the development of unique communities of consumers and decomposers having high respiratory activity. The abundance of pennate diatoms may be due to several factors, among which the ability to migrate is one means to escape from being covered by sediment. They might also benefit from the release of nutrients in the course of the utilization of the epiphytic matter by the heterotrophic organisms, or be using certain organic compounds as additional energy source, of which the physiological ability is well known (Hellebust & Lewin 1977; Admiraal & Peletier 1979).

Specific oxygen consumption rates in the dark of epiphyton reached 3 mg O_2.(g org. dry wt.)$^{-1}$.h^{-1} or more in summer (Fig. 26), about two thirds of which was due to heterotrophic organisms. This gives a decomposition rate of approximately 3.5% per day. The value for the epiphyton from the upper plant parts was up to 1.5 times as high, the matter of this fraction being more recent and easily degradable. These rates are substantially higher than those reported for tree-leaf organic matter, but within the range of those for 'mixed dead phytoplankton and detritus in lake water' (Saunders *et al.* 1980).

It is not known to what extent the epiphytic matter was utilized by either animals or bacteria and moulds. It is generally believed that epiphytic matter forms a readily accessible source of food for consumers (Hutchinson 1975). The protein content especially is considered to be a decisive factor (Russell-Hunter 1970). The C:N ratio may therefore serve as an index of nutritional value (McMahon *et al.* 1974), a ratio of 17 being critical. The average epiphytic and epipelic organic C:N ratios in the present study were 9.0 and 10.2, being close to those reported for periphyton on glass slides in lentic systems by McMahon *et al.* (1974), but lower than those calculated by these authors from the data of Stockner & Armstrong (1971). The potential role of macro-invertebrates in the littoral of Lake Vechten is discussed by Dvořák & Best (1982). The abundance of ciliates and rotifers (Gons, unpublished) implies that this micro-fauna may have contributed significantly to the oxygen exchange (Fig. 28B).

The summertime % organic matter and R-value (Table 2), chlorophyll content (Fig. 11) and algae number (Fig. 17) of epipelon were lower than those of epiphyton. Although the photosynthetic capacity reached similar values to those of the epiphyton, the epipelon had considerably lower specific respiration in the dark (Fig. 26B). The latter may be explained by the epiphyton containing more detritus in early stages of decomposition, especially during early summer when the vegetation was still young. From August onwards the respiration rates of epiphyton and epipelon became similar.

On an areal basis, the littoral exhibited great quantitative differences during summer and autumn (Table 7). The epiphyton reached considerably higher concentrations of organic carbon and chl *a*, as well as higher photosynthesis and respiration than the epipelon. This was also true for the value of Δ C, which greatly exceeded that of P_g. That the net increase of organic carbon was for the greater part due to import, appeared also from a comparison with the periphyton on the artificial substrate (Table 7B), which, being vertically placed, trapped far less sedimenting matter. This also resulted, of course, to the lower maxima of organic carbon on the perspex, although chl *a* was more like that of the natural substrates. The seasonal maxima of integral chlorophyll, photosynthesis and respiration of the epipelon are comparable to the values compiled by Moss (1968) and Westlake *et al.* (1980). Information on integral values of epiphyton associated with submerged macrophytes is too scanty for comparison. The chlorophyll maximum of the 'perspex' periphyton has a mid range value among those reported for the periphyton upon rocks and artificial substrates (e.g. Stockner & Armstrong 1971; Allen 1971; Hooper-Reid & Robinson 1978).

Further, the timing of maxima of epiphyton and epipelon differed markedly; this appeared to be due to the greater capacity of littoral with dense vegetation to collect and retain sedimenting matter.

The role of redistribution of detrital matter in the metabolism of lakes was emphasized by Wetzel *et al.* (1972). Gulati *et al.* (1982) studied the seasonal changes in sestonic organic carbon in Lake Vechten. Sedimentation rates were measured in order to link the processes in the productive and anaerobic strata (Cappenberg & Verdouw 1982), but little is known about sedimentation and resuspension in the littoral zone. There is evidence, however, that the following general phenomena (see Wetzel 1975) apply in the case of the littoral of Lake Vechten: 1) depth dependent sedimentation rates, which explains the increase of epipelon with depth (Fig. 15), and 2) vertical and horizontal differential distribution, as indicated by the differences in the chemical composition of epipelon, respectively, with depth (Table 2) and on the southern and northern lake side (Table 4).

Verdouw & Dekkers (1982) measured sedimentation in the limnetic region of Lake Vechten during 1977–1979. The rates observed in March-April were consistently high and those in September-October were even higher. The spring values of about 6 g dry wt. $m^{-2}.day^{-1}$ agreed well with the increase in dry weight of epipelon during the spring of 1981 (Fig. 15). Since resuspension, or at least downward shifts along the bottom, may occur more or less throughout the year, the seasonal trends in sedimentation and in epipelic density were comparable until in July-August when the latter had its maximum.

For reasons not yet understood, from these months the epipelon already decreased almost continuously until winter (Figs. 15 and 30). Perhaps this was associated with the increased turbulence along the bottom due to the start of cooling after the temperature maximum in July (Fig. 19A). The decrease rates generally could not be accounted for by the data on the oxygen exchange. For instance, the decrease in organic matter from mid August to September was about twice as high as predicted from the oxygen consumption rates in the dark.

With the submerged vegetation, in contrast, the accumulation of organic matter continued (Fig. 30)

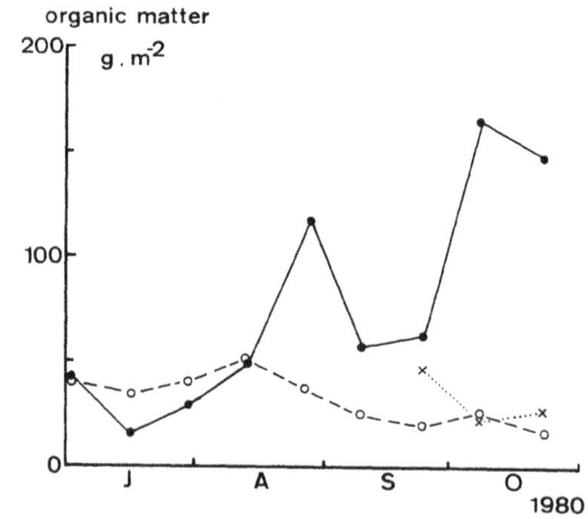

Fig. 30. Influence of dense vegetation on the accumulation of organic matter in the littoral region (epiphyton, dots; epipelon in vegetation, crosses; epipelon at 4 m depth, circles).

and was fastest in the time Verdouw & Dekkers found the highest sedimentation rates.

On the basis of eq. (3), the carbon flow of the epiphyton was generalized for the May-October 1980 growing season (Table 8). The net import of organic carbon, i.e. I–E, was on the average about 4 times as high as the gross photosynthetic fixation of carbon. A similar proportion was also reported by Pieczyńska (1975) for emergent vegetations, but including the macrophytic production.

This result sheds new light on the functioning of the lower littoral of lakes in general and of the part with submerged vegetation in particular. In the case of Lake Vechten, where the region from the shore to the 5 m depth contour occupies nearly 40% of the total lake area, significant proportions of the seston are diverted from sedimenting to the anaerobic strata, and consequently decomposed in aerobic conditions at comparatively favourable temperatures. No doubt this has far-reaching consequences, both in the regeneration of nutrients and energy flow in the higher trophic levels. Since especially during autumn large amounts of organic matter reach the lower littoral, the decline of the sub-

Table 8. Generalized carbon flow in epiphyton of station 2 m for May-October 1980.

ΔC	P_g	R	I–E
(mg $C.m^{-2}.day^{-1}$)			
400–550	150–300	600–700	700–1 100

merged vegetation during the last few years may have great consequences for the seemingly unrelated organisms and processes in the lake.

Summary

The periphyton associated with the submerged macrophytes *Ceratophyllum demersum, Myriophyllum spicatum* and *Elodea* sp., and that on the sandy bottom in the lower littoral of Lake Vechten (The Netherlands) was quantitatively collected from the substrates. The periphyton from the plants is denoted as epiphyton and that from the bottom as epipelon, and includes live micro-organisms, detritus and inorganic matter.

On a dry weight basis the epiphyton was maximal in autumn and often exceeded its substrate. The organic matter, which was mainly detrital, comprised generally about 20% of the dry weight; $CaCO_3$ contributed up to 50% in summer.

During exponential growth of the macrophytes, filamentous green algae, notably species of *Oedogonium, Spirogyra* and *Mougeotia,* predominated. From the end of June freely motile diatoms, belonging mainly to the genera *Nitzschia* and *Navicula,* were the most abundant. Except for June and July, the planktonic species were abundant and, often, predominated in winter and spring.

The chlorophyll *a* in the organic matter was low, about 5 $mg.g^{-1}$ being the upper limit in summer. Phaeophytin and other chlorophyll derivatives contributed about 30% to the total chl *a*. Green algae and diatoms were quantified by linear regression of the pigment ratio chl *b*: chl *a* vs. fucoxanthin: chl *a*. In the summer months the diatom pigments contributed 50% or more to the light absorption by all the photosynthetic pigments.

The epiphyton from the three macrophyte species did not exhibit striking differences.

Epipelon was measured along a fixed transect in the SE part of the lake. Temporal changes were great: at 5 m depth the dry weight varied from about 30 $g.m^{-2}$ in winter to almost 1 900 $g.m^{-2}$ in summer 1981.

In general the epipelon increased with depth. There are indications that epipelon is transported and redeposited along the steep slopes of the lake bottom. Epipelon collected from opposite lake sides differed in chemical composition. The percentages of organic matter and $CaCO_3$ in the epipelon were considerably lower than in the case of epiphyton; also, the energy content of epipelic organic matter was lower.

Epipelic algae comprised mainly pennate diatoms and blue-green algae of the same species as in the epiphyton. Compared with the epiphyton, the epipelon had low numbers of filamentous greens, but more blue-green algae and representatives of the planktonic species. Weight specific densities of algae in epipelon in summer were only a fifth of those in epiphyton. The chl *a* content in the epipelic and epiphytic organic matter was similar in winter, but, contrary to that in the epiphyton, the epipelic chlorophyll remained low in summer.

The epiphyton on *Ceratophyllum* apices received maximally 75% of the surface irradiance. However, for the greater part of epiphyton the light was low, and in dense vegetation part of the epiphyton was virtually in the dark. For the epipelon light was low as well, ranging from 1 to 8% of the surface irradiance at 5 m depth. In summer, the gradients in oxygen concentration and pH were sharp, particularly in the parts with submerged vegetation. In general these gradients indicated the predominance in the lower littoral of heterotrophic processes.

Artificial substrates were exposed at four depths ranging from 0.5 to 4 m. That light was a key factor was demonstrated by both the inhibition of growth near the surface and limitation in the deeper part. In July, however, little variation with depth was found. Filamentous green algae were more abundant at the lesser depths, while diatoms, mostly species other than on the natural substrates, were more numerous at greater depth. The periphyton from the artificial substrates differed both quantitatively and qualitatively from the epiphyton and epipelon on the natural substrates.

The light gradient in dense stands of *Ceratophyllum* caused marked changes in the composition of the epiphyton. In the lower parts of the plant bed, the organic matter and $CaCO_3$ contents of epiphyton relative to ash were low, and the pigments were almost all chlorophyllous breakdown products. The epiphyton of the upper parts exhibited a fairly high photosynthetic capacity, but the respiration was also intense, so that oxygen consumption exceeded its production in the vegetation over the 24 hours. This can be attributed to import of organic matter.

The maximum rate of gross photosynthesis per unit organic matter (p_g^{max}) of epiphyton ranged from 0.3 in winter to 8 mg $O_2.g^{-1}.h^{-1}$ in summer. For epipelon the maximum was 5 mg $O_2.g^{-1}.h^{-1}$. The maximum of specific respiration rate in the dark (r) was 5 for the epiphyton, and 2 mg $O_2.g^{-1}.h^{-1}$ for the epipelon. Except for a relatively brief period in summer, the specific photosynthesis and respiration rates of epiphyton and epipelon were comparable. The p_g^{max}: r ratios indicate that only in very favourable conditions of light does diel photosynthesis exceed diel respiration.

In conclusion, the redistribution processes mainly determine the changes in organic carbon of epiphyton and epipelon. Although integrated photosynthesis may reach high values in the lower littoral of Lake Vechten, the heterotrophic utilization of matter produced elsewhere is more important. The parts covered by vegetation are especially effective in collecting and retaining sedimenting matter, which is thus diverted from anaerobic decomposition in the deeper strata.

Acknowledgements

Without the assistance of Ronny van Keulen – under water and above – in the laboratory and at the drawing board, the work program of 1980 and 1981 would have been much 'diluted' in time and space. To my colleagues in Nieuwersluis thanks are due for their supply of background information and eye-openers. P. H. Best is thanked for introducing me to the littoral and for lots of samples. H. Roon provided unerring transport and assistance in the field. H. J. Korthals and G. Wiegers are thanked for pigment chromatography; M. D. M. Trommel, Th. H. Frank and A. G. Frank-Landman for determination and counting algae; E. M. J. Dekkers for all the CNH-analyses; J. H. A. Dassen and H. Visser for data on the oxygen in the limnetic region (Fig. 21); N. G. Hafkamp and his crew for the construction of the epipelon sampler and assemblages for the artificial substrates; E. M. Mariën for photography and J. H. M. van Kleinwee for typing. Acknowledgements are due to P. Roos (Department of Aquatic Ecology, University of Amsterdam) for the scanning electron photomicrographs. R. D. Gulati and S. Parma are thanked for many improvements on the manuscript. And thank you, Jeanet.

References

Admiraal, W., 1977. Salinity tolerance of benthic estuarine diatoms as tested with a rapid polarographic measurement of photosynthesis. Mar. Biol. 39: 11–18.

Admiraal, W. & Peletier, H., 1979. Influence of organic compounds and light limitation on the growth of estuarine benthic diatoms. Br. phycol. J. 14: 197–206.

Allanson, B. R., 1973. The fine structure of the periphyton of Chara sp. and Potamogeton natans from Wytham Pond, Oxford, and its significance to the macrophyte-periphyton metabolic model of R. G. Wetzel and H. L. Allen. Freshwat. Biol. 3: 535–541.

Allen, H. L., 1971. Primary productivity, chemo-organotrophy and nutritional interactions of epiphytic algae and bacteria on macrophytes in the littoral of a lake. Ecol. Monogr. 41: 97–127.

Best, P. H., 1979. Ecophysiological studies on growth and development of the aquatic macrophyte Ceratophyllum demersum L. Thesis, State University of Utrecht. 104 pp.

Best, P. H., 1982. The aquatic macrophytes of Lake Vechten. Species composition, spatial distribution and production. Hydrobiologia 95: 65–77.

Blaauboer, M. C. I., 1982. The phytoplankton species composition and the seasonal periodicity in Lake Vechten from 1956 to 1979. Hydrobiologia 95: 25–36.

Buscemi, P. A., 1958. Littoral oxygen depletion produced by a cover of Elodea canadensis. Oikos 9: 239–245.

Cappenberg, Th. E. & Verdouw, H., 1982. Sedimentation and breakdown kinetics of organic matter in the anaerobic zone of Lake Vechten. Hydrobiologia 95: 165–179.

Castenholz, R. W., 1960. Seasonal changes in the attached algae of freshwater and saline lakes in the lower Grand Coulee, Washington. Limnol. Oceanogr. 5: 1–28.

Cattaneo, A., 1978. The microdistribution of epiphytes on the leaves of natural and artificial macrophytes. Br. phycol. J. 13: 183–188.

Cattaneo, A., Ghittori, S. & Vendegra, V., 1975. The development of benthic phytocoenosis on artificial substrates in the Ticino River. Oecologia 19: 315–327.

Cattaneo, A. & Kalff, J., 1980. The relative contribution of aquatic macrophytes and their epiphytes to the production of macrophyte beds. Limnol. Oceanogr. 25: 280–289.

Cholnoky, B. J., 1968. Die Ökologie der Diatomeen in Binnengewässern. Cramer, Lehre, 699 pp.

Dvořák, J. & Best, P. H., 1982. Macro-invertebrate communities associated with water vegetation in Lake Vechten. Hydrobiologia 95: 115–126.

Eminson, D. & Phillips, G., 1978. A laboratory experiment to examine the effects of nutrient enrichment on macrophyte and epiphyte growth. Verh. int. Verein. Limnol. 20: 82–87.

Fitzgerald, G. P., 1969. Some factors in the competition or antagonism between bacteria, algae and aquatic weeds. J. Phycol. 5: 351–359.

Gons, H. J. & Mur, L. R., 1980. Energy requirements for growth and maintenance of Scenedesmus protuberans Fritsch in light-limited continuous cultures. Arch. Microbiol. 125: 9–17.

Gulati, R. D., Siewertsen, K. & Postema, G., 1982. The zooplankton: its community structure, food and feeding, and role in the ecosystem of Lake Vechten. Hydrobiologia 95: 127–163.

Hager, A. & Stransky, H., 1970a. Das Carotinoidmuster und die Verbreitung des lichtinduzierten Xantophyll-cyclus in verschiedenen Algenklassen. III. Grünalgen. Arch. Mikrobiol. 72: 68–83.

Hager, A. & Stransky, H., 1970b. Das Carotinoidmuster und die Verbreitung des lichtinduzierten Xanthophyll-cyclus in verschiedenen Algenklassen. V. Einzelne Vertreter der Cryptophyceae, Euglenophyceae, Bacillariophyceae, Chrysophyceae und Phaeophyceae. Arch. Mikrobiol. 73: 77–89.

Hallegraeff, G. M., 1976. Pigment diversity, biomass and species diversity of phytoplankton of three Dutch lakes. Thesis, University of Amsterdam. 177 pp.

Harris, G. P., 1978. Photosynthesis, productivity and growth: The physiological ecology of phytoplankton. Arch. Hydrobiol. Beih. 10: 1–171.

Hellebust, J. A. & Lewin, J., 1977. Heterotrophic nutrition. In: Werner, D. (Ed.) The Biology of Diatoms. Blackwell, Oxford. 288 pp.

Hoek, C. van den, 1978. Algen. Einführung in die Phykologie. Thieme, Stuttgart. 481 pp.

Hooper, N. M. & Robinson, G. G. C., 1976. Primary production of epiphytic algae in a marsh pond. Can. J. Bot. 54: 2810–2815.

Hooper-Reid, N. M. & Robinson, G. G. C., 1978. Seasonal dynamics of epiphytic algal growth in a marsh pond: productivity, standing crop, and community composition. Can. J. Bot. 56: 2434–2440.

Hutchinson, G. E., 1975. A Treatise on Limnology. III. Limnological Botany. Wiley, New York. 660 pp.

Iwamura, T., Nagai, H. & Ichimura, S., 1970. Improved methods for determining contents of chlorophyll, protein, ribonucleic acids and deoxyribonucleic acid in planktonic populations. Int. Rev. ges. Hydrobiol. 55: 131–147.

Jeffrey, S. W., 1961. Paper chromatographic separation of chlorophylls and carotenoids from marine algae. Biochem. J. 80: 336–342.

Jeffrey, S. W., 1968. Quantitative thin-layer chromatography of chlorophylls and carotenoids from marine algae. Biochim. Biophys. Acta 162: 271–285.

Jensen, A. & Sakshaug, E., 1973. Studies on the phytoplankton ecology of the Trondheimsfjord. II. Chloroplast pigments in relation to abundance and physiological state of the phytoplankton. J. exp. mar. Biol. Ecol. 11: 137–155.

Jones, R. C., 1980. Primary production, biomass, nutrient limitation, and taxonomic composition of algal communities epiphytic on the submersed macrophyte Myriophyllum spicatum L. in a hardwater, eutrophic lake. Thesis, University of Wisconsin-Madison. 200 pp.

Jørgensen, E. G., 1969. The adaptation of plankton algae. IV. Light adaptation in different algal species. Physiol. Plant. 22: 1307–1315.

Kajak, Z., Hillbricht-Ilkowska, A. & Pieczyńska, E., 1972. The production in several Polish lakes. In: Kajak, Z. & Hillbricht-Ilkowska, A. (Eds.) Productivity Problems of Freshwaters. PWN Warszawa, 918 pp.

Kloet, W. A. de, 1982. The primary production of phytoplankton in Lake Vechten. Hydrobiologia 95: 37–57.

Madsen, B. L., 1972. Detritus on stones in small streams. Mem. Ist. ital. Idrobiol. 29: 385–403.

Mann, K. H., 1975. Patterns of energy flow. In: Whitton, B. A. (Ed.) River Ecology. Blackwell, Oxford. 725 pp.

Mason, C. F. & Bryant, R. J., 1975. Periphyton production and grazing by chironomids in Alderfen Broad, Norfolk. Freshwat. Biol. 5: 271–277.

McMahon, R. F., Douglas-Hunter, R. & Russell-Hunter, W. D., 1974. Variation in Aufwuchs at six freshwater habitats in terms of carbon: nitrogen ratio. Hydrobiologia 45: 391–404.

Milner, H. W., 1953. The chemical composition of algae. In: Burlew, J. S. (Ed.) Algal Culture from Laboratory to Pilot Plant. Carnegie Inst., Washington, D. C. 357 pp.

Moss, B., 1968. The chlorophyll a content of some benthic algal communities. Arch. Hydrobiol. 65: 51–62.

Odum, E. P., 1971. Fundamentals of Ecology. Saunders, Philadelphia. 574 pp.

Parsons, T. R., Takahashi, M. & Hargrave, B., 1977. Biological Oceanographic Processes. Pergamon, Oxford. 332 pp.

Perkins, M. A. & Kaplan, L. A., 1978. Epilithic periphyton and detritus studies in a subalpine stream. Hydrobiologia 57: 103–109.

Phillips, G. L., Eminson, D. & Moss, B., 1978. A mechanism to account for macrophyte decline in progressively eutrophicated freshwaters. Aquat. Bot. 4: 103–126.

Pieczyńska, E., 1975. Ecological interactions between land and the littoral zones of lakes (Masurian Lakeland, Poland). In: Hasler, A. D. (Ed.) Coupling of Land and Water Systems. Springer, Berlin. 309 pp.

Round, F. E., 1973. The Biology of the Algae. Edward Arnold, London. 278 pp.

Russell-Hunter, W. D., 1970. Aquatic Productivity. Macmillan, New York. 306 pp.

Saunders, G. W., Cummins, K. C. & Gak, D. Z. et al., 1980. Organic matter and decomposers. In: Le Cren, E. D. & Lowe-McConnell, R. H. (Eds.) The Functioning of Freshwater Ecosystems. Cambridge University Press, Cambridge. 588 pp.

Siver, P. A., 1977. Comparison of attached diatom communities on natural and artificial substrates. J. Phycol. 13: 402–406.

Sládečková, A., 1962. Limnological investigation methods for the periphyton ('Aufwuchs') community. Bot. Rev. 28: 286–350.

Sládeček, V. & Sládečková, A., 1964. Determination of periphyton production by means of the glass slide method. Hydrobiologia 23: 125–128.

Spodniewska, I., Pieczyńska, E. & Kowalczewski, A., 1975. Ecosystem of the Mikolajskie Lake, primary production. Pol. Arch. Hydrobiol. 22: 17–37.

Spoehr, H. A. & Milner, H. W., 1949. The chemical composition of Chlorella: effects of envirionmental conditions. Plant Physiol. 24: 120–149.

Steemann Nielsen, E., 1975. Marine Photosynthesis, with Special Emphasis on the Ecological Aspects. Elsevier, Amsterdam. 141 pp.

Steenbergen, C. L. M. & Verdouw, H., 1982. Lake Vechten: aspects of its morphometry, climate, hydrology and physicochemical characteristics. Hydrobiologia 95: 11–23.

Stockner, J. G. & Armstrong, F. A. J., 1971. Periphyton of the experimental lakes area, Northwestern Ontario. J. Fish. Res. Bd Can. 28: 215–229.

Strathmann, R. R., 1967. Estimating the organic carbon content of phytoplankton from cell volume or plasma volume. Limnol. Oceanogr. 12: 411–418.

Tippett, R., 1970. Artificial surfaces as a method of studying populations of benthic microalgae in freshwater. Br. phycol. J. 5: 187–199.

Verdouw, H. & Dekkers, E. M. J., 1982. Nitrogen cycle of Lake Vechten (The Netherlands); role of sedimentation. Arch. Hydrobiol. 94: 251–263.

Weber, C., 1973. Research development in the measurement of the response of plankton and periphyton to changes in their environment. In: Glass, G. E. (Ed.) Bioassay Technique and Environmental Chemistry. Ann Arbor Science Publ. 499 pp.

Westlake, D. F., Adams, M. S. & Bindloss, M. E. et al., 1980. Primary production. In: Le Cren, E. D. & Lowe-McConnell, R. H. (Eds.) The Functioning of Freshwater Ecosystems. Cambridge University Press, Cambridge. 588 pp.

Wetzel, R. G., 1975. Limnology. Saunders, Philadelphia. 743 pp.

Wetzel, R. G. & Allen, H. L., 1972. Functions and interactions of dissolved organic matter and the littoral zone in lake metabolism and eutrophication. In: Kajak, Z. & Hillbricht-Ilkowska, A. (Eds.) Productivity Problems of Freshwaters. PWN, Warszawa. 918 pp.

Wetzel, R. G., Rich, P. H., Miller, M. C. & Allen, H. L., 1972. Metabolism of dissolved and particulate detrital carbon in a temperate hard-water lake. Mem. Ist. ital. Idrobiol. 29: 185–243.

Macro-invertebrate communities associated with the macrophytes of Lake Vechten: structural and functional relationships

Jan Dvořak[1] & Elly P. H. Best[2]
[1] *Institute of Botany, Czechoslovakian Academy of Sciences, 25243 Průhonice u. Prahy, Czechoslovakia*
[2] *Limnological Institute, 'Vijverhof' Laboratory, Rijksstraatweg 6, 3631 AC Nieuwersluis, The Netherlands*

Keywords: macro-invertebrates, macrophytes, species inventory, trophic relationships, standing crop

Abstract

The macro-invertebrates associated with eight species of emergent, floating-leaved and submerged macrophytes were recorded in May–June 1979. The plant species were: *Phragmites australis, Sparganium erectum, Typha angustifolia, Polygonum amphibium, Ceratophyllum demersum, Elodea* sp., *Myriophyllum spicatum* and *Nitella mucronata.*

All macrophytes had a high number of macro-invertebrate species in common; however, some animals were also species-specific, particularly those occurring on the emergent plants. As regards the food source and feeding mechanism, the detritivores, periphytiscrapers and omnivores predominated, followed by predatory carnivores. The seston-filtrators also form a significant part of the community biomass. The macrofaunal distribution depended strongly on colonizable plant surface area and vegetation density. The macrofaunal communities of the *Elodea, Ceratophyllum* and *Phragmites* stands contributed mainly to the lake's carbon cycle. Being poorly represented by herbivores, the macro-invertebrate communities will affect the lake's primary production only slightly.

Introduction

In Lake Vechten, The Netherlands, integrated studies are in progress on the biological characteristics, energy flow and nutrient recycling (Best *et al.* 1978). In this context much attention has been paid to the primary producers (Best 1982; Blaauboer 1982; Gons 1982; De Kloet 1982), zooplankton (Gulati *et al.* 1982) and bacteria (Steenbergen 1982). On the macrofauna only limited knowledge exists. A preliminary inventory of the bottom fauna was made and an autecological study was carried out on the formerly most important benthic larvae of *Chaoborus flavicans* (Parma 1971). As to the dwellers of the littoral zone, no information is available so far.

Therefore a study was started on the structure and size of the invertebrate communities of Lake Vechten. This paper reports the data on the macrophyte-associated invertebrates. The bottom fauna, which was far less rich, will be described elsewhere. The littoral zone, i.e. the area occupied by macrophytes, covered at the end of 1978 about 19% of the total lake area (Best, unpubl.). The study pertains to only a short period, i.e. May–June 1979. However, since it is a time of high production rates, both for plants and invertebrate littoral communities, the main lines are likely to become obvious.

Material and methods

Sampling

The physico-chemical characteristics and morphometry of Lake Vechten are described by Steenbergen & Verdouw (1982).

The macrophyte-associated invertebrates were

Hydrobiologia 95, 115–126 (1982). 0018–8158/82/0951–0115/$02.40.

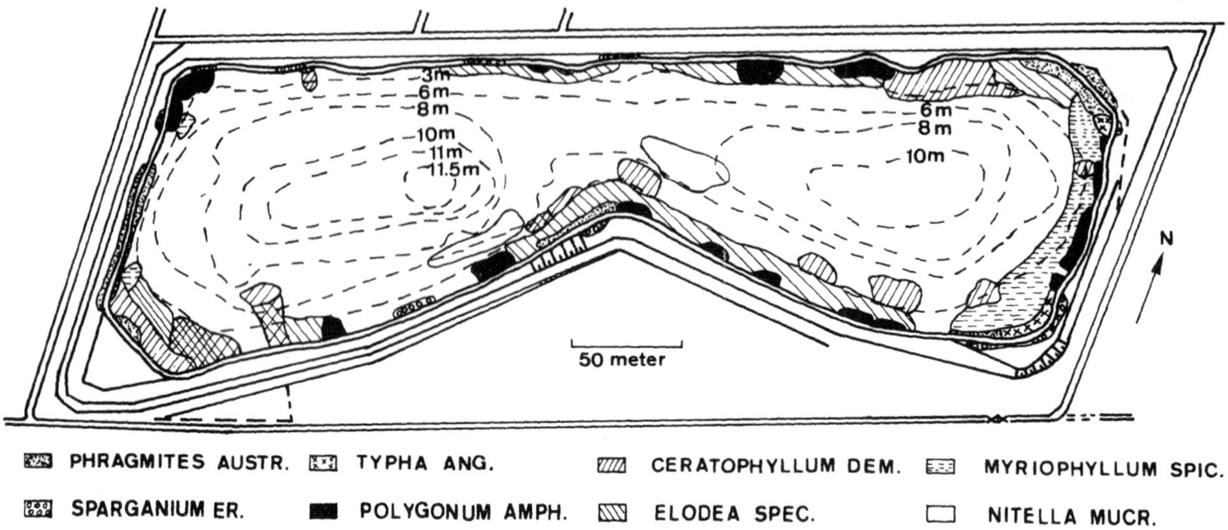

Fig. 1. Schematic representation of the main vegetation types of Lake Vechten at the end of 1978. As criterion for a vegetation type, a cover percentage higher than 10% was used.

collected from the macrophyte samples. For this all macrophytic species which formed a distinct vegetation were sampled (see Fig. 1). These were: *Phragmites australis* (Cav.) Trin. ex Steud., *Sparganium erectum* L., *Typha angustifolia* L., *Polygonum amphibium* L., *Ceratophyllum demersum* L., *Elodea* sp. (mixture of *E. canadensis* Michx., and *E. nuttallii* (Planck) St. John), *Myriophyllum spicatum* L. and *Nitella mucronata* (A. Br.) Miq. (f Leiopitys (Wheld.) R.D.W.).

Most macrophytic species including the emergent and floating-leaved plants were sampled using SCUBA diving since they were rooted fairly deeply in the water. The rootstocks of the emergent and floating-leaved plants were first separated from the main system and subsequently the lower plant parts were caught in a specially constructed handnet (Higler 1981) and brought ashore. Thus, losses of macrofauna were minimized. The submerged plants were collected from the lake bottom in plastic bags. For *Ceratophyllum* sp. three depth classes were distinguished, rooting at depths ranging from surface to 1 (I), 1 to 2.5 (II) and from 2.5 to 4 m (III), respectively. Only *Sparganium* was sampled from the lake shore.

Sampling was carried out three times. Every time a substantial quantity of plant biomass was taken. The first and third samplings on 11 May and 21 June 1979, respectively, included all the plant species. The main purpose of the second sampling (6 June) was to distinguish invertebrate communities in dead and living shoot material of the emergent plants. *Sparganium* was only sampled in June. On 21 June the plants were sampled quantitatively, i.e. quadrats of 0.25×0.25 m were harvested from the different vegetation types. The other plant samples were collected only qualitatively (5 May and 6 June).

Sample treatment

The plant material was immersed in 4% formalin solution, which was changed several times. The animals were removed from the formalin solution using a metal sieve (mesh size 0.5 mm).

The plant material was dried at 70 °C until constant weight; ash was determined at 550 °C.

All animals were preserved in 4% formalin at 5 °C prior to identification. In a few cases, some specimens were cultivated for identification.

Invertebrate biomass was determined as formalin ash-free dry weight. For this the animals were handled like the plant material.

Results and discussion

The invertebrate communities were analyzed qualitatively as well as quantitatively. They were

Table 1. The predominant macro-invertebrate species characteristic for the macrophytes of Lake Vechten.

CNIDARIA
- *Hydra* sp.

MOLLUSCA
- *Acroloxus lacustris* L.
- *Armiger crista* L.
- *Dreissena polymorpha* (Pallas)
- *Gyraulus albus* (O. F. Müller)

OLIGOCHAETA
- *Limnodrilus hoffmeisteri* Claparede
- *Ophidonais serpentina* (Müller)
- *Stylaria lacustris* L.

INSECTA

EPHEMEROPTERA
- *Caenis horaria* L.
- *C. moesta* Bengtsson

TRICHOPTERA
- *Cyrnus flavidus* Mac Lachl.
- *Cyrnus crenaticornis* (Kolenati)
- *Ecnomus tenellus* (Ramb.)
- *Oxyethira flavicornis* (Pict.)
- *Oxyethira* sp.

DIPTERA

Chironomidae
- *Ablabesmyia monilis* L.
- *A. longistyla* Fittkau
- *Endochironomus albipennis* (Meig.)
- *Glyptotendipes* sp.
- *Parachironomus gr. arcuatus*
- *Phytotendipes gripekoveni* Kieff

Ceratopogonidae
- *Bezzia* sp.
- *Palpomyia lineata* (Meig.)

classified into different groups according to species structure, trophic role and biomass.

Specific structure

The majority of the predominant invertebrate species occurred frequently on most of the plant species, but some showed preference for certain plant species. Thus, in spite of some specificity in distribution, these invertebrates can be considered as characteristic for the whole macrophytic vegetation of Lake Vechten. They are listed in Table 1.

In addition, several invertebrates occurred in smaller numbers specifically on some emergent plants and *Polygonum*. Specific for both *Phragmites* and *Typha* was *Potamopyrgus jenkinsi* Smith (Gastropoda, Prosobranchiata). The caddisfly larvae *Tinodes waeneri* L. and *Lype phaeopta* (Steph.) occurred on *Phragmites, Typha* and *Polygonum*. Caterpillars or pupae of *Nonagria* sp. (Insecta, Lepidoptera, Noctuidae) were found only on *Phragmites* on 11 May.

The remaining species occurred only in low numbers and they did not substantially influence the structure and diversity of the macrofauna associated with the plants. Some of them, however, notably those belonging to the Hirudinea and Odonata, affected the biomass of the macrofauna considerably. This remaining group is composed of representatives of Turbellaria, Hirudinea, Hydracarina, Insecta (Odonata, Hemiptera, Lepidoptera, Coleoptera) and the rest of the Mollusca, Oligochaeta and Insecta (Trichoptera and Chironomidae).

The structure of the invertebrate communities is described by a) the share of the characteristic groups of animals in the total abundance and b) the predominant species. Thus, similarities and differences between the communities associated with different plant species become apparent.

A rather high degree of similarity was found in the communities of the emergent *Phragmites* and *Typha,* and the floating-leaved *Polygonum* (Table 2). The macrofauna consisted mainly of Mollusca, Oligochaeta, Ephemeroptera, Trichoptera, Chironomidae and Ceratopogonidae. The main characteristics were the high numbers of Ceratopogonidae, particularly in June, scarcity of mayfly nymphs and the decrease in abundance of the Chironomidae in June.

A quite different community structure occurred on the emergent *Sparganium* sp. (Table 2). Its composition was similar to that on the submerged *Myriophyllum* in June (Table 4). The Chironomidae contributed 56% to the total numbers, followed by Oligochaeta (12.5%) and Mollusca (5.4%). Rather poor were the Ceratopogonidae, caddis larvae and mayfly nymphs.

The macrofaunal community of *Ceratophyllum* was investigated most thoroughly. Here zonation and variability in fauna of one single plant species growing at three different depths could be studied (Table 3). The macrofauna of all depth classes of *Ceratophyllum* was characterized by the relatively constant high share of Oligochaeta and Chironomidae, variable occurrence of mayflies and by the very low numbers of Palpomyiinae. Some specific differences were found between the communities of *Ceratophyllum* I and those of *Ceratophyllum* II

RELATIVE CONTRIBUTION TO TOTAL ABUNDANCE (%)	PHAGMITES, TYPHA, POLYGONUM: MAY (scale 0–100)	PHAGMITES, TYPHA, POLYGONUM: JUNE	SPARGANIUM: JUNE
MOLLUSCA	→	→	→
Acroloxus lacustris	•	•	
Bythinia tentaculata L.			•
Dreissena polymorpha	•	•	
Gyraulus albus			
Physa fontinalis L.			•
Potamopyrgus Jenkinsi		•	
OLIGOCHAETA	→	→	→
Limnodrilus hoffmeisteri			•
Stylaria lacustris	•	•	•
EPHEMEROPTERA	→	→	→
Caenis horaria	•	•	•
TRICHOPTERA	→	→	→
Cyrnus flavidus	•	•	
Ecnomus tenellus	•	•	
Lype phaeopta (Steph)	•	•	
Mystacides longicornis			•
Oxyethira sp.	•	•	•
Tinodes waeneri	•	•	
DIPTERA			
CHIRONOMIDAE	→	→	→
Endochironomus albipennis	•		•
Glyptotendipes sp.			•
Phytotendipes gripekoveni	•		
CERATOPOGONIDAE	→	→	→
Bezzia sp. diversa		•	•
Palpomyia lineata	•		

Table 2. Structure characterization of the macro-invertebrate communities associated with the emergent and floating-leaved macrophytes of Lake Vechten. Parameters are a) the relative contribution to the total abundance of the characteristic macro-invertebrate groups, and b) the predominant species (occurrence marked by •). Zero values for abundance are omitted.

RELATIVE CONTRIBUTION TO TOTAL ABUNDANCE (%)	CERATOPHYLLUM I: MAY (scale 0–100)	I: JUNE	II: MAY	II: JUNE	III: MAY	III: JUNE
CNIDARIA						→
Hydra sp.						•
MOLLUSCA	→	→	→	→	→	→
Armiger crista		•		•		•
Dreissena polymorpha	•	•	•	•	•	•
Gyraulus albus		•		•		•
OLIGOCHAETA	→	→	→	→	→	→
Limnodrilus hoffmeisteri	•		•	•	•	•
Ophidonais serpentina						
Stylaria lacustris	•	•		•		•
EPHEMEROPTERA	→	→	→	→	→	→
Caenis horaria	•		•	•	•	•
C. moesta Bengts.		•		•		•
C. robusta Eaton						
TRICHOPTERA	→	→	→	→	→	→
Cyrnus flavidus	•		•	•	•	•
Cyrnus crenaticormis		•		•		•
Oxyethira sp.						
DIPTERA						
CHIRONOMIDAE	→	→	→	→	→	→
Endochironomus albipennis	•	•	•	•	•	•
Parachironomus gr. arcuatus						•
CERATOPOGONIDAE	→	→	→	→	→	
Bezzia sp. diversa		•			•	
Palpomyia lineata	•		•	•		

Table 3. Structure characterization of the macro-invertebrate communities associated with the submerged *Ceratophyllum* of Lake Vechten. Parameters are a) the relative contribution to the total abundance of the characteristic macro-invertebrate groups, and b) the predominant species (occurrence marked by •). Zero values for abundance are omitted.

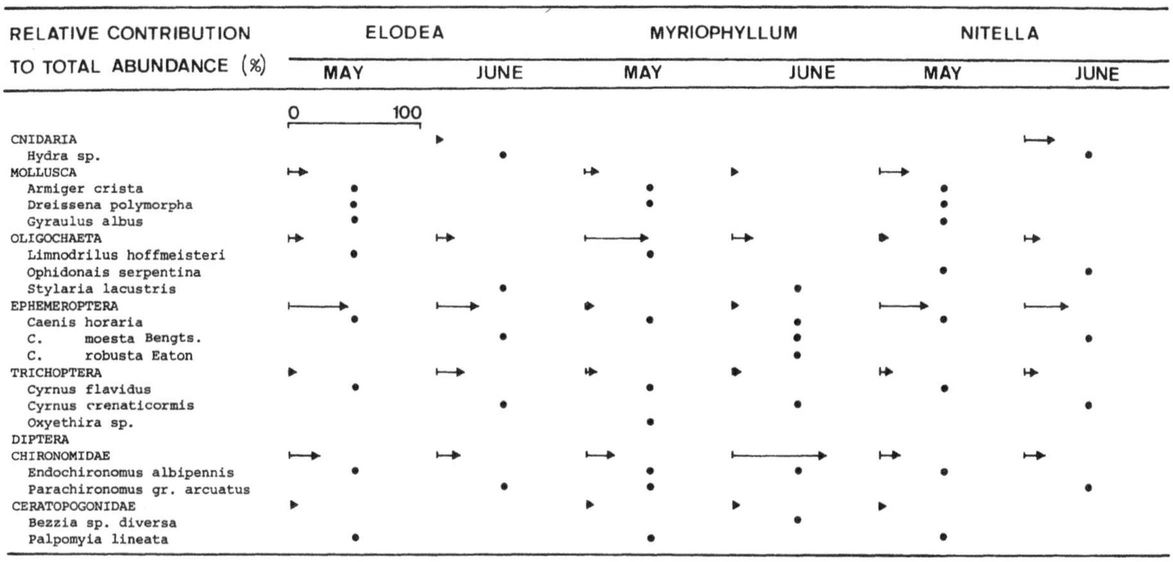

Table 4. Structure characterization of the macro-invertebrate communities associated with submerged macrophytes of Lake Vechten. Parameters are a) the relative contribution to the total abundance of the characteristic macro-invertebrate groups, and b) the predominant species (occurrence marked by ●). Zero values for abundance are omitted.

and III. In May the macrofauna of *Ceratophyllum* I was rather similar to that of *Elodea*; both these species occurred in about the same depth range and were dominated by the Chironomidae and Ephemeroptera, followed by molluscs and caddis larvae. The structure of macrofauna on *Ceratophyllum* II and III was similar to that on *Myriophyllum; Myriophyllum* occurs normally at a rooting depth of 2.5 m. The communities of both *Ceratophyllum* II and III depth classes were characterized by high shares of Chironomidae and Oligochaeta, followed by Mollusca and Trichoptera. In June the communities of the three depth classes of *Ceratophyllum* were similar, in that the animals colonizing consisted mainly of Oligochaeta, followed by Ephemeroptera and caddis larvae. However, the share of the chironomids to the community of *Ceratophyllum* I was far less than that of *Ceratophyllum* II and III. The depth class I had the highest number of Palpomyiinae of all submerged plant species.

The invertebrate communities of *Elodea* and *Nitella* were similar to a certain degree (Table 4). Both these plants were characterized by the high proportions of mayfly nymphs and chironomid larvae and the scarcity or absence of Palpomyiinae. In May the mayfly nymphs prevailed on both plants, followed by Mollusca, Chironomidae, and Trichoptera and Oligochaeta. In June the mayfly nymphs still prevailed, but caddis larvae became abundant, followed by Chironomidae and Oligochaeta. The *Nitella* community differed from that of *Elodea* in having a high percentage of *Hydra* sp.

The macrofaunal community on *Myriophyllum* was characterized by relatively high numbers of Chironomidae and Oligochaeta and a very low percentage of mayflies and Palpomyiinae. In May the Oligochaeta predominated, followed by chironomid larvae, Mollusca and caddis larvae: mayflies and Palpomyiinae formed a minority. The structure was rather similar to that of the community on *Ceratophyllum* II and III. In June the Chironomidae prevailed strongly, followed by Oligochaeta, Trichoptera and Ephemeroptera.

The macrofaunal structure in May and June differed; this was caused evidently by the succession in generations of the predominant invertebrate species. In May the macrofauna consisted partly of representatives of the overwintering generations, particularly of species having only two or three generations per year, e.g. *Endochironomus albipennis, Phytotendipes gripekoveni, Cyrnus flavidus,* etc. In the samples of 21 June new summer generations were encountered. On this day diversity of the macrofauna on the various macrophytes was high because of the differentiation in ecological conditions. In a similar study on the seasonal changes in the structure of macrofauna on submerged plants, an even greater species diversity than in June was found in autumn (Higler 1981). Since autumn samples were lacking in the present study a similar comparison could not be made.

Trophic structure

The macrophyte-associated macrofauna in the ecosystem was grouped according to the food source and feeding mechanism into trophic categories (Cummins 1973; Higler 1977; Dvořák *et al.* 1982) corresponding roughly to various trophic levels and representing a means by which the primary production in the ecosystem can be used. These categories are:

1) Species directly consuming living plant tissue or, occasionally, dead plants (DC). These are typical herbivores, also called shredders and grazers.
2) Miners, consuming partly vascular plant tissue and partly detritus and algae (M). This group consists of herbivores and detritivores.
3) Detritivores including collectors (D), periphytiscrapers (P) and species tending to omnivory (O). These animals collect their food from the plant surface.
4) Filtrators (F), feeding on fine detritus (seston) gained from the water around the macrophytes like filter-feeding zooplankton.
5) Predating carnivores (C_{pred}), feeding on various invertebrates associated with macrophytes and on zooplankton.
6) Parasitizing carnivores (C_{par}), leeches living on fishes, frogs, etc.

Figures 2 and 3, giving the distribution of the macrofaunal trophic categories on the macrophytes, are used as follows to evaluate the trophic structure.

The direct consumers (DC) occurred in 60% of the samples, being represented by caterpillars of *Nymphula nymphaeata* L., *Cataclysta lemnata* L., *Nonagria* sp., by larvae of weevils (on *Myriophyllum* only) and by *Donacia* sp. They contributed 0.6% (range 0.2–1.2%), to the total abundance and 3.5% (range 0.02–15.8%) to the total macrofaunal biomass. The values in June were mostly lower than those in May. They were absent on *Nitella.*

The miners (M) occurred in 90% of the samples, being represented mainly by larvae of the chironomids *Phytotendipes gripekoveni, Glyptotendipes* sp., *Endochironomus tendens* (Fabr.) and by *Limnochironomus* spp. The numbers of miners

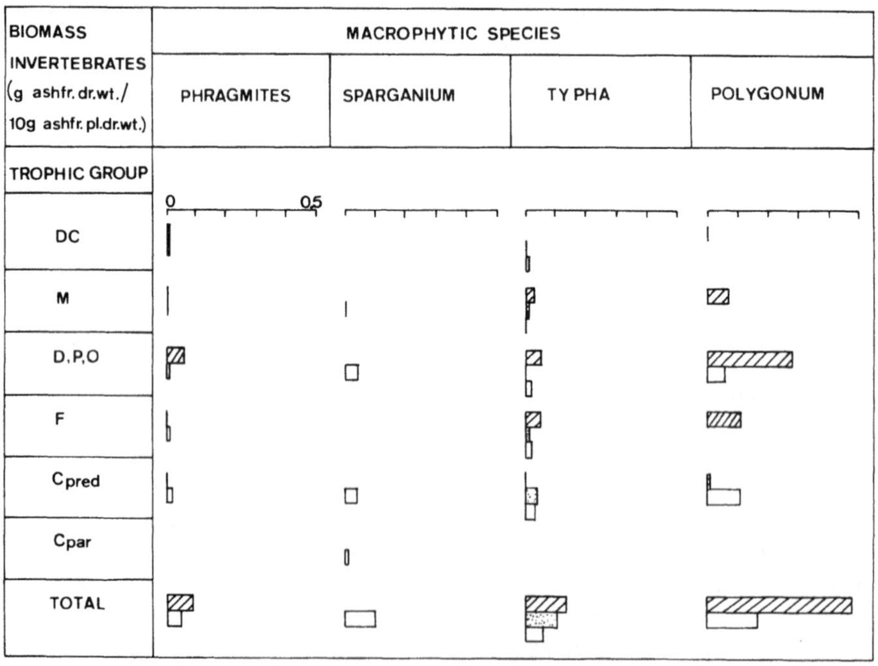

Fig. 2. Distribution of the macro-invertebrate trophic groups on the emergent and floating-leaved macrophytic species in May and June. DC–direct plant consumers; M–miners; D, P, O–detritivores, periphytiscrapers, omnivores; F–filtrators; C_{pred}–predating carnivores; C_{par}–parasitizing carnivores. ▨ May; ⁚⁚ June–dead plant material. Zero values are omitted.

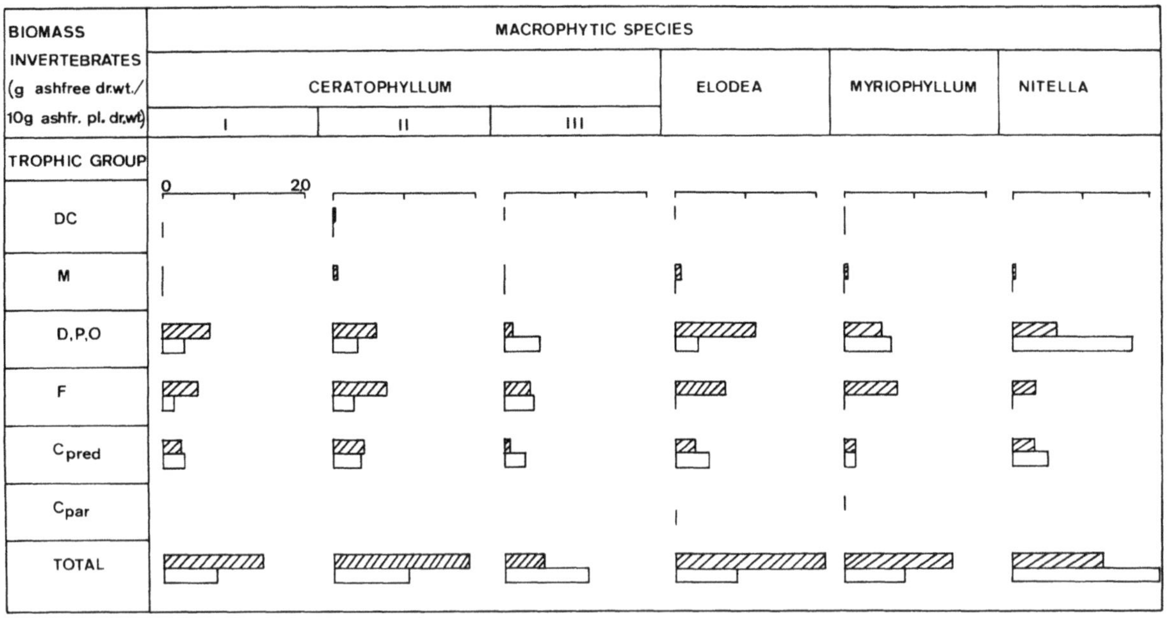

Fig. 3. Distribution of the macro-invertebrate trophic groups on the submerged macrophytic species in May and June. DC–direct plant consumers; M–miners; D, P, O–detritivores, periphytiscrapers, omnivores; F–filtrators; C_{pred}–predating carnivores; C_{par}–parasitizing carnivores. ▨ May; □ June. Zero values are omitted.

were highest on the emergent plants and *Polygonum,* which provide suitable conditions for the building of mines. Here the miners contributed 5.5% (range 0.2–15.8%) to the total numbers, and 7.9% (range 0.8–21.6%) to the total macrofaunal biomass. In the submerged plants the mining animals contributed only 1.5% (range 0.2–4.1%) to the total number and 1.2% (range 0.03–3.8%) to the total biomass.

The biomass of DC and M together was very low compared to that of their host plants, forming, respectively, 0.08–0.7% and 0.2–0.8% of the emergent and submerged plants. These values indicate a very low rate of direct utilization of the vascular plants by the invertebrates. This is supported by experimental evidence of Soszka (1975) who found a maximum macrofaunal consumption rate of living plant material of 10% of their own body weight per day (on a fresh weight basis). As to the miners, only part of their food consists of living plant material. Most DC and M species of Lake Vechten have low turnover rates, characterized by one or two generations per year (Dvořák *et al.* 1982).

The detritivores, periphytiscrapers and omnivores (D, P, O) occurred in all samples. They were represented by Oligochaeta, mayfly nymphs (mainly *Caenis*), non-predatory caddis larvae, Coleoptera and Chironomidae (mainly *Endochironomus albipennis*). In May D, P, O predominated all the trophic categories in most of the plants. They contributed 73.7% (range 60.8–83.8%) to the total numbers and 44.2% (21.9–68.9%) to the total biomass. From May to June, however, the detritivores and periphytiscrapers decreased substantially most of the plants, namely from 77.6 to 39.0% in numbers and from 51.9 to 26.6% in biomass. They occurred in relatively shallow water, notably on *Phragmites, Typha, Polygonum, Ceratophyllum* I and *Elodea; Sparganium* was the only exception in this respect. In contrast, on the deeper rooted submerged plants, *Ceratophyllum* II, III, *Myriophyllum* and *Nitella* in May to June, the numbers of D, P, O increased from 68.6 to 75.1% and the total biomass from 34.5 to 56.7%.

The filter-feeders (F) occurred in 90% of the samples. This category was represented by *Dreissena polymorpha* (Pallas) and *Pisidium* sp. These animals contributed only 3.7% (range 0.3–9.7%) to the total numbers but 27.5% (range 0.5–64.3%) to the total biomass.

The predatory carnivores (C_{pred}) also occurred in all the samples. They were represented by *Hydra* sp., Turbellaria (*Dugesia* sp.), Oligochaeta (*Chaetogaster diaphanus* Gruithuisen), Hirudinea (*Helobdella stagnalis* L., *Glossiphonia heteroclita* L.), Hydracarina, Odonata (*Ischnura elegans* v.d. Linden and Agrioninae), Trichoptera (*Cyrnus* sp., *Ecnomus tenellus* Ramb.) and Diptera (*Parachironomus* gr. *arcuatus* Goetgh, *Cryptochironomus* sp., *Ablabesmyia* sp., *Palpomyia* sp., *Bezzia* sp.). The food requirements of this category in Lake Vechten were covered mainly by zooplankton living within the vegetation and by minute invertebrates dwelling at the surface of the vegetation, e.g. Oligochaeta and some chironomid larvae. The share of these carnivores in the invertebrate communities of all the plants increased substantially from May to June. The contribution of the predatory carnivores to the total numbers increased from 16.7% (range 8.4–25. 6%) in May to 42.8% (range 13.2–86.7%) in June. As to the biomass, their share increased from 13.0% (range 3.9–24.6%) in May to 39.8% (range 20.8–69.7%) in June.

The parasitizing carnivores (C_{par}) occurred on *Sparganium, Elodea* and *Myriophyllum.* They were found in only 15% of the samples, and were represented by the leeches *Piscicola geometra* L. and *Hemiclepsis marginata* (O. F. Müller). These animals contributed only 0.8% (range 0.1–1.8%) to the total numbers and 4.2% (range 0.4–11.6%) to the total macrofaunal biomass.

The substantial changes in the trophic structure of the macrofauna were caused by two main factors: 1) the fluctuations in the numbers of the main representatives of the categories D, P, O and F, mostly due to changes in life cycle and migration; and 2) the strong development of new summer generations of carnivores which occurred on all plants.

Quantitative characteristics

The invertebrate communities were quantified both as numbers and as biomass. The biomass, expressed in ash-free dry weight, was related to the macrophytic biomass, the colonizable plant surface area and the vegetation area. Also a conservative carbon budget was made, based on data of June only.

These ranges in total number of animals associated with the macrophytes were roughly estimated. On an ash-free plant dry weight basis, the highest number of individuals occurred on the submerged plants. The numbers for all submerged plants, except *Nitella* sp., ranged from 800 to 6 700 per 10 g ash-free dry weight; *Nitella* sp. had the highest numbers of all plant species, namely 10 450 individuals per 10 g ash-free plant dry weight, and the emergent plants the lowest numbers with 140 to 300 ind.; the floating-leaved *Polygonum* sp. took an intermediate position with 700 to 1 050 ind.

A similar tendency as in animal numbers was observed also in the animal biomass expressed in ash-free dry weight per 10 g ash-free plant dry weight. The biomass characteristics of the invertebrate communities are depicted in Figs. 3 and 4.

The biomass ranged from 0.58 to 2.30 g in the submerged plants, 0.17 to 0.49 in *Polygonum* and 0.05 to 0.14 g in the emergent plants. Certain dis-

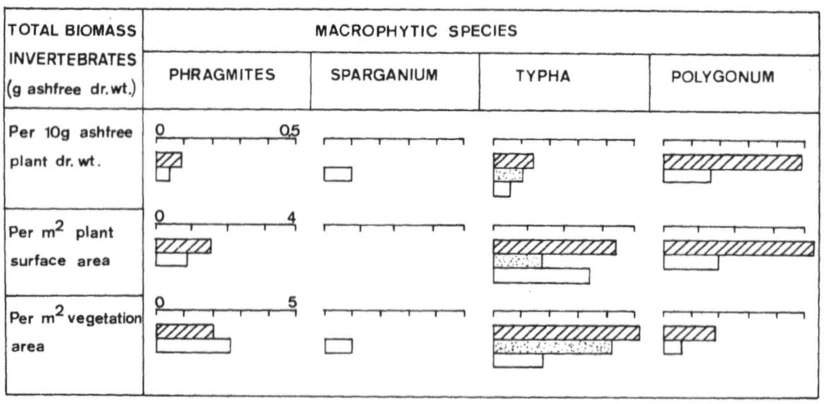

Fig. 4. The total biomass of macro-invertebrates on the emergent and floating-leaved macrophytes in May and June, per 10 g ash-free plant dry wt., m² plant surface area and m² vegetation area. [hatched] May; [open] June; [stippled] dead plant material June.

Table 5. The ratio plant surface area (m^2): biomass (ash-free dry wt.) in the macrophytes in June.

Plant species	Ratio
Phragmites	0.0055
Typha	0.0040
Polygonum	0.0113
Ceratophyllum	0.1206
Elodea	0.1333

crepancies between the animal numbers and biomass were caused by the mass appearance of small larvae of some insects (Palpomyiinae, the youngest instars of *Caenis, Cyrnus* and *Endochironomus)* and by the unequal distribution of some relatively heavy animals such as *Dreissena,* nymphs of the Agrionidae, and larvae of some Trichoptera and Diptera.

The quantitative differences in the macrofaunal communities depended strongly on the colonizable plant surface area of the various life forms of the aquatic vegetation. For most plant species the ratio between plant surface area (m^2) and biomass as ash-free dry weight was determined (see Table 5). The submerged plants had the highest ratio, the emergent species the lowest one, and *Polygonum* showed an intermediate value. The animal biomass of the macro-invertebrate communities of the emergent and floating-leaved plants was much lower on the basis of plant dry weight rather than plant surface area, pointing to a relatively large part of the animal community living inside the plants (Fig. 4). However, a similar tendency was observed in the animal communities of the submerged plants between animal biomass on the basis of plant dry weight and plant surface area (Fig. 5). In the latter case the quantity of macro-invertebrates was proportionally more related to the plant surface area than to plant biomass. Submerged plants with finely dissected or small and densely packed leaves offer better shelter and a more generous food source (periphyton) for the macrofauna than do the massive, sclerenchymatous and simply built emergent macrophytes. Similar relationships were found earlier by Krecker (1939) and Korinkova (1971).

When the macrofaunal biomass is expressed per m^2 vegetation area, a picture quite different from that based on ash-free plant dry weight and per m^2 surface area emerges (Figs. 4 and 5). The highest value was found for the dead *Typha* stems (5.23 g m^{-2} in May). This one was followed by *Polygonum, Phragmites* and *Ceratophyllum* (1.41–4.35 g m^{-2}) and *Sparganium, Polygonum, Elodea* and *Nitella* (0.58–1.67 g m^{-2}). The lowest value occurred in a *Myriophyllum* vegetation (0.006–0.01 g m^{-2}). This very low biomass on *Myriophyllum* might be explained by the bad physiological condition of the plants; the whole vegetation disappeared completely from the lake in the summer of 1979. No correlation was found between the macrofauna per m^2 vegetation area and the different life forms of the plants.

Carbon budget

For a carbon budget the biomass data of June were used. Since in June the macrophytic substrates were sampled quantitatively, and the plant-occu-

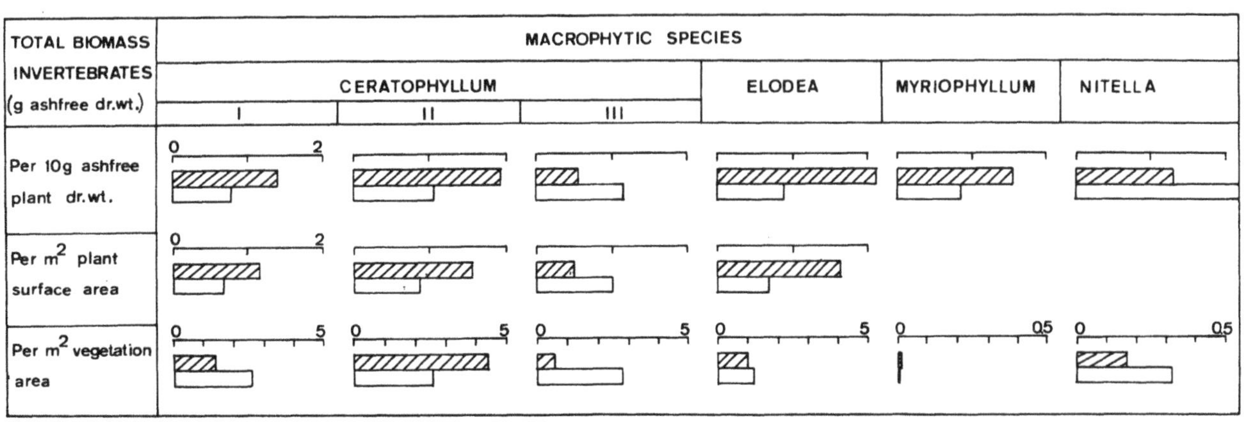

Fig. 5. The total biomass of macro-invertebrates on the submerged macrophytes in May and June, per 10 g ash-free plant dry wt., m^2 plant surface area and m^2 vegetation area. [hatched] May; [open] June.

Table 6. Calculation of the macro-invertebrate biomass on m² lake basis in June, expressed in carbon. An average carbon concentration of 55% ash-free animal dry wt. is assumed.

Plant species		Plant biomass (g ash-free dry wt./ m² vegetation area)	Cover stand area (m²)	Invertebrate biomass (g ash-free dry wt./ 10 g plant ash-free dry wt.)	Invertebrate biomass (g C/m² lake area)
Phragmites		497.1	768	0.052	0.023
*Sparganium**		98.6	168	0.103	0.002
Typha		329.0	72	0.056	0.002
Polygonum		33.5	218	0.173	0.001
Ceratophyllum	I	33.8	622	0.775	0.019
	II	24.9	692	1.043	0.021
	III	23.5	602	1.199	0.020
Elodea		13.9	4 185	0.877	0.060
Myriophyllum		0.07	1 232	0.865	0
Nitella		1.4	212	2.300	0.001

* Shoot material only.

pied areas were also known (Best, unpubl.), the macrofaunal biomass could be calculated fairly accurately.

The submerged vegetation consisting of *Elodea* and *Ceratophyllum,* and the *Phragmites* stands contributed most significantly to the macrofaunal carbon (Table 6). In June the average macro-invertebrate standing crop was 0.015 g C m^{-2} lake area. If at that time the total biomass of the animals, which consume living plant material (macrophytes, or algae or both, i.e. DC + M + F), is calculated for the different vegetation types, an average of 0.00131 g C m^{-2} lake area is found. Assuming a maximum daily consumption rate equivalent to 10% of their own biomass (Soszka 1975) this would indicate the use of 0.000131 g C m^{-2} lake area day^{-1}. Since the total primary production of the lake amounts to about 21.2 kg C m^{-2} day^{-1}, of which 7.4% is contributed by macrophytes, this would mean the use of 0.03% of the total daily primary production, and if only macrophytes were used, 0.39% of the macrophytic production. The plant-associated macro-invertebrate biomass was small compared to that of the filter-feeding zooplankton, 0.015 vs. 0.73 g C m^{-2} lake day^{-1} (Gulati *et al.* 1982); however, the latter value is an annual average. The biomass of several macro-invertebrate species in macrophytic stands similar to those in Lake Vechten may increase between June and November by a factor of 3 for chironomids to 100 for *Dreissena* (Glowacka *et al.* 1976; Geerlings *et al.* 1980). Assuming a similar increase in Lake Vechten of the macro-invertebrate biomass, in November carbon values close to those of the filter-feeding zooplankton in the lake may be obtained. However, since the plant consumption rate of the macro-invertebrates is much lower than that of the zooplankton the upper range of their grazing pressure on the lake's total primary production would also be much lower. Since, during the last few years the macrophytes have declined, particularly *Phragmites* and all the submerged species (Best 1982), this grazing pressure would become even less due to the disappearance of macrophyte-specific macro-invertebrates.

Acknowledgements

The stay of J. Dvorák at the Limnological Institute was supported by the Beyerinck-Popping Foundation.

Mr. J. H. A. Dassen is gratefully acknowledged for the technical assistance, and Dr. L. W. G. Higler for critical reading of the text.

Summary

The macrophyte-associated macro-invertebrate communities in the littoral zone of Lake Vechten, The Netherlands, were studied during early summer.

The macro-invertebrates could be divided into three groups, a) characteristic for all macrophytes,

b) associated with emergent and floating-leaved macrophytes only, and c) others. As regards the share of the characteristic groups of animals in the total abundance, and the predominance of the species, several similarities and differences became clear. The communities on the emergent *Phragmites* and *Typha* and on the floating-leaved *Polygonum* showed a striking resemblance, while those on the emergent *Sparganium* resembled those on the submerged *Myriophyllum* more than those on the other emergent plants. Among the communities associated with the submerged plants, the plants' rooting depth played an important role, while species-specific differences also occurred. The macrofauna on all the depth classes of *Ceratophyllum* showed a rather high similarity. The influence of rooting depth was clear in that a) the community on the shallowest *Ceratophyllum* depth class resembled that on *Elodea,* and b) those on the two other depth classes of *Ceratophyllum* resembled that on *Myriophyllum.* However, there was also some concurrence between the communities on *Elodea* and on *Nitella,* the latter plant species occurring at greater rooting depth. The seasonal differences in macrofaunal structure were mainly caused by the succession in generations of the predominant invertebrate species.

The macro-invertebrates' trophic role was described in terms of food source and feeding mechanism. The detritivores, periphytiscrapers, omnivores and carnivores occurred with the highest frequency, the first three predominating strongly, mainly in May. In June they decreased in the shallow regions but increased in the deeper parts of the lake. Of the carnivores, the predators fed mainly on zooplankton and small invertebrates and formed the main bulk. The parasitizing carnivores were of minor importance. The total biomass of the macrophyte-consuming animals was small compared to that of their host plants, indicating little use of the host plants. The algae-filtrating invertebrates formed a significant part of the whole community.

Quantitatively, the distribution of the macro-invertebrates in the littoral zone depended strongly on the colonizable surface area of the macrophytes and on the density of their respective vegetation types. On a unit ash-free plant dry weight basis, the highest animal numbers and their biomass were found on the submerged plants, and the lowest on the emergent plants. However, the *Typha, Phragmites* and *Ceratophyllum* stands had the highest animal biomass per m^2 vegetation due to high density of the plants, while the remaining stands were far less rich.

Based on a conservative carbon budget on lake basis in June *Elodea, Ceratophyllum* and *Phragmites* stands contributed the most to the macrofaunal carbon. Although the seasonal average of macrofaunal biomass might be in the same range as the other secondary producers, i.e. the filter-feeding zooplankton, utilization of the primary production by macrofauna is probably small due to the relatively poor representation among them of herbivores.

References

Best, E. P. H., 1982. The aquatic macrophytes of Lake Vechten. Species composition, spatial distribution and production. Hydrobiologia 95: 65–77.

Best, E. P. H., Blaauboer, M. C. I., Cappenberg, Th. E. et al., 1978. Towards an integrated study of the ecosystem in Lake Vechten. Hydrobiol. Bull. 12: 107–118.

Blaauboer, M. C. I., 1982. The phytoplankton species composition and the seasonal periodicity in Lake Vechten from 1956 to 1975. Hydrobiologia 95: 25–36.

Cummins, K. W., 1973. Trophic relations of aquatic insects. Ann. Rev. Ent. 18: 183–206.

Dvorák, J. et al., 1982. The role of animal communities in wetlands. In: Ecology of Wetlands, Chap. 5. Cambridge Univ. Press.

Geerlings, R., Lee, A. v.d. & Schouten, T., 1980. Investigations on the quantitative and qualitative development of macrofaunal coenoses in three different vegetation types in the Lake Maarsseveen I. Internal Report, Research Institute for Nature Management, Leersum. 64 pp. (in Dutch).

Glowacka, I., Soszka, G. J. & Soszka, H., 1976. Invertebrates associated with macrophytes. In: Pieczynska, E. (Ed.) Selected Problems of Lake Littoral Ecology, pp. 97–122. Warschau.

Gons, H. J., 1982. Structural and functional characteristics of epiphyton and epipelon in relation to their distribution in Lake Vechten. Hydrobiologia 95: 79–114.

Gulati, R. D., Siewertsen, K. & Postema, G., 1982. The zooplankton: its community structure, food and feeding, and role in the ecosystem of Lake Vechten. Hydrobiologia, 95: 127–163.

Higler, L. W. G., 1977. Macrofauna-coenoses on Stratiotes plants in Dutch broads. Proc. Research Institute for Nature Management 11. V. Beek, Hooglanderveen. 86 pp.

Higler, L. W. G., 1981. Bottom fauna and littoral vegetation fauna in lake Maarsseveen. Hydrobiol. Bull. 15: 82–87.

Kloet, W. A. de, 1982. The primary production of phytoplankton in Lake Vechten. The Netherlands. Hydrobiologia 95: 37–57.

Korinkova, J., 1971. Sampling and distribution of animals in submerged vegetation. Vest. csl. zool. Spol. 3: 208–211.

Krecker, F. H., 1939. A comparative study of the animal population of certain submerged aquatic plants. Ecology 20: 553–562.

Parma, S., 1971. Chaoborus flavicans (Meigen) (Diptera, Chaoboridae): an autecological study. Ph.D. thesis, Groningen University. 128 pp.

Soszka, G. J., 1975. Ecological relations between invertebrates and submerged macrophytes in the lake littoral. Ekol. pol. 23: 393–415.

Steenbergen, C. L. M., 1982. Contribution of photosynthetic sulphur bacteria to primary production in Lake Vechten. Hydrobiologia 95: 59–64.

Steenbergen, C. L. M. & Verdouw, H., 1982. Lake Vechten. Aspects of its morphometry, climate, hydrology and physico-chemical characteristics. Hydrobiologia 95: 11–23.

The zooplankton: its community structure, food and feeding, and role in the ecosystem of Lake Vechten

R. D. Gulati, K. Siewertsen & G. Postema
Limnological Institute, 'Vijverhof' Laboratory, Rijksstraatweg 6, 3631 AC Nieuwersluis, The Netherlands

Keywords: zooplankton, filter-feeders, filtering rates, seston, feeding, assimilation, methodological errors of ^{14}C-technique, physiological efficiencies, turnover rates

Abstract

The structure, feeding and metabolism of the filterfeeders community of Lake Vechten (The Netherlands) were investigated for seven years in relation to the functioning of the lake's ecosystem. The ^{14}C-technique used in the grazing and assimilation study is discussed in detail with a critical analysis of the methodological errors.

The three major species which contributed to the annual density, biomass and grazing maxima in spring are: *Bosmina longirostris, Daphnia* spp. and *Eudiaptomus gracilis.* The rise in grazing pressure in recent years, particularly in May, was accompanied by a corresponding decrease in the seston (<33 μm) biomass, and in increase of inedible algae, especially *Ceratium hirudinella,* in late summer. The means of daily grazing ranged from 3% in March to 34% in June. The mean annual ratio ingestion: phytoplankton production varied from 70 to 230%. The specific filtering rate, SFR (ml $\cdot$ day^{-1} $\cdot$ mg^{-1} $\cdot$ zoop $\cdot$ C), was related directly to water temperature but inversely to the food concentrations.

The main errors in the ingestion and assimilation rates were related to the leaching of the isotope from the animals in the preservation fluid. The loss of tracer was 42 and 26%, respectively, for the two rates.

In spring, the food removed by the grazers per day was equivalent to 125–400% of the daily primary production. This caused a sharp decrease in the seston concentrations and a recurring 'clear water' phase because of a sharp increase in the Secchi depth. The zooplankton assimilatory removal of carbon and the sedimentation loss rates to the hypolimnion exceeded the primary production rates. The inconsistencies in the carbon budget are possibly due to our lack of knowledge of the horizontal transport of material from the littoral, bacteria as an alternative food source for zooplankton, and the DOC dynamics.

The grazers' activity as SFR in deep, stratifying lakes like Vechten is 3 to 4 times that in the shallow, mixed and more eutrophic Dutch lakes. In the former category of lakes the crustacean herbivores serve as an important link in the food chain in the limnetic region.

Introduction

Small, stratifying freshwater bodies are common in the Netherlands, and their numbers are increasing due to increased 'sand-winning' operations. Despite this, their zooplankton has been poorly investigated. The studies on Lake Vechten (Gulati 1975, 1976, 1978a, b) and on Lake Maarsseveen (Butter 1981; Hulsmann & Ringelberg, in prep.; Ringelberg 1981) are, however, notable exceptions.

The first phase of zooplankton studies on Lake Vechten dates back to 1956 when Leussink (1956) examined the vertical distribution of both phytoplankton and zooplankton in spring and summer in relation to temperature and oxygen changes. Salomé (see Parma 1971) made an inventory of phytoplankton and zooplankton species in both the limnetic and the littoral regions of the lake.

In 1972 a study on the community structure and grazing, assimilation and respiration rates of zoo-

Hydrobiologia 95, 127–163 (1982). 0018-8158/82/0952-0127/$07.40.

plankton was started (Gulati 1975, 1976, 1978a, b). The food chain research, especially that concerning herbivore zooplankton, was intensified in the period 1976–1978, together with a more coordinated, ecosystem approach to the production and mineralization studies on the lake (Best *et al.* 1978; Parma 1982). The vertical differences in the grazing activity of the herbivore-zooplankton community and the role therein of different species were examined (Gulati 1978a; Postema & Siewertsen 1981). Also, phytoplankton and its primary production were studied (Blaauboer 1982; De Kloet 1982) simultaneously with zooplankton grazing during 1977 and 1978 to provide information on the food of zooplankton.

The current zooplankton studies on Lake Vechten are concerned mainly with the role of zooplankton excretion products in the nutrient dynamics and availability for the limnetic primary production (Gulati *et al.* 1981).

The present paper is the first endeavour to collate the existing information and to provide a comprehensive review of both the methods and our knowledge to date of the community structure, food relations and metabolism of the predominantly crustacean herbivores in the lake. These aspects are examined with emphasis on the impact of grazing in the functioning of the lake's ecosystem.

The lake

Lake Vechten is a small, deep, stratifying and clear water body, has no surface drainage and is free from external pollution. The lake was created in 1941 by sand digging. Its morphometry and bathymetry have been investigated in detail (Parma 1971, 1980). The physiological features, waterbalance, light climate, thermal regime of the lake and meteorological data are described in Steenbergen & Verdouw (1982). The morphometric data relevant to the present study are summarized:

surface area, 4.71 ha
max. depth, 11.9 m
mean depth, 6.0 m
relative depth (Z_r), 4.9%
shoreline, 1 130 m
shoreline development D_L, 1.47

A ridge running NW to SE across the breadth divides the lake into an eastern and a western basin that are almost equal in area, 2.3 and 2.4 ha, respectively. The maximum depth along the ridge is 6 m, so that the separation into two basins is confined to the water strata below 6 m. The eastern basin where zooplankton and other related research was mainly carried out has maximum and mean depths of 10.5 and 5.6 m, respectively, compared with 11.9 and 6.5 m in the western basin.

Methods

Field methods

The sampling techniques, sampling frequencies, depths of water layers sampled and the nature of the parameters measured differed during the different periods in the course of studies (see Table 1).

The water was generally sampled between 9.30

Table 1. Summary of sampling programme and parameters measured.

Year/Worker	Sampling depth (m)	Period/frequency	Parameters measured	Remarks
1. 1956 (Leussink)	0–10	March–July, f	p, z	First study
2. 1962–1967 (Salomé)	s	Jan.–Dec., f	p, z	Seasonal changes
3. 1972–1979 (Present study)				
a. 1972–1974	s–5	March–Feb., f	z, gr	Mixed samples
b. 1974–1976	s–O_2 zero depth	March–Feb., f	z, gr	Mixed samples
c. 1976–1977	s–O_2 zero depth	March–Feb., w	z, gr	4 or 5 strata from s to O_2 zero depth
d. 1977	s–O_2 zero depth	March–Nov., f	z, gr	4 or 5 strata from s to O_2 zero depth
e. 1977–1979	s–O_2 zero depth	Dec. '77–Feb. '79, f	z, gr	2 strata from s to O_2 zero depth

w, weekly; f, fortnightly; s, surface water; z, zooplankton; p, phytoplankton; gr, grazing.

and 10.30 a.m., with a 5-l Friedinger sampler (cylinder length, 60 cm). The sampling procedure in the March 1972–Feb. 1974 period differed from that in March 1974–Feb. 1979 in two main respects: first, the depth of the sampling column was limited to the epilimnetic water (about 5 m column); and secondly the successive 60 cm sampler hauls, from the surface to 5 m, were mixed. During March 1974–Feb. 1976 the entire water column, up to the depth where oxygen could no longer be detected with a Mackereth (1964) electrode, was sampled; these samples were pooled too. But from March 1976 to February 1979, 30 l of water was sampled from each of four or five 1.8 m strata, or from two strata of 3.6 m each, extending from the surface to the depth where the oxygen concentration became zero (see also Table 1). These strata are subsequently called I, II, III and IV and correspond, respectively, to the following depths: surface–1.8 m, 1.8–3.6 m, 3.6–5.4 m, and 5.4–7.2 m or from 5.4 m to depth of zero oxygen.

Biomass measurements

The biomass of both zooplankton (>105 μm) and 'seston' (<105 μm), i.e. phytoplankton, detritus and bacteria colonizing the latter, was measured. The seston was fractionated into two size classes, 3–15 μm and 15–105 μm up to Feb. 1976 (Gulati 1975), or 3–33 μm and 33–105 μm from March 1976 onwards.

The biomass was measured after fractionation as COD (Golterman 1969). The samples were digested for 2 h in a pressurized oven (2.5 atmospheres) at ±140 °C. The technique was modified in 1973 when checks using glycine standards indicated 20–30% underestimation of carbon (Siewertsen 1979). Almost complete recovery was obtained, even in 2 h, if the volume of distilled water in the digesting flasks was reduced from the recommended 70 ml (Golterman 1969) to about 30 ml. Thus we reduced the total volume in the flasks – including that of the reagents and the sample – from 100 to 60 ml, and this resulted in a more complete oxidation. The total volume was made up to 100 ml with ±40 ml distilled water, after digestion, i.e. before measuring the extinction values colorimetrically. Even though the COD data of 1972 were corrected using a correction factor based on the modified COD technique, the data of 1972 are less reliable, and therefore excluded from the annual comparisons involving seston mass. The COD values were converted to carbon according to Winberg (1971).

During a comparative study of the biomass parameters, nitrogen, biochemical composition, organic dry weight and heat of combustion of seston and zooplankton were also measured (Siewertsen 1979).

Plankton counting

For zooplankton density estimation, 1 or 2 l of water was subsampled from the 30-l sample for each of the four strata, filtered through an 85 μm metal gauze and the zooplankton preserved in 4% formalin.

Usually, the entire zooplankton sample was counted under a microscope. Recently, a comparison of 33 μm and 85 μm mesh size filters for concentrating zooplankton indicates that losses of small sized rotifers and nauplii through the 85 μm mesh size are substantial. However, considering the very small share in the zooplankton biomass of these relatively smaller animals, neither the zooplankton biomass nor the herbivore grazing is assumed to be much underestimated.

For details of the phytoplankton subsampling, preservation and counting techniques, see Blaauboer (1982). Netplankton (>33 μm) and nanoplankton (<33 μm) were fractionated as for the seston for carbon measurements.

Grazing rates

The grazing and assimilation rates of zooplankton were measured in the laboratory using the ^{14}C-tracer technique (Nauwerck 1959). The size of the *in situ* assemblage of suspended particulates used as tracer was below 15 μm until Feb. 1976 (Gulati 1975) but below 33 μm in the period thereafter.

The size and form of food particles (Gliwicz 1969a) together with the mesh size of the filtering apparatus (Geller & Müller 1981) determine the clearance rates of the filterfeeders. Thus, any shift due to laboratory manipulations in the frequency distribution of size spectrum and in composition of the algal species or food particles in suspension is likely to influence the filtering rates of the herbivores. The published literature on the subject is scanty, though it has been the topic of some discussion during the Plankton Ecology Group (PEG) meetings in the recent years. There is undoubtedly a need to describe the techniques used elaborately and to compare critically the results obtained using different techniques. Our procedures of labelling the *in situ* food and carrying out the grazing experiments are thus described in some detail.

Preparation of ^{14}C-tracer. The lake water for preparing the tracer-food was sampled 5 to 10 days prior to the grazing experiment. The scheme of sampling depth for the tracer-food was as follows:

Period	*Stratum*	*Tracer prepared from*
1972–1975	S–5 m	Mixed sample
1976–1979	I & II	Mixed sample from S–3.60 m for both strata
	III & IV	Mixed sample from 3.6 m–O_2 zero depth for both strata

One litre of the sample was pre-filtered with a 15 or 33 μm mesh sieve to remove the zooplankton and coarse sestonic particulates. This pre-filtration step was omitted for grazing experiments with coarser algae. To the filtrate 100 μCi $NaH^{14}CO_3$ was added. The water was incubated at the mean temperature of the column of lake water sampled, and at a light intensity higher than that prevailing in the field, but not exceeding 30 W · m^{-2}. Above this intensity photosynthetic inhibition usually occurred in the lake (Gulati, unpublished data). In periods of low primary productivity (de Kloet 1982) the filtered lake water was mixed in the ratio 1:1 with a synthetic medium identical to that of Gorham but devoid of carbonates and bicarbonates, in order to accelerate assimilation of $^{14}CO_2$.

The incubated sample was allowed to settle overnight in a refrigerator at about 5 °C; the supernatant was gently decanted thereafter to obtain a few millilitres of the concentrated tracer food. This was then diluted to about 100 ml with 0.5 μm membrane filtered lake water, filtered through a 15 or 33 μm mesh sieve and allowed to settle for at least 4 h in a long, narrow 150 ml measuring cylinder. The decantation and dilution steps were repeated to remove as far as possible the $H^{14}CO_3$ in solution. The coarse fraction retained on the filter was also used as tracer in experiments on size selective feeding.

The ^{14}C-uptake by the fine seston fraction was maximally 20% of the radioactivity added, but usually around 10%. Thus, about 10–20 μCi of the ^{14}C added was assimilated. The remaining 80–90 μCi was mainly in the inorganic phase.

Changes in food concentration due to tracer addition. The addition of the tracer food to the grazing chamber will not normally alter the *in situ* food concentration. This is especially true if the tracer has a high specific activity as is the case in highly labelled single-cell cultures of algae (Haney 1971, 1973; Lampert 1977a), yeast or bacteria (Haney 1973). Therefore, the effects of using an '*in situ* tracer' on *in situ* food concentration in a lake like Vechten, which has a low concentration of food, need to be examined carefully.

The per cent increase of food concentration due to the addition of the tracer in the grazing chamber can be derived from the following equation:

$$S_i = V_a \cdot P_a \cdot I_c \cdot C_f \cdot V_g^{-1} \cdot S_l^{-1} \cdot 10^2 \qquad (1)$$

where S_i = % increase of seston (<33 μm; μg C · ml^{-1}) due to the addition of tracer;
V_a = volume in ml of tracer added to the grazing vessel;
P_a = proportion of the added radioactivity (μCi) assimilated by seston;
I_c = dissolved inorganic carbon (μg C · ml^{-1}) in lake water used for tracer preparation;
C_f = ratio initial volume of lake water incubated for preparing the tracer to the volume of tracer left after decanting, etc. Here, losses in the supernatant are ignored;
V_g = volume of *in situ* lake sample in grazing chamber; and
S_l = seston concentration (μg C · ml^{-1}) in lake water.

Using the average values of the terms in equation (1) (V_a, 5 ml; I_c, 30 μg C · ml^{-1}; C_f, 10; S_l, 0.5 μg C · ml^{-1}; V_g, 5 000 ml), and the maximal value of P_a, 0.2, the S_i, i.e. the increase in seston concentration in the grazing vessel, was 12%. Since this S_i value is based on a P_a of 0.2, which, as already mentioned, is generally the maximal value in our tracer preparation, it appears high. Mostly, however, P_a was lower than 0.2, and so S_i was lower too.

Grazing experiments. The methods employed in the 1972–1975 period to measure grazing are summarized in Gulati (1975). From 1976 onwards, the procedure of concentrating zooplankton was abandoned in favour of *in situ* concentration since this reduced cladoceran mortality, particularly of *Bosmina* sp. Five litres of lake water from each of the four strata was poured into graduated, translucent plastic beakers and placed in a 'conditioned' cabinet present at the lake water temperature and light conditions. After acclimation for 30–45 min (see Geller, 1975) a known volume of tracer food, generally 5 ml, was added to the grazing vessel, mixing gently while adding. The radioactivity in the grazing vessel (R_F) varied between 125 and 225 pCi $\cdot$ ml^{-1}, respectively 275 and 500 DPM $\cdot$ ml^{-1}. This could be approximately calculated beforehand from the equation:

$$R_F = R_a \cdot P_a \cdot V_a \cdot V_t^{-1} \cdot V_g^{-1} \quad (2)$$

where R_F = is radioactivity of food in the grazing vessel in DPM $\cdot$ ml^{-1};
R_a = radioactivity (DPM) in μCi $\times$ 2.2 $\times$ 10^6 (1 μCi = 2.2 $\times$ 10^6 DPM) added to prepare the tracer;
V_t = volume in ml of the concentrated tracer stock; and
P_a, V_a and V_g are the same terms as defined for equation (1).

The time between the addition of tracer-food and the end of the grazing experiment was usually 10 min. The gut-filling time for the dominant filter feeders in Lake Vechten generally varies between 8 and 18 min (Koole & Vulto 1976).

The experiment was terminated by filtering the zooplankton on a 85 μm mesh sieve. The zooplankton was killed in hot water (90 °C) and thereafter rinsed with 0.5 μm-filtered lake water, preserved in 1% formalin and stored at a low temperature (about 5 °C). The preservation and storage steps are indispensable because of the time-consuming procedure of sorting the species to determine their share in the community grazing (Gulati 1978a).

For examination the animals were poured into a petri dish, and sorted to species under stereo binoculars using a fine forceps to pick the animals holding on to the antenna, tail spine or valve. They were transferred to scintillation vials containing a few drops of distilled water. The species were sorted into length size classes when feasible, but this was limited to three categories, viz. large, medium and small. Usually, 10 to 20 individuals of each size class were needed per replicate to get reliable estimates of radioactivity. The rest of the sample, which contained the unsorted individuals, was poured over a 30 μm nylon filter, back-washed with 10–15 ml distilled water, and tipped into a scintillation vial. Replicate sub-samples of the filtrate were also prepared in order to measure 'losses' related to the method of killing, preservation, storage, handling, etc. (Holtby & Knoechel 1981). This is a serious but as yet unresolved problem with our study.

Radioactivity counting. The animals in the scintillation vials were dried at 55 °C in an oven. Thereafter, they were wetted with 100 μl distilled water before 0.25 ml of Soluene 350 (Packard Instruments) was added; the wetting improves the dissolving action of Soluene. The vials, with screw caps closed, were kept in the oven (55 °C) for at least 4 h to solubilize. Ten millilitres of scintillation cocktail (5.5 g secondary scintillator containing PPO and bis-MSB in the ratio of 10:1 per litre toluene) was added per sample.

The radioactivity of the samples was counted in a Packard Liquid Scintillation Spectrometer (Model 2425).

Quench curve correction was based on External Standard Ratio (ESR) and counting efficiency, employing a computer program to calculate the DPM via the 3rd degree polynomial regression. The above procedures of sample preparation and counting were started in late 1975. For the preceding period, the gas-flow counting technique was used (Gulati 1975).

Assimilation experiments. The assimilation rates of the zooplankton were measured simultaneously with and in the same way as the grazing rates; but the feeding time after tracer addition lasted 45 min or 1 h, after which the animals were transferred to the natural non-active lake food, and allowed about 45 min to clear their guts of the undigested labelled food.

Calculation of filtering, consumption and assimilation rates. The individual filtering rate per day (F) for the different species was derived from:

$$F = R_z \cdot R_F^{-1} \cdot 60 \cdot 24 \cdot t^{-1} \quad (3)$$

where R_z and R_F are the radioactivities per individual and per ml food in the grazing vessel in DPM (see also equation 2), respectively. The daily community filtering (F_{total}), i.e. filtering rate in ml. · day^{-1} l^{-1}, was also calculated in the same way as in equation (3):

$$F_{total} = \Sigma R_z \cdot R_F^{-1} \cdot (60 \cdot 24 \cdot t^{-1}) V_g^{-1} \cdot 10^3 \quad (4)$$

where ΣR_z is the radioactivity of all animals of all the species in the community, including the unsorted sample, and V_g is the volume in ml of water containing zooplankton in the grazing vessel (as in equation 1). The daily grazing (Gr) or per cent food consumed is:

$$Gr\ (\%\ day^{-1}) = F_{total} \times \frac{100}{1000}. \quad (5)$$

The ingestion (I) in μg C · day^{-1} · l^{-1} (or mg C · day^{-1} · m^{-3}) is calculated as follows:

$$I = F_{total} \cdot S_l \quad (6)$$

where F_{total} is the volume of water filtered in ml · day^{-1} · l^{-1} (equation 4) and S_l is the seston concentration (3–33 μm) in μg C · ml^{-1} (equation 1).

The terminology and calculation of assimilation (A) data used in this paper correspond to those set out in equations (3)–(6). The assimilation efficiency (%) is computed from A/I · 100 where A and I are, respectively, assimilation and ingestion rates. The specific daily filtering (SFR), specific ingestion or daily ration (DR) and specific daily assimilation are the weight specific rates calculated respectively from F_{total}:B, I:B and A:B, where B is zooplankton biomass.

Other measurements

The respiration rates of the zooplankton were measured during 1973–1975 potentiometrically, employing the Winkler technique (Gulati 1975). For the years 1976–1978, the rates were computed from the exponential regression equation: $Y = ae^{bx}$, where Y = respiration in mg C per mg body carbon day^{-1}; a and b are constants, being 0.0303 and 0.0977, respectively; and X is water temperature in °C. The correlation was highly significant ($r = 0.85$; $n = 61$; and $P < 0.001$).

The work on nitrogen and phosphorus excretion of zooplankton has been in progress since 1980; the techniques used have been reported briefly by Gulati *et al.* (1981) and will be described in detail elsewhere.

Calculation of mean depth, areal rates, etc.

The data on zooplankton grazing and assimilation rates, and on biomass densities and seston food of the different strata were integrated on a weighted basis to get the areal values. For this, percentage volumes of the four strata were calculated for each sampling date. These volumes are relative and based on Z_{mo}, i.e. depth in metres below which the water in the eastern depression of the lake is anaerobic; thus Z_{mo}, unlike the mean depth Z_m, varies seasonally (see Fig. 1) and depends directly on the ratio volume of aerobic water:volume of anaerobic water.

The Z_{mo} that equals Z_m, during most of the circulation period mid-Nov–April (5.55 m), decreases to 4.7 m in July when the summer stratification is fully developed, but increases again as the epilimnion extends downwards (Fig. 1).

The areal values (m^{-2}) thus obtained are assumed to be valid for the whole lake area, and when divided by Z_{mo} give the rates or concentrations per m^3. This approach to calculating the weighted areal values has three main advantages in that: 1) the grazing study in the lake was confined to the strata in which oxygen could still be measured with a Mackerath electrode; 2) the increase or decrease in the share of certain strata in zooplankton grazing owing to factors such as migration, temperature, food, etc., could thus be taken care of; and 3) the grazing and other relevant data can easily be converted to values for the whole lake and fitted to the production–consumption scheme of the lake's ecosystem (see Gons 1981).

Results

The seasonal changes in zooplankton densities during 1973–1978 with particular reference to the grazer species are described in brief first, followed by those in the zooplankton biomass and in average biomass per individual in the community. The variations in the zooplankton food, namely, seston 3–33 μm, grazing, assimilation and different physiological and ecological efficiencies form the main content of the results and discussion that follows.

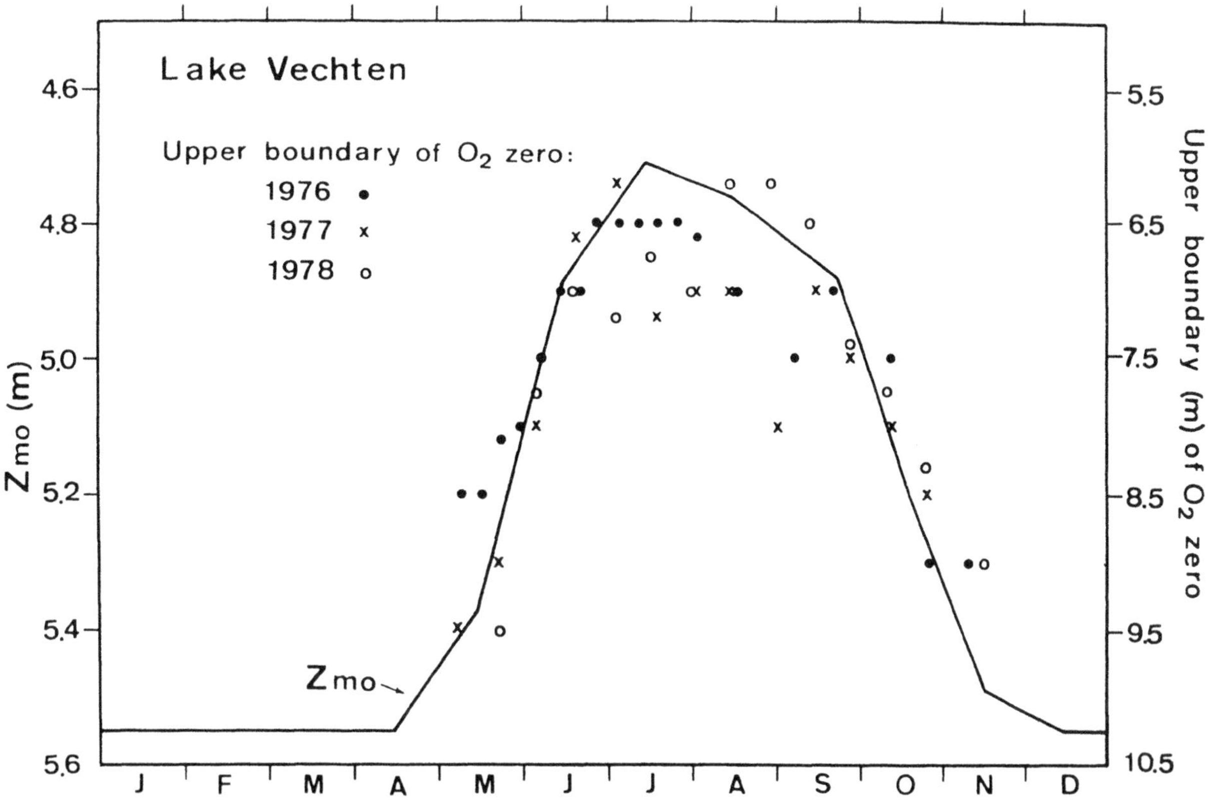

Fig. 1. Changes in the depth of upper boundary of anoxic layer derived from the monthly means and the corresponding decrease in the mean depth, Z_{mo}, of water stratum above this layer.

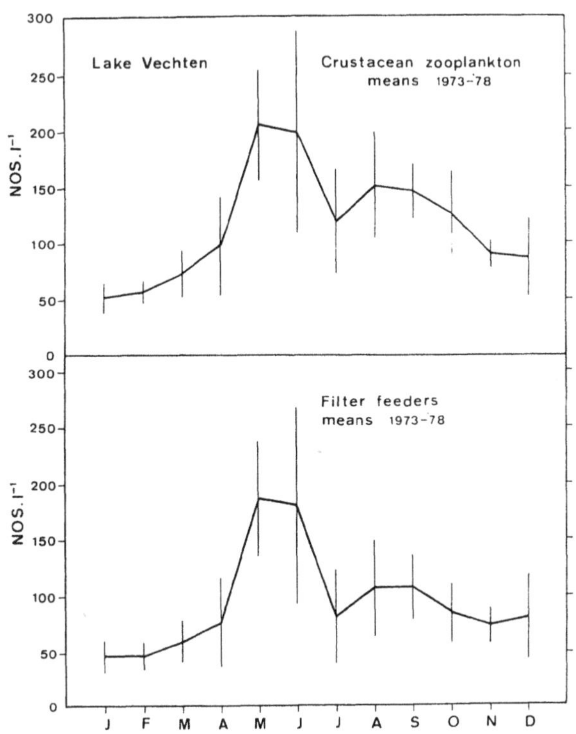

Fig. 2. The weighted monthly means ± SD of the total densities (nos · l^{-1}) of crustacean zooplankton and of the filterfeeders during the study period. The means were computed from the weighted integral data (nos · m^{-2}) of strata above the oxygen zero depth divided by Z_{mo} depth (m) · 10^3.

Zooplankton densities and biomass

The filterfeeders comprise the bulk of the total zooplankton density in the limnetic region, i.e. from 76 to 84% (Fig. 2). Important species with their depth of maximum abundance are listed in Table 2. The annual means for total zooplankton varied between 109 ind. · l^{-1} in 1973 and 128 ind · l^{-1} in 1978. The grand annual and summer (May–Sept.) means were 117 ± 7 and 164 ± 24 ind · l^{-1}, respectively.

The zooplankton community in the period Dec.–Feb. is dominated mainly by *Eudiaptomus gracilis.* The first steady increase in March–April is owing to nauplii which, together with *E. gracilis,* constitute the main filterfeeders. The first major

Table 2. The crustacean zooplankton species in Lake Vechten, their depths (S = surface) and periods of abundance and grand annual means with standard deviation (ind · l^{-1}) for the years 1973–1978. The depth of stratum below 5.4 m is variable depending on the upper boundary of O_2 zero (see Fig. 1).

Species	Abundance		Mean ± SD
	Depth (m)	Period	
Daphnia spp.	S -5.4	May–June	21.3 ± 3.9
Bosmina longirostris	5.4–7.2	May–June	15.1 ± 3.7
Diaphanosoma brachyurum	3.6–7.2	Aug.–Sept.	6.9 ± 2.9
Ceriodaphnia pulchella	5.4–7.2	Aug.–Sept.	2.2 ± 1.7
Eudiaptomus gracilis	S -3.6	Nov.–June	20.2 ± 5.3
Nauplii	S -7.2	March–May Sept.–Oct.	29.4 ± 2.5
Cyclopoids	3.6–7.2	July–Oct.	21.8 ± 2.3
Total	3.6–7.2	May–June	116.9 ± 6.8

increase in zooplankton density occurs from April to May, and is from 97 ± 44 to 205 ± 49 ind · l^{-1} (Fig. 2). The three major filterfeeders which contribute to this spring and annual maximum in May in their order of importance are: *Bosmina longirostris; Daphnia* spp., predominantly *D. cucullata;* and *Eudiaptomus gracilis.* Though the mean for the month of June during the six years is similar to that of May, the fluctuations in June from year to year are the highest (Fig. 2) and caused mainly by the timing of oscillations in *Bosmina* populations, but in *Daphnia*, as well.

The decline of zooplankton populations in July (Fig. 2) is owing to all three major species of filterfeeders. However, *Diaphanosoma brachyurum* and species of cyclopoids and their nauplii increase in numbers in this period. The zooplankton numbers in August and September are constant at around 150 ind · l^{-1}, about two-thirds of which are filterfeeders (Fig. 2). In late summer, important species are: *D. hyalina* and *D. brachyurum,* in the depth below 3.6 m; and *Ceriodaphnia pulchella,* below 5.4 m; also, cyclopoids have their annual maximum in this period.

The autumn (Oct.–Nov.) is marked by a further decrease in zooplankton concentrations, but for *Eudiaptomus* sp. which exhibits a recurring increase. This species is dominant in winter (Dec.–Feb.) when it contributes significantly to both the zooplankton numbers and the biomass.

Filterfeeders. The grazers dominate the zooplankton of Lake Vechten. They comprised annually 76–84% (mean, 81 ± 3%) of the total zooplankton densities (Table 3). This is equally true also if the data of summer periods only (May–Sept.) are considered even though the variation in the summer means is higher. This, together with narrow fluctuations in numbers, indicates a constancy in regard to zooplankton during the years 1973–1978 (Table 3).

The monthly variations, as expected, are more pronounced and depend on a variety of factors, among which seasonal changes in water temperature, food quality and quantity, and predation may be primarily important. The recurrent abundance of filterfeeders in May and June followed by a sharp decline in July wil be examined in relation to the variations in food concentration and feeding intensity.

Table 3. Annual and summer means of weighted densities (ind l^{-1}) of the crustacean filterfeeders, and their percentage share in the total crustacean zooplankton in Lake Vechten (1973–1978) (see also Fig. 2 for monthly means).

Year	Annual		Summer	
	Mean	% share	Mean	% share
1973	83	76	102	78
1974	93	81	122	79
1975	98	84	164	83
1976	97	82	136	83
1977	89	80	122	81
1978	106	83	154	84
Grand mean (SD)	94(8)	81(3)	133(23)	81(2)

Individual biomass. The zooplankton biomass does not always keep pace with the changes in zooplankton densities but follows a somewhat different course depending on the species composition and mean size of individuals in the community. The nauplii, being very small, will contribute negligibly to the community biomass. Therefore, for comparing the seasonal variations in numbers with those in biomass, nauplii were excluded. This was done using the data on monthly means of zooplankton (nos. l^{-1}) and their biomass ($\mu g\ C \cdot l^{-1}$) for the period 1973–1978 (see Fig. 3).

From January to March the total numbers remain constant but the biomass increases (Fig. 3), owing mainly to an increase in the individual biomass (Fig. 4) of *E. gracilis*. During April and May the increase in community biomass does not keep pace with that in the numbers. The pronounced decrease in the mean individual biomass is caused by sharp increase in *Bosmina longirostris* and *Daphnia cucullata* populations. In short, these species cause not only the most important increase in community density and biomass but also a substantial decrease in community size structure.

The monthly mean weight of the individuals varied from a high of 4.3 ± 0.8 in March to $1.1 \pm 0.1\ \mu g\ C \cdot ind^{-1}$ in September. In the stratification period (May–Oct.) it did not exceed $1.7\ \mu g\ C \cdot ind^{-1}$ and was rather constant except for some increase in July (Fig. 4) owing mainly to an increase in size of cyclopoids in the deeper strata. The rise in individual weight, especially from October to November, is again due to the dominance of *E. gracilis*.

Community standing crop. The zooplankton standing crop biomass (>105 μm) was measured throughout the study period (March 1972–Feb. 1979), parallel with that of seston food (3–33 μm)

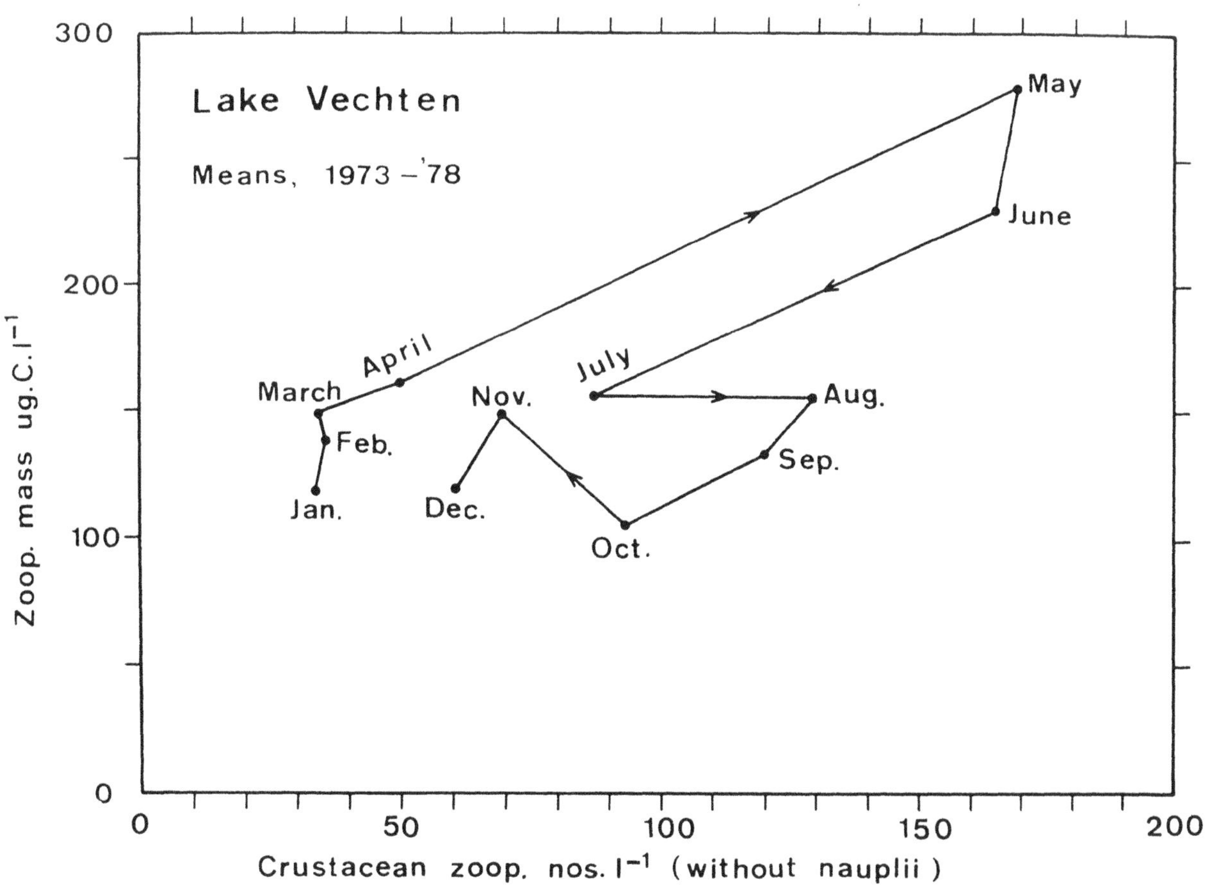

Fig. 3. The monthly means as in Fig. 2 but excluding the nauplii in relation to the monthly changes in the weighted zooplankton mass.

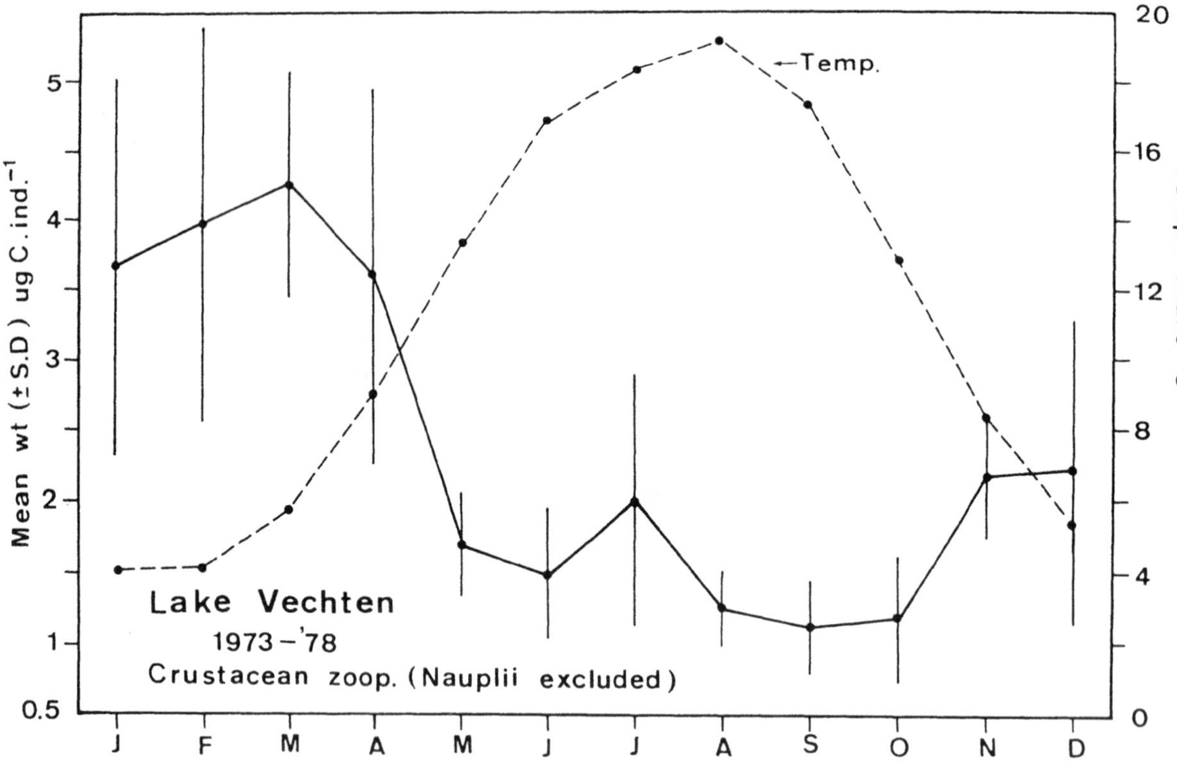

Fig. 4. Changes in the monthly means of individual weights ± SD of the crustacean community, excluding nauplii; and weighted means of water temperature (°C) in the oxic layers.

and total seston (3–105 μm). The monthly areal means for seven years are illustrated in Fig. 5; they ranged from 0.69 ± 0.13 g C · m^{-2} in December to 1.56 ± 0.37 g C · m^{-2} in May. The striking increase in the zooplankton biomass from April to May, from 0.97 ± 0.26 to 1.56 ± 0.37 g C · m^{-2}, is caused mainly by the filterfeeders. The decrease in the standing crop that follows the annual maximum in May continues until July; the monthly means during the rest of the year vary narrowly, from 0.7 to 0.9 g C · m^{-2}.

In comparing the annual means of zooplankton standing crop of different years (Table 4) it is seen that:

1) the means of the years 1976, 1977 and 1978 are higher than those in the preceding three years and
2) the increase from 1975 to 1976 is most striking and from 1976 onwards the annual means are stable.

A further scrutiny of the differences in the zooplankton standing crop (Fig. 6B) between the two

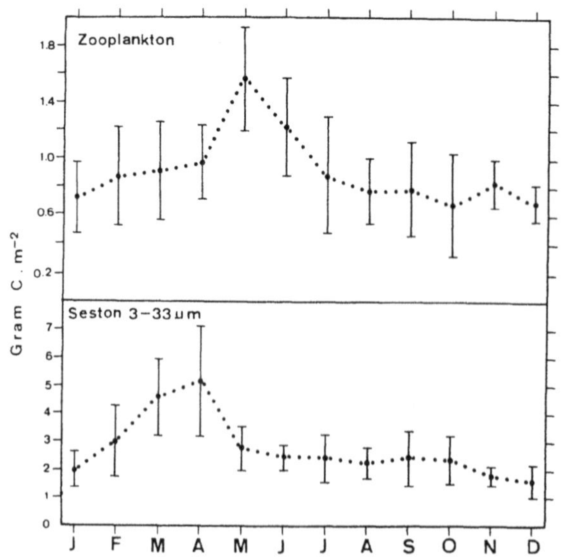

Fig. 5. Changes in the monthly means of the mass and their SD of total zooplankton including cyclopoids, and of the seston fraction used as a measure of zooplankton food.

Table 4. Annual (1973–1978) and three-year period means with SD of zooplankton mass (g C · m^{-2}*) in Lake Vechten. The data are derived from the monthly means (see Fig. 5).

Year	1973	1974	1975	1976	1977	1978
Annual mean	0.81	0.66	0.76	0.93	1.00	0.96
Three year period mean (SD) (1973–1975 and 1976–1978)		0.75(0.08)			0.96(0.04)	

* The concentrations in g C · m^{-3} can be approximated by dividing by 5.25 (Z_{mo} depth).

periods (1973–1975 and 1976–1978) shows that the monthly means in the second period, except for those of October and November, were higher than those of the first period. This increase in 1976–1978 was most conspicuous during March–June, when it was between 31 and 111%. The mean biomass in these months was on the average 60% higher than

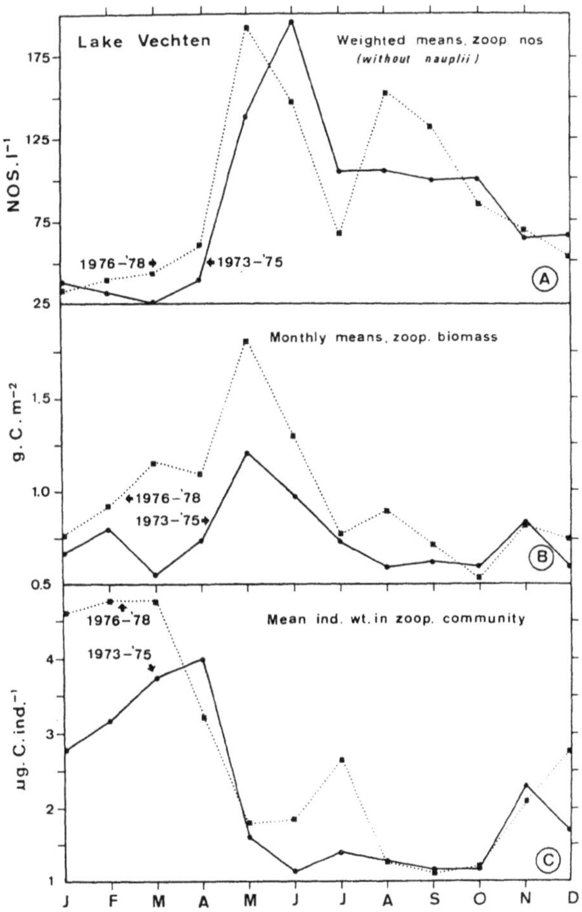

Fig. 6. A comparison of the monthly weighted means of zooplankton densities (A), zooplankton mass (B) and weight per individual (C) in the periods 1973–1975 and 1976–1978.

during March–June of 1973–1975. Interestingly, the rise in community biomass is caused more by the increase in the average annual biomass of individual organisms, namely from 2 ± 0.2 µg C · ind^{-1} in 1973–1975 to 2.6 ± 0.2 µg C · ind^{-1} in 1976–1978 (Fig. 6C), than by zooplankton densities (Fig. 6A). Nevertheless, mean densities have also exhibited changes. In the second period the spring maximum occurs in May instead of June, and densities in June and July are lower, whereas those in August and September are higher compared with those in 1973–1975 (Fig. 6A).

The aforementioned observations indicate that whereas the community composition has not changed, the size structure and densities have.

Zooplankton food

The zooplankton food composed of phytoplankton and detritus with bacteria colonizing it is called seston. The size fractions of the seston used to compute the zooplankton ingestion rates are: 3–15 µm until 1976, and 3–33 µm thereafter; the biomass of coarser fraction (>33 µm) was also monitored in connection with the food selectivity experiments (Postema & Siewertsen 1981).

Phytoplankton. The information on phytoplankton species composition and abundance in the lake is extracted from Blaauboer (1982).

In January the phytoplankton community in Lake Vechten is dominated by *Rhodomonas minuta* – a perennial species –, *Mallomonas akrokomos, Trachelomonas volvocina* and several *Cryptomonas* spp. The growth conditions of phytoplankton in the lake are apparently already quite optimal by late winter (late Jan.–Feb.). The species abundant in this period are *Chlorella pyrenoidosa* (its systematics is not yet resolved) and *Stephanodiscus hantzschii;* the former species has a diameter of about 2 μm and is thus optimal zooplankton food. It attains a maximum of 20–40 · 10^6 cells · l^{-1} in winter, and persists the whole year round.

In spring (March–April), *S. astraea* – which replaces *S. hantzschii* – *Cyclotella comta* and 2-celled coenobia of *Scenedesmus* sp. are important; the first-named species contributes the most to phytoplankton biomass.

In May, with the onset of summer stratification and a great increase in the density of filterfeeders, both the number and the densities of phytoplankton species are reduced. From June to early August the important species in their chronological sequence are: *Dinobryon divergens; Closteriopsis* sp.; *Ankyra* sp.; *Sphaerocystis schroeteri; Gloeocystis vesiculosa;* and several *Oocystis* species.

In August all the early summer species decline, but the population of *Ceratium hirundinella* continues to increase; *Mallomonas caudata* and filamentous blue green-algae, e.g. *Oscillatoria* sp. and *Lyngbya* sp., form the important late summer phytoplankton in the metalimnetic waters. The changes in the summer mass of coarse seston (>33 μm) are governed mainly by *C. hirundinella.* This species has in recent years become increasingly more prevalent.

The phytoplankton densities decrease in October and as the autumnal circulation gradually sets in the detrital component in the seston becomes predominant.

Seston. The seston concentrations (<33 or 15 μm) in Lake Vechten tend to increase already in Jan.–Feb. despite the low water temperature and poor light climate. The rise is uninterrupted until April when the annual maximum, 5.1 ± 2.0 g C · m^{-2} (or 0.92 ± 0.35 g C · m^{-3}; Z_{mo}, 5.55 m), is reached (Fig. 5).

The standing crop of seston and its concentration are reduced in May to almost half of those prevalent in April. The decrease of diatom populations, namely *S. astraea* and *C. comta,* in this period coincides with the peaks of zooplankton biomass and grazing. Other important, concomitant changes in the lake concern light climate and primary production. A sharp increase occurs in depth of Secchi from about 2 m in April to 4 m or more in May and early June – the period of high water transparency or the so-called 'clear water phase' (see p. 151 under 'Discussion'). Secondly, the primary production study (de Kloet 1982) indicates that the annual maxima of daily assimilation index, i.e. production in mg C · mg^{-1} chlorophyll, occur mostly in May and June.

The changes in seston biomass in the period June–October are minor, and, despite the variations in phytoplankton composition, monthly means have a narrow range, viz. 2.2–2.7 g C · m^{-2}. However, measurements of seston concentration in different strata in 1976, 1977 and 1978 indicated significant increases with depth during May–September. The mean ranges of concentrations (mg C · m^{-3}) for the strata were:

Stratum	*Concentration*
I	0.34–0.44
II	0.38–0.42
III	0.37–0.51
IV	0.51–0.73

In Lake Vechten, increase of seston concentrations with depth is caused mainly by accumulation of sedimenting matter (Cappenberg & Verdouw 1982) produced in the epilimnetic layers. However, production of algae (de Kloet 1982) and phototrophic bacteria (Steenbergen 1982) also contribute to the higher concentration in the metalimnion and hypolimnion.

The food levels in November and December decrease further (Fig. 5) because of low primary production rates (de Kloet 1982).

In contrast with the densities and biomass of zooplankton, mean annual concentration of zooplankton food, namely seston (<15 or 33 μm), decreased in the course of the study period. The seasonal averages of seston biomass were higher in the years 1973–1975 than in 1976–1978, especially during March–Nov. (Table 5). Comparing the monthly means of different years, it becomes clear that this decrease is most noteworthy in April and May. In both these months, the food level has steadily declined since 1975. This is indirectly related to an increase in both the numbers and the biomass of zooplankton as well as their grazing (see also p. 143 and Figs. 6 and 8).

Even though the coefficient of variation of annual average of total seston biomass based on seven years was only 10% (mean, 3.7 ± 0.37 g C · m^{-2}; range 3.3–4.3), the relative shares of fine and coarse fractions in the total seston did change notably. The

Table 5. Seasonal means (SD) of seston (<33 μm; zooplankton food) mass* (g C · m^{-2}) during the years 1973–1975 and 1976–1978 in Lake Vechten; the mass is also expressed as mean (SD) per cent of total seston mass (<125 μm).

Year/season		Dec.-Feb.	March-April	May-Sept.	Oct.-Nov.
1973–1975	Mass	2.2(0.6)	5.4(2.0)	2.5(0.7)	2.1(0.6)
	%	83 (5.5)	91 (2.5)	64 (14.8)	80 (10.7)
1976–1978	Mass	2.3(1.2)	3.8(0.5)	2.1(0.5)	1.7(0.3)
	%	87 (1.3)	88 (2.3)	53 (7.8)	63 (12.5)

* For conversion to g C · m^{-3}, see Table 4.

proportion of the latter fraction has increased from about 20% in 1973 and 1974 to between 29 and 38% in the subsequent four years (Fig. 7).

This increased dominance of coarse seston fraction since 1975 is caused mainly by *Ceratium hirudinella,* which has become important in two respects. First, it has been showing a rise in its densities in summer periods; secondly, both its growth and period of abundance start earlier in the summer and extend later, i.e. up to the start of autumnal turnover (Blaauboer 1982). The accompanying decrease in primary production rates (de Kloet 1982) in the years 1977 and 1978 is significant.

Grazing

The monthly and seasonal means of the zooplankton community grazing (%) and ingestion and assimilation rates during 1972–1978 are described first. Thereafter, changes in the grazing activity in the periods 1973–1975 and 1976–1978 are examined in relation to zooplankton densities, biomass and food.

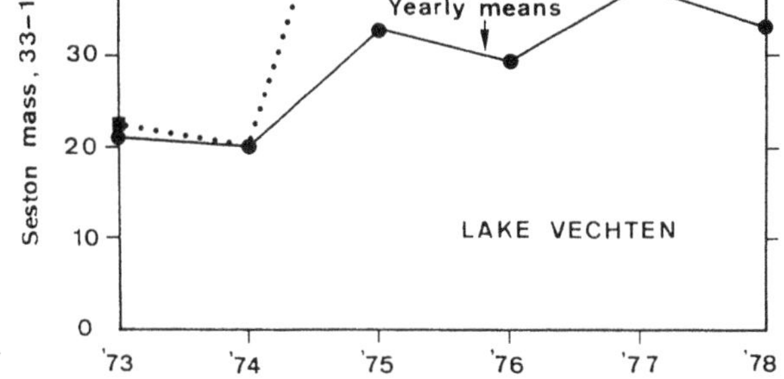

Fig. 7. The mean annual and seasonal (Aug.-Oct.) ratios (%) net seston:total seston mass in the years 1973–1978.

Community grazing rates. The fluctuations in community grazing were sharp both from season to season and from year to year. Also, spatial variations were great, as reported earlier (Gulati 1978a). The monthly means of daily grazing (Fig. 8) ranged from 2.9 ± 2.1 in March to 33.8 ± 16.3% day^{-1} in June. However, the ranges for individual months during different years were much wider. For example, the extreme values for these months were: March, 0.4 and 7.4% day^{-1}; and June, 12 and 73% day^{-1}.

The decrease in community grazing which starts in December continues until March. The factors governing the winter minimum (Table 6) are: low densities of the grazers other than *E. gracilis;* low water temperature; and to some extent increase in the food concentration. In March and April the grazing activities are still low though the feeding rates (see Fig. 8 and Table 6) do rise considerably, particularly in April when food concentration is at its annual maximum.

The sharp rise in the grazing rates from April to May, caused by the strong increase in the populations of *Bosmina, Daphnia* and *Eudiaptomus* species, is one of the most important aspects of the functioning of the lake's ecosystem. Also, warming up of the surface water and decreasing food concentration as a result of phytoplankton wax and wane and increased grazing pressure are the environmental changes contributing chiefly to the increased grazing activity. Interestingly, the zooplankton grazing rates in the month of May have steadily

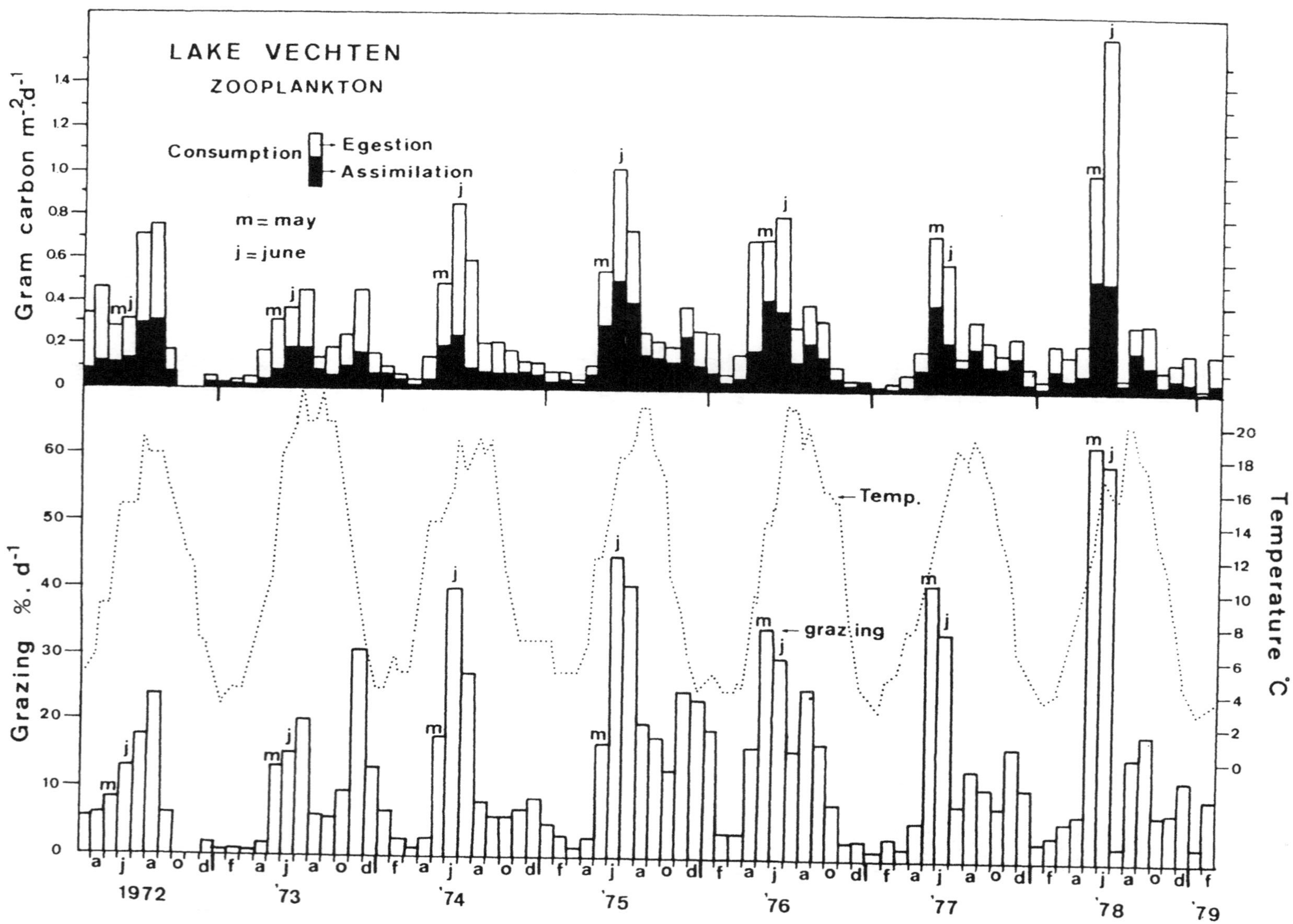

Fig. 8. Monthly means of daily grazing (%) and water temperature (lower) and the daily consumption and assimilation rates (upper) from March 1972 to Feb. 1979. Note the consistent increases in May (m) in the course of the years.

Table 6. Seasonal means (± SD) of zooplankton grazing and assimilation in Lake Vechten during March 1972–February 1979; n = number of observations.

Season	Grazing (% day^{-1})	Ingestion (mg C · m^{-2} · day^{-1})	Assimilation (mg C · m^{-2} · day^{-1})	Ass. efficiency (A/I (%))	n
March–April	4.4 ± 3.1	216 ± 136	70 ± 37	35 ± 11	34
May–Sept.	21.5 ± 6.7	473 ± 106	209 ± 66	44 ± 10	83
Oct.–Nov.	11.0 ± 5.5	193 ± 92	103 ± 54	53 ± 15	25
Dec.–Feb.	6.5 ± 5.5	107 ± 65	48 ± 29	45 ± 8	33
Annual	13.3 ± 3.4	291 ± 57	129 ± 35	44 ± 9	175

increased since 1972, but more so since May 1975 (Fig. 8). This is caused by a shift in spring maximum from June to May since 1976, similar to that in the grazers' densities and biomass (Fig. 6). The decline in grazing activity which sets in in July continues until the fall overturn in early November. But in some years in November an autumnal peak of grazing (15 ± 11% day^{-1}) occurred, caused mainly by *E. gracilis*, but also contributed to by *Daphnia* sp.

Ingestion rates. The course of seasonal fluctuations in the feeding rates is broadly similar to that of the grazing, but for the effects of changes in food concentration. The community ingestion rates ranged between 23 and 1632 mg C · m^{-2} · day^{-1}, respectively, in January 1973 and June 1978. Also the monthly mean minima and maxima fell in these months, viz. 85 ± 88 and 768 ± 471 mg C · m^{-2} · day^{-1} (Fig. 9). On a seasonal basis 67% of the annual consumption occurred in the period May–September (Table 6); the mean daily rate of 473 mg C · m^{-2} · day^{-1} in this period is more than twice the daily rate of phytoplankton production in the limnetic region.

On an annual basis the daily means for seven years varied from 218 in 1973 to 368 mg C · m^{-2} · day^{-1} in 1978, the grand study period mean being 291 ± 57 mg C · m^{-2} · day^{-1}. The ratio of daily zooplankton ingestion to daily phytoplankton production, for the years in which both parameters were measured, has significantly increased:

Year	1972	1973	1974	1977	1978
Ratio	1.38	0.63	0.68	1.80	2.26

Assimilation rates. The pattern of changes in the assimilation rates was similar to that of the ingestion rates (Fig. 9). Also, in this case spring is an important period, and a shift of maxima from June (1972–1975) to May occurred in the years 1976–1978. The assimilation efficiences (A × 100/I) fluctuated between 35 ± 11% in March–April and 53 ± 15% in October–November (Table 6; see also Fig. 9), and thus related indirectly to food concentration. The food in the spring period is predominated by nanoplankton but in autumn by detritus.

The annual means of assimilation efficiency ranged from 35% in 1974 to 56% in 1975 (study period mean, 44 ± 9%; see Table 6). In 1975, the monthly means (43–67%) were the highest compared with those in the other years; incidently, 1975 marked a transition to increase in zooplankton biomass and their grazing activities, etc. Because the mean assimilation efficiencies are often lower than approx. 50%, daily zooplankton egestion (I–A), would generally amount to more than half of the food ingested daily (unshaded area, Fig. 8, top). Depending on the time of the year, the amount of detritus daily generated by the zooplankton will fluctuate, the mean range being from 59 in Dec.–Feb. to 260 mg C · m^{-2} · day^{-1} in May–Sept. (computed from I–A in Table 6). Although a still unknown part of this detritus will directly enter the dissolved organic carbon (DOC) pool, there is some evidence that the principal part of it is in particulate form and enters the seston pool (Gulati, unpublished data). Especially in summer, when detritus daily generated is high (260 mg C · m^{-2} · day^{-1}), it will – despite 'losses' to the DOC pool – form an important part of seston food available to the herbivores.

Changes in grazing activities. The grazing patterns changed in accordance with the shift in the maxima of zooplankton densities from June in 1973–1975 to

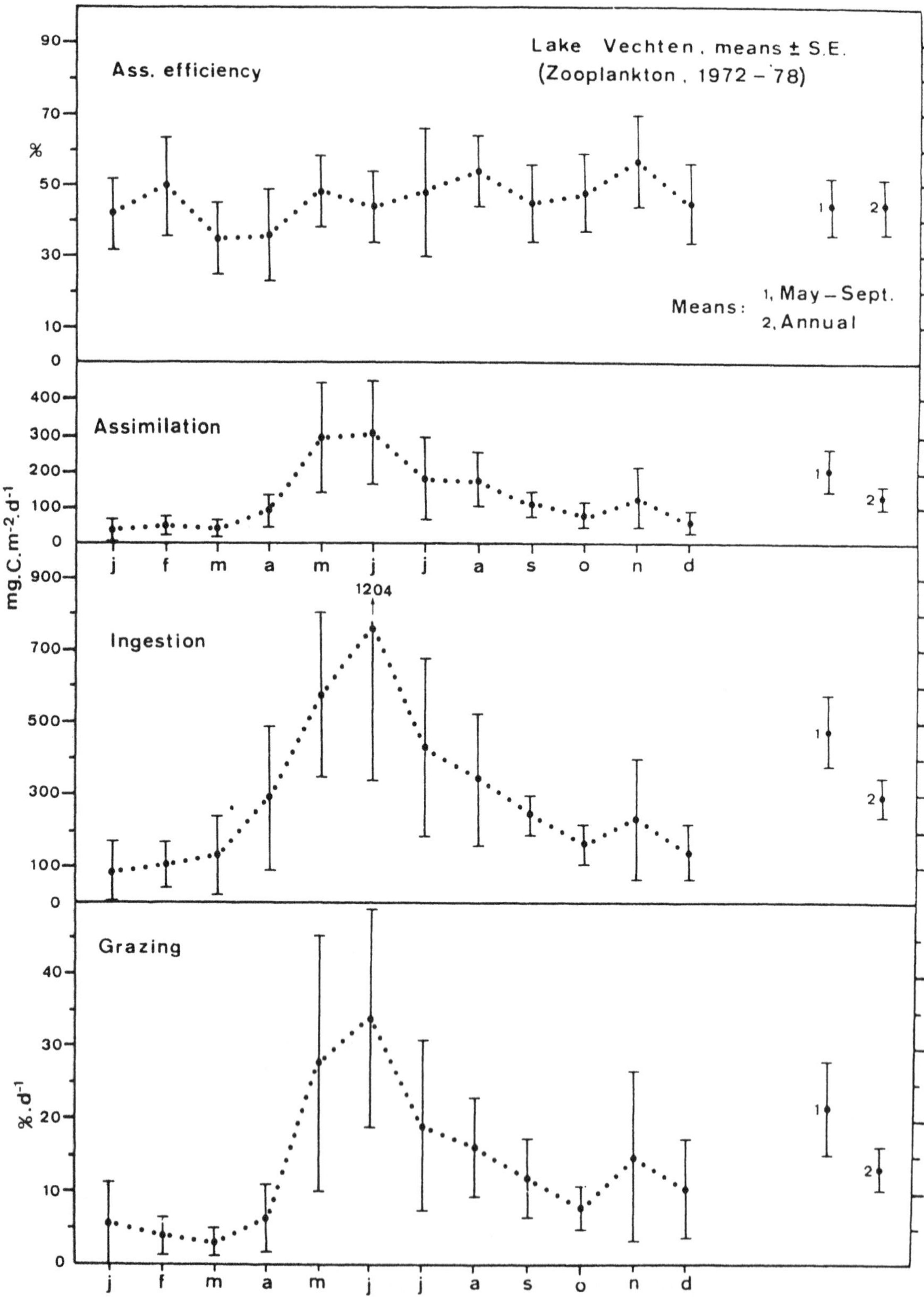

Fig. 9. Grand daily means with S.E. (95% confidence limits) (n = 7) in the different months; the data are derived from Fig. 8. Summer period and daily means are also shown.

May in 1976–1978, and with the increases in standing crop biomass and mean individual weights (Table 4; Fig. 6) during 1976–1978.

A comparison of the monthly means of the zooplankton grazing activity during 1973–1975 and 1976–1978 reveals increases in Feb.–June and Aug.–Sept. during 1976–1978. A three-fold rise in zooplankton grazing pressure in May is highly significant ($t = 3.2$; $P < 0.01$; $n = 15$), reflected also in the consumption and assimilation rates. These changes are owing mainly to the increase in the densities of major grazer species, namely, *E. gracilis, Daphnia* spp. and *B. longirostris*. Contrary to these upward trends in the zooplankton grazing pressure, in July and during Oct.–Dec. (Fig. 8) the grazing rates decreased. Here again, the decline in herbivore densities or the increase of biomass per individual (Fig. 6C) or both played a role.

The opposite patterns of changes noted in the course of different months during 1973–1975 and 1976–1978 do offset the differences in daily means computed from the annual data. Therefore, the overriding influence of increased zooplankton grazing (Gr.), ingestion (I) and assiilation (A) rates in the spring periods of 1976–1978 is only partly recognizable from the comparison of the annual means of two periods as follows:

Period	*1973–1975*	*1976–1978*
Gr (% day^{-1})	13 ± 3	15 ± 2
I (mg C · m^{-2} · day^{-1})	271 ± 57	308 ± 73
A (mg C · m^{-2} · day^{-1})	120 ± 55	143 ± 12

Specific filtering rates. The daily grazing rates (%) of the zooplankton community may be expressed as F_{total}, i.e. filtering rate in ml · day^{-1} · l^{-1} (equation 4, in Methods). However, the seasonal variations in F_{total} caused by water temperature and food concentration can only be examined by transforming the F_{total} into specific filtering rate (SFR), i.e. into ml · day^{-1} · mg^{-1} zoop. C; this is done by dividing the F_{total} by the filterfeeders biomass in mg · C · l^{-1}.

The SFR data thus obtained were subdivided into two categories, i.e. one for the food concentrations below 0.5 mg C · l^{-1} (mean 0.34) and the other for those above 0.5 mg C · l^{-1} (mean 0.71). Each category was split further into five temperature intervals of 5 °C, from 0 to 25 °C (Fig. 10).

The SFR at the lower food concentration rises from 683 at 0–5 °C to 1 730 ml · day^{-1} · mg^{-1} · zoop · C at 15–20 °C, but falls as the temperature exceeds 20 °C. This is even more so at the higher food concentrations; and, as expected, over the entire temperature range SFR is lower, but shows the sharpest increase as temperatures increase from 0–5 to 5–10 °C. This reflects the late-April to early-May situation in the lake when food concentration is still higher than 0.5 mg C · l^{-1} but the temperature rises. The decrease in SFR if the temperature exceeds 20 °C – at both food concentrations – is conspicuous.

The increase in water temperature especially in spring promotes the development of grazers and rise in their densities so that food concentrations sharply decline causing a sharp increase in SFR. Thus, in late April–May both temperature and food changes are 'conducive' to increased filtering; contrariwise, in March both low temperature and high food concentration depress such an activity. This explains the order of magnitude difference between the minimum and maximum SFR (120–1 730 ml day^{-1} · mg^{-1} · zoop · C; see also Fig. 10).

SFR versus temperature and food. As expected, in Lake Vechten a combination of low temperature and higher food concentration results in much lower SFR than if the temperature is high and food concentration low. On the other hand, simultaneous decrease of both variables – or their increase – is likely to nullify each other's effect on SFR. Thus, it is not possible to distinguish between the effect of temperature and that of food, especially in the field situation. In this regard, multiple regression statistics of log-transformed data of temperature, food concentration and SFR was done according to the equation: $Y = a \cdot X_1^{b_1} \cdot X_2^{b_2}$, where Y is SFR; X_1 and X_2 are temperature and food variables; a is the intercept; and b_1 and b_2 are, respectively, the partial regression coefficients of X_1 and X_2.

The sets of data used for the regression analysis are (Table 7):

A) seven years (1972–1978) separately;
B) seven years pooled ($n = 166$); and
C) the averaged monthly means of seven years ($n = 12$).

In case A the variance ratio (F) was significant in all seven years ($P < 0.001$, but $P < 0.002$ in 1972 and 1974). The beta weights calculated according to Davis (1971) showed the relative importance of b_1 and b_2 in controlling the SFR. In addition, keeping

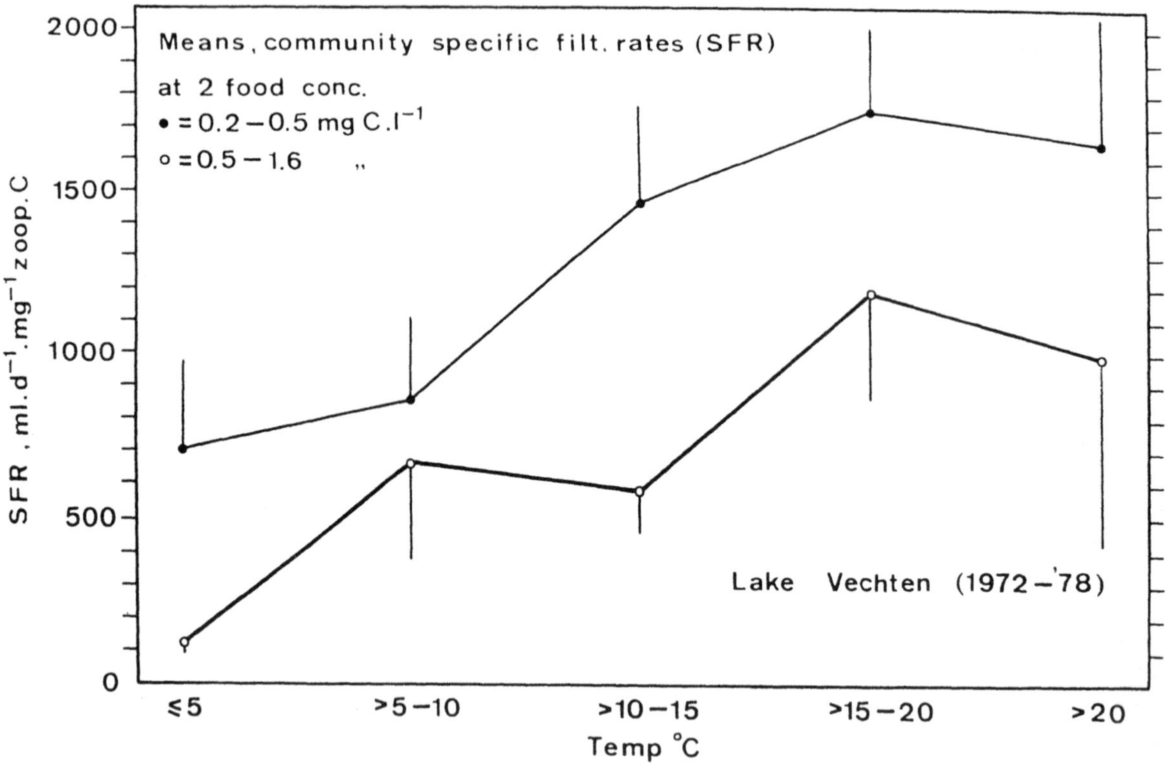

Fig. 10. Effect of food concentration and water temperature on the community SFR; the prevalent food concentrations were split into two arbitrary categories, namely, one below, and the other above 0.5 mg C · l^{-1}. The water temperature was also divided into five temperature categories of 5 °C intervals. The standard errors of the mean SFR on the lower side of the upper curve and upper side of lower curve are omitted for clarity.

Table 7. Results of multiple regression of log-transformed data ($Y = a \cdot X_1^{b_1} \cdot X_2^{b_2}$) on water temperature ($X_1$) and food ($X_2$) against SFR (Y); a is the intercept, and b_1 and b_2 are, respectively, the partial regression coefficients of the water temperature (°C) and food concentration in mg C · l^{-1}. See also Table 8.

Data sets	Year	a	b_1	b_2	R^2	F	D.F.	P<
A	1972	10.52	1.454	0.104	0.63	11.1	13	0.002
	1973	17.33	1.247	-1.157	0.61	17.5	22	0.001
	1974	8.68	1.540	-1.011	0.51	10.8	21	0.002
	1975	52.88	0.627	-1.447	0.62	17.0	21	0.001
	1976	23.11	1.403	0.091	0.76	42.1	27	0.001
	1977	20.15	0.950	-1.334	0.81	46.2	21	0.001
	1978	19.81	1.054	-1.340	0.52	11.6	20	0.001
B	1972–1978	19.16	1.1261	-1.106	0.56	104.2	163	0.001
C	1972–1978	21.04	1.1437	-1.451	0.96	106.9	9	0.001

A: All the years separately.
B: Pooled data of the seven years.
C: Pooled monthly means (n = 12) derived from the monthly means of the seven years.

the food level at 0.5 mg C · l^{-1} and varying the temperature from 10 to 20 °C or vice versa, the changes in SFR were examined. The coefficients of determination (R^2) show that the changes in SFR explained by those in temperature and food ranged from 51% in 1974 to 81% in 1977.

In case B, R^2 explained 56% of the SFR variations. However, in case C, R^2 accounted for 96% of the changes in SFR. Apparently, the striking reduction in the variations is caused by pooling the seven-years' monthly observations which ranged from 11 to 17 per month.

The partial regression coefficients, b_1 and b_2, varied from year to year, too. The beta weights show that in 1972 and 1976, changes in the temperature and not in food concentration appeared to govern SFR. In 1972 it was related to high food levels, but in 1976 to the highest water temperatures, 1976 being the warmest year in the seventies.

Using the data in Table 7, the effects of temperature and food were examined in some detail and are summarized in Table 8. For this, first, the food was kept constant at 0.5 mg C · l^{-1}, but the temperature varied from 10 to 20 °C (Q_{10}), and, secondly, food was varied from 0.5 to 0.25 mg C · l^{-1}, but temperature kept constant at 15 °C. The Q_{10} ranged from 1.5 to 2.9. Interestingly, the SFR more than doubled in five of the seven cases if the food level was halved (temperature 15 °C), i.e. the community consumed more food despite the decrease in food concentration. Thus the daily ration increased even at a constant temperature. This, however, possibly reflects the May–June situation when the increases in SFR are accentuated by small grazers, e.g. *Daphnia cucullata* and *Bosmina longirostris*.

On the other hand, although the food concentrations in Lake Vechten are constant at approx. 0.5 mg C · l^{-1} from May to September, the SFR does keep on increasing at the same time as the temperature until August, and decreases in September concomitantly with temperature. Thus using the monthly means of temperature and food, this would give us the best predictions for the SFR (Fig. 11). However, for these predictions the role of changes in composition and size of the animals may be important, but was not considered.

Table 8. Computed factors showing increase in specific filtering rates (SFR): I), if water temperature was increased from 10 to 20 °C (Q_{10}) but food was constant at 0.5 mg C · l^{-1} and II), if food concentration was decreased from 0.5 to 0.25 mg C · l^{-1}, keeping water temperature constant at 15 °C. The data were derived from the regressions in Table 7.

Data sets	Year	I (Q_{10})	II
A*	1972	2.74	0.93
	1973	2.37	2.23
	1974	2.91	2.02
	1975	1.54	2.73
	1976	2.64	0.94
	1977	1.93	2.52
	1978	2.08	2.53
B*	1972–1978	2.04	2.16
C*	1972–1978	2.16	2.35

* See footnotes to Table 7.

Daily ration (DR), and effect of temperature. Although the SFR does give insight into the potential feeding activities of the filterfeeders, the daily ration (DR) – the ratio between ingestion rate and filterfeeders biomass (I × 100/B) – is a useful index of the food consumed. Besides, the DR is a useful mean for comparing growth potential of zooplankton in different lakes.

The annual means of DR ranged from 26 to 61%; but those during May–Sept. were higher (45–128%) and almost three times the ones in the period Oct.–April (Table 9). The differences between the summer and rest of the year are pronounced and most likely due to higher temperatures in summers.

The effect of water temperature (X) on the DR (Y) in the different strata – both individually and integrated – was analysed in 1976 and 1977. For this log-log ($Y = aX^b$) and exponential ($Y = ae^{bx}$) regressions were performed on the two years' data, separately and pooled. In both types of treatment, there was a significant positive correlation ($P < 0.001$) between DR and water temperature (t). The regression equations for the integrated, pooled data

Table 9. Seasonal ranges and means of daily ration per cent during the period 1972–1978. The data are based on averages of seven years (n = 7).

Period	May–Sept.	Oct.–April	Annual
Range	45–128	16–36	26–61
Mean (SD)	68(29)	24(7)	41(11)

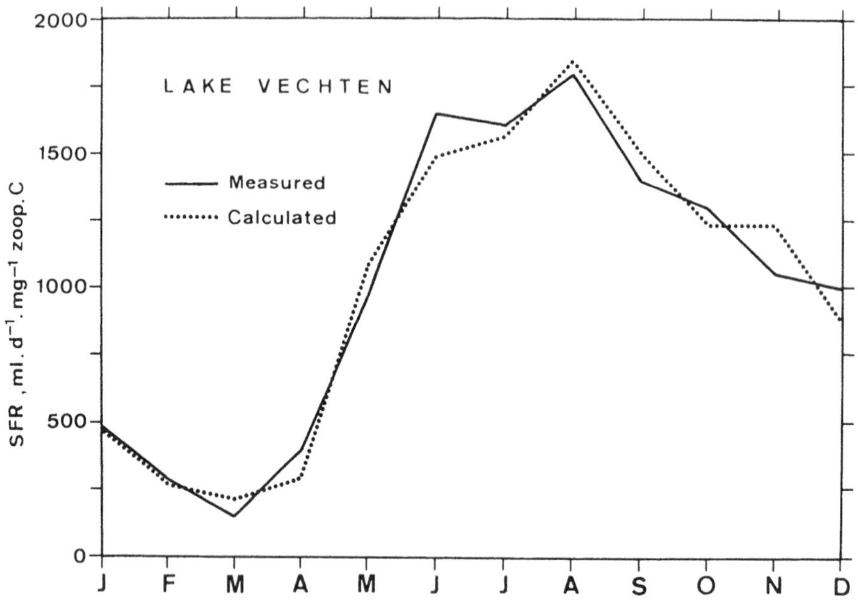

Fig. 11. The measured mean monthly values of SFR during the seven-year study and those calculated via the regression equation as in Table 7 under C.

of 1976 and 1977 are:

Power function, $DR = 2.0162\ t^{1.094}$ (r, 0.75; DF, 53; $P < 0.001$)

Exponential, $DR = 7.299\ e^{0.105\ t}$ (r, 0.71; DF, 53; $P < 0.001$).

In both cases, the unexplained variation ($r^2–1$) is rather high, possibly because the seasonal changes in the filterfeeders species composition are not accounted for in the regressions. In addition, the striking decrease in food concentration in the spring period in the lake is apparently compensated by the increased community filtering. Lastly, the quality of the food particles certainly may influence ingestibility. The Q_{10} values of integral data of 1976 and 1977 were, respectively, 2.21 and 2.00, the pooled mean being 2.13. In 1976 the mean Q_{10} value of 2.21 for DR is substantially lower than that (Q_{10} 2.64) for SFR (Table 8). Examination of raw data shows that at temperatures of about 10 °C (April) SFR was relatively low but the daily ration high, because of the high food concentration in April. The Q_{10} for DR was therefore reduced. In 1977 the Q_{10} for the SFR (1.93; see Table 8) and DR were quite similar.

Assimilation index, a ratio similar to DR in which assimilation (A) substitutes for ingestion, is influenced directly by the assimilation efficiency. The temperature effect on the assimilation index was more profound than on DR, particularly in the deeper layers in which the Q_{10} was 2.53 as against 1.80 in surface waters; the mean Q_{10} for the integrated, pooled data was 2.47 in contrast with 2.13 for DR.

Discussion

One of the important problems in our grazing study concerned errors relating to the ^{14}C-tracer technique. A critical analysis of the methods employed is essential and, therefore, precedes the discussion on the impact of grazing in the lake's ecosystem. The physiological efficiencies and turnover rates obtained from our results are compared with those in the literature. The grazing is compared with data on other Dutch lakes and finally summed up in the light of energy balance of the filterfeeders.

Methodological problems and errors

In spite of the high reproducibility of the ^{14}C-tracer technique, several errors in the methods employed are unavoidable. Besides, it is almost impossible to improvise and simulate the field situation under laboratory conditions. The changes in the behaviour of the organisms related to diel migratory rhythms may also introduce errors in extrapolations based on a single daily measurement. Some errors especially concerning the ^{14}C-methodology have come to light during the present study; the others, we assume, are likely to compensate for each other in a long-term study, such as the present one.

The techniques employed and the data collected are critically analysed. This is important, both in comparing the results with those from other temperate lakes and in examining the role of the zooplankton in the ecosystem of the lake itself. The errors in the data of one parameter will have a bearing on another, and so on; an under- or overestimation of grazing, for example, may respectively over- or underestimate the assimilation efficiency. Consequently, errors in both the assimilation and the respiration rates will be reflected in the calculated production, and thus in the gross and net growth efficiencies and in the turnover rates. Based on these errors, the data on grazing, assimilation and respiration rates are assessed and re-evaluated.

Errors in grazing rates. Important among the likely errors in the grazing methodology are:

1) The direct effects using tracer food concern: a) the shift in the species composition and in the size of particles in the tracer during its preparation; b) differences in the specific activity of the particles of different sizes; c) changes in the *in situ* food concentration caused by the tracer addition; and d) the size fraction used as labelled food and bacteria as food source.

2) Leaching or loss of isotope associated with killing and preservation of the labelled organisms before handsorting of the individuals of the different species.

3) Diurnal changes in the grazing patterns.

Among the errors listed under 1), because of the long exposure period for tracer preparation, there was indeed a shift in species composition (R. Pot, pers. comm.). Some less frequent and rare species in the lake, e.g. *Scenedesmus,* increased in numbers; the species normally abundant, however, remained dominant, even after the labelling period, e.g. *Chlorella pyrenoidosa* and the smaller flagellates. There was also a slight shift after labelling in the size spectrum especially in the particles ranging from 4–7 μm to 3–5 μm. Haney (1973) found no significant differences using tracer particles in the size ranges varying from 12–14 μm in algae, to 3.5 μm in yeast, and to 0.5–1.2 μm in bacteria. A possible over- or underestimation of grazing due to use of tracer in the 3–33 μm size class during 1976–1978, and 3–15 μm up to 1975, could have occurred. However, the size classes used are well within the range of particle size, namely, 1–15 μm, that the planktonic herbivores compete for (Hall *et al.* 1976; see also Brooks & Dodson 1965), even more so in Lake Vechten because most particles in the 3–33 μm fraction fall within the 3–15 μm range (pers. unpublished data).

The increase in the *in situ* food concentration due to the addition of tracer was maximally 12%, but mostly much lower. A correction for this is relatively more important at or below the incipient limiting level of food concentration, i.e. for short periods in May–June in some of the years only. In the rest of the year, when the *in situ* food concentrations are above the critical food level, the ingestion rates would be somewhat underestimated. However, a slight decrease in *in situ* food concentration during the acclimation period of about one hour, i.e. before the addition of tracer, would partly nullify the effects of tracer addition. Thus the net increase in the *in situ* food concentration would not to any great extent underestimate the feeding rates.

The role of net seston >33 μm as a food source was examined simultaneously with the routine studies using the smaller seston fraction (<33 μm) during June–October 1977 (unpublished data; Postema & Siewertsen 1981). The mean ratio of filtering rates on small: large fractions was 1:0.60; and that of the seston biomass in these fractions was 1:0.57. Thus, the fractions were consumed in the ratio of 1:0.34 (0.60 × 0.57), and the consumption rates based on the total seston will be 1.34 times the ones based on the smaller fraction alone. However, these food selectivity experiments were restricted to the adults of the different filterfeeders, the biomass of which is roughly one half of the total community biomass. It has become evident that usually the potential to feed on the larger food particles increases with age because of the increase in filtermesh size (Geller & Müller 1981). Assuming that juveniles feed solely on the smaller seston fraction the ratio ingestion of nanoseston: net seston by the community will be 1:0.17 in contrast with the 1:0.34

Table 10. The measured and corrected daily rates (mg C · m^{-2} · day^{-1}) of the different food budget parameters; CF: correction factors; for details see text.

Parameter	Measured	Corrected	CF_1	CF_2	CF_3	CF_4
Ingestion (I)	291	586	1.72	1.17	–	–
Assimilation (A)	130	218	1.35	1.09	1.14	
Respiration (R)	77	102	–	–	–	1.33
Production (P)	53	116	–	–	–	–

CF_1: Leakage of ^{14}C in preservation medium (for I, 42% for A, 26%).
CF_2: Net-algae eaten (for I, 14.5% and for A 7.7%).
CF_3: Loss of $^{14}CO_2$ during assimilation, 12.5%.
CF_4: Metabolic losses of $DO^{14}C$ during respiration, 25%.

based on adults alone. Thus the daily grazing and ingestion rates based on the small fraction alone underestimate the community ingestion by 15% (see Table 10).

As the bacterial biomass in the lake (1977–1979) varies roughly between 0.2 and 0.8 mg C · l^{-1} (recalculated from densities measured by R. van Keulen), the role of bacteria as a potential food source cannot be neglected (Geller & Müller 1981). The community grazing rates would be, therefore, higher than the ones measured in the present study.

The second category among the errors relating to radio-tracer technique is the loss or leaching of isotope into the preservation fluid (Holtby & Knoechel 1981). We found this to be the principal source of error and have since 1976 kept a full inventory of the preservation losses in formalin. Since one of the aims of the grazing study in Lake Vechten was to examine the role of individual species (Gulati 1978), preservation and storage of the labelled organisms was indispensable (hand-sorting being time-consuming). The average isotope loss as a percentage of the total radioactivity, i.e. in the animals plus that in the preservation fluid, was 42 ± 10% in the summer periods of 1976 and 1977. A linear regression relationship between the loss of radioactivity in DPM (X) and total radioactivity (Y), i.e. loss plus radioactivity in the animals, gave a highly significant correlation ($r = 0.99$; $n = 58$). The regression equation was:

$$Y = 244 + 1.57X.$$

Holtby & Knoechel (1981) reported an even higher loss, namely 57%, in the formalin preserved animals, and a stabilization time of 26 h even though about 90% of this loss occurred in the initial 4 to 5 h. The ^{14}C loss of 42% in our study also stabilized in about 4–6 h. Both dislodgement of radioactive food from the food groove (Rigler 1971) and expulsion of the gut contents are probably the most important constituents of the material lost. However, the losses of the assimilated organic material are also highly significant (see later). Therefore, the third most likely component in the ingested ^{14}C-material lost is the dissolved organic carbon. The 'actual' ingestion rates can be computed by multiplying by a factor of 1.72 (100/{100–42}). The corrected data in Table 10 indicate grazing pressure to be much higher than estimated so far.

The third type of error concerns possible differences in day and night grazing rates; it needs to be especially checked in our study in which the conversion into daily rates is based on the single day time measurements. From the 26 incidental, diurnal measurements in the course of study in the lake, the evidence of a diel grazing rhythm as reported by Haney & Hall (1975) was equivocal. In another deep Dutch lake, namely Maarsseveen, which has a filterfeeder composition similar to that in Lake Vechten, a diurnal grazing study – stretching over three consecutive days – also did not reveal a clear or consistent increase in the grazing activity at night (Gulati in Ringelberg 1981). In Lake Vechten the daily ration (%) calculated by integrating the specific ingestion rates measured after sunset, before sunrise and during the day was 38% compared with 44% based on the daytime measurement alone. This means a decrease of 15% in the grazing activity at night, which finding is contrary to the findings of Haney (1973) on Heart lake but in agreement with Gliwicz's unpublished data (cited by Haney 1973) on two Polish lakes. The discrepancy in Lake Vechten in the zooplankton biomass – invariably higher at night than during the day – is as yet unresolved. A correction for the grazing differences between day and night at this stage is, therefore, premature.

The above discussion demonstrates that the corrected or 'true' grazing and other food budget parameters in Lake Vechten are apparently much higher than mentioned in the results (see Table 10).

Errors in assimilation rates. The errors in the assimilation rates concern the entry of freshly assimilated ^{14}C-isotope first into a short-term or metabolic pool with a much faster turnover rate than that of the much larger, long-term structural pool (Conover & Francis 1973; Lampert 1975). Lampert (1975) found that about 25% of the freshly assimilated material in *Daphnia pulex* was lost in the metabolic process in the first hour. The rate of loss, however, appears to depend on the respiration rates which in Lake Vechten zooplankton had an annual mean of 0.4% h^{-1} of body carbon compared with 0.8% h^{-1} in Lampert's animals. Therefore, the respiratory loss of $^{14}CO_2$ during the assimilation experiments in our case will be about 12% (25 × 0.4/0.8) on an annual basis and the corrected assimilation rates 1.14 (12/(100–12) × 100) times higher; see Table 10.

The other two sources of error, namely the loss in the preservation fluid of ^{14}C and the one concerning macro-algae (>33 μm), were discussed under the errors in the grazing rates. The per cent loss in the preservation medium was, as in the ingestion samples, also highly significant (r = 0.99) and fairly constant, but markedly lower, viz. 26 ± 8.5, the correction factor being 1.35. The underestimation due to not accounting for netplankton was 8.3% (= a correction factor of 1.09). The grand total correction for the assimilation was 68% (1.14 × 1.09 × 1.35 = 1.68), resulting in a revised mean assimilation rate of 218 mg C · m^{-2} · day^{-1}, compared with the uncorrected 130 mg C · m^{-2} · day^{-1} (Table 10).

Errors in respiration rates. The respiration rates were measured in the years 1973–1975 (Gulati 1975). For the years 1976–1978, the respiration data were computed from the temperature and zooplankton using a regression equation ($Y = ae^{bx}$) based on the observations (see under Methods, p. 132). The mean respiration rate calculated for the filterfeeder community during the entire study period was 77 mg C · m^{-2} · day^{-1} (Table 10). Almost identical results were obtained in 1980, when respiration rates were measured in relation to nutrient regeneration by zooplankton (Gulati *et al.* 1981).

The studies on nutrient regeneration by zooplankton indicate that approx. 80% of P and 50% of N excreted may be in inorganic form. One can therefore assume that losses as dissolved organic carbon are substantial. The works of Dagg (1976) and Copping & Lorenzen (1980) clearly demonstrate the significance of these losses in constructing the carbon budgets. The losses as DOC found by these workers vary respectively from about 25 to 100% of the respiration losses, measured as CO_2 produced or O_2 consumed. Thus a minimal correction of our respiration results based on the literature data and on our own preliminary findings, would be an increase of 25%. Multiplying the uncorrected respiration rates of 77 mg C · m^{-2} · day^{-1} with 1.33 (1 + 25/100 – 25) gives a mean daily rate of 102 mg C · m^{-2} (Table 10).

Impact of grazing in the lake ecosystem

The data on daily grazing, assimilation and metabolic rates formed the main bulk of the information collected. A substantial part of the grazing study 1977–1979 was conducted simultaneously with the investigations on phytoplankton and on primary production, mineralization and sedimentation rates. The most recent studies on zooplankton concern the role of nutrient regeneration by zooplankton in meeting the demands of phytoplankton. The data are essential in interpreting the impact of grazing in its totality (see p. 139). Based on grazing pressure, the food situation and the climatic factors, the seasonal events in Lake Vechten can be divided into six different periods:

a) Dec.–Feb., the period of low daily grazing, especially in terms of daily ingestion and assimilation rates;
b) March–April, low grazing pressure but increasing ingestion and assimilation rates due to high food levels;
c) May–June, the period of 'clear water' related to highest grazing, ingestion and assimilation rates, highest densities of filterfeeders, sharp decrease in food concentration;
d) July–Sept., high grazing pressure, but moderate ingestion and assimilation rates; and
e) Oct.–Nov., the transition to low grazing.

a) Dec.–Feb., the period of low activity. The low grazing pressure, 6.5 ± 5.5% day^{-1}, and high fluctuations seem to be caused by low water temperature and low densities of filterfeeders. However, in view of the lowest annual primary production rates and low biomass of phytoplankton (de Kloet 1982), the pressure on the latter appears to be high. The phytoplankton daily turnover rate, k_i (Frost 1980), calculated from the primary production and the carbon data of 1977–1978 derived from chlorophyll following Golterman & Kouwe (1980), varied between 0.03 and 0.10 day^{-1}. In comparison, the grazing rate coefficient g_i (daily grazing %/100; cf. Frost 1980) ranged from 0.01 to 0.17 and the ratio (%) g_i:k_i from 40 to 380 (average 135). Thus, despite the low grazing activity in Dec.–Feb., the food consumed by the grazers was far greater than produced by the phytoplankton. This was especially true in December and early January. The mean egestion rate during Dec.–Feb., i.e. the input of detritus via zooplankton was about 60 mg C · m^{-2} · day^{-1} (Table 6) compared with the primary production of about 30 mg C · m^{-2} · day^{-1} in early winter (de Kloet 1982). Therefore, the enrichment of the seston pool in December to early January via detritus will exceed that via phytoplankton production. The lake being in circulation, except for short spells of ice cover (Steenbergen & Verdouw 1982), resuspended detritus from the lake sediments and from the littoral region (Gons 1982) apparently prevents the food concentration falling to below 1.6 ± 0.6 g C · m^{-2} or 0.18–0.40 g C · m^{-3} (Z_m, 5.55 m, Fig. 1). The food levels are roughly around the incipient limiting level for the filtering activity (Lampert 1977b). *Eudiaptomus gracilis,* the most dominant filterfeeder in the lake during winter, appears to be able to exploit the conditions in early winter with success. The ability of this calanoid to utilize detritus (Nauwerck 1963) may contribute to its success. In late January, but usually in February, the nutrients regenerated during winter mixing and increasing light and temperature promote nanoplankton increased algal growth in the lake, as in other temperate lakes (Hutchinson 1967).

Both the primary production (de Kloet 1982) and the food concentration start increasing rather early in the year. Here, cryptomonads, small flagellates and *Stephanodiscus hantzschii* (Blaauboer 1982) are important as zooplankton food since they have optimal size to graze upon (Porter 1977). The grazing activity in February, however, seems to be still low probably because of low water temperature.

b) March–April, the period of high food concentration. In March and April, the seston concentrations increase sharply because of the reasons already mentioned. Although the grazing rate mean of 4.4 ± 3.1% day^{-1} is the lowest compared with the other periods, the consumption and assimilation rates are not so low (Table 6). The average g_i in March and April during the years 1977 and 1978, 0.037 and 0.063, compared with the k_i of 0.072 and 0.128 in these months for the phytoplankton gives g_i:k_i ratios (%) of 51 and 49 respectively. Thus the mortality caused by the grazers is equivalent to 50% of the material daily produced. This, however, does not imply that grazers consume *per se* 50% phytoplankton and the rest detritus. The proportions in the grazers' menu of phytoplankton and detritus will depend directly on the pool sizes of these two food sources, provided there is no selective feeding on one of the pools. The assimilation efficiencies of 35 ± 11% (Table 6, Fig. 9) in this period are the lowest, reflecting on the food quality and possibly on the concentration maxima. The high food levels may lead to short gut passage time or superfluous feeding. This latter has been disputed by Conover (1966), and for this the data on cladocerans are controversial (Lampert 1977b). The food levels in March and April, respectively, of 4.6 ± 1.4 and 5.1 ± 2.0 g C · m^{-2}, or 0.83 ± 0.25 and 0.92 ± 0.36 g C · m^{-3} are well above the incipient limiting levels. One would, therefore, expect the assimilation rate to reach a plateau, i.e. follow the same pattern as ingestion rates. Relatively greater underestimation of assimilation rates than those of ingestion, owing to the methodological errors, does not seem to have occurred (Table 10).

The doubling of daily grazing % and of consumption and assimilation rates in April compared with those in March (Fig. 9) does not result in a decrease in seston concentrations. Both the phytoplankton densities (Blaauboer 1982) and primary production rates (de Kloet 1982) are higher in April than in March.

The increased food availability in the spring every year most probably forms the main basis for high fecundity (Wetzel 1975). Also, the clutch size of *E. gracilis* attains maximal value under the optimal food conditions (Elster 1954; Comita & Anderson 1959; see also Hutchinson 1967). The rise in temperature in this period (Fig. 4) perhaps contributes significantly in causing the increased rates of egg production and instantaneous growth (Hall 1964) and decreased duration of egg development (Vijverberg 1980). This latter holds particularly true for *Bosmina longirostris* and *Daphnia* spp. both of which increase sharply in density in May. Since the average individual weight of the dominant filterfeeders is much lower in May than in April (Fig. 6), the generation time would be considerably shorter too (Weglеńska 1971).

c) May–June, the period of highest activity. The optimal food situation and increasing temperatures in early spring result after a time lag (Hutchinson 1967) in a sharp increase in May and June in the densities of filterfeeders. The community is dominated by the species of *Bosmina, Daphnia* and *Eudiaptomus.* This is, however, the most crucial period in regard to the timing of several events which perhaps follow each other in quick succession and certainly have a great bearing on the zooplankton grazing intensity. Important events in the lake are: 1) the start of summer stratification and warming up of the upper water layers (Steenbergen & Verdouw 1982); 2) a sharp incease in N:P ratio, due to the depletion of PO_4-P because of relatively high algal production in early spring (de Kloet 1982); 3) a pronounced reduction in the densities of dominant nanoplankters, less due to natural mortality but more due to the grazing pressure; this results in a noticeable decrease of seston biomass, which would contain increasingly more detritus and less live algae, as evident also from the very low chlorophyll values (de Kloet 1982); and 4) an increase with depth of seston concentrations, possibly because of increased sinking, and the most significant increase in grazing losses and low food availability (Gulati 1978a) in the upper layers.

A consequence of warming up, nutrient depletion and of reduction in nanoplankton and seston biomass in response to the mortality caused by the grazers is increase in water transparency to 4–4.5 m in contrast with 2–3 m normally, and a decrease in extinction coefficient (E Phar, see also de Kloet 1982; and Steenbergen & Verdouw 1982). This period of improved light climate, called the 'clear water phase', has been observed in other deep, stratifying temperate lakes during late May or June (Nowak 1975; Lampert 1978). Despite the marked decreased in nanoplankton and chlorophyll concentration in this period, in May the primary production rates are in the range of annual means, and those in June even higher. The primary production: chlorophyll concentration ratios are the highest in this period (de Kloet 1982), and significantly correlated with zooplankton assimilation maxima which in turn are related with zooplankton excretion rates (unpublished data). The mean daily zooplankton excretion rates in May and June 1978 were: NH_3-N, 5 $\mu g \cdot l^{-1} \cdot day^{-1}$; and PO_4-P, 1 $\mu g\, l^{-1} \cdot day^{-1}$, and covered about 70 and 100% of the daily nutrient demands of phytoplankton (Gulati *et al.* 1981). Besides, the indirect effect of grazing will be a reduction in the producers' standing crop, leading to their increased turnover rates. The producers may, therefore, respond independently of the increased nutrient regeneration brought about by grazing (Cooper 1973). These feedbacks in the grazed–grazer relations are likely to provide a considerable stability to the apparently grazer-dominated system in this period.

Although the mean daily grazing rates in the months of May and June were 27.7 ± 19.1 and 33.8 ± 16.3%, respectively, the individual values in these months varied widely, from 2.3% day^{-1} in May 1975 to 72% day^{-1} in May 1978 (see also Fig. 8). Corresponding values in June were 12 and 73% day^{-1}, respectively. The grazing maxima in Lake Vechten, though high, are much lower than those of 469 and 114% recorded by Haney (1971, 1973) in the nutrient rich Lake Heart and in a bog lake, respectively. Due to the lack of data on primary production in these lakes, it is difficult to compare the grazing pressures in terms of primary production. However, Haney's rates in Heart Lake are contrary to the concept (Hillbricht-Ilkowska 1972) that grazing pressure decreases as the trophy grade increases.

The annual grazing maxima in May may be attributed mainly to the increase in both the density of filterfeeders and their biomass and the decrease in food concentration and rise in water temperature. The water temperature in May and June is lower than that in the subsequent summer months and is probably less important in causing the annual maxima. The weight specific filtering rate or SFR in ml · $day^{-1} \cdot mg^{-1}$ zoop C (see p. 143) is a useful means of comparing the grazing intensities in relation to food concentration and temperature in the different periods. The mean SFR of 966 in May is about 2.5 times that in April (Fig. 11). In contrast, the grazing % per day in May is about 4.5 times that in April, namely 27.7 against 6.1% day^{-1}. Using the regression equation in Table 7 under C (see also Table 8), it becomes clear that the decrease in mean food concentration from 0.93 mg C · l^{-1} in April to 0.51 mg C · l^{-1} in May alone is responsible for doubling the SFR, and is thus a major factor causing an increase in grazing activity. The effect of temperature increase, however, cannot be neglected. In Lake Vechten the variations in the mean water temperatures in the month of May, from year to year, are no more than a few degrees centigrade. Thus, both decrease of the food concentration and increase of the cladoceran densities, in the course of years, have caused a pronounced increase in the grazing pressure. The increase in the cladoceran densities is due mainly to those of *Bosmina* and *Daphnia* species. The former species was reported by Lewis (1977) to be preyed upon preferentially by *Chaoborus* larvae. A decrease in predation pressure of *Chaoborus,* related to a decrease in its densities, may have resulted in increased cladoceran numbers. Recently, Weekenstroo *et al.* (in prep.) have presented evidence indicating that in 1981 the larval densities of *Chaoborus flavicans* in Lake Vechten were 75% lower than those reported by Parma (1971). In June, on the other hand, the increase in grazing pressure in the course of years is less marked, owing mainly to a decrease in the densities of grazers. The rise in grazing pressure compared with that in May is about 20%, most probably the consequence of warming up alone. Since the grazers' densities decrease and the species composition remains unchanged compared with those in May, the temperature effects may be even more.

The mean daily feeding rates of 580 and 790 mg C · m^{-2} · day^{-1} in May and June (Fig. 9) are roughly in the range of primary production maxima (de Kloet 1982). However, a direct comparison of phytoplankton daily growth rate (k_i) and daily grazing (g_i) is restricted to 1977 and 1978 when both the parameters were measured simultaneously (Table 11). The g_i:k_i ratios varied between 124 and 395% indicating that the removal of food by the grazers far exceeded its replenishment via daily primary production. Because the mean assimilation efficiency in May and June was 45% (Fig. 9), the net daily removal of seston by the grazers is equivalent to 56–178% (g_i:k_i ratio × 0.45) of the primary production; and the egestion products – an important fraction – enter the seston pool as detritus. It is clear that the grazers through their feeding activity, including the subsequent detritus production and its re-utilization, form an important link between the grazer and detritus food chains. Thus, when related to ingestion, assimilation and egestion, the energy transport within the limnetic food chains is likely to be most efficient in May and June. Besides, due to the mineralization of organic matter in the oxic layers (Olie & Cappenberg 1982), the sinking of material to the anoxic hypolimnetic water would be considerably reduced. Although some decrease in sedimentation is evident in May and June (Cappenberg & Verdouw 1982) the losses found in the hypolimnion are higher than expected from the assimilation rates of zooplankton. A monthly input–output budget of phytoplankton primary production and zooplankton assimilation was constructed for the year 1978 to check a possible inverse relationship between zooplankton assimilation in the aerobic water and sedimentation losses found in the anaerobic zone. The equation used is:

$$L = P_p - Z_a - C_b$$

where L = sedimentation loss or elimination, in g C · m^{-2} · $month^{-1}$; P_p = phytoplankton primary

Table 11. Instantaneous coefficients of phytoplankton mortality, g_i, attributable to zooplankton daily grazing (grazing % day^{-1} · 10^{-2}) and of phytoplankton growth, k_i (= daily primary production/ phytoplankton carbon), in Lake Vechten. The data are means of two or three monthly measurements.

	1977			1978		
	g_i	k_i	g_i/k_i (%)	g_i	k_i	g_i/k_i (%)
Jan.	0.012	0.029	41	0.028	0.054	52
Feb.	0.032	0.040	80	0.038	0.100	38
March	0.019	0.057	33	0.058	0.105	55
April	0.056	0.087	64	0.068	0.152	45
May	0.411	0.332	124	0.618	0.313	197
June	0.339	0.160	211	0.589	0.149	395
July	0.080	0.108	74	0.020	0.057	35
Aug.	0.133	0.034	391	0.150	0.050	300
Sept.	0.108	0.052	208	0.186	0.093	200
Oct.	0.080	0.039	205	0.068	0.050	136
Nov.	0.167	0.028	596	0.072	0.041	175
Dec.	0.106	0.028	378	0.119	0.053	224

production, Z_a = zooplankton assimilation and C_b = the change in seston biomass in a month.

The calculated loss is about 1 g C · m^{-2} · $month^{-1}$ in January and February, and approx. 7 and 4.5 g C · m^{-2} · $month^{-1}$ in March and April, but one finds a 'negative elimination' of about 9 g C · m^{-2} · $month^{-1}$ in May and June. In these two months, even if we ignore the sedimentation losses, there should have been an import of material from other parts of the lake to meet the needs of zooplankton assimilation alone. Horizontal transport of organic material from the littoral is a possibility, but has not yet been investigated.

d) July–September, the period of moderately high grazing. The daily grazing is reduced to approximately half that in May and June; the mean rates lie between 19% in July and 12% in September. It is interesting that the summer grazing maximum has shifted from August in 1972 to July in 1973, to June in 1974 and 1975, and to May in subsequent years (Fig. 8, bottom). But this is more due to the increase in grazing pressure in May and June since 1974 than to its decrease in July–August. In July, however, the grazing has consistently dropped since 1975. This may be a consequence of changes in seston size composition and of decrease in primary production rates (de Kloet 1982) in the course of the last six or seven years. The mean concentration of <33 μm seston remains more or less constant around 0.49 mg C · l^{-1}, i.e. the late spring level (Fig. 5). However, its share in the total seston decreases owing to the abundance in this period of the non-edible net plankton (Fig. 7), particularly *Ceratium hirundinella* (Blaauboer 1982). This shift to the dominance of net plankton, particularly during summer and early autumn, started in 1975. Such changes, namely, dominance of 'summer' and 'autumn' associations early in the season, point to a eutrophic situation (Gliwicz 1967a; Gliwicz & Hillbricht-Ilkowska 1975; Porter 1977).

The mean ingestion rates of 430 ± 265 in July and 242 ± 59 mg C · m^{-2} · day^{-1} in September (Fig. 9) are far greater than the daily nanoplankton primary production, particularly in 1977 and 1978. The g_i and k_i ratios in 1977 and 1978 show daily removal by the grazers in July–Sept. ranging from 35 to 390% of the primary production, with means of 224 and 178%. However, a direct comparison of the daily rates of zooplankton ingestion and of nanoplankton primary production rates gives consumption: production ratios of 156 and 193%, respectively. The discrepancy between this direct method of comparing the zooplankton ingestion and primary production and the one based on g_i and k_i ratios is not surprising since the k_i was calculated using a constant carbon:chlorophyll ratio of 35:1 (Golterman & Kouwe 1980) which in the field situation is very unlikely. Undoubtedly, however, the zooplankton ingestion rates far exceeded those of the primary production. The direct comparison of the simultaneous production and consumption rates avoids the inherent errors due to the changes in the chlorophyll content of the algae and conversions to carbon, and thus seems to be more reliable.

Using the mean assimilation efficiencies in July–Sept. during 1977 and 1978 of 63 and 57%, net daily removal of seston by the grazers would amount to 98 (156 × 0.63) and 110% (193 × 0.57) during July–Sept. of the two years, respectively. Again, as in May–June, the grazers as primary consumers appear to play an important part in the lake's food economy. The constancy of seston biomass (<33 μm ϕ) throughout this period hints at an apparent equilibrium between the production and consumption process. However, one cannot account for sedimentation loss (Cappenberg & Verdouw 1982) which in 1978 was 170 mg C · m^{-2} · day^{-1} compared with zooplankton assimilation of 156 mg C · m^{-2} · day^{-1}. Probably, primary production of non-edible algae, particularly *Ceratium* sp., which amounts to about 10–40% of total primary production in July–Sept. (de Kloet 1982), contributes increasingly more to sedimentation in this period.

Furthermore, grazers do not seem to depend solely on limnetic primary production. Their vertical grazing patterns (Gulati 1978a) indicate that during the day-time, the rates in the layers below 3.6 m, but particularly below 5.4 m, are two to four times as high as in the upper 1.8 m, and may reach 100% day^{-1} in early August as in 1976. The quantity of food daily produced in these deeper strata is only a minor fraction of both the total integral primary production (de Kloet 1982) and the daily consumption by zooplankton. Furthermore, transport of seston from the upper strata would be reduced considerably in view of the above discussion, and because the nocturnal grazing in the upper layers removes two to three times more food from these strata than that during the day (Postema & Siewertsen 1981). Information on transport of material from the littoral or from the outer pelagic region, and on the bacteria in suspension, as supplementary sources of food for the herbivores, may help to explain the inconsistencies in the results. Recent work in our laboratory has demonstrated that the food assimilated by *Diaphanosoma brachyurum* in summer may comprise approx. 10% bacteria (pers. comm. R. Pot). This may be also true for both *Daphnia cucullata* (Geller & Müller 1981) and *Ceriodaphnia* sp. (Gophen *et al.* 1974).

About 30% of the total annual ingestion and assimilation occurs in July–September. This was 40% in May–June, together. Thus, more than two-thirds of the energy transfer in the limnetic food chain of the lake occurs in May–September. These results are very similar to those found in Heart Lake (Haney 1973).

The presence in summer of all the five major species of filterfeeding Cladocera, showing only a partial segregation and niche overlap, suggests a very efficient utilization of the spectrum of size particles available. This will be especially so for the seston fraction below 33 μm used as a measure of available food in our study.

e) Oct.–Nov., transition to low grazing activity. The fall in water temperature, leading to a breakdown of the thermal stratification and decrease in the light intensities, results in reduced phytoplankton production (de Kloet 1982), but the fall in seston concentration occurs much later, in November (Fig. 5). The densities of phytoplankton species encountered are relatively low (Blaauboer 1982) and a clear algal periodicity is lacking. Thus, the detritus in suspension should form a major component of the seston. The minor filterfeeders, viz. *C. pulchella* and *D. brachyurum,* become sparse in October. This is generally also true for *B. longirostris.* However, the increase in zooplankton biomass in November (Fig. 5) is caused especially by that in the individual weights of *E. gracilis* which dominates the filterfeeders' community. The filterfeeders biomass:seston biomass (<33 μm) ratio of 0.4 in November is only second highest to that (0.5) in May.

Because the course of both climatic and biological changes is opposite to that noted in March and April, these affect the rate of reproduction and development of the crustaceans in an opposite manner, so that their growth is considerably reduced. *Eudiaptomus gracilis,* however, appears to be an exception, possibly because of its ability to feed on a wider range of food particles (Gliwicz 1969a, b; Hillbricht-Illkowska *et al.* 1979). This may be an added advantage together with an egg duration time of this species of 4–8 days at the temperatures prevailing in October and November (Jacobs & Bouwhuis 1979).

The grazing rate increases from about 8% day^{-1} in October to about 15% day^{-1} in November. The variations from year to year as in winter are rather high. The ingestion and assimilation rates follow an almost identical pattern; their mean values of 193 ± 92 and 103 ± 54 mg C · m^{-2} · day^{-1} are respectively six and three times as high as those of the primary production. The algae, therefore, constitute a minor fraction of the grazers' food menu in this period.

Eudiaptomus gracilis seems to be able to exploit detritus over the entire depth of the lake. This is reflected only partly in the decrease in seston concentration even though in November it drops to one of its lowest in the year – next only to the annual minimum in December. Again, input of decomposing and dead materials from the littoral (Best 1982; Gons 1982) to the pelagic region, and possibly some resuspension from shallow sediments, may help explain the less-than-expected decrease in seston concentration from October to November. This decrease of C_b based on the difference in mean seston mass in October and November was only about 0.5 g C · m^{-2} (Fig. 5), as against zooplankton mean assimilation (Z_a) in one month of 3 g C · m^{-2}. Using the loss equation (p. 152) and P_p, Z_a and C_b values in November of 1.0, 3.0 and -0.5 g C · m^{-2} · $month^{-1}$, we could account only for 50% of the assimilation losses. The input from the littoral may explain the anomalies.

Lastly, in view of the corrections of grazing data (Table 10) the impact of zooplankton on the primary producers would be even greater than so far assessed. The mean primary production rate of about 300 mg C · m^{-2} · day^{-1} appears to be inadequate in meeting the food demand of zooplankton in May–September. Since the primary production data were also measured using the ^{14}C-technique, a critical analysis of the methodological problems and the possible errors is needed. This will throw more light on the importance, direct or otherwise, of the other primary producers, e.g. periphyton (Gons 1982) and macrophytes (Best 1982) as food sources for zooplankton.

Physiological efficiencies and turnover rates

Field studies dealing with physiological efficiencies of the filterfeeder community are scarce. A comparison of field data with those from the laboratory studies (see e.g. Lampert 1977c; Lei & Armitage 1980) is difficult since the latter are restricted to single species populations, and under constant environmental conditions too.

The seasonal mean range of assimilation efficiencies (A/I) in Lake Vechten of 35–57% is well within that of the 7–99% reported by Sushchenya (1969) for several aquatic crustaceans. The mean A/I of 44% in our lake is, however, lower than that of 58% found by Sushchenya and those of 60–80% encountered in the East European literature (Winberg 1972). Hillbricht-Illkowska *et al.* (1972) used an A/I of 60%, together with the assumed P/A (K_2) of 40% (Winberg 1972) to back-calculate the I from P data. Such assumptions will give a constant value of K_1. For example, the assumed K_1 in Lake Vechten will be 24% (0.6 × 0.4) compared with the mean monthly range of 15–30% (Table 12).

Table 12. The study period monthly means of the different physiological efficiencies and turnover rates of the crustacean filterfeeders in Lake Vechten; all data based on the daily rates. The abbreviations used are: I, ingestion; A, assimilation; B, mass of filterfeeders; R, respiration; and P, production (A–R). Only A/I from the period March 1972 to Feb. 1979, the rest from July 1973 to Feb. 1979.

Month	A/I	P/I	P/A	I/B	R/B	P/B
Jan.	0.42	0.15	0.35	0.15	0.041	0.023
Feb.	0.50	0.17	0.37	0.16	0.046	0.027
March	0.35	-0.08	-0.22	0.12	0.054	-0.010
April	0.36	0.12	0.33	0.35	0.078	0.040
May	0.48	0.31	0.59	0.50	0.116	0.157
June	0.41	0.23	0.58	0.87	0.155	0.196
July	0.48	0.14	0.32	0.77	0.196	0.106
Aug.	0.54	0.21	0.38	0.53	0.191	0.114
Sept.	0.45	0.18	0.46	0.62	0.151	0.112
Oct.	0.48	0.30	0.60	0.50	0.095	0.148
Nov.	0.57	0.33	0.61	0.32	0.068	0.106
Dec.	0.45	0.21	0.50	0.26	0.054	0.053
Year	0.44	0.22	0.49	0.43	0.107	0.094

The lowest A/I values were found in March–April; they seem to be related to high food concentration. Surprisingly, the phytoplankton in March–April is dominated by small green algae, flagellates and diatoms that are regarded as good food (Arnold 1971; Schindler 1971). There is, however, no evidence that 'superfluous feeding' occurred and A/I was reduced. The data of Oct.–Nov. are quite contrary, namely, low food level and poor food quality – seston being dominated by detritus which has low assimilability (Bell & Ward 1970) – and relatively high A/I. This was despite high values of DR. It is interesting that in both spring and autumn *Eudiaptomus gracilis* was the dominant filterfeeder, and the water temperature was roughly in the same range, 8–13 °C.

The inconsistencies in the results are probably caused by the differences in food quality related to the tracer technique. Silver & Davoll (1978) reported loss of tracer related to the phytoplankton species composition and cell condition. Both I and A, but particularly A, were presumably affected more seriously in March–April when small algae and more fragile flagellates were dominant. These algae may be digested and absorbed more quickly and enter the metabolic pool of the animal faster than poorly assimilable food, e.g., as in Oct.–Nov. In the one hour period after the animals are transferred from the active to non-active food, the respiratory loss of the assimilated material both as $^{14}CO_2$ and $DO^{14}C$ may start earlier than when food quality is poor. This could give relatively higher losses of A than of I, and thus lower efficiencies. On the other hand, relatively slow food absorption, if the food quality is poor as in Oct.–Nov., may result in lower losses of the assimilated tracer during the measurements and higher A/I.

The mean DR values of 50–57% in summer are about 2.5 times those in the rest of the year. The mean values in Lake Vechten are lower than the ones reported for both laboratory cultured animals and those based on field data, e.g. Gliwicz (1968) and Monakov & Sorokin (cited by Hillbricht-Illkowska *et al.* 1972). However, our corrections of the I data (Table 13) do indicate that the underestimates of DR were rather serious. Since the food concentrations in Lake Vechten are more or less constant in summer, the dependence of DR on SFR and thus on temperature is quite clear. A slight reduction in DR during August, possibly due to

Table 13. The means of physiological efficiencies and turnover rates (day^{-1}) calculated as in Table 12 (I), for the entire study period (II) and after correcting for the various errors (III; see Table 10). All data are in percentages.

Parameter	I	II	III
Ass eff. (A/I)	44	44	37
Daily ration (I/B)	43	41	79
P/I (K_1)	22	18	20
P/A (K_2)	49	41	53
R/B · day^{-1}	10	10	14
R/A	52	59	47
Turnover (P/B · day^{-1})	9	7	16

temperatures above 20 °C, seems to be compensated by an increase in A/I. Our results do not show a negative relation between A/I and K_2 as reported by Welch (1968). The variations in both the environmental factors and the species composition probably mask such effects in the field data (Lei & Armitage 1980). The mean K_1 and K_2 efficiencies in Lake Vechten fall in the range reported for several crustacean species (Sushchenya 1970). The mean K_2 values in May–June, 57–59%, and in Oct.–Nov., 60–61%, were relatively the highest. In the first period the temperature range was 13–17 °C and *Bosmina* sp. and *Daphnia* spp. dominated, whereas in the second period temperature range was 8–13 °C and *Eudiaptomus* sp. dominated. Though the food concentrations are relatively low in both these periods, they generally do not affect the K_2 adversely, thus indicating the absence of food limitation for growth. The effects of the methodological errors on the K_2 are notable from the corrections (Table 13) and its values are, therefore, likely to be somewhat higher than estimated. The negative values of both K_1 and K_2 in March point to a more serious underestimation of A rather than an overestimation of R; the low A/I values, as previously discussed, also support this argument.

Although R/B is related directly to temperature, R/A is not. The differences in the magnitude of seasonal errors in the assimilation rates, related to the ^{14}C-technique, need to be investigated further so that K_2 can be assessed more accurately.

The mean daily turnover rates (P/B day^{-1}) varied seasonally by an order of magnitude, viz. from 0.023 in January to 0.2 in June. These variations are not unexpected considering both the seasonal fluctuations in temperature, food conditions and species composition. These factors are also the likely source of wide scatter among the IBP and other data summarized by Nauwerck *et al.* (1980, see Table 6.2, p. 270). The regression relation between P and B of $P = 9.097\ B^{1.237}$, based on pooled data of herbivores and carnivores given by these authors, was applied to our mean B of 31.4 kJ · m^{-2} (= 0.75 g C · m^{-2}). The computed P of 645 kJ · m^{-2} yr^{-1} gave a value of 42 mg C · m^{-2} · day^{-1}, which is comparable with our study period mean of 50 mg C · m^{-2} · day^{-1}. However, our P data are derived indirectly from the relation P = A – R rather than based on population dynamics which was generally the approach of most IBP studies. Our mean P/B · day^{-1} of 0.094 gives a turnover time of 11 days or about 34 turnovers per year. This value is well in the middle of the range reported in the IBP data. On a monthly basis the seasonal changes in the P/B values are: less than one turnover in Jan. and Feb., due chiefly to *E. gracilis;* 5 or 6 turnovers in May and June corresponding to the abundance and exponential growth period of *B. longirostris* and *D.* spp.; and about 3 turnovers during July–September, in which period *D. brachyurum* is also present among the major species. Since the temperature in May–June is moderate, the P/B maxima seem to be correlated with low R/A ratios.

It may be remarked that methodological errors, especially those concerning A, R or both, will effect the accuracy of the results. In our case the corrected mean P/B ratio (Table 13) appears to be high although the validity of this has not yet been confirmed on the basis of correcting the data of individual measurements.

The grazing: some comparisons and role in the lake's ecosystem

Interactions between phytoplankton and zooplankton have been the subject of several works cited by Serruya *et al.* (1980). The data emanating from the studies are, nevertheless, difficult to compare, particularly in view of the heterogeneity of the approaches and methods used. The assumptions involved in deriving the parameters not measured is another obstacle in the comparisons. Our grazing study on Lake Vechten, in spite of the drawbacks of the ^{14}C-technique, is based on a long-term collection of data of the major carbon budget parameters. Some conclusions on the role of zooplankton in the lake's ecosystem (Fig. 13) may be drawn. Further-

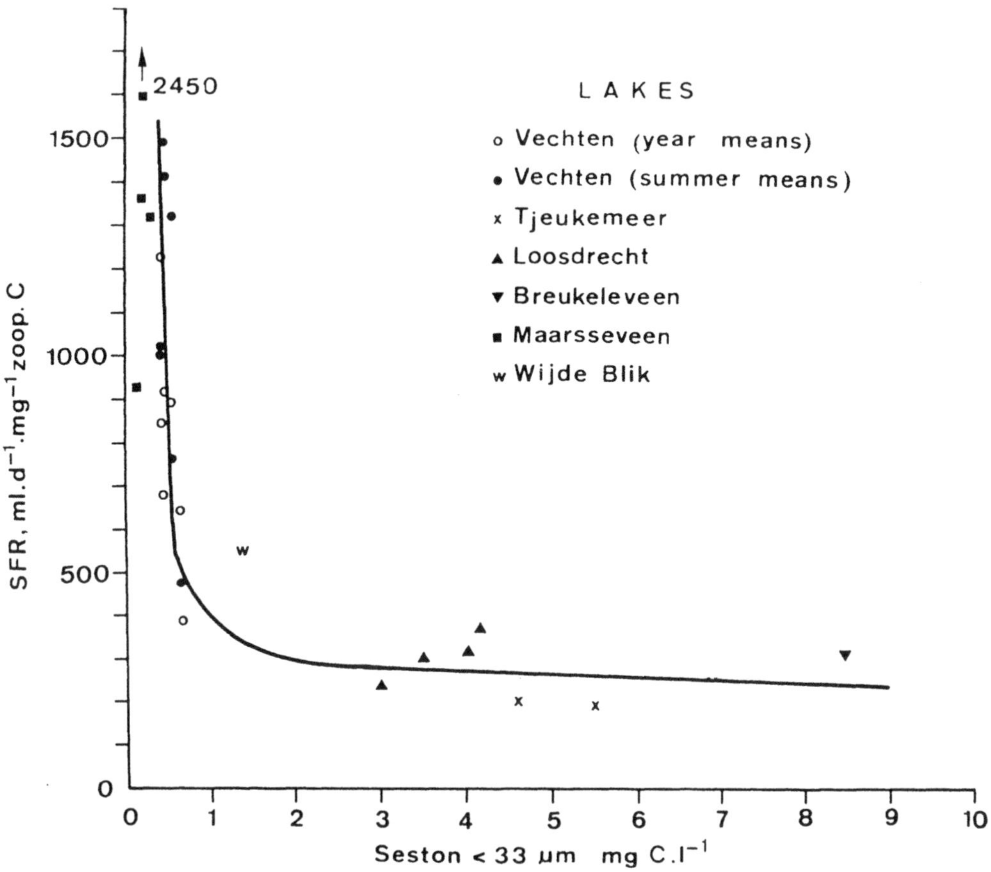

Fig. 12. The effect of food concentrations on SFR in Dutch lakes of varying trophy and depth. The Lake Vechten and Tjeukemeer data pertain to annual and summer means of the seven and three years, respectively, and those of the other lakes are means of several measurements in summer.

more, our studies on other Dutch lakes of varying trophy and morphometry (Gulati *et al.* 1982), but similar composition of the herbivore zooplankton community, have enlarged the scope of the comparisons (Fig. 12). The grazing data on all these lakes were converted to SFR, as explained earlier (p. 143). The inverse relationship between food concentration and SFR is quite clear (Fig. 12) and the course of the SFR curve interesting. The three deep, stratifying and less eutrophic lakes tend to lie on the left side of the curve where SFR declines almost precipitously, as food concentration slightly rises. Lake Vechten takes an intermediate position between Lake Maarsseveen, which has the highest SFR, and Lake Wijde Blik. The SFR in the first two lakes appears to fluctuate enormously, the food concentrations being close to the incipient limiting concentration. The primary production rates in these deep lakes (Gulati 1972; see also Flik & Keyser 1981) are almost an order of magnitude lower than in shallow, mixed and eutrophic lakes. Because of the great mean depth and summer stratification, nutrients appear to limit primary production in the deep lakes. So, the zooplankton grazing becomes a crucial link in both the grazer and the decomposer chains, and possibly in nutrient recycling through production of detritus, thus providing a suitable substrate for bacterial growth. The impact of zooplankton grazing in the internal recycling of N and P in these lakes seems to be an essential feedback link providing stability in the system.

The SFR in the shallower lakes is low and decreases only slowly with increase of food level, indicating a relatively minor role of zooplankton feeding. This is more so because a greater part of the

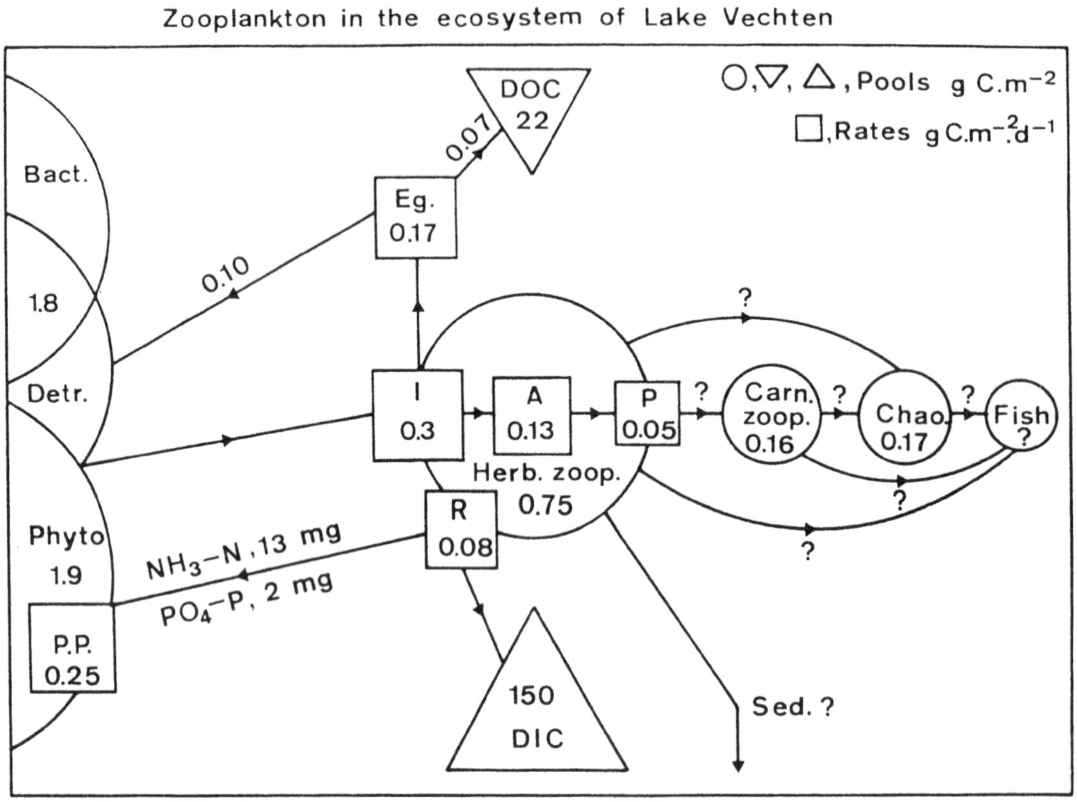

Fig. 13. The crustacean filterfeeders as a link in the grazer and detritus food chain. The rates and pools are based on annual means of 1972–1978. The daily nutrient regeneration (N and P) by the herbivore zooplankton is computed from Gulati *et al.* (1981). For abbreviations and other details, see text.

primary production is as inedible algae that are not available to zooplankton.

Thus, on the basis of SFR the Dutch lakes can be split into two broad categories: 1) high SFR, related to greater mean depth, summer stratification, low nutrient and seston concentration, low primary production, and high filterfeeder: phytoplankton ratios; and 2) low SFR, shallow, continuously mixed, high seston content, with dominance of coarse algae and eutrophic, with grazing playing a relatively minor role.

An illustration in Fig. 13 of the pools and rate processes relating to zooplankton, as a central link in the food chain, is based on the study period means. Admittedly, this is a 'nutshell', simplified and crude representation of a scheme of complex events in which the zooplankton, either directly or indirectly, has an impact on the functioning of a lake's ecosystem. It is abundantly clear that energy involved in the transfer during the grazing (I) is roughly equivalent to the entire carbon input via primary production (p.p.) in the limnetic region. However, during periods of high grazing intensity, namely May and June, the carbon input is grossly inadequate to maintain the flux through the second trophic level. It is equally true in autumn and winter, despite the fact that grazer consumption rates are at a low ebb. That the primary production rates in the open water fall considerably short of the consumption by the herbivores is supported strongly by the corrections for the methodological errors of the grazing data. Two important implications of the above discussion are: first, primary produce in the littoral, not considered in the scheme in Fig. 13, may enter the open water and offset the 'shortages' in the pelagic region. Nevertheless, this does not seem likely in May and June, because most of the material produced in the littoral, i.e. the epipelon, tends to accumulate *in situ* or migrate downwards along the slope of the sediments. A transport to the epilimnion is perhaps unimportant since the lake is stratified. Secondly, the primary production in the limnetic region may have been underestimated. This hypothesis is supported by the overwhelmingly high net output, as zooplankton assimilation and sedimentation together, compared with the input through primary production. The losses as seston mineralization and respiration, as yet unaccounted for, will only accentuate the negative balance between the producers and consumers.

One important shortcoming in our study is lack of information on the dynamics of the DOC pool size, which is immense. A slight daily change in this pool, say 1%, may deliver as much bacterial carbon as the daily production of algal carbon. Such a minor but important output in the DOC pool is difficult to measure with the existing techniques. The inputs to this pool of extracellular products of phytoplankton and those of zooplankton metabolism add further to the difficulties. The direct measurement of the role of bacteria as a food source will provide insight into the discrepancies of the carbon cycle in early summer, and into the role of zooplankton. Similarly, more information is needed on the horizontal transport from the littoral in autumn, when dead and dying epiphytic material may be an important component of seston.

Lastly, the changes in the structure of the filterfeeder community, the increase in the grazing pressure, and in the phytoplankton composition and abundance cannot be restricted to the two trophic levels. That these are a consequence of feedbacks within the ecosystem needs to be confirmed from the studies on the zooplankton predators in Lake Vechten.

Summary

The results of a seven-year study on the filterfeeder community in Lake Vechten (The Netherlands) are summarized. The community structure, food and feeding relations and metabolism are discussed with emphasis on the impact of grazing in the functioning of the lake's ecosystem. Lake Vechten is a small (4.7 ha), deep (mean depth, 6.0 m), stratifying and relatively clear waterbody. The lake was created in 1941 due to sand digging.

The grazing and assimilation rates were measured in the laboratory under *in situ* conditions of temperature and light. The ^{14}C-technique was used to prepare the tracer food, using an *in situ* seston size fraction below 15 or 33 μm. The technique and the drawbacks are discussed critically.

The filterfeeders comprised 76–84% of the total zooplankton densities; the annual means varied between 109 and 128 ind $\cdot$ l^{-1}, with a grand annual summer mean for the 7 years of 164 $\pm$ 24 ind l^{-1}. *Eudiaptomus gracilis* dominated the community from October to March, but *Bosmina longirostris* and *Daphnia* spp. were predominant in May and June when both the density and biomass maxima of the crustacean herbivores occurred. *Diaphanosoma brachyurum* and *Ceriodaphnia pulchella* were important only in late summer.

The pattern of fluctuations in the zooplankton densities and biomass were not strictly parallel. The mean weight of the individuals in the community varied from 4.3 in March to 1.1 μg C $\cdot$ ind^{-1} in September. In April–May, the biomass did not keep pace with the densities. The monthly means of the total community biomass ranged from 0.7 g C $\cdot$ m^{-2} in December to 1.6 g C $\cdot$ m^{-2} in May. The increase in the mean annual biomass in the years 1976–1978, compared with that in 1973–1975, was caused mainly by the increase in the average individual weight of the organisms, rather than their densities.

The phytoplankton community in January–February was domninated by *Rhodomonas minuta, Cryptomonas* spp., *Stephanodiscus hantzschii* and other nanoplankton. The phytoplankton densities were the lowest in May–June, but dominated in the late summer by *Ceratium hirundinella, Mallomonas caudata* and species of blue-greens, *Oscillatoria* and *Lyngbya.* The biomass of phytoplankton and detritus, i.e. seston, increased in Jan.–Feb., despite low water temperature and poor light conditions. The annual maxima of seston, 5.1 $\pm$ 2.0 g C $\cdot$ m^{-2}, usually occurred in April. The minima in May or June coincided with the peaks of filterfeeder biomass. This early spring period was marked by a sharp increase in Secchi depth, i.e. the so-called 'clear water' phase. During June–Oct., the seston biomass was rather constant, viz. 2.2–2.7 g C $\cdot$ m^{-2}, but thereafter decreased to its minimum in November or December. The decrease in seston (<33 μm) mass since 1975 was accompanied by a concomitant, significant increase in grazing pressure, as well as a decrease in the primary production rates.

The community grazing rates fluctuated sharply both from season to season and annually. The mean daily grazing ranged from 3 ± 2 in March to $34 \pm 17\%$ day^{-1} in June. The extreme low and high values in these months were, respectively, 0.4 and 73% day^{-1}. The rise in the grazing rates were sharpest from April to May and could be attributed mainly to species of *Bosmina* and *Daphnia*. The means of community ingestion rates (I) varied from 23 to 1632 mg C · m^{-2} · day^{-1}, respectively, in January 1973 and June 1978. Nearly two-thirds of the annual consumption occurred in the period May–Sept. The zooplankton ingestion: phytoplankton primary production ratio varied between 0.7 and 2.3 on an annual basis. The assimilation rates followed the same pattern as that of the ingestion rates. The assimilation efficiencies (A/I) were relatively low, $35 \pm 11\%$, in March–April and high, $53 \pm 15\%$, in Oct.–Nov. The study period mean was $44 \pm 9\%$. The grazing pressure increased three-fold in May in the period 1976–1978 compared with that in May 1973–1975. This significant rise in the zooplankton grazing was owing to those of densities of the major grazers.

The zooplankton specific filtering rate, SFR (ml · day^{-1} · mg^{-1} · zoop · C), varied between 44 and 2814 on a monthly basis, and was correlated directly with water temperature, but indirectly with the food concentration. This explained the maximum rise in May–June, in which period the temperature rose but food levels decreased. The Q_{10} values for daily ration fluctuated between 2 and 2.2.

The methodological problems mainly concerned the radiotracer technique. The loss or leaching of the label from the labelled animals was the most serious of the errors in our results. This resulted in a 42% underestimation of the ingestion rates, and 26% underestimation of the assimilation. The respiratory loss of $^{14}CO_2$ during the assimilation experiment was computed to be about 12%. The underestimation of the data, due to not accounting for net seston, was about 15% for the ingestion rates and 8% for the assimilation rates. In short, the corrected rates were significantly higher than the measured ones.

The comparison of phytoplankton primary production with that of its elimination via zooplankton assimilation and sedimentation to the hypolimnion revealed serious discrepancies, particularly in the spring but also in summer. The carbon input via production appeared to be inadequate as a food source. The food produced in the littoral, via its horizontal transport to the limnetic region, may be an important supplementary food source. Also, bacteria as a food source may partially explain the deficit in the carbon budget based only on the primary production.

The physiological efficiencies in the lake are in the range of those cited in the literature. The turnover rates of about 9% day^{-1} indicated optimal conditions for growth of zooplankton. In comparison with other Dutch lakes, the SFR in Lake Vechten appears to be higher – as in other deep and stratifying lakes – than in the shallow, mixed lakes of high trophy.

Acknowledgements

We thank all the colleagues who helped in various ways in the field and laboratory. We owe special gratitude to Dr. J. Vijverberg for his critical comments on the manuscript and useful suggestions, and to Drs. B. Z. Salomé for improving the readability. To Miss J. van Kleinwee we are specially indebted for her endurance during the typing work, particularly in view of the last-minute insertions as well as deletions in the typed text and tables. Mr. E. Mariën is thanked for the photography. Dr. S. Parma's help as a 'silent' source of encouragement is appreciated.

Last, but not least, the first author (R.D.G.) learnt a lot from the none-too-easy discussions with the colleagues on the lake's carbon budget. The inconsistencies in the budget are certainly a source of inspiration for the continuity of the grazing studies in Lake Vechten, and their extension to other Dutch lakes.

References

Arnold, D. E., 1971. Ingestion, assimilation, survival, and reproduction by Daphnia pulex fed seven species of blue-green algae. Limnol. Oceanogr. 16: 906–920.

Bell, R. K. & Ward, F. J., 1970. Incorporation of organic carbon by Daphnia pulex. Limnol. Oceanogr. 15: 713–726.

Best, E. P. H., 1982. The aquatic macrophytes of Lake Vechten. Species composition, spatial distribution and production. Hydrobiologia 95: 65–77.

Best, E. P. H., Blaauboer, M. C. I., Cappenberg, Th. E., Gons, H. J., Gulati, R. D., Kloet, W. A. de, Steenbergen, C. L. M. & Verdouw, H., 1978. Towards an integrated study of the ecosystem of Lake Vechten. Hydrobiol. Bull. 12: 107–118.

Blaauboer, M. C. I., 1982. The phytoplankton species composition and the seasonal periodicity in Lake Vechten from 1956 to 1979. Hydrobiologia 95: 25–36.

Brooks, J. L. & Dodson, S. I., 1965. The effects of prey size selection by lake planktivores. Syst. Zool. 17: 272–291.

Butter, M. E., 1981. Estimation of zooplankton in lake Maarsseveen I: Problems, procedures and results. Hydrobiol. Bull. 15: 51–59.

Cappenberg, Th. E. & Verdouw, H., 1982. Sedimentation and breakdown kinetics of organic matter in the anaerobic zone of Lake Vechten. Hydrobiologia 95: 165–179.

Comita, G. W. & Anderson, G. C., 1959. The seasonal development of a population of Diaptomus ashlandi Marsh, and related phytoplankton cycles in lake Washington. Limnol. Oceanogr. 4: 37–52.

Conover, R. J., 1966. Factors affecting the assimilation of organic matter by zooplankton and the question of superfluous feeding. Limnol. Oceanogr. 11: 346–354.

Conover, R. J. & Francis, V., 1973. The use of radioactive isotopes to measure the transfer of materials in aquatic food chains. Mar. Biol. 18: 272–283.

Cooper, D. C., 1973. Enhancement of net primary productivity by herbivore grazing in aquatic laboratory microcosms. Limnol. Oceanogr. 18: 31–37.

Copping, A. E. & Lorenzen, C. J., 1980. Carbon budget of a marine phytoplankton–herbivore system with carbon-14 as a tracer. Limnol. Oceanogr. 25: 873–882.

Dagg, M. J., 1976. Complete carbon and nitrogen budgets for the carnivorous amphipod, calliopius laeviusculus (Krøyer). Int. Revue ges. Hydrobiol. 61: 297–357.

Davis, R. G., 1971. Computer Programming in Quantitative Biology. Academic Press, London. 492 pp.

Elster, H. J., 1954. Über die Populationsdynamik von Eudiaptomus gracilis Sars and Heterocope borealis Fischer in Bodensee-Obersee. Arch. Hydrobiol., Suppl. 20: 546–614.

Flik, B. J. G. & Keyser, A., 1981. Estimation of the primary production in Lake Maarsseveen I with an incubator technique. Hydrobiol. Bull. 15: 41–50.

Frost, B. W., 1980. Grazing. In: Morris, I. (Ed.) The Physiological Ecology of Phytoplankton (Studies in Ecology, Vol. 7), pp. 465–491. Blackwell Scientific, Oxford.

Geller, W., 1975. Die Nahrungsaufnahme von Daphnia pulex in Abhängigkeit von der Futterkonzentration, der Temperatur, der Körpergrösse und dem Hungerzustand der Tiere. Arch. Hydrobiol., Suppl. 48: 47–107.

Geller, W. & Müller, H., 1981. The filtration apparatus of Cladocera: filter mesh-sizes and their implications on food selectivity. Oecologia 49: 316–321.

Gliwicz, Z. M., 1967. The contribution of nannoplankton in pelagial primary production in some lakes with varying trophy. Bull. Acad. pol. Sci., Cl. II 15: 343–347.

Gliwicz, Z. M., 1968. The use of anaesthetizing substances in the studie on the food habits of zooplankton communities. Ekol. pol., Ser. A. 16: 279–295.

Gliwicz, Z. M., 1969a. Studies on the feeding of pelagic zooplankton in lakes with varying trophy. Ekol. pol., Ser. A. 17: 663–707.

Gliwicz, Z. M., 1969b. The share of algae, bacteria and trypton in the food of the pelagic zooplankton of lakes with various trophic characteristics. Bull. Acad. pol. Sci. 17: 159–165.

Gliwicz, Z. M. & Hillbricht-Illkowska, A., 1972. Efficiency of the utilization of nannoplankton primary production by communities of filterfeeding animals measured in situ. Verh. int. Verein. Limnol. 18: 197–203.

Gliwicz, Z. M. & Hillbricht-Illkowska, A., 1975. Ecosystem of the Mikolajskie Lake. Elimination of phytoplankton biomass and its subsequent fate in lake through the year. Pol. Arch. Hydrobiol. 22: 39–52.

Golterman, H. L., 1969. Methods for Chemical Analysis of Freshwater. I.B.P. Handbook, No. 8. Blackwell Scientific, Oxford. 172 pp.

Golterman, H. L. & Kouwe, F. A., 1980. Chemical budgets and nutrient pathways. In: Le Cren, E. D. & Lowe-McConnell, R. H. (Eds.) The Function of Freshwater Ecosystems, I.B.P. No. 22, pp. 85–140. Cambridge University Press, Cambridge.

Gons, H., 1981. Project 'Carbon cycle in Lake Vechten'. In: Progress Report Limnological Institute 1980, Verh. k. ned. Akad. Wetensch., Afd. Natuurk. 2 rks, 77: 30–36.

Gons, H., 1982. Structural and functional characteristics of epiphyton and epipelon in relation to their distribution in Lake Vechten. Hydrobiologia 95: 79–114.

Gophen, M., Cavari, B. Z. & Berman, T., 1974. Zooplankton feeding on differentially labelled algae and bacteria. Nature 247: 393–394.

Gulati, R. D., 1972. Limnological studies on some lakes in the Netherlands. Freshwat. Biol. 2: 37–54.

Gulati, R. D., 1975. A study on the role of herbivorous zooplankton community as primary consumers of phytoplankton in Dutch lakes. Verh. int. Verein. Limnol. 19: 1202–1210.

Gulati, R. D., 1976. Studies on the food, grazing and metabolism of a filter feeder community of zooplankton in the lakes Vechten and Tjeukemeer. Hydrobiol. Bull. 10: 10–12.

Gulati, R. D., 1978a. Vertical changes in the filtering, feeding and assimilation rates of dominant zooplankters in a stratified lake. Verh. int. Verein. Limnol. 20: 950–956.

Gulati, R. D., 1978b. The ecology of common planktonic crustacea of the freshwaters in the Netherlands. Hydrobiologia 59: 101–112.

Gulati, R. D., Siewertsen, K. & Postema, G., 1981. Phytoplankton–zooplankton interactions. In: Progress Report Limnological Institute 1980. Verh. k. ned. Akad. Wetensch., Afd. Natuurk. 2 rks. 77: 12–16.

Gulati, R. D., Siewertsen, K. & Postema, G., 1982. Grazing studies in Loosdrechts lakes. In: Progress Report Limnological Institute 1981, Verh. k. ned. Akad. Wetensch., Afd. Naturk. 2 rks. (in press).

Hall, D. J., 1964. An experimental approach to the dynamics of a natural population of Daphnia galeata mendotae. Ecology 45: 94–112.

Hall, D. J., Threlkeld, S. T., Burns, C. W. & Crowley, P. H.,

1976. The size-efficiency hypothesis and the size structure of zooplankton communities. Annu. Rev. Ecol. Syst. 7: 177–208.

Haney, J. F., 1971. An in situ method for the measurement of zooplankton grazing rates. Limnol. Oceanogr. 16: 970–977.

Haney, J. F., 1973. An in situ examination of the grazing activities of natural zooplankton communities. Arch. Hydrobiol. 72: 87–132.

Haney, J. F. & Hall, D. J., 1975. Diel vertical migration and filterfeeding activities of Daphnia. Arch. Hydrobiol. 75: 413–441.

Hillbricht-Illkowska, A., Spodniewska, I., Weglеńska, T. & Karabin, A., 1972. The seasonal variation of some ecological efficiencies and production rates in the plankton community of several Polish lakes of different trophy. In: Kajak, Z. & Hillbricht-Illkowska, A. (Eds.) Productivity Problems of Freshwaters, pp. 111–127.

Hillbricht-Illkowska, A., Spodniewska, I. & Weglеńska, T., 1979. Changes in the phytoplankton–zooplankton relationship connected with the eutrophication of lakes. Symp. Biol. Hung. 19: 59–75.

Holtby, L. B. & Knoechel, R., 1981. Zooplankton filtering rates: error due to loss of radioisotopic label in chemically preserved samples. Limnol. Oceanogr. 26: 774–780.

Hulsmann, A. D. & Ringelberg, J. (in prep.) In situ graasonderzoek in de Grote Maarsseveense Plas (Chapter, Ph.D. thesis, of A. D. Hulsmann: in Dutch). University of Amsterdam, Amsterdam.

Hutchinson, G. E., 1967. A Treatise on Limnology. II. Introduction to Lake Biology and the Limnoplankton. John Wiley, New York. 1115 pp.

Jacobs, R. P. W. M. & Bouwhuis, A. M. J., 1979. The year cycle of Eudiaptomus vulgaris (Schmeil, 1896) (Copepoda, Calanoida) in a small, acid waterbody during 1973. Development in the natural habitat and relationships between temperature and duration of developmental stages. Hydrobiologia 64: 17–36.

Kloet, W. A. de, 1982. The primary production of phytoplankton in Lake Vechten. Hydrobiologia 95: 37–57.

Koole, E. & Vulto, T., 1976. Een onderzoek naar phyto- en zoöplanktonrelaties en energietransfer in plasje Vechten (west). Student Rep. (in Dutch) Limnol. Inst. 1977-2, 77 pp.

Lampert, W., 1975. A tracer study on the carbon turnover of Daphnia pulex. Verh. int. Verein. Limnol. 19: 2913–2921.

Lampert, W., 1977a. Studies on the carbon balance of Daphnia pulex as related to environmental conditions. I. Methodological problems of the use of ^{14}C for the measurement of carbon assimilation. Arch. Hydrobiol, Suppl. 48: 287–309.

Lampert, W., 1977b. Studies on the carbon balance of Daphnia pulex De Geer as related to environmental conditions. II. The dependence of carbon assimilation on animal size, temperature, food concentration and diet species. Arch. Hydrobiol., Suppl. 48: 310–335.

Lampert, W., 1977c. Studies on the carbon balance of Daphnia pulex De Geer as related to environmental conditions. III. Production and production efficiency. Arch. Hydrobiol., Suppl. 48: 336–360.

Lampert, W., 1978. Climatic conditions and planktonic interactions as factors controlling the regular succession of spring algal bloom and extremely clear water in Lake Constance. Verh. int. Verein. Limnol. 20: 969–974.

Lei, C. & Armitage, K. B., 1980. Energy budget of Daphnia ambigua Scourfield. J. Plankton Res. 2: 261–281.

Leussink, J. A., 1956. Het plankton, de temperatuur, het zuurstofgehalte en hun onderling verband in de plas bij Vechten, gedurende een deel van 1956. Student Report (in Dutch). Univ. Utrecht. 10 pp.

Mackereth, F. J. H., 1964. An improved galvanic cell for determination of oxygen concentrations in fluids. J. sci. Inst. 41: 38–41.

Nauwerck, A., 1959. Zur Bestimmung der Filtrierrate limnischer Planktontiere. Arch. Hydrobiol. Suppl. 25: 83–101.

Nauwerck, A., 1963. Die Beziehungen zwischen Zooplankton und Phytoplankton im See Erken. Symb. bot. uppsal. 17: 1–164.

Nauwerck, A., Duncan, A., Hillbricht-Illkowska, A. & Larsson, P., 1980. Zooplankton. In: Le Cren, E. D. & Lowe McConnell, R. H. (Eds.) The Function of Freshwater Ecosystems, I.B.P. No. 22, pp. 85–140. Cambridge University Press, Cambridge.

Nowak, K. E., 1975. Die Bedeutung des Zooplanktons für den Stoffhaushalt des Schierensees. Arch. Hydrobiol. 75: 149–224.

Olie, J. J. & Cappenberg, Th. E., 1982. Aspects of aerobic mineralization during spring in Lake Vechten with special reference to the ^{14}C-labelling technique. Hydrobiologia 95: 181–190.

Parma, S., 1971. Chaoborus flavicans (Meigen) (Diptera, Chaoboridae): an autecological study. Thesis, Groningen University. 128 pp.

Parma, S., 1980. Historie en morfometrie van 'Vechten' (in Dutch). Intern. Rep., Limnol. Inst., 1980-3. 20 pp.

Parma, S., 1982. The twenty-fifth anniversary of the Limnological Institute, The Netherlands (1957–1982). Hydrobiologia 95: 1–9.

Porter, K. G., 1977. The plant–animal interface in freshwater ecosystems. Am. Scient. 65: 159–170.

Postema, G. & Siewertsen, K., 1981. De consumptie- en assimilatiesnelheden van het herbivore zoöplankton in het plasje 'Vechten' (in Dutch). Intern. Rep. Limnol. Inst. 1981-1. 23 pp.

Rigler, F. H., 1971. Feeding rates: zooplankton. In: Edmondson, W. T. & Winberg, G. G. (Eds.) A Manual on Methods for the Assessment of Secondary Productivity in Fresh Waters. I.B.P. Handbook No. 17, pp. 228–255. Blackwell Scientific, Oxford.

Ringelberg, J., 1981. A diel study in a water column of lake Maarsseveen I. Hydrobiol. Bull. 15: 60–71.

Schindler, J. E., 1971. Food quality and zooplankton nutrition. J. Anim. Ecol. 40: 589–595.

Serruya, C., Gophen, M. & Pollingher, U., 1980. Lake Kinneret: carbon flow patterns and ecosystem management. Arch. Hydrobiol. 88: 265–302.

Siewertsen, K., 1979. Een biochemisch planktononderzoek in het plasje 'Vechten' (in Dutch). Intern. Rep. Limnol. Inst. 1979-3, 29 pp.

Silver, M. W. & Davoll, P. J., 1978. Loss of ^{14}C activity after chemical fixation of phytoplankton: error source for autoradiography and other productivity measurements. Limnol. Oceanogr. 23: 362–368.

Steenbergen, C. L. M., 1982. Contribution of photosynthetic sulphur bacteria to primary production in Lake Vechten. Hydrobiologia 95: 59–64.

Steenbergen, C. L. M. & Verdouw, H., 1982. Lake Vechten: aspects of its morphometry, climate, hydrology and physico-chemical characteristics. Hydrobiologia 95: 11–23.

Suchchenya, L. M., 1969. Quantitative relations of metabolism and transformation of matter and energy in Crustacea. Fish. Res. Bd Can. Trans., Ser. 1374. (Ph.D. thesis).

Suchchenya, L. M., 1970. Food rations, metabolism and growth of crustacean. In: Steele, J. H. (Ed.), Marine Food Chains, pp. 127–141. University of California Press, Berkeley.

Vijverberg, J., 1980. Effect of temperature in laboratory studies on development and growth of Cladocera and Copepoda from Tjeukemeer, The Netherlands. Freshwat. Biol. 10: 317–340.

Weekenstroo, J. E., Parma, S. & Gulati, R. D. (in prep.). The larvae of Chaoborus flavicans (Meigen) in Lake Vechten (The Netherlands): the decrease in the densities and its relation to zooplankton predation.

Węgleńska, T., 1971. The influence of various concentrations of natural food on the development, fecundity and production of planktonic filtrators. Ecol. pol., Ser. A. 19: 427–473.

Welch, H. E., 1968. Relationships between assimilation efficiencies and growth efficiencies for aquatic consumers. Ecology 49: 755–759.

Wetzel, R. G., 1975. Limnology. Saunders Philadelphia. 740 pp.

Winberg, G. G. et al., 1971. Symbols, Units and Conversion Factors of Freshwater Productivity. IBP, London. 23 pp.

Winberg, G. G., 1972. Some results of Soviet IBP investigations on lakes. In: Kajak, Z. & Hillbricht-Illkowska, A. (Eds.). Productivity Problems of Freshwaters, pp. 363–381. Krakow.

Sedimentation and breakdown kinetics of organic matter in the anaerobic zone of Lake Vechten

Th. E. Cappenberg & H. Verdouw
Limnological Institute, 'Vijverhof' Laboratory, Rijksstraatweg 6, 3631 AC Nieuwersluis, The Netherlands

Keywords: sedimentation rates, turnover rates, carbon recycling, anaerobic mineralization

Abstract

The turnover and exchange rates, as well as the sedimentation rates concerning the input and output of carbon in the anaerobic hypolimnion and the sediment in Lake Vechten were studied. The carbon input data were derived from those on sedimentation. By comparison of C-fixation data, sedimentation rates measurements and microbial reduction rates of observed fluxes of electron acceptors a mineralization of 20–30% of the limnetic C-fixation was found in the hypolimnion. The rate-limiting step of methanogenesis in anaerobic mineralization is the breakdown of algal cell wall components to their main product, namely acetate. This intermediate has a relatively fast turnover rate to the end-products, namely the gases CH_4 and CO_2. Finally, from the data on the diffusion of low-molecular weight metabolic intermediates and on the rates of formation of CH_4 and CO_2, the output of carbon and its cycling in Lake Vechten are discussed.

Introduction

The sequence aerobic mineralization in the photogenic zone to anaerobic mineralization in the tropholytic zone are linked by the sedimentation of organic matter. The rates at which the particulate material reaches the anoxic layers in the hypolimnion will determine the share of aerobic and anaerobic breakdown. For an evaluation of the role of this downward transport of organic carbon, fixed in the primary production, sedimentation rates must be measured. In addition, for quantifying the breakdown rates of organic matter in the sediments and in overlying waters, one needs to understand (1) the mechanisms controlling the mineralization rates, and (2) the exchange rates across the interfaces of sediment–water and metalimnion–hypolimnion.

The present views on sedimentation trap technique (see Bloesch & Burns 1980, and Blomqvist & Håkanson 1981) indicate that sedimentation rates measurements are realistic, provided they are carried out in systems with little water turbulence. Also, the use of traps of appropriate design, i.e. cylinders with a favourable ratio of length to internal diameter ('aspect ratio'), is important. In water bodies like Lake Vechten, sedimentation rates can be measured most suitably with traps below the thermocline, to avoid resuspension of bottom material.

In the hypolimnion and the sediments organic substrates are broken down ultimately into their most reduced and oxidized forms, methane and carbon dioxide respectively. This involves many microbial species, while the bacterial food chain starts with hydrolysis of complex organic materials as algal cell wall debris. It is well documented that acetate is the main precursor for methanogenesis in the anaerobic environment: 50–70% of the methane produced is derived from acetate, which is thus one of the key metabolic intermediates in the conversion of organic matter (Cappenberg & Prins 1974; Strayer & Tiedje 1978; Winfrey & Zeikus 1979; Sansone & Martens 1981).

Hydrobiologia 95, 165–179 (1982). 0018-8158/82/0952-0165/$03.00.

This paper describes several aspects of the input and output of carbon in the anaerobic zone of Lake Vechten. The carbon input data are derived from those on sedimentation. The breakdown of algal cell wall debris as the rate-limiting step in the conversion of complex organic matter into low-molecular weight intermediates as lactate and acetate are examined. Besides their turnover rates, the exchange rates by diffusion from the sediments to the overlying water and the fluxes of important electron acceptors for anoxic bacterial reactions of total organic carbon oxidation are considered. Using these data and those on the production rates of CO_2 and CH_4, the output of carbon and its recycling in Lake Vechten are discussed.

Materials and methods

Study area and sampling

Lake Vechten is a slightly eutrophic seepage lake (mean daily C-fixation of 15.4 kg C.day^{-1}, mean total dissolved organic carbon of 4.5 to 6.5 mg C.l^{-1}), situated in the centre of the Netherlands. The lake has a surface area of 4.7 ha and consists of two basins with maximum depths of 10.5 and 11.9 m, separated by a ridge at about 6 m. After the onset of thermal stratification in May the oxygen in the deeper water layers is rapidly depleted. This results in anaerobic hypolimnetic conditions from June to November. A general description and morphometric data of the lake are given by Steenbergen & Verdouw (1982).

The sediment below the 8 m depth contour of the lake is muddy. During the summer stratification the mud is black owing to precipitated FeS. Undisturbed mud cores were sampled using a modified Jenkin surface-mud sampler from the deepest part of the eastern basin of the lake. Sub-samples were drawn by piercing with a syringe through 2 mm holes, covered with Scotch tape no. 471, in the perspex sampling tube. The sediment composition was determined using intact cores taken during SCUBA diving at several places in the deepest area. The percentage dry weight, total carbon, carbonate, nitrogen and organic matter were measured as a function of depth in the sediment.

Sediment traps and procedure for collecting and processing of the samples are described in Verdouw & Dekkers (1982a). Carbon content of the dried particulate matter was determined with a Perkin Elmer NCH Analyzer (type 204). Organic matter was measured as loss on ignition for 2 h at 550 °C. Carbonate-C was estimated by C-determinations after ignition. Carbon sedimentation data could not directly be derived from the carbon content of the collected material, as this was considerably mineralized in the traps. For nitrogen this was accounted for by determination of the total nitrogen content, in both particulate form and dissolved in the water in the traps.For carbon, only data on particulate matter are available. Approximate carbon sedimentation rates were derived by multiplying the N-sedimentation rates with the C:N ratio of the dried particulate material. It is assumed that nitrogen and carbon mineralization in the sediment traps did not differ significantly. The positive correction for edge effects caused by transport along the lake bottom was estimated to be about 10% for measurements at 8 m depth (Verdouw & Dekkers 1982a), but was cancelled by a negative correction of the same magnitude for the presence of carbonate-C in the collected material. Thus, the input of organic carbon to the anaerobic zone of the lake can be roughly calculated.

Analyses of lower fatty acids, methane and carbon dioxide

The analysis of volatile lower fatty acids in the centrifuged (5 min at 10 000 *g*) pore water was performed on a Packard-Becker 417 gas chromatograph, fitted with an FID-detector, using helium (passed over formic acid) as the carrier gas (Cappenberg & Jongejan 1978). The analysis of non-volatile lower fatty acids was performed using high pressure liquid-chromatography, after derivatization (Hordijk & Cappenberg 1982). Carbon dioxide and methane were measured gaschromatographically as described in Cappenberg (1976).

For determining ^{14}C-labelled compounds, a gas chromatography-gas proportional counting system was used (Cappenberg 1980; Cappenberg & Prins 1974). Uniformly labelled *Chlorella* cell walls were obtained from precultured, $NaH^{14}CO_3$ containing, axenic batch cultures. The radioactivity of the cell wall debris was measured after sonifying the algal cells, and filtering and combusting the filters in a Packard Tri-Carb (model B 306) sample oxidizer.

The $^{14}CO_2$ was trapped in a Carbo-Sorb solution and determined using a Packard Tri-Carb (model 2425) liquid scintillation spectrometer. The counting efficiency of the samples was determined by automatic external standardization.

Assay of cofactor F_{420}

Sub-samples of sediment cores of various depths were treated for 10 min at 100 °C directly after sampling and centrifuged (10 min at 9 000 *g*) at 4 °C. Iso-propanol was added in the ratio 3:1 to an aliquot of the supernatant and the mixture was centrifuged again at 9 000 *g* for 10 min at 4 °C. One part by volume of aqueous 1 N KOH was added to the mixture and the amount of cofactor F_{420} was determined in the alkaline extract at an excitation wavelength of 420 nm and an emission wavelength of 470 nm, using a P-204 spectro-fluorimeter (Perkin-Elmer) and pure cofactor F_{420} as a standard. The assays were carried out by M.D. Delafontaine (Laboratoire de Chimie et Biologie de l'Eau, Université Catholique de Louvain).

Counting procedures for bacteria

Sub-samples were incubated after serial dilution using the anaerobic roll-tube technique for counting sulphate reducers, methane producers and algal cell-wall splitting bacteria in the sediments. To inoculate the tubes, about 1 g of wet mud was collected from the mud cores, and the syringe was quickly weighed, emptied into the tube to start the dilution series and weighed again. The initial 10^{-1} dilutions were mixed with a Vortex mixer and appropriate dilutions were made in butylrubber-stoppered roll-tubes. Anaerobic conditions were maintained throughout the whole incubation procedure, by flushing with oxygen-free nitrogen, which had passed through a column of BASF catalyer R 3-11 at 150 °C. For counting sulphate reducers and methane producers, media as described by Cappenberg (1974) were used. For counting algal cell-wall-splitting bacteria, a basal medium (KH_2PO_4, 0.75 $g.l^{-1}$; K_2HPO_4, 0.75 $g.l^{-1}$; NaCl, 1.5 $g.l^{-1}$; $(NH_4)_2SO_4$, 0.75 $g.l^{-1}$; $MgSO_4.7H_2O$, 0.15 $g.l^{-1}$; $CaCl_2.2H_2O$, 0.15 $g.l^{-1}$ and yeast extract, 1 $g.l^{-1}$) was used, in addition to Bacto-Casitone (Difco, 0.1% w/v), Na-resazurine (0.01% w/v), L-cysteine (0.3% w/v) and cell-walls of *Chlorella pyrenoidosa* (0.6% w/v). The gas atmosphere of the anaerobic culture tubes was oxygen-free N_2. An *in situ* incubation temperature of 6 °C was used for at least 4 weeks. Specific colonies were counted in the lowest dilution series (triplicates). The counting procedures were checked with dilution series of pure cultures of *Desulfovibrio desulfuricans, Methanobacterium formicicum* and pure cultures of *Chlorella* algal cell-wall-splitting bacteria, in a range from 10^{-4} to 10^{-8} each in triplicate; the reproducibility as judged from colony counts was excellent (± 5%).

Results and discussion

Sedimentation rates

The seasonal patterns of C-sedimentation at 8 and 10 m depth closely resemble those of nitrogen, due to the fact that the carbon figures were derived from the nitrogen data and the carbon:nitrogen ratios of the collected material did not vary much (C:N values of 9 ± 1). Peaks in sedimentation rates were observed in spring and autumn (Verdouw & Dekkers 1982a). For the evaluation of the role of sedimentation in the carbon cycle and, more specifically, as the input source of organic material for the anaerobic mineralization, the surface areas covering the 8 and 10 m contours were considered for both production and sedimentation.

The input and output data of carbon below the 8 m depth contour in Lake Vechten during summer stratification, together with those on the average planktonic C-fixation in the epilimnion, are given in Table 1. The ratios of surface areas were calculated with the area at 10 m depth as unity, and the carbon data were corrected accordingly. It would be interesting to have these data for the individual collecting periods, in order to study the events occurring between production in the epilimnion and arrival at the sediment traps. For instance, grazing by zooplankton populations in May-June is important (Gulati *et al.* 1982). C-sedimentation as a percentage of C-fixation averaged 88 ± 32 (n = 9) at 8 m and 56 ± 21 (n = 9) at 10 m depth. For the individual periods, the differences between 8 and 10 m depth averaged 30 ± 15%. A sedimentation percentage exceeding 100 occurred in July at 8 m depth, when C-fixation and zooplankton grazing

Table 1. Carbon-fixation*, sedimentation and carbon output rates in Lake Vechten, eastern depression. Average values in the period 3 May–20 September 1977 (with S.D.; n = 9).

	$g\,C \cdot m^{-2} \cdot day^{-1}$	Area ratio	$g\,C \cdot m^{-2} \cdot day^{-1}$ (corrected)
C-fixation	0.17 ± 0.06	10.2	1.7
C-sedimentation at 8 m	0.52 ± 0.09	2.7	1.4
C-sedimentation at 10 m	0.90 ± 0.17	1.0	0.9
C-output as production rate of CH_4 below the 8 m depth contour line	0.24 ± 0.03		
C-output as production rate of CO_2 below the 8 m depth contour line	0.08 ± 0.01		

* Primary production measurements of the limnetic planktonic algae; data provided by W. A. de Kloet.

were low. The fortnightly frequency of primary production measurements, however, is too low to compare the results of individual periods. In addition, only the limnetic production of planktonic algae is taken into account here. The input of carbon by macrophytes and associated periphyton and epipelic algae in the littoral zone, together with allochthonous input (leaf fall) may be considerable and could influence the comparison of sedimentation with that of production.

Direct comparison of C-fixation rates in the production zone with sedimentation rates at a given depth on areal basis (g $C.m^{-2}.day^{-1}$) is difficult, because of the morphometric 'funnel' effect. If sediment focussing is complete, the ratio of surface areas involved can be used to relate production and sedimentation as was done in Table 1 (see also e.g. Kimmel & Goldman 1977).

Whether or not this procedure is justified depends on the degree of focussing. The role of sedimentation in the nitrogen cycle of Lake Vechten has been investigated in some detail by Verdouw & Dekkers (1982a). This study suggested that a precise measurement of sedimentation is impossible, as a part of the particulate material is transported downwards along the sloping lake bottom. Approximate measurements, however, are possible and, in view of the necessity of quantifying separate processes in nutrient cycles, indispensable; cf. Vollenweider (1976).

The suggested mineralization figure of 30 ± 15% between 8 and 10 m depth is rather tentative, since the exact mechanism of particle transport below 8 m is not known. Nevertheless, the C-sedimentation at 8 m depth can be used as the input for anaerobic mineralization, in the water column and at the bottom. For the sedimentation of organic material during the circulation period, the data are not reliable because of resuspension of sedimented material. This is true both for the deep areas and for the littoral region of the lake. Thus a direct comparison of C-fixation and sedimentation during the circulation is not possible. At present, an evaluation of sedimentation is limited to the stratification period.

Mineralization in the hypolimnion

During summer stratification several bacterial reactions may occur in the anoxic hypolimnion; these reactions use organic substrates, e.g. for bacterial reduction of nitrate and sulphate, and fermentative processes in which organic matter is broken down to intermediate metabolites and CO_2. Because the amount of electron acceptors is available to only a limited extent, the anoxic reactions are competitive, i.e., an increase in one occurs at the expense of the other. The relative importance of each activity is a function of thermodynamics, the concentrations of reactants and products at a given moment, and other environmental conditions which might be more suitable for one group than the other. The magnitude of these activities and the relative importance of the different anoxic bacterial reactions was examined, i.e. the amount of organic carbon mineralized by each reaction was computed and compared with the C-fixation data. As nitrate is not available in the hypolimnion of Lake Vechten as a potential electron acceptor, and the amount of reducing equivalents represented by iron and manganese is only 0.5 to 1% in comparison with the amount formed in the primary production (Ver-

Table 2. Relative carbon-mineralizing activity of different hypolimnetic bacterial reactions during summer stratification, 1980. Samples were taken each week at 1.2 m depth intervals in the eastern depression of Lake Vechten.

Reactions	O_2 reduction (1 mol C + 1 mol O_2→1 mol CO_2)		SO_4^{2-} reduction (3 mol C + 1 mol SO_4^{2-}→1 mol S^{2-} + 2 mol C + 1 mol CO_2)		Fermentation (ΔDIC)	
Time intervals[1]	t_1	t_2	t_1	t_2	t_1	t_2
Δ, mol[2]	1 280	1 300	740	200	–	–
Flux, mol[3]	1 340	5 200	600	860	–	–
C-fixation, kg C[4]	504	1 270	504	1 270	504	1 270
Δ, mol CO_2	–	–	–	–	6 350	19 560
% carbon mineralization	6.0	5.7	6.4	2.4	15.1	19.3

[1] t_1 = 20-5-1980 to 17-6-1980; t_2 = 12-8-1980 to 9-9-1980.
[2] Δ, mol = total change in quantity of species for the various water layers in the hypolimnion (total volume of the eastern depression = 24.3×10^6 l).
[3] Flux, mol = total flux in quantity of species computed through the 6 m plane (11 326 m²), with D = 1 ± 10^{-2}.cm².s⁻¹ and the existing gradient of species between the 5 and 7 m water layer.
[4] C-fixation, kg C = total limnetic primary production of oxygen and anoxygenic photosynthesis computed for the various strata and time intervals (see Steenbergen & Korthals 1982).

douw & Dekkers 1980, 1982b), only the reduction rates of oxygen and sulphate were considered.

Because of the presence of strong concentration gradients of O_2 and SO_4^{2-} in the upper region of the hypolimnion, vertical diffusion corrections were necessary across the 6 m plane of the lake and were made using the following equation:

$$F = -D.\ d_c/\ d_x$$

where F is the flux in moles.cm^{-2}.s^{-1}, D is the average eddy diffusion coefficient in cm^2.s^{-1}, and $d_c/\ d_x$, is the existing concentration gradient in moles.cm^{-3}.cm^{-1}. Calculations of the coefficient of eddy conductivity, using observed heating rate and temperature increase as a function of depth gave values of about 1×10^2 cm^2.s^{-1} in the waterlayers from 5 to 9 m (see Steenbergen & Verdouw 1982).

The total change in quantity of each concentration of O_2 and SO_4^{2-} measured was computed for two 29-day periods, i) May–June at the onset of the stratification, and ii) in August–September during optimal stratification conditions (Table 2). To further characterize the hypolimnetic metabolism in Lake Vechten, the CO_2 produced by fermentation (increase of the inorganic carbon pool) was compared with the limnetic C-fixation. In the first period 6% mineralization contributed to O_2 as electron acceptor, 6.4% mineralization contributed to SO_4^{2-} as electron acceptor, and 15.1% mineralization contributed to fermentation was calculated from the primary production, i.e. a total potential mineralization of 27.5% of the C-fixation. For the optimal stratification period, i.e. August–September, these figures were, respectively, 5.7, 2.4 and 19.3, giving a value of 27.4% for potential mineralization of the limnetic C-fixation.

In any lake the relative predominance of each type of hypolimnetic bacterial activity should be a function of 1) the relative concentration of each different electron acceptor (O_2, SO_4^{2-}, etc.), 2) organic loading to the sediments, and 3) the duration of anoxic conditions following a circulation event–O_2 and SO_4^{2-} reduction would be more prevalent just after complete circulation than late in the summer as found in the seepage Lake Vechten. In addition, in late summer the sedimentation rate of particulate organic matter as a source for fermentative processes is almost twice as high in the hypolimnion than at the start of the stratification (Verdouw & Dekkers 1982a). Although the mineralization figures are not realistic for the actual mineralization rate in the hypolimnion of Lake Vechten, they do, however, favourably compare with the suggested mineralization percentage of 30 ± 15 between 8 and 10 m depth computed from the sedimentation rate measurements. This calculated mineralization rate is also comparable with those given by other workers for lakes of similar volume: depth ratio in the hypolimnia as in Lake Vechten

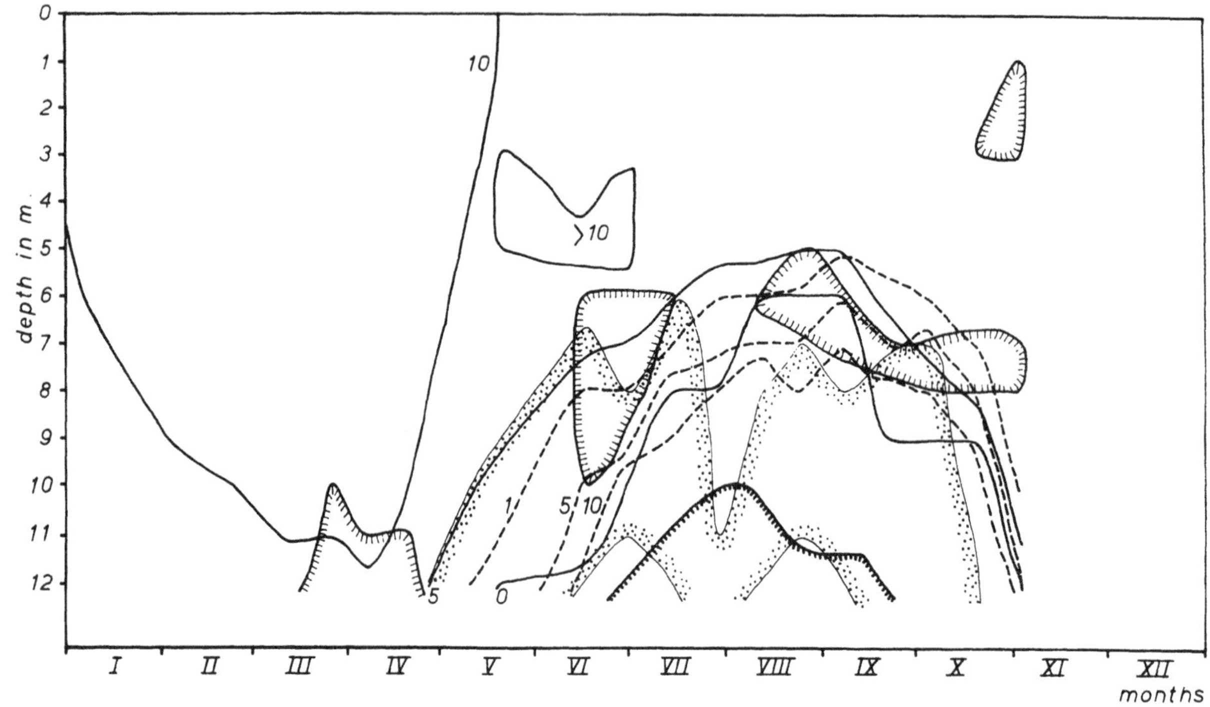

Fig. 1. Seasonal distribution of facultative aerobic heterotrophic (HB), obligate anaerobic sulphate-reducing (SRB) and micro-aerophylic methane-oxidizing bacteria (MOB) in Lake Vechten (1970), together with isopleths of oxygen (10,5 and 0 $mg.l^{-1}$) and methane (10, 5 and 1 $ml.l^{-1}$). For counting procedures and measurements of parameters, see Cappenberg 1972.

numbers of HB, $\geq 10^9.l^{-1}$

numbers of MOB, $\geq 5.10^5.l^{-1}$

numbers of SRB, $\geq 4.10^4.l^{-1}$

concentration of oxygen

concentration of methane.

(Jones 1976; Lastein 1976; Anderson & Jacobson 1979).

Bacterial distribution in the hypolimnion

Our knowledge of the various breakdown pathways and turnover rates of organic matter in the hypolimnion is inadequate. A general picture can be obtained from the total numbers of important bacterial groups, i.e. the facultative aerobic heterotrophic, the obligate anaerobic sulphate-reducing and the micro-aerophylic methane-oxidizing bacteria. The maximal numbers, together with the changes of concentrations of oxygen and methane in time and depth, are given in Fig. 1.

Maxima of the heterotrophs are found usually in the boundary between aerobic and anaerobic zones. During stratification the stagnant water layers became anaerobic owing to the metabolic activity of the bacterial flora. Two gradients play an important role: firstly, the oxygen is depleted in the hypolimnion and circulation in epilimnion supplies oxygen to the metalimnion; and secondly, there is an organic matter gradient. The metalimnion, being a barrier between the epi- and hypolimnion due to the marked viscosity and density differences, traps the downward moving organic matter, resulting in a concentration gradient. A large accumulation of particulate organic matter, viz. 3 000 mg organic-$C.m^{-3}$, has been found at the border of the

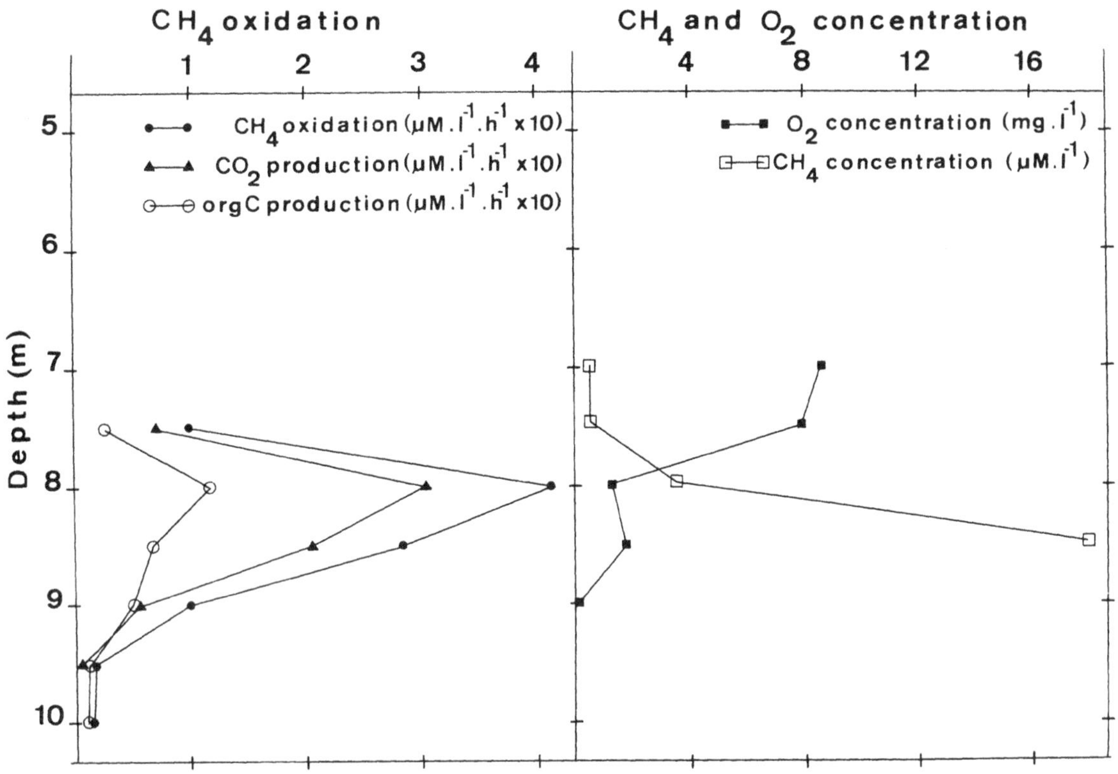

Fig. 2. Typical profile of summer methane oxidation showing very low concentration of methane within zone of methane-oxidizing activity at the meta-hypolimnion border in Lake Vechten (12-9-1976).

metalimnion and the hypolimnion. Besides the formation of CO_2, the fermentation products of the microbial flora can give rise to concentration gradients of low-molecular-weight organics, e.g. as lactate and acetate. Preliminary observations using HPLC methods confirm this. Also, the algal excretion products will contribute to the dissolved organic pool present here (Blaauboer *et al.* 1982) and may supply suitable electron donors for anoxygenic photosynthesis by green and purple sulphur bacteria in the border between the meta- and hypolimnion (Steenbergen & Korthals 1982). Further information is needed about the actual mineralization rates of the DOC-pool and the possible link between the carbon and sulphur cycles in the anaerobic zone of Lake Vechten.

Both the maximal numbers of methane oxidizers and the maximal methane concentration occur at the same depth (Fig. 1). Preliminary data indicated that these organisms are facultative methane oxidizers and can be regarded as micro-aerophylics (Cappenberg 1972). As methane oxidation proceeds, residual oxygen must be consumed and the hypolimnion depleted of oxygen. Using the methods proposed by Rudd & Hamilton (1978) ^{14}C-CH_4 measurements indicate the oxidation rate of CH_4 to be about 0.4 μmol $CH_4.l^{-1}.h^{-1}$ which is about 8 mol $CH_4.h^{-1}$ for the water layers in the meta-hypolimnion borderline. On the other hand, the production of CH_4 in the sediments below the 8 m contour line is about 9 mol.h^{-1}, indicating that most of the methane produced in Lake Vechten will be oxidized in the water phase of the lake (Fig. 2).

In the meta-hypolimnion border, the maximum concentration of bacteria found is $2 \times 10^{10}.l^{-1}$; presumably, the bacteria here utilize the CH_4 and the excretion products of phytoplankton as substrates. In addition, bacterial colonization on particulate detritus is abundant here (Blaauboer *et al.* 1982). The importance of bacteria, particularly those colonizing detritus, as food for the zooplankton feeders in these strata, remains to be investigated.

From June until September decreasing sulphate

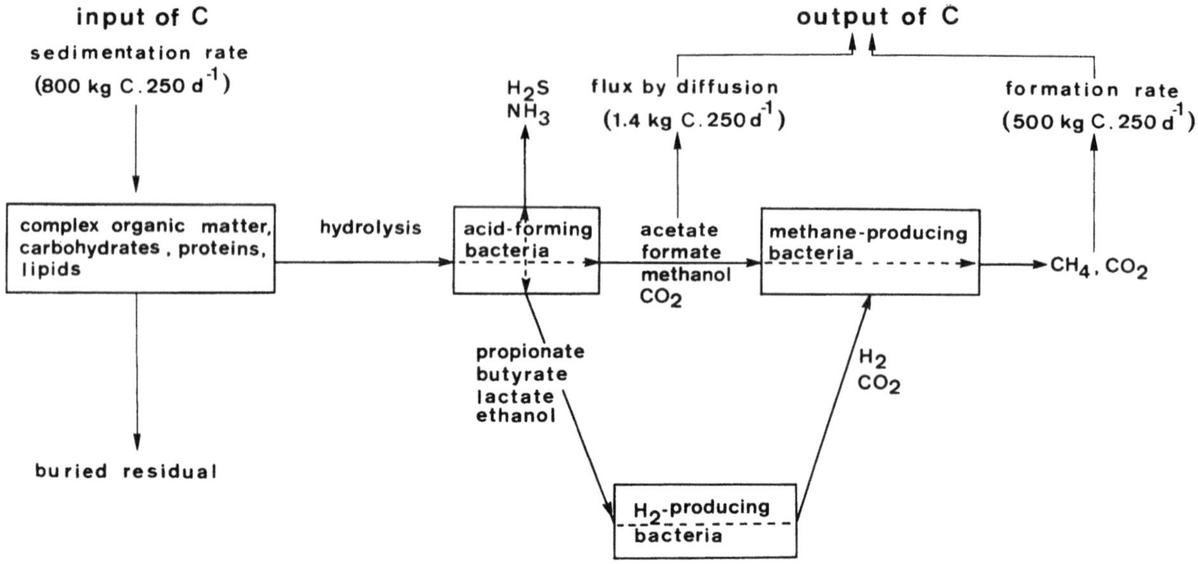

Fig. 3. Schematic representation of the anaerobic mineralization processes in the sediments, together with the sources of in-and output of carbon.

concentrations in the hypolimnion of Lake Vechten coincide with an increasing number of sulphate-reducing bacteria (van Gemerden 1967; Cappenberg 1972). Although sulphate reduction occurs usually in the sediment, the stagnant water layers, with their favourable reducing capacity, may form a suitable biotope for sulphate reducers; this requirement is partly fulfilled by the activity of methane oxidizers which remove the residual O_2 below the metalimnion. It can be assumed that *in situ* CO_2-fixation by green and purple sulphur bacteria and the oxidation of H_2S as an electron donor, produced by sulphate-reducing bacteria, are coupled. Oxidizable carbon, as energy-yielding substrate for the sulphate reducers, may be, as stated before, obtained from fermentative processes in the hypolimnion. Apparently, further studies using the ^{35}S-method are needed to obtain conclusive evidence for the above assumptions.

Mineralization in the sediments

The input and output of carbon in Lake Vechten below the 8 m depth contour during summer stratification in 1977 are given in Table 1. The total input of C for the area below this contour (= 6300 m^2) calculated using the sedimentation data was about 800 kg C during a stratification period of 250 days. The output of C calculated similarly from the production rates of CH_4 and CO_2, respectively, was about 500 kg C.250 day^{-1}. Though these are average figures based on average sedimentation and production rates and the fluctuations of these rates during the stratification period are great, they do indicate the importance of the anaerobic breakdown of the sedimented organic matter. Thus a figure about 62% mineralization (500/800 × 100) in the sediments is comparable with other lakes in the temperate zone (Gasith 1976; Kimmel & Goldman 1977; Kelly & Chynoweth 1980; Molongoski & Klug 1980). The remaining 13% of the sedimented particulate organic matter (that is, after deleting the 62% mineralization in the sediments and 25% mineralization in the hypolimnion) may be buried more or less permanently by the subsequent deposits or may be partly released in exchange processes of lower fatty acids at the sediment–water interface.

Our knowledge of the various anaerobic breakdown pathways and of turnover rates of carbon between its input and output in the sediments is inadequate. A scheme of anaerobic mineralization processes is given in Fig. 3. The quantity and quality of sedimenting particulate organic matter are linked closely to the production dynamics of the autotrophs in the pelagic zone. Chemical analysis indicated that this organic matter is planktonic in

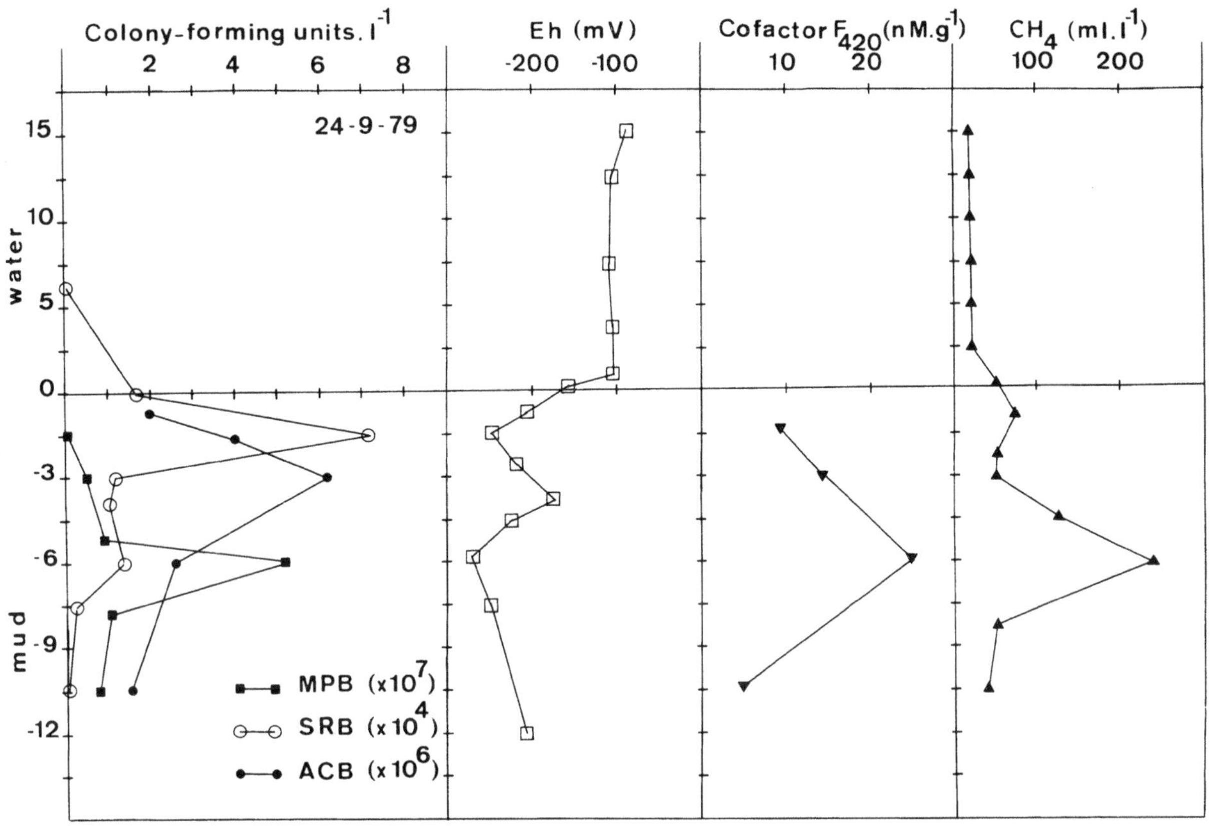

Fig. 4. Distribution of methane-producing (■—■), sulphate-reducing (○—○) and algal cell-wall-splitting bacteria (●—●) in bottom deposits of Lake Vechten, together with the redox potential values, concentrations of cofactor F_{420} and methane concentration as a function of depth in the mud core. Core was taken 24-9-1979 at maximal stratification of the lake.

origin and is dominated by cellulose. Organic-rich sediment was present in the upper 20 or 30 cm, depending on the sampling station. Below these depths the layer is firmly clayey and mixed with sand. The gradients of the measured parameters were confined to the upper 20 cm; thus, it is considered to be the active layer for studying nutrient exchange with the overlying water. A detailed study of this in respect in nitrogen exchange processes is in progress.

Organic matter as a percentage of dry weight varied between 20 and 30 at the surface and 10 and 20 in the deeper sediment layers. The solids by volume in the mud ranged from approximately 7% in the upper few mm to 70% at a depth of 4 cm in the sediment. Concentration of total dissolved carbon (TDOC, $< 0.45\ \mu$m) ranged in the sediment from 12.4 to 15.3 mg C.l^{-1} during the stratification. The C:N ratio increased with depth in the mud from 11:1 to 14:1, indicating a relatively greater increase in organic C-compounds than organic N-compounds with depth. The sedimented material is dominated by the cell wall and debris of phytoplanktonic origin, especially of green algae. In Lake Vechten the green alga *Chlorella pyrenoidosa* dominates with maximal numbers reaching 20×10^6.l^{-1} (Blaauboer *et al.* 1982). The breakdown of *Chlorella* cell walls was, therefore, studied in detail.

To indicate the importance of this breakdown some data on the distribution of algal cell-wall-splitting bacteria in the sediment of Lake Vechten are given (Fig. 4), together with that of sulphate-reducing and methane-producing bacteria. Despite seasonal effects, the maxima of both methane producers and the concentration of methane and cofactor F_{420} tend to coincide with depth in the mud of minimum redox potential. Cofactor F_{420} is an electron carrier specifically linked to the reduction pathway of HCO_3^- into CH_4 and can be used as a potential parameter for methanogenic activity. The maximum number of the methane producers occurs at a redox potential of about -275 mV at depths from 5 to 7 cm in the mud, that of the sulphate reducers at about -200 mV at depths from 1 to 3 cm, and that of algal cell-wall-splitting bacteria at intermediate value, i.e. -150 mV, and at depths from 3 to 5 cm in the mud.

It was shown earlier (Cappenberg 1974) that sulphate in the interstitial water of the sediments is reduced by the sulphate reducers, which are also limited by sulphate concentration. Methane producers occur deeper in the mud at lower concentration of hydrogen sulphide, and therefore the depth segregation of the methane producers (MPB), sulphate reducers (SRB) and algal-splitting bacteria (ACB) may be partly due to sensitivity to sulphide. In addition, an interrelationship depending on the level of substrate flow between the groups could be the formation of acetate and lactate by algal-splitting organisms and their subsequent utilization by the two other groups, as acetate is the main precursor for methanogenesis and lactate the main carbon source for sulphate reduction in the sediment of Lake Vechten (Cappenberg & Prins 1974; Cappenberg & Jongejan 1978).

Breakdown kinetics in the sediments

To study the breakdown kinetics of organic carbon into its important intermediate metabolites one needs to assume that the pool sizes of the metabolites do not change appreciably in the mud, and therefore that their rates of dissimilation must equal those of their formation. The dissimilation rate of an intermediary is the product of its turnover rate constant (k-value) and its pool size. The magnitude of these dissimilations can be estimated by the disappearance from mud samples of ^{14}C-labelled substrates. The turnover rate constants are calculated using first-order kinetic equations, after the log of the radioactivity retained in the samples is plotted against time.

The kinetics of breakdown of algal cell walls in mud were studied with U-^{14}C-labelled *Chlorella* cell wall material (specific activity about 5.10^5 DPM. mg^{-1}). To the mud samples of 4 to 5 g wet weight from various depths in the core, radioactive substrate (200 μl of ^{14}C cell suspension) was injected, rapidly mixed and the samples were incubated under strict anaerobic conditions at the *in situ* temperature of 6 °C. The aliquots of the mud were removed at appropriate time intervals for measurements of the radioactivity. When plotted against time in days the $^{10}\log$ of radioactivity in the cell wall fraction gave a straight line ($r = -0.955$). From the regression equation, turnover rate constants (k-values) calculated for the various mud layers were: 0 to 1 cm, 0.18 day^{-1}; 0 to 4 cm just under the interface, 0.32 day^{-1}; and in the 6 to 10 cm, 0.21 day^{-1}.

The average concentration of the pool of cell wall material obtained by common extraction methods was: the surface layer, 66 $\mu g.g^{-1}$ wet mud; under the surfaces layer 80 $\mu g.g^{-1}$ wet mud; and in the deeper layers, 37 $\mu g.g^{-1}$ wet mud. The computed turnover rates for the three layers were, respectively, 11.8, 25.6 and 7.7 $\mu g.g^{-1}.day^{-1}$ of cell wall material. The derived value per cm^2 of sediment area based on the respective rates in the three layers, is about 23 $\mu g.g^{-1}.day^{-1}$, giving for the whole sediment area (=6 300 m^2) and calculated from the gross formula $(C_6H_{10}O_5)_n$, a turnover of about 360 kg C.250

Table 3. Production rate of metabolic intermediates from algal cell wall material of *Chlorella*.

Composition of cell walls of *Chlorella pyrenoidosa:* hemicellulose 59%, cellulose 33%, lignin 2.6%, polysaccharides 2.8%, pectin 1.8% (general composition of 27.79% C, 3.34% N, 4.92% H). Production rate of $\mu mol.h^{-1}$ from the isolates (gross ratio of 6:3:1):

Intermediates	Sporeforming rods ($Gram^+$)	Non-sporeforming rods ($Gram^-$)	Cocci ($Gram^+$)
Acetic acid	2.8	3.2	1.2
Propionic acid	0.08	1.1	2.3
Isovaleric acid	–	–	0.006
Butyric acid	0.003	0.04	–
Lactic acid	0.5	0.6	–

Production of the gases CO_2 (8.5% v/v) and H_2 (0.08% v/v); traces of capronic and isobutyric acid.

days^{-1}. Although this value is based on laboratory studies and thus not realistic for the lake, the breakdown of cell wall material is however favourably comparable with the input and output data of carbon as mentioned earlier for the stratification period.

From the mud samples incubated with several components of cells walls as the only energy and carbon source, numerous anaerobic bacteria were isolated (Table 3). These isolates consisted of a gross ratio of 6:3:1 of sporeforming, non-sporeforming and cocci; they were cultured using convential roll-tube methods and the rate of breakdown products monitored in time. From the ratio of the gross formula of cell wall material and the acetic-carbon produced, it can be calculated that almost 60% of the carbon is channeled into this metabolic intermediate; therefore, acetate appears to be the most important breakdown product. In addition, incubation of mud samples with cell wall material gave a stoichiometric increase of breakdown products which also indicates that the rate-limiting step in anaerobic mineralization of algal components is cell wall material.

The breakdown kinetics of acetate was studied in the same way as that of the algal cell walls, using a gas chromatography–gas proportional counting system (Cappenberg 1980). For the 5 cm depth layer in the mud, the turnover rate constant was 0.24 h^{-1}; the average concentration of the endogenous acetate pool was 0.30 $mg.l^{-1}$ of wet mud, giving a turnover rate of 0.07 $mg.l^{-1}.h^{-1}$ in this layer. At a depth of 2 cm in the mud a k-value was 0.07 h^{-1} and a pool size 0.60 $mg.l^{-1}$, giving a turnover rate of 0.04 $mg.l^{-1}.h^{-1}$. The relatively higher rates of acetate in lower mud layers of Lake Vechten confirms our earlier observation on the maximum of acetate-fermenting methanogenic bacteria being at 5–7 cm depth. The incubation of samples, under environmental conditions, just above the sediment–water interface gave a turnover rate constant of 0.31 h^{-1}; with average pool size of 0.60 $mg.l^{-1}$ of acetate, the turnover rate was found to be 0.19 $mg.l^{-1}.h^{-1}$.

The breakdown studies of acetate in relation to sediment depth, incubating the mud at the environmental temperatures, and analyses of $^{14}CH_4$ and $^{14}CO_2$ (calculated from the ratios

$$F_{CO_2} = \frac{^{14}CO_2}{^{14}CO_2 + {}^{14}CH_4}$$

) produced from 1-^{14}C-acetate, 2-^{14}C-acetate or U-^{14}C-acetate, show clearly that oxidation of acetate to CO_2 is more important in the upper layers; this is due to anaerobic, sulphur-reducing, acetate-oxidizing bacteria (Cappenberg & Jongejan 1978). On the other hand, the acetate-splitting methanogenic processes are more important in the lower layers in the mud (Fig. 5). Thus, the breakdown of acetate in mud may be illustrated as follows:

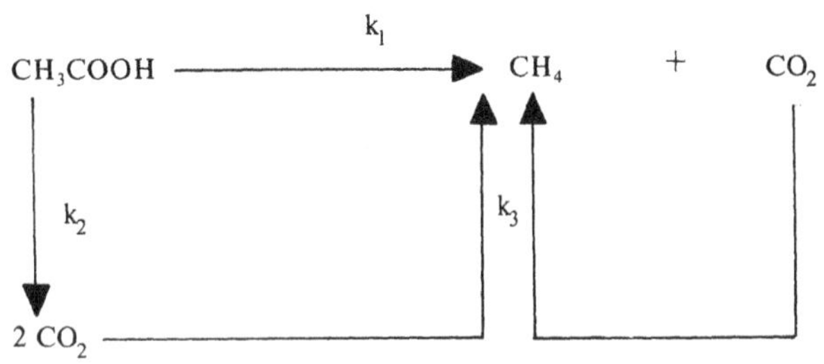

From the F_{CO_2}-ratios obtained by incubating mud with specifically labelled acetates, the k-values deduced for the deeper mud layers (4 to 6 cm) are: k_1, 0.18 h^{-1}; and k_3, 0.24 h^{-1}. Similarly, for the mud in the upper layers (0 to 2 cm), these values are: k_1, 0.038 h^{-1}; k_2, 0.27 h^{-1}; and k_3, 0.66 h^{-1}.

Diffusion of intermediate metabolites

Since acetate is the most important intermediate in anaerobic mineralization, its diffusion from the sediment to the overlying water may be used to calculate the carbon flow. For computing the flux of acetate, one needs to know the existing acetate gradients at the mud–water interface. Such a gradient can be obtained using small sample sizes only, so that the gradients are not disturbed. An analytical method with high resolution was developed using a special HPLC-technique (Hordijk & Cappenberg 1982); care was taken while sampling that the mud cores were not disturbed. For this, the concentration gradients of acetate in two cores of mud–water interface were compared (Fig. 6); the agreement between the two cores was good. The main observations regarding this gradient are: a point source of acetate lies at approximately 2.5 cm in the mud layer and an acetate sink exists both in the deeper layers and at the sediment interface.

If we assume that only molecular diffusion in the deeper parts of Lake Vechten sediments occurs, a diffusion rate can be calculated using the formula of

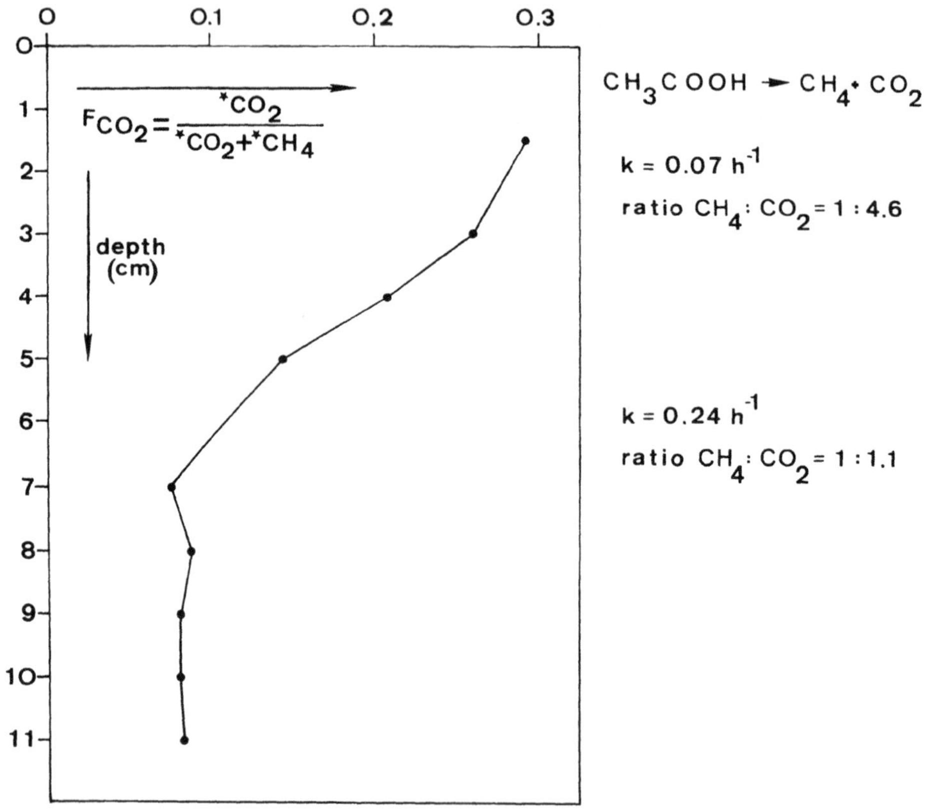

Fig. 5. Label distribution between CH_4 and CO_2 ($F_{CO_2} = \frac{^{14}CO_2}{^{14}CO_2 + ^{14}CH_4}$) in relation to the depth in the mud, together with turn-over rate constants (k-values) and ratios of CH_4:CO_2 formed by the breakdown of acetate at a depth of 2 and 6 cm in the mud, respectively.

Fick's law: flux = – D. ($^dc/^dx$). Since the diffusion of organic molecules in the sediments is dependent largely on the sediment porosity, one needs to calculate the effective diffusion coefficient (D_s) in the system under study. Using sterilized mud cores, incubated under *in situ* circumstances and adding a point source of acetate, D_s computed for acetate is 2.2×10^{-6} $cm^2.s^{-1}$ (Cappenberg *et al.* 1982). As the concentration gradient of acetate in the mud layer of 2.5 cm is about 20 μmol, the flux of acetate is about 0.06 $\mu mol.cm^2.h^{-1}$. The total flux calculated, using $D_s.^dc/^dx$.area.time, for the 250 days and for the whole sediment area, was about 1.4 kg acetate-carbon, and this simple molecular diffusion flux is not important for comparing the figures of input of carbon with its output as given before.

Bioturbation of *Chaoborus flavicans* larvae may positively affect the diffusion. Parma (1971) found a larval abundancy in winter and early spring of about 5 000.m^{-2} in the eastern depression of the lake. Recently, however, this number decreased to about 1 000.m^{-2} (Weekenstroo, pers. comm.). Although bubble formation of methane near the sediment–water interface has not been demonstrated in Lake Vechten so far, high production of methane during summer may cause saturation concentrations and subsequent bubble ebullition may increase the apparent sediment diffusivity as in organic-rich marine ecosystems (Martens & Klump 1980; Klump & Martens 1981). *In situ* flux measurements, placing chambers on the sediments, might give a more realistic flux of carbon from the sediments to the overlying waters.

With other important metabolic intermediates in the carbon flux, i.e. lactate, the concentration gradient was found to be much less steep and its con-

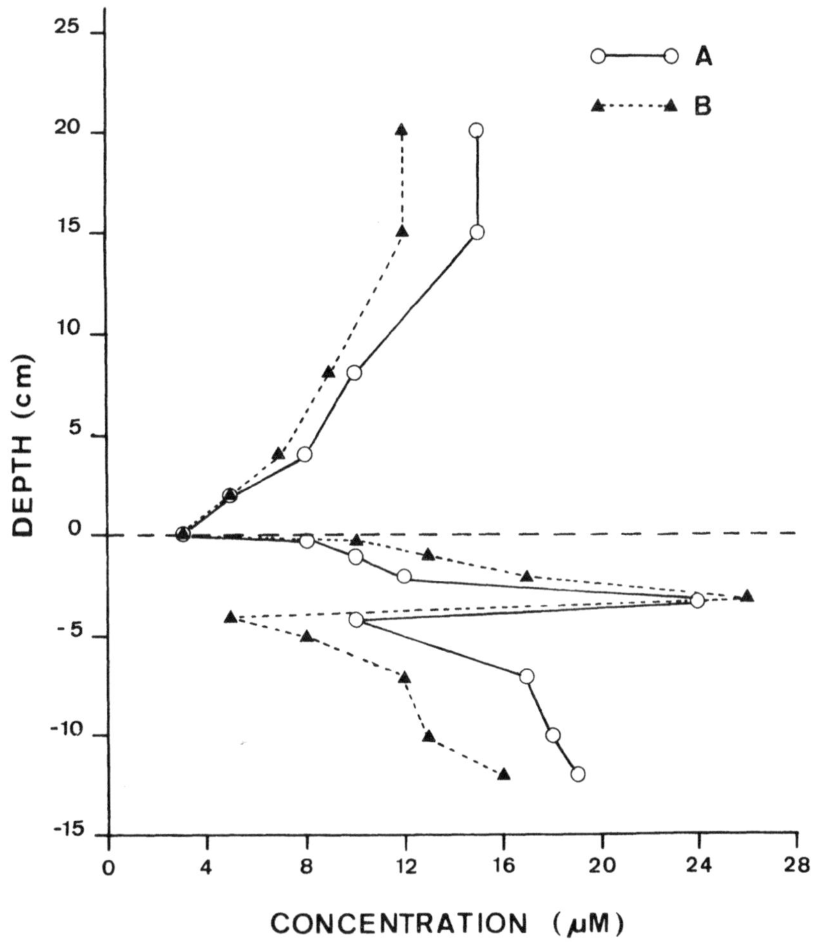

Fig. 6. Concentration gradient of acetate (μmol.l^{-1}) at the mud–water interface. A and B are two different mud cores taken at the same time from the deepest point of the eastern depression in Lake Vechten.

centration ten-fold lower than that of acetate (Cappenberg *et al.* 1982). Also, lactate being a larger molecule, its effective diffusion coefficient should be considerably smaller. The calculations employing first-order kinetic equations and experiments with uniformly and non-uniformly ^{14}C-labelled lactate show that lactate is broken down into propionic, butyric and acetic acids and carbon dioxide simultaneously, with different values of turnover rate constant (Cappenberg 1980). Thus, these intermediates, which are less oxidized than acetate, will play a minor role compared with acetate in the carbon flow from the more complex organic matter towards acetate and methane, according to the concept of interspecies electron transfer by hydrogen gas with advantages of coupled (electron transfer by H_2) over uncoupled (reduced fermentation end products) metabolism.

Summary

The turnover and exchange rates, as well as the sedimentation rates concerning the in- and output of carbon compounds in the anaerobic hypolimnion and the sediments, were studied. In addition, field observations of numbers of important bacterial groups in breakdown processes of organic matter are given. The carbon input rates were derived from the sedimentation rates of particulate organic matter (about 800 kg C.250 days^{-1}) during the stratification period. The nature of the sedimented POC, and its breakdown pathways and turnover rates towards important metabolic intermediates in methanogenesis were examined.

In the hypolimnion a potential mineralization of 20–30% of the limnetic C-fixation could be computed from sedimentation rate measurements and from microbial reduction rates of observed fluxes in electron acceptors. Seasonal *in situ* measurements of carbon fluxes, simultaneous with those of the actual turnover rates of important carbon intermediates, are needed to elucidate the rates and mechanisms of chemical and biological exchange and their importance in the hypolimnion.

The breakdown kinetics of *Chlorella* cell walls, a dominant green alga in Lake Vechten, was studied using U-^{14}C-labelled cell walls. The breakdown of the cell walls (about 360 kg C.250 days^{-1}) appears to be the rate-limiting step in anaerobic mineralization. Using first-order kinetic equations, turnover rate constants (k-values) of between 0.18 and 0.32 day^{-1} and pool sizes of algal cell walls of 37 to 88 μg.g^{-1} wet mud were found, giving turnover rates of 7.7 to 25.6 μg.g^{-1}.day^{-1} of cell wall material. The turnover rates (k-values between 0.07 and 0.31 h^{-1}) of acetate, the most important breakdown product, and its concentration profile (between 5 and 30 μM), diffusion coefficient ($D_s = 2.2 \times 10^{-6}$ cm^2.s^{-1}) and breakdown pathways to CH_4 and CO_2 in the sediments were quantified.

Finally, from the flux data of acetate (about 1.4 kg C.250 days^{-1}), and those on the rates of formation (about 500 kg C.250 days^{-1}) of CH_4 and CO_2, it may be concluded that the major part of the input of particulate organic matter is mineralized and channeled as output via gases into the overlying waterlayers and the remaining is buried more or less permanently into the sediments, while precise data are needed to elucidate the importance of this last factor in the carbon cycling in Lake Vechten. Last, most of the produced CH_4 is oxidized in the meta–hypolimnion border, where maximum concentration of methane-oxidizing and heterotrophic bacteria (about 2×10^{10}.l^{-1}) are located during the stratification period.

Acknowledgements

Thanks are particularly due to E. M. J. Dekkers and C. A. Hordijk for skilful technical assistance. The authors are indebted to R. D. Gulati and S. Parma for reading the manuscript.

References

Andersen, J. M. & Jacobsen, O. S., 1979. Production and decomposition of organic matter in eutrophic Frederiksborg Slotssö, Denmark. Arch. Hydrobiol. 85: 511–542.

Blaauboer, Marianne C. I., Keulen, R. van, Cappenberg, Th. E., 1982. Extracellular release of photosynthetic products by freshwater phytoplankton populations, with special reference to the algal species involved. Freshwat. Biol. (in press.)

Bloesch, J. & Burns, N. M., 1980. A critical review of sedimentation trap technique. Schweiz. Z. Hydrol. 42: 15–55.

Blomqvist, S. & Håkanson, L., 1981. A review on sediment traps in aquatic environments. Arch. Hydrobiol. 91: 101–132.

Cappenberg, Th. E., 1972. Ecological observations on heterotrophic, methane oxidizing and sulfate reducing bacteria in a pond. Hydrobiologia 40: 471–485.

Cappenberg, Th. E., 1974. Interrelations between sulfate-reducing and methane-producing bacteria in bottom deposits of a fresh-water lake. I. Field observations. Antonie van Leeuwenhoek 40: 285–295.

Cappenberg, Th. E., 1976. Methanogenesis in the bottom deposits of a small stratifying lake. In: Schlegel, H. G., Pfennig, N., Gottschalk, G. (Eds.) Microbial Production and Utilization of Gases, H_2, CH_4, CO, pp. 125–134. E. Goltze-Verlag, Göttingen.

Cappenberg, Th. E., 1980. Use of radio gas chromatography in studying breakdown processes of organic matter in aquatic ecosystems. In: Agrochemical Residue–Biota Interactions in Soil and Aquatic Ecosystems, pp. 55–66. International Atomic Energy Agency, Vienna.

Cappenberg, Th. E. & Prins, R. A., 1974. Interrelations between sulfate-reducing and methane-producing bacteria in bottom deposits of a fresh-water lake. III. Experiments with ^{14}C-labelled substrates. Antonie van Leeuwenhoek 40: 457–469.

Cappenberg, Th. E. & Jongejan, E., 1978. Microenvironments for sulfate reduction and methane production in freshwater sediments. In: Krumbein, W. E. (Ed.). Environmental Biogeochemistry & Geomicrobiology, Vol. 1, pp. 129–138. Ann Arbor sci. Publ.

Cappenberg, Th. E., Hordijk, K. A., Jonkheer, G. J., Lauwen, J. P. M., 1982. Carbon flow across the sediment–water interface in Lake Vechten, The Netherlands. In: Sly, P. G. (Ed.). Sediment/Freshwater Interaction, pp. 161–168. Proc. 2nd. Int. Symp., Kingston, Ontario. Junk, The Hague.

Gasith, A., 1976. Seston dynamics and tripton sedimentation in the pelagic zone of a shallow eutrophic lake. Hydrobiologia 51: 225–231.

Gulati, R. D., Siewertsen, K. & Postema, G., 1982. The zooplankton: its community structure, food and feeding, and role in the ecosystem of Lake Vechten. Hydrobiologia 95: 127–163.

Hordijk, C. A. & Cappenberg, Th. E., 1982. Quantitative HPLC-fluorescence determinations of some important lower fatty acids in lake sediments. Appl. envir. Microbiol. (submitted).

Jones, J. G., 1976. The microbiology and decomposition of seston in open water and experimental enclosures in a productive lake. J. Ecol. 64: 241–278.

Kelly, C. A. & Chynoweth, D. P., 1980. Comparison of in situ and in vitro rates of methane release in freshwater sediments. Appl. envir. Microbiol. 40: 287–293.

Klump, J. V. & Martens, C. S., 1981. Biochemical cycling in an organic-rich coastal marine basin – II. Nutrient sediment-water exchange processes. Geochim. cosmochim. Acta 45: 101–121.

Kimmel, B. L. & Goldman, C. R., 1977. Production, sedimentation and accumulation of particulate carbon and nitrogen in a sheltered subalpine lake. In: Golterman, H. L. (Ed.) Interactions between Sediments and Freshwater, pp. 148–155. Junk, The Hague.

Lastein, E., 1976. Recent sedimentation and resuspension of organic matter in eutrophic Lake Esrom, Denmark. Oikos 27: 44–49.

Martens, C. S. & Klump, J. V., 1980. Biochemical cycling in an organic-rich coastal marine basin – I. Methane sediment-water exchange processes. Geochim. cosmochim. Acta, 44: 471–490.

Molongoski, J. J. & Klug, M. J., 1980. Anaerobic metabolism of particulate organic matter in the sediments of a hypereutrophic lake. Freshwat. Biol. 10: 507–518.

Parma, S., 1971. Chaoborus flavicans (Meigen) (Diptera, Chaoboridae): an autoecological study. Ph.D. thesis, Rijksuniversiteit Groningen. 128 pp.

Sansone, F. J. & Martens, C. S., 1981. Methane production from acetate and associated methane fluxes from anoxic coastal sediments. Science 211: 707–709.

Steenbergen, C. L. M. & Korthals, H. J., 1982. Distribution of phototrophic microorganisms in the anaerobic and microaerophilic strata of Lake Vechten (The Netherlands). Pigment analysis and role in primary production. Limnol. Oceanogr. (in press).

Steenbergen, C. L. M. & Verdouw, H., 1982. Lake Vechten: aspects of its morphometry, climate, hydrology and physicochemical characteristics. Hydrobiologia 95: 11–23.

Strayer, R. F. & Tiedje, J. M., 1978. Kinetic parameters of the conversion of methane precursors to methane in a hypereutrophic lake sediment. Appl. envir. Microbiol. 36: 330–340.

Rudd, J. W. M. & Hamilton, R. D., 1978. Methane cycling in a eutrophic shield lake and its effects on whole lake metabolism. Limnol. Oceanogr. 23: 337–348.

Van Gemerden, H., 1967. On the bacterial sulfur cycle of inland waters. Ph.D. thesis, Rijksuniversiteit Leiden. 110 pp.

Verdouw, H. & Dekkers, E. M. J., 1980. Iron and manganese in Lake Vechten (The Netherlands): dynamics and role in the cycle of reducing power. Arch. Hydrobiol. 89: 509–532.

Verdouw, H. & Dekkers, E. M. J., 1982a. Nitrogen cycle of Lake Vechten (The Netherlands): role of sedimentation. Arch. Hydrobiol. 94: 251–263.

Verdouw, H. & Dekkers, E. M. J., 1982b. Nitrogen cycle of Lake Vechten: concentration patterns and internal mass-balance. Hydrobiologia 95: 191–197.

Vollenweider, R. A., 1976. Rotsee, a source, not a sink for phosphorus? A comment to and a plea for nutrient balance studies. Schweiz. Z. Hydrol. 38: 29–34.

Winfrey, M. R. & Zeikus, J. G., 1979. Microbial methanogenesis and acetate metabolism in a meromictic lake. Appl. envir. Microbiol. 37: 213–221.

Aspects of aerobic mineralization during spring in Lake Vechten with special reference to the ^{14}C-labelling technique

J. J. Olie & Th. E. Cappenberg
Limnological Institute, 'Vijverhof' Laboratory, Rijksstraatweg 6, 3631 AC Nieuwersluis, The Netherlands

Keywords: aerobic mineralization, respiration, photo-assimilation, extracellular release, carbon recycling

Abstract

Aerobic mineralization, i.e. seston respiration, microbial breakdown of detritus and microbial assimilation–dissimilation of photosynthetically derived D(issolved) O(rganic) C(arbon) was measured in concentrated samples from the pelagic zone of Lake Vechten. The samples were described by cell numbers of dominant algae prior to concentrating by centrifuge with continuous rotor. The concentrated samples were incubated in the laboratory at *in situ* temperature and a light intensity of 30 $W \cdot m^{-2}$ for measuring primary production and photosynthetically derived DOC. After filtration the particulate fraction was incubated in unlabelled lake water for measuring respiratory production of $DI^{14}C$. Portions of the same particulate fraction were sterilized and incubated in unlabelled sample concentrate for measuring microbial breakdown of detritus. The seston respiration amounted to 19–30% loss of the particulate fraction over 15–45 h. Microbial breakdown of detritus amounted to 28–40% loss of the particulate fraction over 24–168 h. In both cases P(articulate) O(rganic) ^{14}C(arbon) was transformed to mainly $DI^{14}C$. Microbial assimilation–dissimilation of photosynthetically derived DOC could not be measured reliably.

Introduction

In a comprehensive study on the carbon cycle in the stratified Lake Vechten (for details see Steenbergen & Verdouw 1982) the production of organic matter by autotrophs, its consumption by herbivores and its aerobic and anaerobic mineralization by bacteria are being studied simultaneously. Aerobic mineralization, i.e. seston respiration, microbial breakdown of detritus and microbial assimilation–dissimilation of photosynthetically derived D(issolved) O(rganic) C(arbon) forms an essential part of this carbon cycle study.

Studies on mineralization in Lake Vechten have been focussed on respiration of zooplankton (>104 μm, Gulati 1977) and on anaerobic mineralization of sedimented organic matter in the hypolimnion and in the sediments (Cappenberg & Verdouw 1982). The work on aerobic mineralization was started recently to complete the C-budget studies. To date it had been assumed that maximally 50% of the total breakdown was aerobic.

Since the oxygen measurement techniques have proved to be insensitive in the pelagic zone of the lake autotrophic production of organic matter, including photosynthetically derived DOC (Blaauboer *et al.* ... in press), was measured with ^{14}C-tracer techniques. However, measuring aerobic mineralization with ^{14}C-tracer technique in this relatively low-productive lake system was found to be inadequate in respect to *in situ* incorporation of ^{14}C by autotrophs. The present paper describes aerobic mineralization in samples which were concentrated by centrifuging with continuous rotor (Wiebe & Smith 1977b) during late winter–early spring 1980, a period characterized by increasing biological activity. Consequences for the carbon cycle are discussed.

Hydrobiologia 95, 181–190 (1982). 0018-8158/82/0952-0181/$02.00.

Materials and methods

Sampling

From 6 February to 8 April 1980, 30 l water was sampled from the surface–2.4 m depth at weekly intervals using a 5 l Friedinger sampler. The sampling station was located in the middle of the eastern depression of the lake (Steenbergen & Verdouw 1982). The water was filtered through a 104 μm net to remove zooplankton and transported to the laboratory taking precautions against exposure to light and rise of water temperature. In the laboratory, i.e. within 3 h of sampling, a 5 l subsample was taken for phytoplankton counting following Utermöhl (1958) and the remaining 25 l was concentrated in an MSE 18 highspeed (15 000 r.p.m.) refrigerated (4 °C) centrifuge with continuous rotor at a flow rate of 12 $l \cdot h^{-1}$. From the supernatant 2 l was filtered simultaneously through 3 and 0.2 μm membrane filter (Sartorius, GMBH) used as incubation medium in the ^{14}C-label experiments. The seston pellet was resuspended in the filtrate of supernatant to 400 ml (Wiebe & Smith 1977b) so that the seston was concentrated about 60 times. The sampling and the sample treatments are illustrated in Fig. 1.

Analysis of pigments in concentrated seston

Part of the concentrated seston was used to determine chlorophyll *a* and phaeophytin by the ethanol extraction–acidification procedure (Steenbergen & Korthals, in press). Spectra of the extracts were recorded with a Unicam SP 1800 UV spectrophotometer coupled to a Unicam AR 25 recorder.

^{14}C-label experiments

From every sample 10 ml was filtered simultaneously through 5 and 0.2 μm Sartorius membrane filter under partial vacuum (300 mm Hg). Cell damage was minimal at this reduced pressure (Blaauboer *et al.*, in press). The filters were prewetted with filtered supernatant to prevent ^{14}C-adsorption onto the filters. After the filtration the filters were rinsed with 5 + 20 ml filtered supernatant to flush out the D(issolved) I(norganic) ^{14}C(arbon). The first 5 ml was collected in the sample filtrate. The second 20 ml was collected in a waste erlenmeyer.

From every sample the particulate fraction, total label in filtrate and the photosynthetically derived DOC was determined. Also, at the start of the primary production measurement, the total supplied label was determined. The radioactive filters were combusted when still wet in a Packard Tricarb sample oxidizer B 306 trapping the produced $^{14}CO_2$ in Carbosorb. POPOP + bisMSB in toluene scintillator (22 $g \cdot l^{-1}$) was added as fluor. Total label in filtrate was determined by taking two subsamples of 5 ml directly after filtration and by adding them to scintillation vials containing 10 ml Instagel (Packard, Brussels) + 0.2 ml N NaOH (Blaauboer *et al.*, in press) to prevent loss of $^{14}CO_2$. The remaining filtrate was acidified with 0.16 N HCl to pH 2.5 and stripped of DI^{14}C by bubbling with CO_2-free air for 30 min. The $^{14}CO_2$ escaping from the samples was trapped in washing bottles containing 5 N NaOH. After bubbling, two subsamples were added to scintillation vials containing 10 ml Instagel for determination of photosynthetically derived DOC.

The radioactivity of the combusted filters and the filtrates (i.e. subsamples for total label and for photosynthetically derived DOC) was counted by a Packard Tricarb liquid scintillation spectrometer 2425 using the automatic external standard ratio for quench correction. The counts of the filtrates were corrected for dilution by rinsing of filters and acidifying of the filtrates. The DI^{14}C produced is the difference between total label in the filtrate and photosynthetically derived DOC. Percentage extracellular release (PER) is the fraction of organic carbon passing the 0.2 μm filter of the total ^{14}C assimilated (i.e. the sum of the PO^{14}C and DO^{14}C). This is only used after photosynthetic ^{14}C-fixation.

All samples were placed in an incubator with *in situ* temperature (measured at sampling date) and when light was required at a light intensity of 30 $W \cdot m^{-2}$ using 4 TL 30 fluorescent tubes (Philips). The light incubation was performed in a Fernbach flask; the controls and all dark incubations were done in scintillation vials wrapped in aluminium foil. Before measuring primary production, the concentrated seston was diluted with fil-

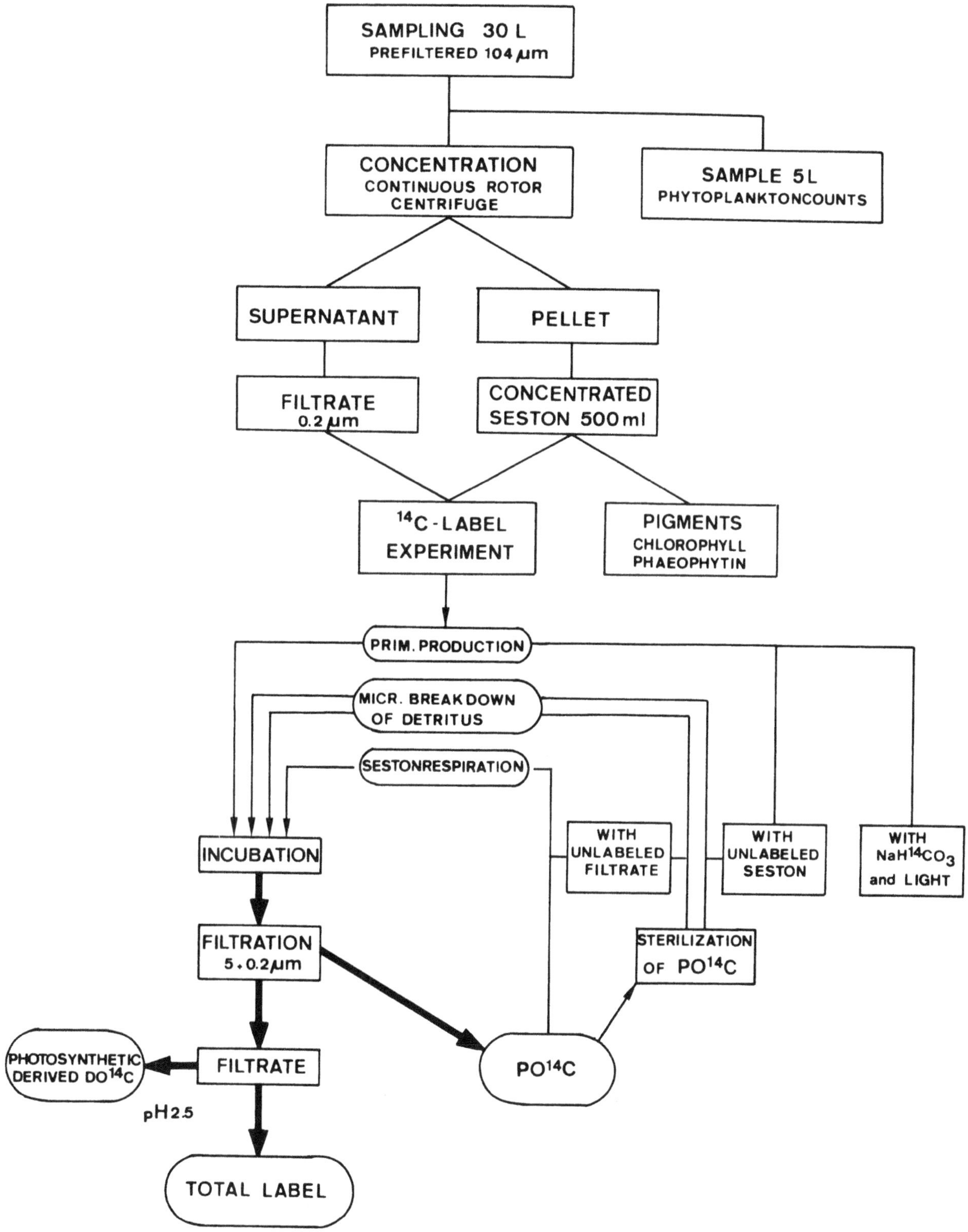

Fig. 1. Scheme of sampling and successive experimental procedures.

tered supernatant to a concentration factor of about 30.

The photoassimilation of $^{14}CO_2$ was measured using 50 μCi $NaH^{14}CO_3$ (Radiochemical Centre, Amersham, specific activity 55 mCi·mg^{-1}) per 130 ml of concentrated seston. Total radioactivity supplied was determined by taking duplicate aliquots of 100 μl and by adding them to scintillation vials with fluor as used in determination of total label in filtrate. Dark ^{14}C-fixation was measured in duplicate samples of 10 ml. Zero time (t_0) fixation of ^{14}C was determined in a 10 ml sample to which 2.5 ml 40% formaldehyde was added. After 4–6 h incubation from the rubber stoppered Fernbach flask, t_0 and dark ^{14}C-fixation samples were filtered for determination of $PO^{14}C$-fixation and extracellular release and for experiments of seston respiration and microbial breakdown of detritus.

Additional sets of filters with labelled seston were incubated in 10 ml unlabelled filtrate of supernatant in the dark for measurement of respiratory loss. After incubation of 15–45 h the loss of $PO^{14}C$ was measured concomitantly with $H^{14}CO_3^-$ (bi^{14}Carbonate) production and photosynthetically derived DOC. The loss of $PO^{14}C$ must be equalled by the produced bi^{14}Carbonate and the photosynthetically derived DOC.

In order to prepare detritus additional sets of filters with labelled seston were frozen (–20 °C), or organically destructed in acetone, or heat sterilized (115 °C for 15 min). The microbial breakdown of detritus was followed according to Iturriage & Hoppe (1977) and Cole & Likens (1979). A duplicate set of filters with labelled detritus was added to 10 ml of unlabelled seston concentrate. Separately, a duplicate set of filters with labelled detritus was added to 10 ml of unlabelled filtrate of supernatant for checking of microbial contamination as a result of filtration (Cole & Likens 1979). Both treatments were incubated in the dark for 26–168 h. Zero time controls were included for checking the rinsing procedure of filter and for measuring the organic fraction passing the filter due to preparation of detritus. Again the loss of $PO^{14}C$ must be equalled by the produced $DI^{14}C$ and the photosynthetically derived DOC.

Chemical analysis

Dissolved inorganic carbon (DIC) was determined potentiometrically with a Quéré automatic titrator (Golterman *et al.* 1978).

Results and discussion

Algae numbers and pigments

In the surface–2.4 m depth samples, a succession from centric diatoms to *Asterionella* and *Chlorella* was observed. In early February centric diatoms were found in relatively large numbers, i.e. up to 14×10^5 cells·l^{-1}, whereas the numbers of *Asterionella* sp. and *Chlorella* sp. were relatively low ($0.5–1 \times 10^6$ cells·l^{-1}, Fig. 2A). Both these latter algae increased gradually in March, i.e. *Asterionella* from 0.5×10^5 cells·l^{-1} to 23×10^5 cells·l^{-1} and *Chlorella* to 250×10^5 cells·l^{-1}, followed by a decrease in cell number of *Chlorella* in early April. Blaauboer (1982) reported for different years a dominance in the spring phytoplankton of centric diatoms together with *Asterionella* sp. and *Chlorella* sp. for mixed samples from the water column.

Chlorophyll *a* in the concentrated samples followed mainly the cell numbers of the dominant algae (Fig. 2C). In February chlorophyll *a* decreased concomitantly with the decrease in cell numbers of centric diatoms. The increase of chlorophyll *a* in early March coincided with an increase in cell numbers of *Asterionella* and *Chlorella.* However, towards the end of March, while the cell numbers of the latter algae were still increasing, the chlorophyll *a* decreased.

The changes in the phaeophytin concentration in the concentrated samples (Fig. 2C) were almost opposite to those of chlorophyll *a*. Especially in February, when the concentration of chlorophyll *a* declined, a maximum of phaeophytin was observed, probably due to enhanced mineralization. During the succession of *Asterionella* and *Chlorella,* when chlorophyll *a* reached its maximum, phaeophytin reached its minimum. Afterwards, when chlorophyll *a* decreased, phaeophytin was a little enhanced to a steady low concentration, and the mineralization was probably at a steady low rate.

Primary production and extracellular release

The response of the concentrated samples to the light intensity of 30 W·m^{-2}, expressed as $PO^{14}C$-

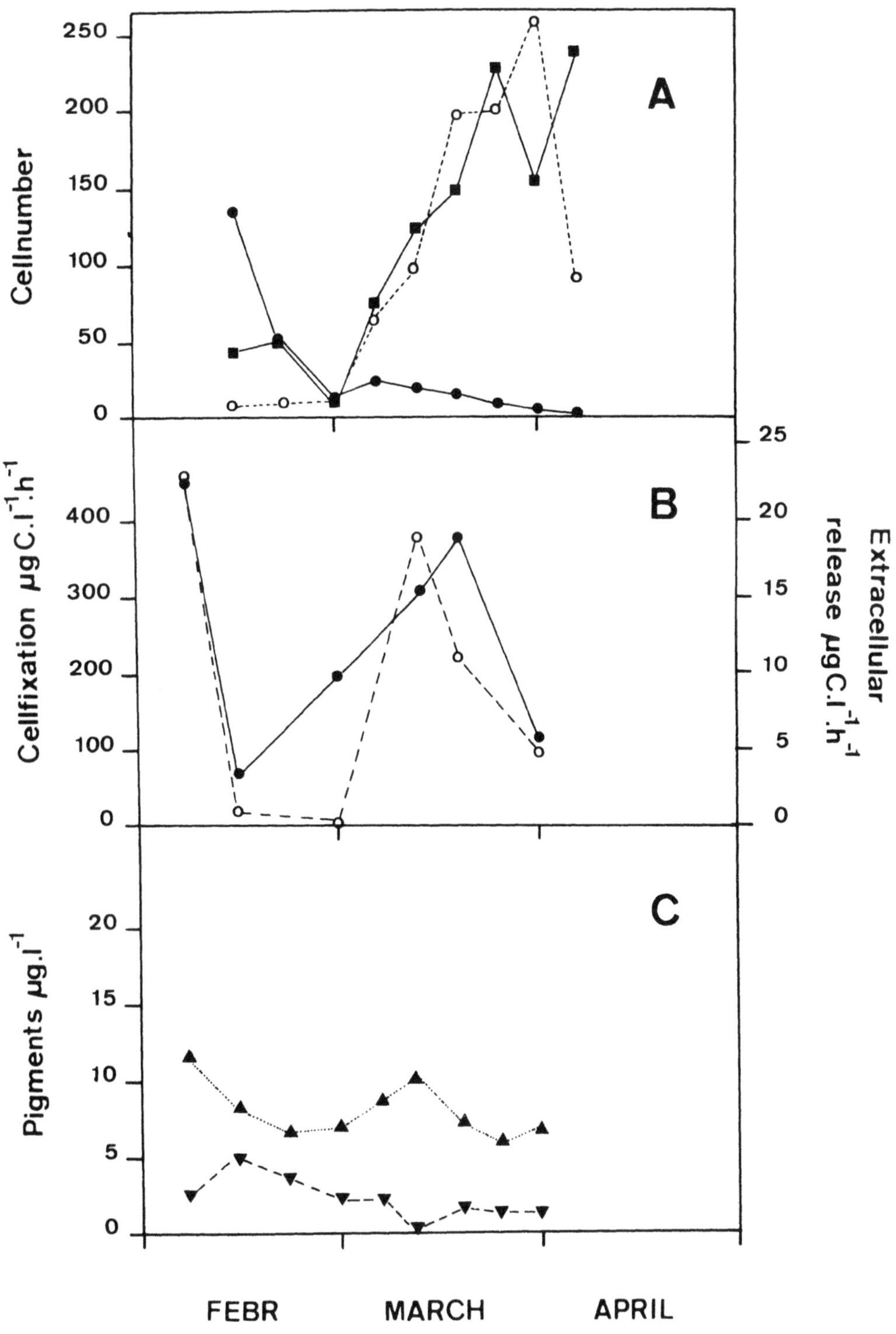

Fig. 2. A: Cell number of dominant algae in mixed samples (surface–2.4 m depth) from pelagic zone of Lake Vechten, spring 1980: centric diatoms $\times 10^4$ cells$\cdot l^{-1}$ (●---●), *Asterionella* sp. (■---■) $\times 10^4$ cells$\cdot l^{-1}$ and *Chlorella* sp. $\times 10^5$ cells$\cdot l^{-1}$ (○---○). B: Rates of $PO^{14}C$-fixation (●---●) and extracellular release (○---○), measured in concentrated samples of 0–2.4 m depth from Lake Vechten, spring 1980. C: Concentrations of chlorophyll *a* (▲---▲) and phaeophytin (▼--▼) measured in the same concentrated samples.

fixation, and extracellular release mainly followed the chlorophyll *a* concentration (Fig. 2B). In early February, within 1 week the photosynthetic ^{14}C-fixation rate declined together with the cell numbers of centric diatoms and chlorophyll *a* concentration, from 470 to 70 $\mu g\ C \cdot l^{-1} \cdot h^{-1}$ in the case of the $PO^{14}C$-fixation and from 23 to 0 $\mu g\ C \cdot l^{-1} \cdot h^{-1}$ in the case of extracellular release. The PER decreased from 6 to 0, presumably lowered by bacteria (Bell & Sakshaug 1980) active in mineralization, as indicated by the increase of phaeophytin (Fig. 2C). Starting in early March the photosynthetic fixation rates increased together with chlorophyll-*a* concentration: $PO^{14}C$-fixation to 380 $\mu g\ C \cdot l^{-1} \cdot h^{-1}$ and the extracellular release to 19 $\mu g\ C \cdot l^{-1} \cdot h^{-1}$, although the maximum for the $PO^{14}C$-fixation was found 1 week later than that of extracellular release. The PER again reached a maximum of 6. Bell & Sakshaug (1980) suggested bacterial inhibition by a rapid diatom bloom in Norwegian Fjords. This could explain the increase of the PER and the decrease of the phaeophytin. The difference of 1 week between the maxima of chlorophyll *a* and the $PO^{14}C$-fixation could be owing to the more efficient use of the available chlorophyll *a* by one of the dominant algae. In early April both chlorophyll *a* and the photosynthetic ^{14}C-fixation rates were declining together with the cell numbers of *Chlorella*. Still phaeophytin did not reach a new maximum, probably caused by the increase of *Asterionella*, which may partially inhibit the uptake of extracellular products by bacteria as indicated by Bell & Sakshaug (1980) for *Skeletonema* blooms in a Norwegian Fjord.

Blaauboer *et al.* (in press) investigated *in situ* extracellular release in relation to primary production in Lake Vechten for three successive years. In the period February–April, when the primary production rate ranged from 4 to 20 $\mu g\ C \cdot l^{-1} \cdot h^{-1}$, they reported a PER of 0 to 15. The primary production of 1 to 20 $\mu g\ C \cdot l^{-1} \cdot h^{-1}$ and a PER of 0 to 6 found by us in the laboratory with 30 $W \cdot m^{-2}$ compare favourably with their data. During this period, on 5 occasions $PO^{14}C$-fixation was measured at 1 m depth *in situ* in Lake Vechten ranging from 6 to 8 $\mu g\ C \cdot l^{-1} \cdot h^{-1}$. Again these results compare favourably with our results.

With regard to the C-cycle, two periods of high carbon input can be distinguished, one in early February caused by high numbers of centric diatoms, and another in the middle of March caused by high numbers of *Asterionella* and *Chlorella*. During the period between these maximal inputs mineralization is enhanced as pointed out by phaeophytin concentration and the lack of PER. After the second maximal input the situation is more complicated by the lack of a new phaeophytin maximum and by partial depression of PER.

We were not able to measure the microbial assimilation–dissimilation of the extracellular products. We could not recover the loss of $DO^{14}C$ into microbial $PO^{14}C$ and $DI^{14}C$.

Respiration of the concentrated seston

In order to avoid problems of dark ^{14}C-fixation of $DI^{14}C$ during the incubation for measuring seston respiration the filters with labelled seston were rinsed with and incubated in unlabelled filtrate of supernatant. Blaauboer *et al.* (in press) reported a dark ^{14}C-fixation less than 5% of primary production. We measured a similar proportion of dark ^{14}C-fixation in 4 h incubation but in some cases dark ^{14}C-fixation increased if the incubation period was longer. Refixation of the respired $^{14}CO_2$ can be excluded by dilution of $DI^{12}C$ (Wiebe & Smith 1977a). By incubating in $DI^{12}C$ containing lake water, we measured the isotopic dilution of the organic pools used for respiration purposes.

Respiration is expressed as the percentage of $PO^{14}C$ lost during incubation, while this loss of $PO^{14}C$ should be recovered as $DI^{14}C$ and as photosynthetically derived DOC (Table 1). This recovery is expressed as a percentage of the $PO^{14}C$ loss. We found 19–30% of the $PO^{14}C$ was lost after 15–45 h incubation. The recoveries of $PO^{14}C$ as $DI^{14}C$ and photosynthetically derived DOC were 69–97% and 0–7% respectively. The technical difficulties in measuring extracellular release according to Blaauboer *et al.* (in press) sometimes prevented complete recovery of the $PO^{14}C$ loss, but the $PO^{14}C$ loss is mainly transferred to $DI^{14}C$.

In marine environments, Derenbach & Williams (1974) attributed a loss of 30% of the primary production to respiration of the population in the size fraction $>3\ \mu m$, while the population in the size fraction $<3\ \mu m$ showed no loss of label during 15 h incubation. Eppley & Sharp (1975) reported a loss of 50% after 12 h incubation for both size fractions. In Lake Kinneret, Berman & Gerber (1980) found

Table 1. Respiratory losses of concentrated seston.

Sampling date	Incubation period (h)	Loss of $PO^{14}C$		Recovery loss of $PO^{14}C$	
		$\mu g\ C \cdot l^{-1} \cdot h^{-1}$	%	% $DI^{14}C$	% photosynthetically derived DOC
6 February	15	545	21	69	3
12 February	18	269	16	92	0
27 February	19	187	19	92	0
5 March	45	362	21	97	0
19 March	44	520	30	69	1
1 April	18	208	28	82	7

losses from 3 to 20% after 21 h incubation in the two fractions as used by Derenbach & Williams. These results are comparable with our results for Lake Vechten. In addition, we found that the percentage of $PO^{14}C$ loss did not increase notably if the incubation time increased to more than 15 h. Although the experiments did not have a time course character the results seemed to indicate a slowdown of label transfer from $PO^{14}C$ to $DI^{14}C$ after 15 h incubation in regard to incubation in $DI^{12}C$ containing lake water.

Mineralization of detritus

The method of preparing detritus, namely freezing (−20 °C), organic destruction in acetone and heat sterilization (115 °C, 15 min) of filters with labelled seston did not affect the percentage of $PO^{14}C$ loss. Zero time controls irrespective of the method of detritus preparation showed roughly 20% of the $PO^{14}C$ passed 0.2 μm filter as photosynthetically derived DOC. In the filtrate less than 1% of the $PO^{14}C$ was recovered as $DI^{14}C$. Therefore when seston turned into detritus following the three methods, roughly 20% of the $PO^{14}C$ was leached out of the cells as photosynthetically derived DOC, presumably the same organic ^{14}C-pool as used for respiration in living cells.

The percentage loss of $PO^{14}C$ in case of incubation with filtered supernatant and with concentrated seston (Table 2) did not differ. In Mirror Lake Cole & Likens (1979) reported similar results for filtrates passing a 0.4 μm filter. We used 0.2 μm filter; apparently, this pore size is large, because Cole & Likens

Table 2. Microbial breakdown of detritus by filtered supernatant (0.2 μm) and by concentrated seston.

Sampling date	Incubation period (h)	Loss of $PO^{14}C$		Recovery loss of $PO^{14}C$	
		$\mu g\ C \cdot l^{-1}$	%	% $DI^{14}C$	% photosynthetically derived DOC
Filtered supernatant (0.2 μm)					
6 February	26	996	40	ND	ND
12 February	28	708	40	ND	ND
5 March	167	521	31	77	2
19 March	168	619	36	63	3
1 April	167	226	31	81	8
Concentrated seston					
6 February	26	716	28	ND	ND
12 February	28	792	46	88	3
5 March	167	625	37	66	0.2
19 March	168	625	36	56	0.2
1 April	167	211	29	86	5

ND = not determined.

(1979) reported passing of mineralizing agent through a 0.1 μm Sartorius filter. They found that the activity of this mineralizing agent was inhibited by addition of formaldehyde prior to incubation. In addition, we found in the filtrates almost no $DI^{14}C$ and almost unchanged photosynthetically derived DOC if formaldehyde was added prior to incubation. Therefore filtration as a method of separating particulate and dissolved fractions should be avoided, whereas the mineralizing agent found in the filtered supernatant is probably microbial in origin.

We found the microbial breakdown of detritus amounted to 28–46% of the $PO^{14}C$ irrespective of the incubation time (Table 2). The recoveries of the $PO^{14}C$ loss were respectively 56–88% as $DI^{14}C$ and 0–8% as photosynthetically derived DOC. The rest can be attributed to escape of $^{14}CO_2$ to the air, since the special attention needed for recovery of the photosynthetically derived DOC according to Blaauboer *et al.* (in press) is in conflict with trapping of $^{14}CO_2$ as proposed by Iturriaga & Hoppe (1977) and Cole & Likens (1979). Again the $PO^{14}C$ was mainly transferred to $DI^{14}C$. Prior to incubation as derived from t_0 controls, roughly 20% of the $PO^{14}C$ was present as photosynthetically derived DOC. After incubation the photosynthetically derived DOC was almost depleted and the $PO^{14}C$ loss is increased to 28–46%. The 46% of $PO^{14}C$ loss at 12 February coincided with the enhanced mineralization at that time as indicated by phaeophytin maximum and PER minimum.

In marine environments 6–20% (Iturriaga & Hoppe 1977) or 30% (Herbland 1975) of the $PO^{14}C$ was recovered as $DI^{14}C$ in about 24 h incubation. Herbland (1975) demonstrated in time course experiments that the maximal rates of microbial assimilation–dissimilation of solubilized detritus (comparable with photosynthetically derived DOC after sterilization) were obtained after 24 h incubation. This could explain why the $PO^{14}C$ loss did not increase much after longer incubation times, although our experiments did not have a time course character and in our experiments we could not discriminate between assimilated microbial $PO^{14}C$ and the remaining detrital $PO^{14}C$. In Mirror Lake Cole & Likens (1979) recovered only 7% as $DI^{14}C$ within 72 h incubation and pointed out that the losses due to mineralization will be higher in more productive lakes. In short, our results are in the range reported by the authors cited.

With respect to the carbon cycle, the respiratory loss of 19–30% of POC within a night (15 h) indicates that a pool of organic carbon respired rapidly, although so far none of the authors cited (including us) have investigated shorter incubation times. If it is assumed that the preparation of detritus as described in this paper is simulating the death of seston organisms, roughly 20% of the assimilated carbon is leaching out of the cells and is probably the same organic carbon pool which is used for respiratory purposes (this paper). Still, the problem of isotopic equilibrium of all pools is not completely solved, although we assume that it existed in the experiments, but further work in this respect is in progress. If the detritus is mineralized by the heterotrophic part of the seston population (assumed to be microbial) the leachate of the starving seston is almost completely assimilated–dissimilated and the POC loss is increased to 28–46% after 24 h incubation.

With regard to the isotopic equilibrium between pools, the experiment of microbial breakdown of detritus is difficult to interpret. The first action of microbial breakdown is the assimilation–dissimilation of the leachate. As shown by Herbland (1975) it can be regarded as saturated within 24 h incubation, in which microbial POC is formed. The second action is the active breakdown of detritus POC by bacteria. Again it seemed to be saturated within 26 h incubation (this paper), although Herbland (1975) demonstrated a slow assimilation of extracellular products as of solubilized detritus, after longer incubation times. In our case this resulted in a remaining 65% of the POC which is subject to grazing by zooplankton (including living seston), particularly in the epi- and metalimnion (Gulati 1977) and to sedimentation transport to the anaerobic zone of the lake where it is mineralized further anaerobically (Cappenberg & Verdouw 1982).

By comparison of data on C-fixation, sedimentation and on microbial reduction rates of electron acceptors a mineralization of 20 to 30% of the primary production was calculated for the meta- and hypolimnion (Cappenberg & Verdouw 1982), while from the consumption rates of zooplankton in early spring about 20% of the daily primary production is assimilated by the dominant crustacean filter feeders in Lake Vechten (Gulati 1977). From flux data of low-molecular-weight organics and those on the rates of formation of CO_2 and CH_4, an anaerobic mineralization of about 20% in the profundal sediments could be calculated from the C-fixation data, while an as yet unquantified percentage of the sedimented particulate organic matter will be buried more or less permanently by subsequently deposited material (Cappenberg 1976). Although the value of 65% of organic carbon remaining after microbial breakdown in the aerobic zone was obtained in spring, effects on the aerobic mineralization of temperature, increase in primary production and of autumnal circulation need to be investigated further. Lastly, simultaneous studies on actual turnover rates of important carbon intermediates will elucidate the rates and mechanisms of biological carbon transformations and their importance in the aerobic zone of Lake Vechten.

Summary

Aspects of aerobic mineralization were studied in concentrated samples from the pelagic zone of Lake Vechten during late winter–early spring 1980. The concentrated samples were incubated at *in situ* temperature using a light intensity of 30 W · m^{-2}. The cell numbers of dominant algae in the unconcentrated samples revealed a succession from centric diatoms to *Asterionella* and *Chlorella*. The chlorophyll *a* concentration in the concentrated samples followed the succession in cell numbers of the dominant algae. The phaeophytin concentration, however, was opposite to chlorophyll *a* concentration. The response of the concentrated samples to a light intensity of 30 W · m^{-2} expressed as $PO^{14}C$-fixation and as extracellular release mainly followed the chlorophyll *a* data. With regard to the C-cycle, two periods of high carbon input were measured. The respiratory losses of the seston expressed as loss of $PO^{14}C$ amounted to 19–30% of the POC, while the POC loss was mainly transferred to $DI^{14}C$. After 15 h incubation the respiratory loss did not increase much. During preparation of detritus, irrespective of the method, roughly 20% of $PO^{14}C$ passed the filter as photosynthetically derived DOC. Microbial breakdown of detritus accounted for 28–46% loss of the POC, this loss being mainly transferred to $DI^{14}C$. After incubation, the available photosynthetically derived DOC was almost completely assimilated–dissimilated by the microbial part of the seston. We were not able in measuring microbial assimilation–dissimilation of extracellular products. With regard to the C-cycle, roughly 19–30% of the POC is respired in a night period by the living seston. When seston is dying, 20% of the POC is leaching out of the cells. In microbial breakdown of this detritus the leachate is almost complete assimilated–dissimilated. Furthermore the POC loss is increased to 28–46%, which indicates 65% of the POC is subject to grazing by zooplankton and to sedimentation transport to the anaerobic zone of the lake, where the POC is mineralized anaerobically.

Acknowledgements

We would like to thank M. J. B. Bär-Gilissen for pigment determinations, P. J. Boesewinkel-de Bruyn for counting of the phytoplankton, E. M. J. Dekkers for drawing the figures, and H. Roon for technical field assistence. Thanks are also due to W. A. de Kloet for providing data on *in situ* primary production. Special gratitude is expressed to M. A. C. I. Blaauboer, J. G. Kuenen (Technical University, Delft) and C. L. M. Steenbergen for comments.

References

Bell, W. H. & Sakshaug, E., 1980. Bacterial utilization of algal extracellular products. 2. A kinetic study of natural populations. Limnol. Oceanogr. 25: 1021–1033.

Berman, T. & Gerber, C., 1980. Differential filtration studies of carbon flux from living algae to microheterotrophs, microplankton size distribution and respiration in Lake Kinneret. Microb. Ecol. 6: 189–198.

Blaauboer, M. A. C. I., 1982. The phytoplankton species composition and the seasonal periodicity in Lake Vechten from 1956 to 1979. Hydrobiologia 95: 25–36.

Blaauboer, M. A. C. I., van Keulen, R. & Cappenberg, Th. E. (in press). Extracellular release of photosynthetic products by freshwater phytoplankton populations, with special reference to the algal species involved. Freshwat. Biol.

Cappenberg, Th. E., 1976. Methanogenesis in the bottom deposits of a small stratifying lake. In: (Schlegel, H. G., Pfennig, N., Gottschalk, G., (Eds.). Microbial Production and Utilization of Gases, H_2, CH_4, CO, E. Goltze-Verlag, Göttingen. pp. 125–134.

Cappenberg, Th. E. & Verdouw, H., 1982. Sedimentation and breakdown kinetics of organic matter in the anaerobic zone of Lake Vechten. Hydrobiologia 95: 165–179.

Cole, J. J. & Likens, G. E., 1979. Measurements of mineralization of phytoplankton detritus in an oligotrophic lake. Limnol. Oceanogr. 24: 541–547.

Derenbach, J. B. & Williams, P. J., 1974. Autotrophic and bacterial production: fractionation of plankton populations by differential filtration of samples from the English Channel. Mar. Biol. 25: 263–269.

Eppley, R. W. & Sharp, J. H., 1975. Photosynthetic measurements in the Central North Pacific: the dark loss of carbon in 24 h incubations. Limnol. Oceanogr. 20: 981–987.

Golterman, H. L., Clymo, R. S. & Ohnstad, M. A. M., 1978. Methods for Physical and Chemical Analysis of Fresh Waters. IBP Handbook No. 8. Blackwell Scientific, Oxford.

Gulati, R. D., 1977. Vertical changes in the filtering, feeding and assimilation rates of dominant zooplankters in a stratified lake. Verh. int. Verein. Limnol. 20: 950–956.

Herbland, A., 1975. Utilisation par la flore héterotrophe de la matière organique naturelle dans l'eau de mer. J. exp. mar. Biol. Ecol. 19: 19–31.

Iturriaga, R. & Hoppe, H. G., 1977. Observations of heterotrophic activity on photoassimilated organic matter. Mar. Biol. 40: 101–108.

Steenbergen, C. L. M. & Korthals, H. J. (in press). Distribution of phototrophic microorganisms in the anaerobic and microaerophilic strata of Lake Vechten. Pigments analysis and role in primary production. Limnol. Oceanogr.

Steenbergen, C. L. M. & Verdouw, H., 1982. Lake Vechten: aspects of its morphometry, climate, hydrology and physicochemical characteristics. Hydrobiologia 95: 11–23.

Utermöhl, U., 1958. Zur Vervollkommnung der quantitativen Phytoplankton methodik. Mitt: int. Verein. Limnol. 9: 1–38.

Wiebe, W. J. & Smith, D. F., 1977a. Direct measurement of dissolved organic carbon release by phytoplankton and incorporation by microheterotrophs. Mar. Biol. 42: 213–223.

Wiebe, W. J. & Smith, D. F., 1977b. ^{14}C-labelling of the compounds excreted by phytoplankton for employment as a realistic tracer in secondary productivity measurements. Microb. Ecol. 4: 1–8.

Nitrogen cycle of Lake Vechten: concentration patterns and internal mass-balance

H. Verdouw & E. M. J. Dekkers
Limnological Institute, 'Vijverhof' Laboratory, Rijksstraatweg 6, 3631 AC Nieuwersluis, The Netherlands

Keywords: nitrogen cycle, hypolimnetic accumulation, mass-balance, Lake Vechten

Abstract

Concentration patterns of ammonia, nitrate and particulate- and dissolved organic nitrogen in Lake Vechten (The Netherlands) are given. The main processes playing a role in the lake's nitrogen cycle are inventorized. During spring and summer, the uptake of ammonia and nitrate by primary producers, accompanied by sedimentation of organic matter, result in significant nitrogen losses from the 0–6 m stratum. Ammonia accumulates during the stratification in the hypolimnion and after the autumn overturn is recirculated over the whole water column. In winter, release of ammonia from the sediments and nitrification further restore the lake's nutrient supply. The changes in nitrogen content of the different compartments are quantified, giving a basis for more detailed research.

Introduction

The work on the nitrogen cycle is a part of the integrated ecosystem research in Lake Vechten, as outlined in Best *et al.* (1978). For a broad understanding of the lake's functioning in terms of nitrogen dynamics and for background data for other studies, the concentration patterns of nitrogen in different forms were followed during the years 1977–1980. These patterns give information on important processes within the lake's nitrogen cycle (Brezonik 1972). In our work, nutrient regeneration in hypolimnion and sediments is emphasized, so processes leading to ammonia release (e.g. anaerobic mineralization during the stratification in the hypolimnion) and nitrate formation (nitrification during the circulation periods) are of particular interest.

Also, the nitrogen content of different water strata has been calculated, especially during the summer stratification period, to examine the mass flux within the lake. Based on these data, a simple mass-balance could be drafted, quantifying a number of processes, e.g. nutrient assimilation, sedimentation, denitrification, ammonification and nitrification.

The lake

Lake Vechten is a small (4.7 ha), deep (mean depth 6 m) sandpit, consisting of an eastern and a western basin, with maximum depths of 10.5 and 11.9 m, respectively. The basins are separated by a ridge across the breadth in the middle of the lake at about 6 m depth. The lake can be classified as warm monomictic, with a stable summer stratification from May till the end of October. The hypolimnetic water strata below 6.5 m are anaerobic during most of the stratification period. The ridge, dividing the lower water strata of the lake, enables separate study of the two hypolimnia. The work described in this paper was conducted in the eastern basin. The absence of a surface inlet or outlet is an important hydrological feature of the lake, facilitating calculations on the relations between the nitrogen com-

Hydrobiologia 95, 191–197 (1982). 0018–8158/82/0952–0191/$01.40.

partments. However, seepage does play a role. A general description of the lake is given by Parma (1971) and Steenbergen & Verdouw (1982).

Methods

Water samples were taken with a 5 l Friedinger sampler (cylinder length 60 cm) in the deepest area of the eastern basin of the lake. The consecutive sample hauls (e.g. 0–60 and 60–120 cm) were mixed and analysed for different N-compounds. Similarly, 1.20 m strata down to 9.6 m depth were sampled. Total Kjeldahl-N (= NH_3-N + DON + PON) was determined on unfiltered samples. For determining NH_3-N, NO_2-N, NO_3-N and dissolved Kjeldahl-N (= NH_3-N + DON), water was filtered over 0.45 μm membrane filters.

Kjeldahl-N was converted to NH_3-N following Scheiner (1976); NH_3-N was determined according to Verdouw *et al.* (1978). NO_3-N was reduced to NO_2-N following Wood *et al.* (1967); NO_2-N was measured according to Freier (1964).

Results and discussion

The concentration patterns of the measured N-compounds in 1977 are given in Figures 1, 2 and 3. No substantial differences were noticed in the other years. Horizontal variations of concentrations in the lake are minor; however, between the submersed macrophytes in the littoral zone, significantly lower values (up to 40%) of DON compared with the normal sampling site were encountered, but these do not influence our general considerations. The vertical concentration profiles did not exhibit diurnal variations.

From the annual patterns of concentrations and vertical distribution, general conclusions can be

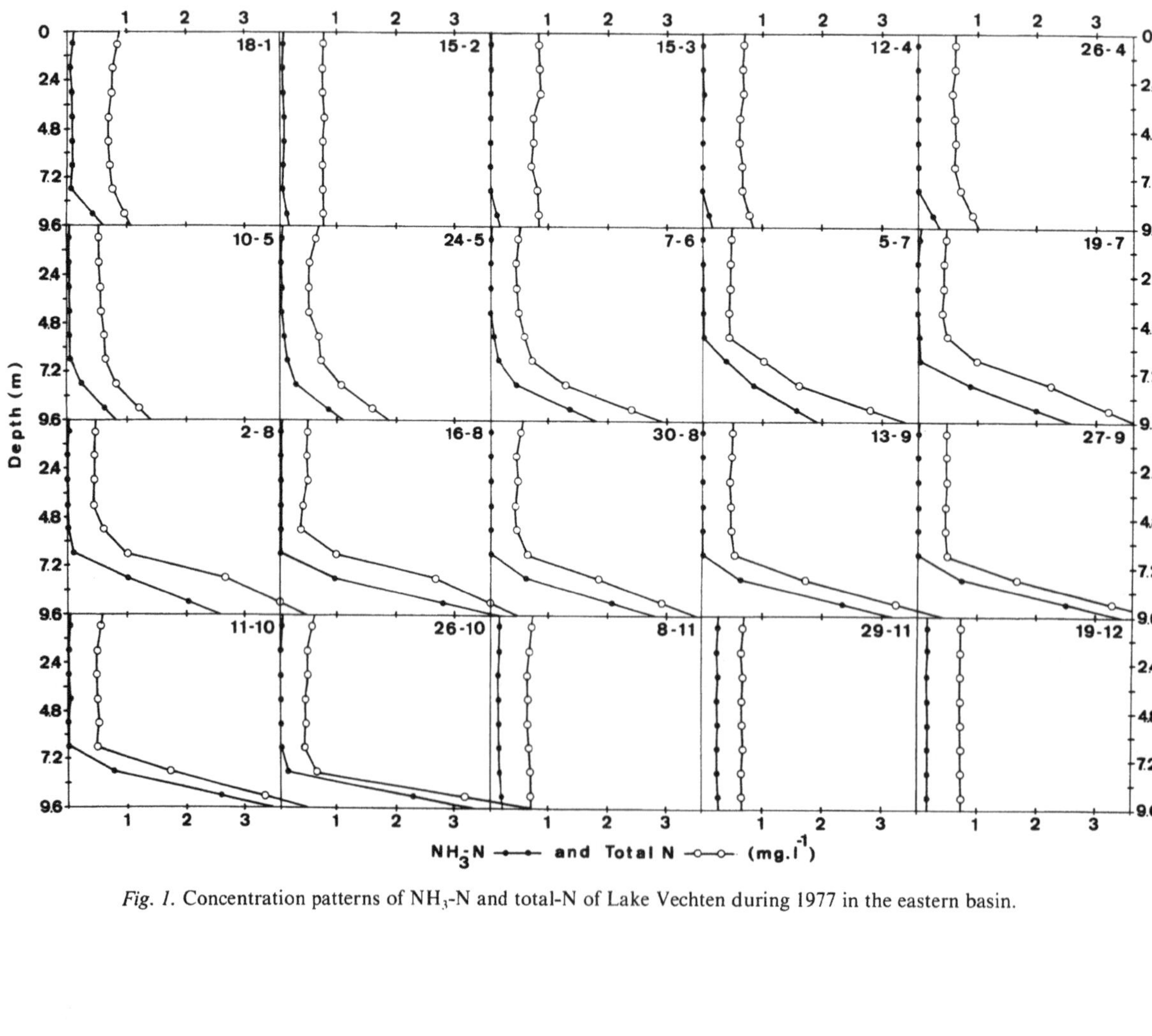

Fig. 1. Concentration patterns of NH_3-N and total-N of Lake Vechten during 1977 in the eastern basin.

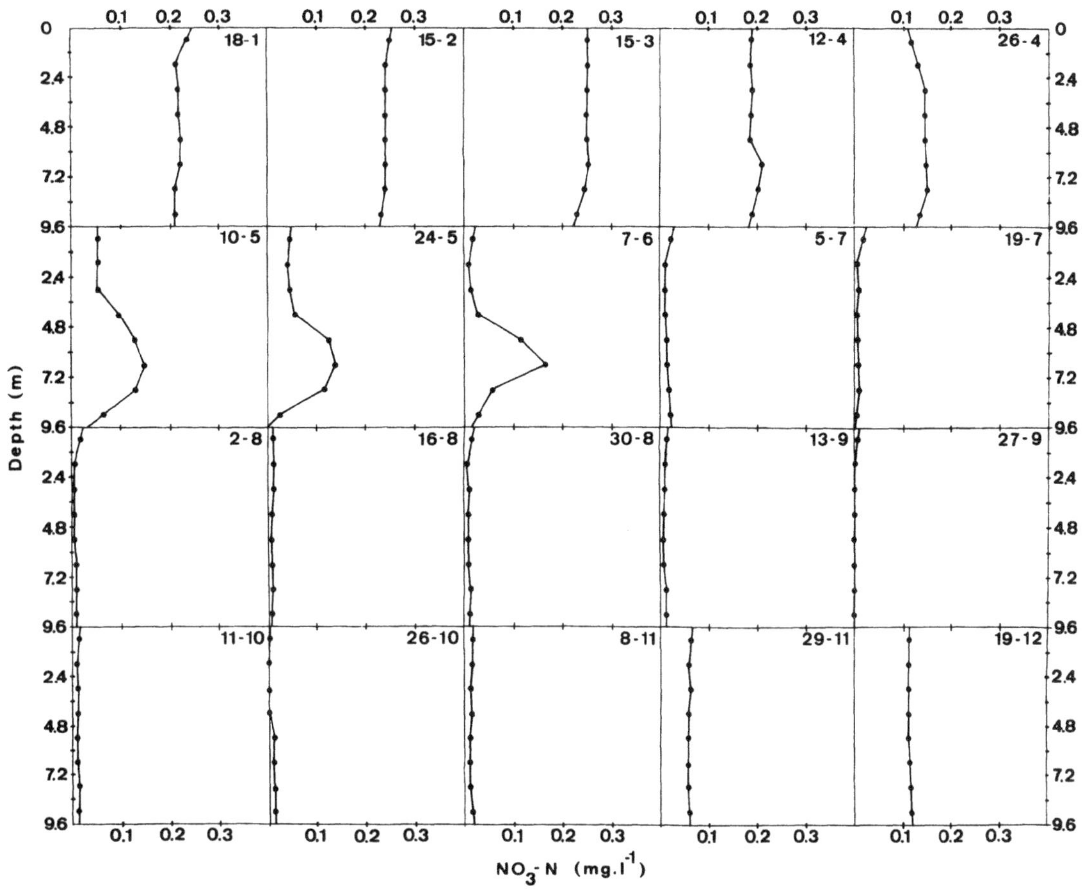

Fig. 2. Concentration patterns of NO_3-N of Lake Vechten during 1977 in the eastern basin.

drawn concerning the dynamics of nitrogen in the lake: concentration differences, due to stratification phenomena, first become visible at the end of April and persist until the time of autumnal circulation in early November. Concentrations in the upper water strata decrease and are relatively uniform during the stratification period, whereas, near the bottom, accumulation of nitrogen compounds (except for nitrate) is observed, with the development of strong concentration gradients.

Based on these differences, the water column can be divided into two strata: one from surface to 6 m depth and the other from 6 m to the bottom. This division is used in our calculations on the internal mass-balance. During the circulation periods, when the concentration profiles from surface to bottom are uniform, the ratio of the nitrogen contents of the strata is the same as that of their water volumes, namely 80: 20, both being the percentages of the total volume of the eastern basin. At the normal sampling site maximum depth, depending on water level of the lake, is 10.5 m. In our calculations on the nitrogen content below 6 m depth, the concentrations from 9.6 to the bottom were estimated by extrapolating the values of the overlying strata. The error involved is insignificant because of the small volume of the water strata below 9.6 m.

The total-nitrogen content showed a clear seasonal pattern in the years 1977–1980. A typical example of this is given in Fig. 4. Maxima are observed in the late winter and during the stratification period. The upper 6 m stratum exhibits only a winter peak. Starting in spring, almost 50% of the total-N content disappears from this compartment, occurring mainly in the period half of March to mid May; by this time stratification is stable and total-N starts accumulating in the hypolimnetic water strata.

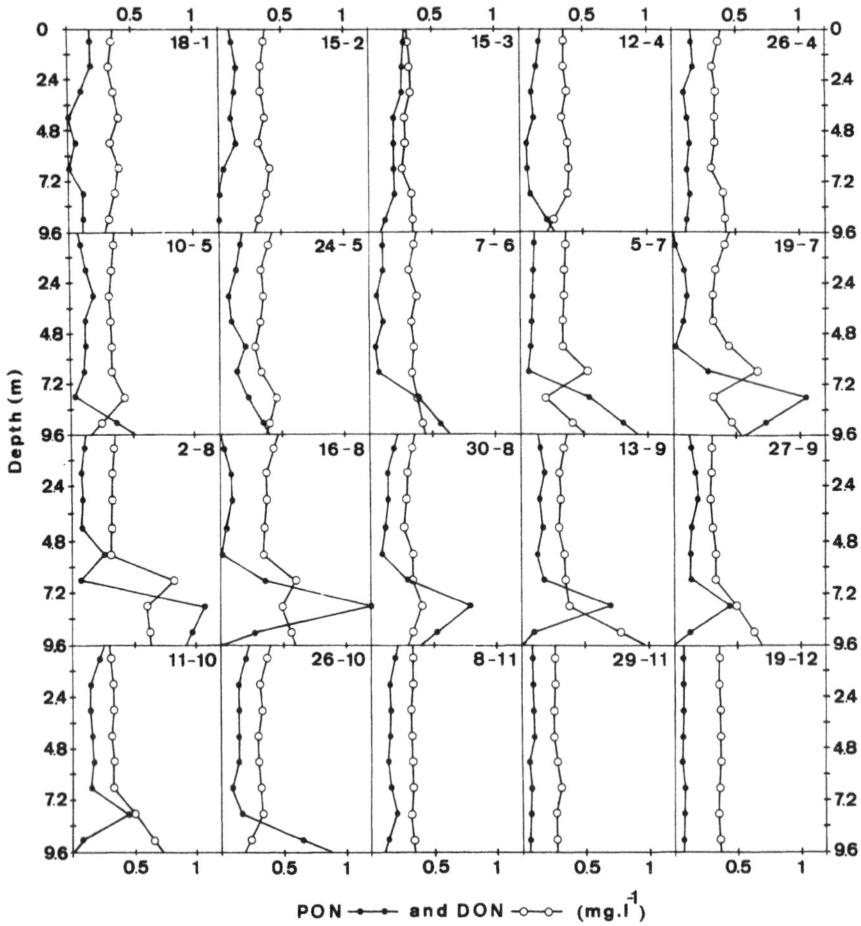

Fig. 3. Concentration patterns of PON and DON of Lake Vechten during 1977 in the eastern basin.

The two nitrogen fractions involved most obviously in these fluctuations are nitrate and ammonia. Nitrate decreases in spring and early summer, whereas after the autumn overturn it reappears. At least three processes are involved. Nitrification is responsible for the regeneration of nitrate in the circulation period. Most of the ammonia in the hypolimnion, distributed evenly in the water column at overturn, plus a certain amount that comes from the sediments in the course of the winter, is converted to nitrate. Nitrate is assimilated by primary producers in the upper water strata and in the littoral zone. A decrease of about 25 kg NO_3-N was measured in the 0–6 m layer in the period March to July. This was not compensated by an increase in the concentrations of other N fractions. Thus, removal of nitrogen by sedimentation of organic matter might be important. Nitrate uptake by macrophytes in the littoral zone can also contribute to such a loss of nitrogen from the water, but is probably of minor importance (see Best 1979). From the start of stratification in May onwards the bottom waters become anaerobic and nitrate concentrations decrease. The regeneration of nitrate by nitrification at the sediment surface is not possible. The denitrification, therefore, is most probably responsible for removal of nitrate, which is exhausted by the end of June. About 6 kg N is involved in this change. The typical changes in the vertical concentration profiles in May and June (Fig. 2), due to uptake by primary producers in the upper water layers and to denitrification near the bottom, have been observed in other stratifying lakes as well (Hutchinson 1957; Tirén 1977).

For ammonia the processes involved in the two compartments were also clearly different. Ammo-

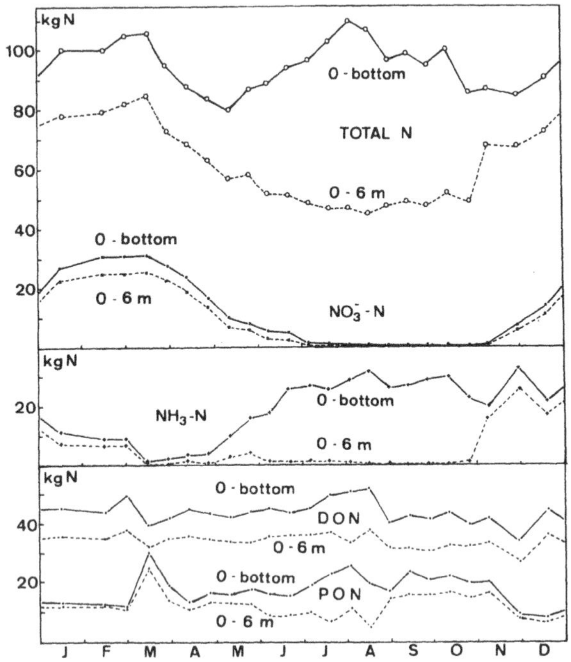

Fig. 4. Nitrogen content of the eastern basin of Lake Vechten in 1977. Solid lines: in total water column. Dashed lines: in water strata from surface to 6 m depth.

nia, present in winter, maximally about 300 $\mu g.l^{-1}$, is quickly used by primary producers in the spring; e.g. the decrease of NH_3-N in March was clearly correlated with the increase in PON concentration (Fig. 4) and with a peak in primary production (Blaauboer *et al.*, in press). This was apparently followed also by sedimentation of organic matter, leading to the observed loss from the water phase.

During summer stratification, the ammonia concentrations decrease to less than 5 $\mu g.l^{-1}$ in the epilimnion and NH_3-N accumulates in the deeper water strata, leading to the development of strong concentration gradients. Ammonia dominates the concentration profile of total-N during that time (Fig. 1). The generally observed tendency of ammonia to accumulate after the onset of anaerobic conditions near the bottom (Hutchinson 1957) is explained by the loss of adsorptive capacity, when the oxidized microzone of the upper sediment layer is reduced; the mineralization of sedimenting particulate material also contributes to this. The direct source of the ammonia in the hypolimnion of Lake Vechten is not quite clear. Part of it seems to come from the sediments, as the increase in the ammonia content of the water near the bottom is greater than the total-N decrease in the 0–6 m compartment. Apparently, there is a delay in the release of ammonia from the material deposited in a preceding period, which might well be due to adsorption of ammonia formed. However, during the winter period, the total-N content of the lake generally increases due to an increase in ammonia, obviously coming from the sediments, followed by nitrification. This effect was most clearly seen in 1978 and 1979. The influence of adsorption capacity in the aerobic sediment top layer, therefore, is questionable and is certainly not the only factor involved in Lake Vechten. It is likely that both ammonia production and uptake rates in the surface sediments are important, i.e. bacterial activity under different conditions of redox potential and input of organic matter might regulate the observed ammonia dynamics.

PON and DON generally show their greatest variations in the hypolimnion (Fig. 3). The patterns observed, especially for PON, are rather irregular. There is, however, a tendency for both PON and DON to accumulate during stratification. The peaks of PON in all the years of the investigations were observed invariably in the layer between 7.2 and 8.4 m, suggesting active breakdown during sedimentation in the water below this stratum. The concomitant increase of NH_3-N and, on some occasions DON, support this, although, as mentioned before, sedimentary nitrogen compounds might also be involved. Comparable patterns were found in hypereutrophic Wintergreen Lake (Molongoski & Klug 1980a, b). DON is an important constituent of the nitrogen pool of Lake Vechten. In the epilimnetic water during the stratification period, when nitrate and ammonia are virtually exhausted, it is by far the most abundant form of nitrogen. Despite its rather constant concentrations, DON might well be a dynamic fraction, playing an important role in the epilimnetic N-cycling.

Internal mass-balance

It is clear that the seasonal changes in the N-content of the two compartments we distinguished in Lake Vechten are important (Fig. 4). This picture, based on the concentrations measured, only gives indications of processes involving net mass changes of nitrogen. As a first approximation to quantifying single processes within the nitrogen cycle, two

Table 1. Summary of the important fluctuations of nitrogen-mass and processes involved, based on data of 1977–1980 from the eastern basin of Lake Vechten.

Period	ΔN (main fractions)	Processes involved
March–May (start of growing season)	Decrease 10–40 kg (NO_3-N, NH_3-N, PON)	Nutrient assimilation, sedimentation, denitrification
May–October (stratification)	0–6 m: decrease about 6 kg NO_3-N)	Nutrient assimilation, sedimentation
	6 m – bottom: increase 10–20 kg (NO_3-N, NH_3-N, PON)	Denitrification, NH_3-N release from sediments and sedimenting material
October–March (circulation)	Increase 30–40 kg (NH_3-N, NO_3-N)	NH_3-N release from sediments, nitrification

phases are distinguished, one of nutrient uptake (whole lake in spring, epilimnion during stratification) and the other of nutrient regeneration (hypolimnion during stratification, whole lake during circulation). For this, the data for 1977–1980 are summarized in Table 1. Though the quantities of nitrogen involved differed from year to year, the qualitative aspects were clearly recognizable each year. The data for the layer 6 m to the bottom represent the net change from May to October. The total-N content generally showed a peak in this period, giving a temporary higher accumulation value than that given in the table (see Fig. 4).

In this inventory of processes involved in nitrogen transformations, nitrogen fixation has not been mentioned. Organisms capable of fixing molecular nitrogen, e.g. blue-green algae and photosynthetic bacteria, are present in the lake (Blaauboer 1982; Steenbergen 1982). Their quantitative role in the nitrogen cycle is unknown, but is probably of minor importance. Only in the metalimnion and the upper part of the hypolimnion, where they sometimes form considerable populations in rather thin strata, might they locally contribute to the lake's nitrogen metabolism.

The data given in Table 1 and Fig. 4 represent a description of nitrogen mass fluxes, which are observed in many lakes, although the quantitative aspects may differ considerably. Generally, there is a net loss of nitrogen in spring by sedimentation and denitrification, followed by an N-gain in summer, due to ammonification and ammonia release from the bottom deposits and in winter, caused by ammonia and nitrate supply from the sediments (Hutchinson 1957; Liao & Lean 1978; Birch & Spyridakis 1981).

The data of Lake Vechten discussed so far form a basis for more extensive investigations, particularly those concerning the mineralization aspects of the nitrogen cycle. Among the processes involved (Table 1), the sedimentation of organic matter (Verdouw & Dekkers, 1982) is a very important transport mechanism between the trophogenic and tropholytic zones of the lake. By the downward flux of particulate material the epilimnion becomes virtually depleted of readily available nitrogen compounds for the primary producers.
ers.

Regeneration of ammonia as zooplankton excretion appears to be of great importance. Laboratory experiments in 1980, extrapolated to 1978 data of primary production, suggest that nutrient demand could be covered substantially by N-excretion of the zooplankton community of the lake (Gulati, in Parma & Gulati 1981; Gulati *et al.* 1982). Supply by zooplankton as a percentage of phytoplankton demand ranged from 36 to 162%.

Apart from this internal cycling in the epilimnion, resupply of ammonia from the deeper water strata, where it is abundant (Fig. 1), might be important. Tentative calculations of the ammonia flux in the hypolimnion in relation to the sedimentation rates suggest that eddy diffusion is a key factor in the nitrogen cycle during stratification (Verdouw, in Parma & Gulati 1981).

The nitrogen dynamics in the hypolimnion and at the sediment-water interface is being investigated. The turnover rates of ammonia are being measured during the stratification period, using a ^{15}N labelling technique and calculations are being made of the nitrogen fluxes (sedimentation and eddy diffusion) in the hypolimnion. It is expected that this approach will clarify to some extent the complex system of ammonia production, consumption and diffusion in relation to the nitrogen input by sedimentation.

Summary

The annual concentration patterns of ammonia, nitrate, PON and DON in the whole water column in Lake Vechten (Figs. 1, 2 and 3) show that a number of processes are quantitatively important in the lake's nitrogen cycle. During stratification, ammonia and nitrate uptake by primary producers and sedimentation of organic matter cause a significant decrease in the nitrogen content of the upper water strata (0–6 m). Accumulation of nitrogen, mainly as ammonia and to some extent as PON and DON, is observed in the lower strata (6 m–bottom). At the autumnal overturn this ammonia is distributed over the whole water column. During the circulation period ammonia is released from the sediments, while nitrification at that time restores the lake's nitrate content.

The study of nitrogen fluctuations in the upper and lower strata during the annual cycle (Fig. 4, Table 1) forms the basis of more detailed research, notably on the role of sedimentation and the nitrogen dynamics at the sediment-water interface and in the hypolimnion.

Acknowledgements

We are indebted to Dr. R. D. Gulati and Dr. S. Parma for reviewing the manuscript and to Mr. E. M. Mariën for photographic work.

References

Best, E. P. H., Blaauboer, M. C. I., Cappenberg, Th. E., Gons, H. J., Gulati, R. D., de Kloet, W. A., Steenbergen, C. L. M. & Verdouw, H., 1978. Towards an integrated study of the ecosystem of Lake Vechten. Hydrobiol. Bull. 12: 107–118.

Best, P. H., 1979. Ecophysiological studies on growth and development of the aquatic macrophyte Ceratophyllum demersum L. Ph. D. thesis, Utrecht.

Birch, P. B. & Spyridakis, D. E., 1981. Nitrogen and phosphorus recycling in Lake Sammamish, a temperate mesotrophic lake. Hydrobiologia 80: 129–138.

Blaauboer, M. C. I., 1982. The phytoplankton species composition and the seasonal periodicity in Lake Vechten from 1956 to 1979. Hydrobiologia 95: 25–36.

Blaauboer, M. C. I., van Keulen, R. & Cappenberg, Th. E. Extracellular release of photosynthetic products by freshwater phytoplankton populations, with special reference to the algal species involved. Freshwat. Biol. (in press).

Brezonik, P. L., 1972. Nitrogen: sources and transformations in natural waters. In: Allen, H. E. & Kramer, J. R. (Eds.) Nutrients in Natural Waters. John Wiley, New York.

Freier, R. K., 1964. Wasser Analyse. Walter de Gruyter, Berlin.

Gulati, R. D., Siewertsen, K. & Postema, G., 1982. The zooplankton: its community structure, food and feeding, and role in the ecosystem of Lake Vechten. Hydrobiologia 95: 127–163.

Hutchinson, G. E., 1957. A Treatise on Limnology. I. Geography, Physics and Chemistry. John Wiley, New York.

Liao, C. F.-H. & Lean, D. R. S., 1978. Seasonal changes in nitrogen compartments of lakes under different loading conditions. J. Fish. Res. Bd. Can. 35: 1095–1101.

Molongoski, J. J. & Klug, M. J., 1980a. Quantification and characterization of sedimenting particulate matter in a shallow hypertrophic lake. Freshwat. Biol. 10: 497–506.

Molongoski, J. J. & Klug, M. J., 1980b. Anaerobic metabolism of particulate organic matter in the sediments of a hypertrophic lake. Freshwat. Biol. 10: 507–518.

Parma, S., 1971. Chaoborus flavicans (Meigen) (Diptera, Chaoboridae): an autecological study. Ph. D. thesis, Groningen.

Parma, S. & Gulati, R. D., 1981. Progress Report 1980. Limnological Institute Verh. Kon. Ned. Akad. Wet. Afd. Natuurkunde, Tweede Reeks, 77.

Scheiner, D., 1976. Determination of ammonia in Kjeldahl nitrogen by indophenol method. Wat. Res. 10: 31–36.

Steenbergen, C. L. M., 1982. Contribution of photosynthetic sulphur bacteria to primary production in Lake Vechten. Hydrobiologia 95: 59–64.

Steenbergen, C. L. M. & Verdouw, H., 1982. Lake Vechten: aspects of its morphometry, climate, hydrology and physicochemical characteristics. Hydrobiologia 95: 11–23.

Tirén, T., 1977. Denitrification in sediment-water systems of various types of lakes. In: Golterman, H. L. (Ed.) Interactions between Sediments and Freshwater. W. Junk and PUDOC, The Hague and Wageningen.

Verdouw, H., van Echteld, C. J. A. & Dekkers, E. M. J., 1978. Ammonia determination based on indophenol formation with sodium salicylate. Wat. Res. 12: 399–402.

Verdouw, H. & Dekkers, E. M. J., 1982. Nitrogen cycle of Lake Vechten (The Netherlands): role of sedimentation. Arch. Hydrobiol. 94: 251–263.

Wood, E. D., Armstrong, F. A. J. & Richards, F. A., 1967. Determination of nitrate in seawater by cadmium-copper reduction to nitrite. J. mar. biol. Ass. U.K. 47: 23–31.

Hydrology of Tjeukemeer

J. D. Leenen
Laboratory for Hydraulics and Catchment Hydrology, Agricultural University, Nieuwe Kanaalweg 11, Wageningen, The Netherlands.
Present address: Haskoning, Consulting Engineers and Architects, P.O. Box 151, 6500 AD Nijmegen, The Netherlands

Keywords: boezem, water motion, tracer, mixing process

Abstract

The hydrology of Tjeukemeer is described in relation to a more complex system. A model is developed – based on fluctuating chloride concentrations – to estimate in- and outflow of the lake. Results are given for actual data.

Introduction

Tjeukemeer, with an area of 20 km^2, is part of a much larger and complex water system, the Frisian 'boezem'. A 'boezem' in the hydrological sense is a system of interconnected canals and reservoirs (lakes) into which superfluous polderwater is pumped and from where it is drained to the sea. The area drained by the Frisian 'boezem' amounts to 3 060 km^2. It includes nearly the whole province of Friesland (2 939 km^2) and small parts of the provinces of Groningen (90 km^2), Drenthe (12 km^2) and Overijssel (19 km^2). This large drainage area is unique in the Netherlands. Moreover, the 'boezem' surface of 140 km^2 makes up 4.6% of the total drainage area. This very high ratio is caused by the many lakes and is advantageous in that a relatively small discharge capacity is needed to maintain the indicated water level of the 'boezem'.

A surplus of polderwater occurs when precipitation exceeds losses due to evaporation and to other causes, e.g. groundwater seepage. The surplus water is, via the 'boezem' system, drained to the North Sea at Harlingen and Lauwersoog and to the IJsselmeer at Lemmer and Stavoren (Fig. 1). When the sea level at Harlingen is high, due to heavy western winds, a free discharge to the sea may be hindered by high tides. Together with heavy rainfall difficulties may arise in the maintenance of the 'boezem' level. In case of a water deficit, however, extra water has to be supplied from the IJsselmeer (Fig. 1) and is pumped into the 'boezem' system.

Water surplus and deficit depend on climatological conditions and, therefore, show a seasonal pattern (de Haan 1982). In Tjeukemeer, as a part of the 'boezem' system, the period of water deficit runs usually from May to October, and that of water surplus from October until May. This periodicity is reflected in the chloride content of the water.

The limnological research in Tjeukemeer concerns the availability of nutrients for algal growth and the population dynamics of the most common fish species related to their food organisms. Both these aspects depend largely on the water movements. Thus, for a full understanding of the hydrobiology of the lake hydrologic and hydrodynamic studies, specifically focussed on the inflow of water into the lake system, are needed.

Water balance

Introduction

A simple water balance equation is:

Hydrobiologia 95, 199–203 (1982). 0018-8158/82/0952-0199/$01.00.

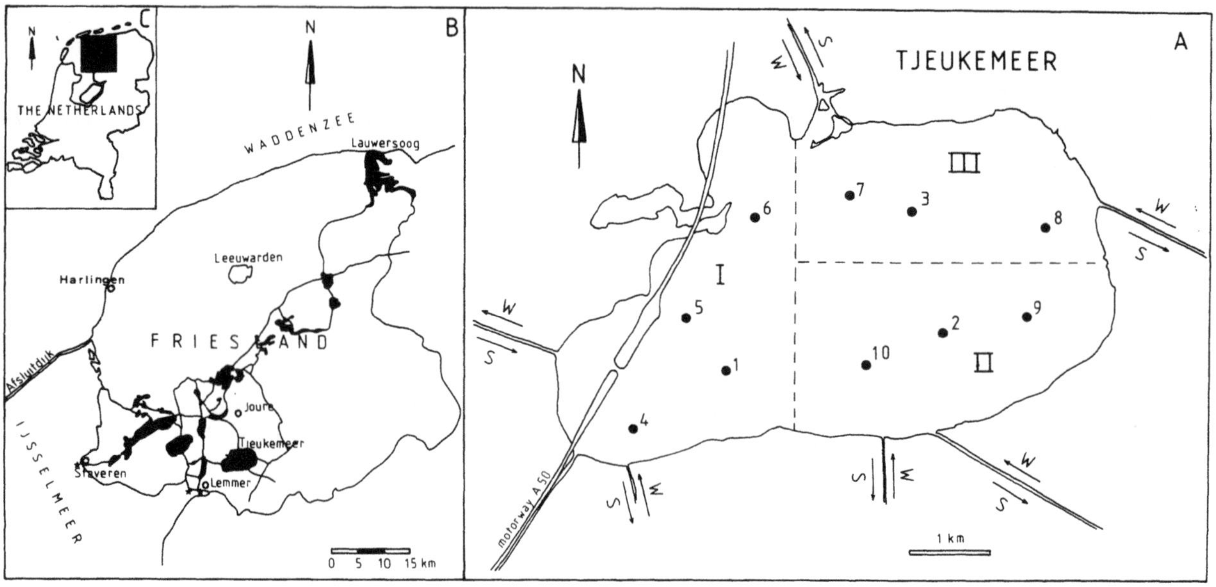

Fig. 1. Map of Tjeukemeer and its situation.

$$I = O + \frac{dB}{dt} \qquad (1)$$

where I = inflow (l^3T^{-1}),
O = outflow (l^3T^{-1}),
B = storage (l^3).

When the time interval is large a change in storage may be ignored. For the determination of inflow and outflow in Tjeukemeer, different methods are available.

Net rainfall input may be calculated from precipitation and evaporation. Inflow and outflow can be measured with flow meters, placed in the various inlets and outlets of the lake. This method is very laborious because of the large amount of measurements necessary to obtain reasonable results.

Knowing the total catchment area of a lake, the inflow may be estimated from net rainfall. This method needs less effort, but experience has shown that the catchment area of Tjeukemeer is variable and depends upon both wind direction and velocity.

Both methods seem to be unattractive. Therefore, a new method has been developed, based on the fluctuating chloride concentrations in the lake. The chloride concentration will be described with a mass-balance equation.

The chloride model

In periods of water deficiency in the 'boezem' system, Tjeukemeer receives IJsselmeer water through the Follegasloot (Fig. 1). During wet periods, polderwater with a lower chloride concentration is drained into the opposite direction. Thus the chloride concentrations range from about 50 mg · l^{-1} in winter to about 250 mg · l^{-1} at the end of the summer (de Haan 1982). The conservative chloride ion is a reliable tracer for studying flow processes in and around Tjeukemeer.

Simplifying the lake to a single reservoir with one inflow and one outflow (Fig. 2), and assuming instantaneous mixing of the water, the chloride balance per time interval Δt reads:

$$q_i \cdot c_i \cdot \Delta t = V \frac{\partial c}{\partial t} \cdot \Delta t + c \cdot \frac{\partial V}{\partial t} \cdot \Delta t + q_u \cdot c \cdot \Delta t. \qquad (2)$$

When Δt is large enough for $c \frac{\partial V}{\partial t}$ to be negligible compared to the other terms, it follows that $q_i\, \Delta t \approx q_u\, \Delta t = Q$ (total inflow equals total outflow during

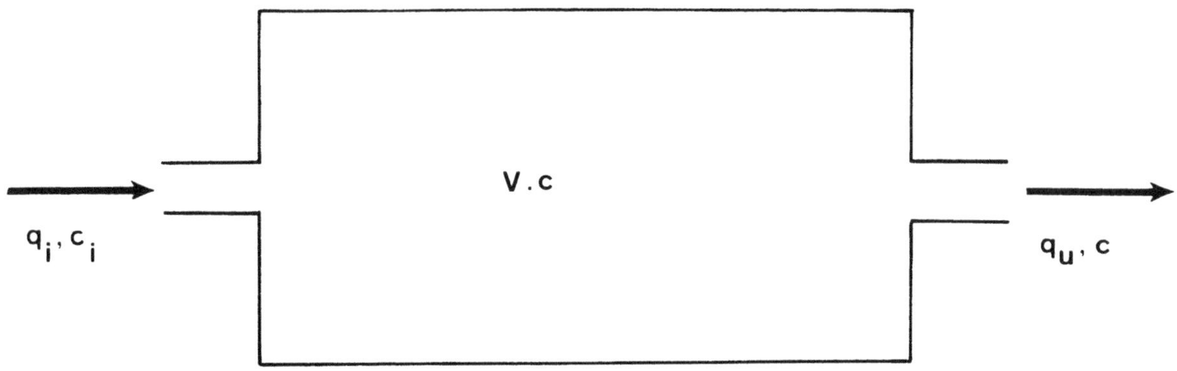

Fig. 2. Schematization as a single reservoir.

Δt). An average value for c over the interval Δt may be taken to be

$$c = \frac{c(t_1) + c(t_2)}{2} .$$

Discretizing the time derivative $\frac{\partial c}{\partial t}$, equation (2) becomes

$$Q \cdot c_i = V \{c(t_2) - c(t_1)\} + Q \frac{c(t_1) + c(t_2)}{2} \qquad (3)$$

or

$$Q = V \frac{c(t_2) - c(t_1)}{c_i - \frac{1}{2}\{c(t_1) + c(t_2)\}} . \qquad (4)$$

Equation (4) gives the total inflow during a period Δt. c_i was assumed to be constant, which is not in fact the case. In the same way as for c, an average value over Δt can be used with data at the beginning and at the end of the considered time interval.

Equation (4) then becomes

$$Q = V \frac{c(t_2) - c(t_1)}{\frac{1}{2}\{c_i(t_1) + c_i(t_2)\} - \frac{1}{2}\{c(t_1) + c(t_2)\}} . \qquad (5)$$

Results

Using weekly chloride samples (Limnological Institute unpublished data), and estimating the lake volume to be 3.5×10^7 m^3, equation (5) permits the calculation of inflow per week. The results for 1971 (Fig. 3) show large fluctuations during the summer period. These will be related to the fluctuations in total inflow of IJsselmeerwater at Lemmer and Stavoren.

The total inflow per 14 days is approximately linearly related to the inflow through the Follegasloot (Fig. 4).

The formula for the regression line in 1971 is

$$Q_F = 0.45\ Q_{L+s} - 6.,0, \qquad (6)$$

where Q_F = inflow in m^3 · 14 days^{-1} through Follegasloot,

and Q_{L+s} = inflow in m^3 · 14 days^{-1} via Lemmer and Stavoren.

The explained variance was 70%.

Discussion

An equation similar to (5) has been used by Golterman *et al.* (1980). However, they took $c(t_1)$ as an estimate for c, instead of an average value over Δt. Consequently this results in an underestimation of Q for the summer period when the chloride concentration increases, and to an overestimation for the winter period with a decreasing chloride concentration.

Our results show that the chloride concentration can be used to estimate the in- and outflow of Tjeukemeer. If the nutrient concentrations of the inflow are known, the nutrient loadings of the lake can be calculated.

In the chloride model, seepage of groundwater, although it does occur, has been neglected; the level of Tjeukemeer is about 1 m higher than the sur-

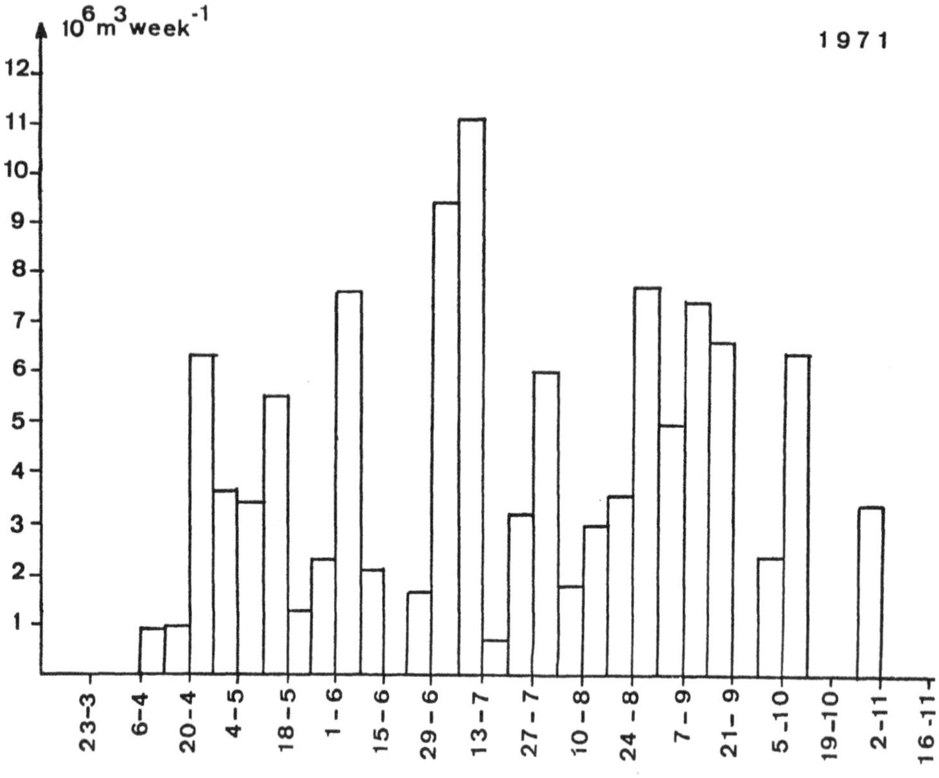

Fig. 3. Inflow through Follegasloot into Tjeukemeer as calculated by means of the chloride model. Data of 1971.

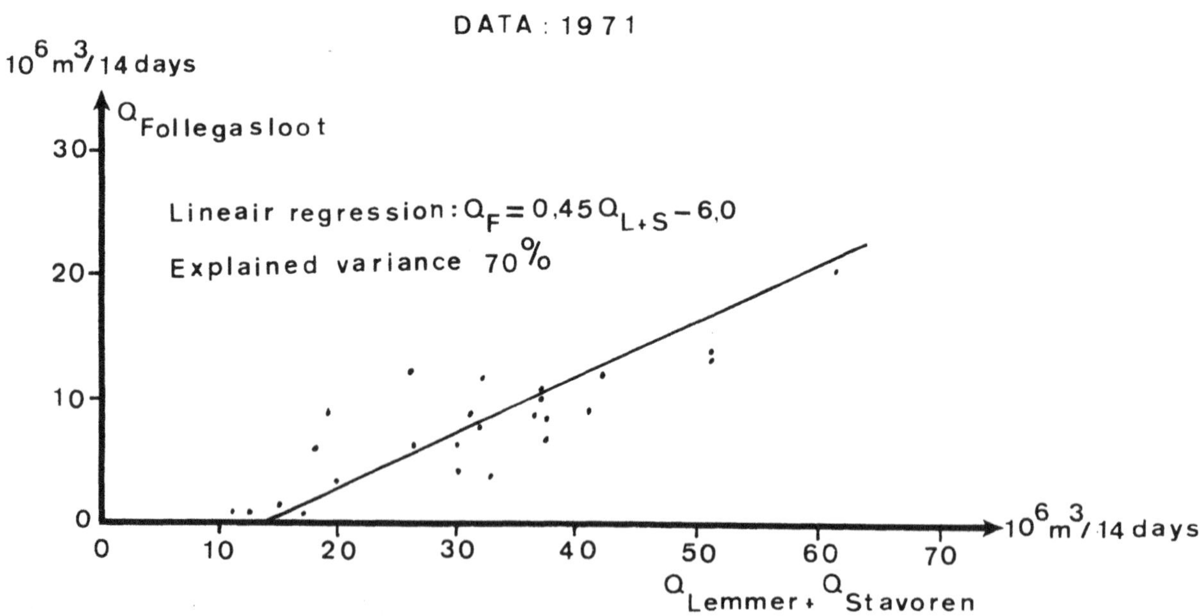

Fig. 4. A linear regression estimate for the relation between the inflow into Tjeukemeer, $Q_{Follegasloot}$ (as calculated with the chloride model) and the inlet at the intake points, Q_{Lemmer} and $Q_{Stavoren}$. Data of 1971.

rounding polder levels. However, it is assumed that the average chloride concentration in the seepage water is the average of that in the lake water, so there is no influence on the calculations.

Although the average chloride concentration in the lake may be used to estimate the inflow, the assumption of instantaneous mixing is only a theoretical tool. In reality the chloride concentration during the summer period is higher in the western part than in the eastern part. In the winter period the situation is the opposite. It takes some time for a given chloride concentration to reach the other side of the lake. Transport of chloride and other dissolved particles occurs through the main flow (convection) and through a mixing process caused by wind action. The latter may be considered as a diffusion process. Thus, the 'transportation time' for a certain concentration at one lake end to reach the other end is smaller than the mean resident time. This can be observed from the 1971 data (Fig. 5). The mean resident time T_r for the observed period (April–October) was about 2 months, to be calculated from the total inflow $\bar{Q}$, the period of inflow $\bar{T}$, and the lake volume V:

$$T_r = \frac{V}{\bar{Q}/\bar{T}}. \quad (7)$$

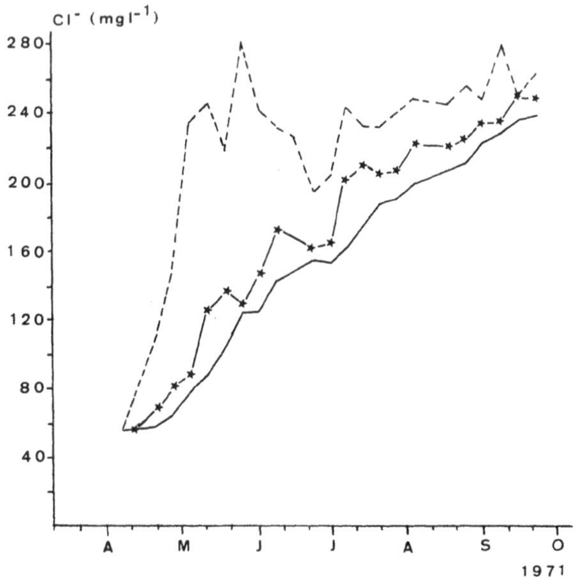

Fig. 5. Chloride concentration of inflow, west side and east side of Tjeukemeer in 1971 (---- = inflow concentration Follegasloot; ★-★-★ = average concentration west side; — = average concentration east side).

From Fig. 3, however, the 'transportation time' appears to be between 2 and 4 weeks.

A dimensional estimate for a longitudinal diffusion coefficient may be obtained from

$$D \sim \frac{L^2}{T}, \quad (8)$$

where D = diffusion coefficient,
L = length scale (horizontal dimension),
T = time scale for the diffusion process.

Taking L = 5 km and T = 3 weeks, equation (8) gives $D \sim 1.2\ km^2 \cdot day^{-1}$. The convection-diffusion equation that describes the mixing process has been solved numerically, following which the diffusion coefficient was calibrated. The results agree favourably with the dimensional estimate from equation (7). The modelling of this mixing process will be reported elsewhere (Leenen, in press).

Summary

The hydrology of Tjeukemeer is described in relation to its environment. Depending upon climatological conditions, the chloride concentration in the lake fluctuates. This fluctuation is the basis for a mathematical model to calculate the in- and outflow of the lake.

A relationship is established with the known intake of seawater at the coast. The chloride data of the lake indicate that the assumption of instantaneous mixing, often used in mathematical models, is only a mathematical tool and not realistic.

Acknowledgement

The author wishes to thank the Limnological Institute, Tjeukemeer Laboratory, Oosterzee, and in particular Dr. J. Moed, for the pleasant cooperation.

References

Golterman, H. L., Voerman, J. & de Nie, J. W., 1980. Fosfaatbelasting van het Tjeukemeer. H_2O 13: 116–121.

Haan, H. de, 1982. Physico-chemical environment in Tjeukemeer, with special reference to specification of algal nutrients. Hydrobiologia 95: 205–221.

Korf, J. F., 1973. Ons water II. Het water van Friesland. Tijdschrift Kon. Ned. Heidemij. 84: 103–126.

Leenen, J. D., in press. Wind induced diffusion in a shallow lake, a case study. Hydrobiol. Bull.

Physico-chemical environment in Tjeukemeer with special reference to speciation of algal nutrients

H. De Haan
Limnological Institute, Tjeukemeer Laboratory, De Akkers 47, 8536 VD Oosterzee, The Netherlands

Keywords: morphometry, hydrology, turbidity, colour, salinity, alkalinity, pH, conductivity, speciation, major, minor and micro elements, fulvic acids

Abstract

This paper attempts to generalize the major controlling factors and their impacts on the physico-chemical environment in Tjeukemeer in relation to the speciation of algal nutrients.

Morphometry, man-made hydrology, climate, chemistry of drainage area and biology of the lake appear to control its physico-chemistry the most. Thus the shallow Tjeukemeer is vulnerable to wind and therefore almost always completely mixed and often turbid owing to resuspended sediments. In normal summers the lake is oligohaline, hard, eutrophic and alkaline, and in winter it is hard, eutrophic, alkaline and humus-rich. The implications of these properties on the following physico-chemical components and their relevance for the speciation of algal nutrients are discussed: turbidity, colour, halinity, pH, alkalinity, salinity, conductivity, ionic composition, ionic balance and organic matter.

Because of the relationships between the physico-chemical environment on the one hand and the speciation and bio-availability of nutrients on the other hand, special attention is given to the speciation of some major and minor elements as derived from ultra filtration experiments. In this context, the metal binding capacity of the organic matter (mainly fulvic acids) is also considered.

Introduction

The physico-chemical conditions in natural waters control to a large extent the qualitative and quantitative aspects of aquatic life. This is particularly so for the primary producers on which the consumers of subsequent trophic levels depend. Factors controlling primary production in Tjeukemeer such as light and major and minor nutrients have been intensively studied (Golterman 1975). However, recently Frey & Small (1980) reported that the effects of micro nutrients on growth and composition of marine phytoplankton species are fundamentally different from those of major nutrients. In natural waters, the availability of micro nutrients, mainly metals, depends strongly on the physico-chemical environment (Florence 1977; Mantoura *et al.* 1978). Since we are interested mainly in the relationships between the quantity, physico-chemical form (species), and availability of major, minor and micro nutrients on the one hand and the algal periodicity in Tjeukemeer on the other hand, knowledge of the physico-chemical environment is a prerequisite. A description of the algal periodicity is given by Moed & Hoogveld (1982). In this paper generalizations based on physico-chemical data with special reference to the speciation of algal nutrients in Tjeukemeer are attempted. Thus the gaps in our knowledge of the relationships between the physico-chemistry and algal periodicity are demonstrated. The major controlling factors of the physico-chemical conditions in the lake are described first, followed by a discussion of the components relevant to the speciation of the algal nutrients. Among these the most important are: windcaused turbidity, the relatively high pH and

Hydrobiologia 95, 205–221 (1982). 0018-8158/82/0952-0205/$03.40.

alkalinity, and the high concentration of fulvic acids.

Major factors controlling the physico-chemical environment

Morphometry

Tjeukemeer is situated in the north of The Netherlands (Fig. 1). Before the construction of the A 50 motorway its area was 21.5 km² (Beattie 1978; Beattie *et al.* 1978), the eastern part now being 20 km² (Leenen 1980). The shoreline development is 1.3 and the relative depth only 0.04, meaning that the lake is almost round and very shallow. The lake is enclosed by a dike as its water level is above the surface of the surrounding polder country. The flat open landscape of the polders and the high water level both make Tjeukemeer very vulnerable to the influence of wind. Beattie (1978) gives a mean daily wind speed of about 10 km h^{-1} above the lake. However, at any time of the year, the wind speed can increase from a gentle breeze to more than 50 km h^{-1} within half an hour. The exposure of the shallow ($\bar{z}$ = 1.5 m) Tjeukemeer to wind speeds exceeding 30 km h^{-1} results in a complete mixing of the water column and resuspension of the top layer of sediment. This may happen several times a year and may significantly influence the vertical and horizontal distributions of particulate matter, temperature, turbidity and the water transparency. Thus the lake is almost always thermally homogeneous and saturated with O_2. The resuspended particles may interact with dissolved and colloidal matter. Therefore the morphometry of Tjeukemeer in particular, and that of shallow wind exposed lakes in general, forms an important controlling factor in the speciation of nutrients.

Man-made hydrology

Beattie *et al.* (1970, 1978) have given a general introduction to the hydrology of the Friesian lakes, and Leenen (1982) describes the hydrology of Tjeukemeer in detail. The relevant aspects for the physico-chemical environment are only broadly considered again.

Tjeukemeer is part of the Friesian boezem, i.e., the inter-connected system of lakes and canals in Friesland (Fig. 1). After the Afsluitdijk was completed, it became possible to introduce fresh water into Friesland via the IJsselmeer which is fed by the river IJssel, a branch of the river Rhine. Via locks

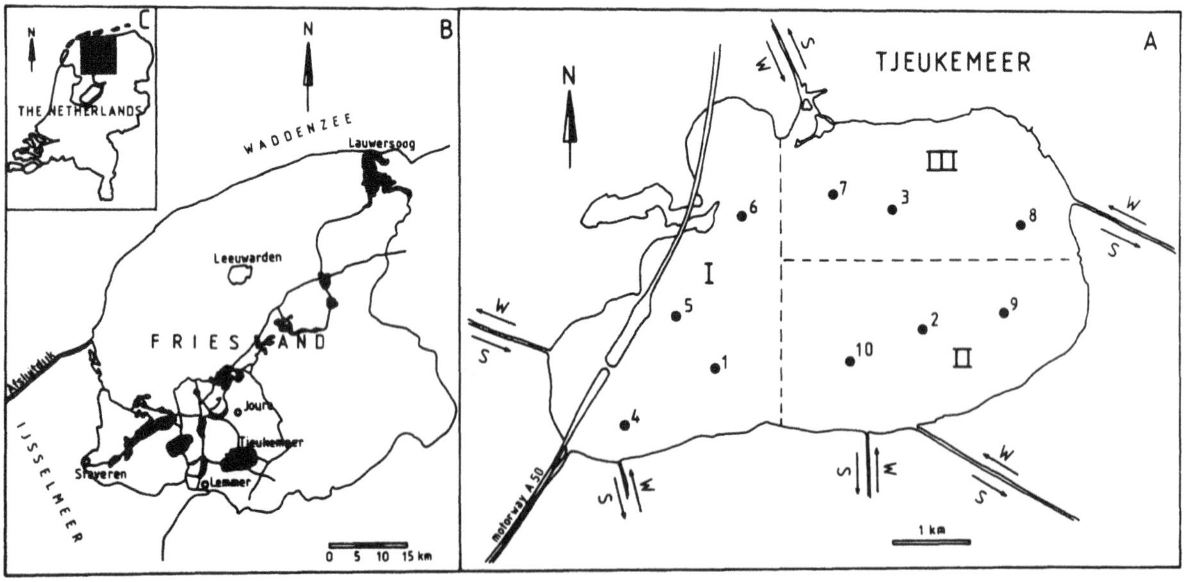

Fig. 1. Map of Tjeukemeer (A) and its situation in Friesland (B) and in the Netherlands (C). The main water movements in summer (S) and in winter (W) are indicated by arrows. The sampling stations are marked by dots; the sections are indicated schematically and marked with roman figures. For further details see text.

at, among other places, Lauwersoog, and water-pumps at Staveren and Lemmer, water can be released into the Waddenzee and IJsselmeer respectively (Fig. 1). Thus the water authorities are able to offset shortages and surpluses of water in Friesland which are determined mainly by agricultural needs. During water shortages, usually in summer, the incoming IJsselmeer water replaces the polder water in the lakes of the Friesian boezem. In winter the surplus of polder water is pumped into the lakes. Consequently, in summer the water flow in Tjeukemeer is the reverse of that in winter.

Climate

Besides the wind (as discussed earlier), other climatological factors which have great impact on the physico-chemical environment in Tjeukemeer are air temperature, rainfall and evaporation. Owing to its shallowness and exposure to wind the water column is not thermally stratified and is sensitive to changes in air temperature. During cold winters, the surface water freezes and ice covers of 10–30 cm may last several weeks (Beattie 1978). The O_2 concentration of the water column under ice rapidly decreases and may reach saturations of less than 10%. Only in these periods reduced metal species can occur in the otherwise always-oxygenated water column.

Rainfall and evaporation mainly control the water table of the polders. Thanks to the man-made hydrology as discussed earlier, changes in the water table can be offset by the inlet of IJsselmeer water or the release of polder water. Thus the nature of the water in Tjeukemeer, originating in the polders or IJsselmeer, is dependent on climatic conditions and the man-made hydrology of the boezem.

Chemical composition of polder and IJsselmeer water and sediments

The chemical composition of the lake water depends mostly on the nature of its sediments, drainage area and in some cases also on the composition of rain water. As already mentioned, the drainage area of Tjeukemeer is vast, extending over several hundreds of square kilometers of polderland and IJsselmeer. So the chemical composition of the lake water depends also on the drainage from these sources.

The sediments of Tjeukemeer are almost unknown. We do know, however, that they consist of a mosaic of peat (60%), sand (35%) and mud (5%) (Beattie *et al.* 1970). The top layer of the sediment is almost always wave-agitated, which has discouraged extensive study of the sediment. This makes the distinction and fractionation of (re)suspended detritus and live phytoplankton difficult. However, the application of not too large enclosures may enable us to study the undisturbed sediments. But extrapolation of the kinetic results thus obtained to the field may seriously underestimate the intensified interactions between sediments and water owing to wind.

The polders surrounding Tjeukemeer consist mainly of peaty soil; thus the run-off water from the polders contains high concentrations of yellow-brown metal-binding fulvic acids, which may give the polder water a colour as high as 200 mg l^{-1} of Pt (De Haan 1975). The polder water is hypertrophic; Tot-N and Tot-P concentrations may exceed 10 and 0.5 mg l^{-1} respectively. To date it is not exactly known how far the intensive agricultural use of the polders contributes to these high nutrient concentrations. However, more insight into the impact of the use of fertilizers in the polders on the nutrient load of Tjeukemeer could help in establishing the role of agriculture in the eutrophication of surface water in the peaty pasture lands of Friesland. From data on fulvic acids, major and minor elements in polder water (Table 1), it is not surprising that polder water has a drastic influence on the hydrochemistry of Tjeukemeer. This is noticeable usually in autumn and winter.

IJsselmeer water, which is normally introduced into the boezem in summer, replaces the polder water so that the chemical composition of Tjeukemeer changes again (Table 1).

Biology

Other major factors governing physico-chemical aquatic environments are the uptake and release of nutrients and metabolic products by micro and macro organisms. Thus, actively metabolizing algae control the concentrations of organic and inorganic nutrients, of which CO_2 and HCO_3^-, for example, determine the pH. Almost nothing is known about the release of organic matter by algae in Tjeukemeer, important for metal speciation because of its

Table 1. Concentration of major elements (meq l^{-1}) in Tjeukemeer in summer (s) and winter (w), IJsselmeer, and polder water after Golterman (1970), and those of some minor elements ($\mu g\ l^{-1}$) in 1978.

	Major elements							
Lake	pH	Ca^{2+}	Mg^{2+}	Na^{+}	K^{+}	HCO_3^-	Cl^-	SO_4^{2-}
IJsselmeer	8.6	3.14	1.24	4.00	0.21	2.02	5.07	1.38
Tjeukemeer (s)	8.5	2.55	0.84	2.76	0.21	1.89	3.17	1.08
Tjeukemeer (w)	7.7	1.80	0.56	1.00	0.25	1.37	1.18	0.50
Polder water	7.6	2.27	0.52	0.97	0.28	1.91	1.14	0.24

	Minor elements				
	PO_4-P	NH_3-N	SiO_2-Si	Tot-Fe	Colour (mg Pt l^{-1})
IJsselmeer	–	–	–	–	–
Tjeukemeer (s)	0	254	815	269	15.4
Tjeukemeer (w)	85	1235	4748	1787	63.7
Polder water	94	3897	7304	3812	102.9

metal complexing properties. However, nutrient uptake has been extensively studied. In spring the diatoms deplete SiO_2-Si in Tjeukemeer (Moed *et al.* 1976), while *Oscillatoria* and *Scenedesmus* species are mainly responsible for the almost simultaneous depletion of NH_3-N, $NO_2 + NO_3$-N and PO_4-P (Golterman 1974; De Haan *et al.* 1982a). The role of herbivorous zooplankton has hardly been investigated. A study by Vijverberg (1976) on the effect of food quality and quantity on the growth of *Daphnia,* gives some insight into preferred algal size but does not contain information about the relationships between zooplankton grazing on phytoplankton and their role in nutrient regeneration in Tjeukemeer. Also, little is known of the contributions of macro organisms which, by feeding on phyto and zooplankton, benthos, excretion of metabolites and bioturbation, may affect nutrient levels. The latter process is probably not important in Tjeukemeer since it may be negligible compared with wind-induced resuspension of the sediment.

The study of the role of microbes in contributing to the size of the organic and inorganic nutrient pools in Tjeukemeer is restricted to that of the mineralization of fulvic acids. *Arthrobacter* spp. appear to be able to degrade fulvic acids slowly and partly (De Haan 1972a, 1977).

Since the bio-availability of micro metals is one of our main interests, special attention has been paid to the metal complexing ability of the fulvic-acid-rich Tjeukemeer water. There is evidence already that the metal binding capacity is also partly controlled by organic matter released by waxing and waning *Oscillatoria* and *Aphanizomenon* populations (De Haan *et al.* 1982b).

Physico-chemical components relevant to speciation of algal nutrients

Turbidity and transparency

In Fig. 2A the changes in Secchi-disc transparency of Tjeukemeer water during 1977 are illustrated. Since the course of the fulvic acid curve (Fig. 2B) indicates the presence of polder water in autumn and winter and of IJsselmeer water in summer, 1977 can be regarded as representative of the usual hydrological situation in the lake. The Secchi-disc

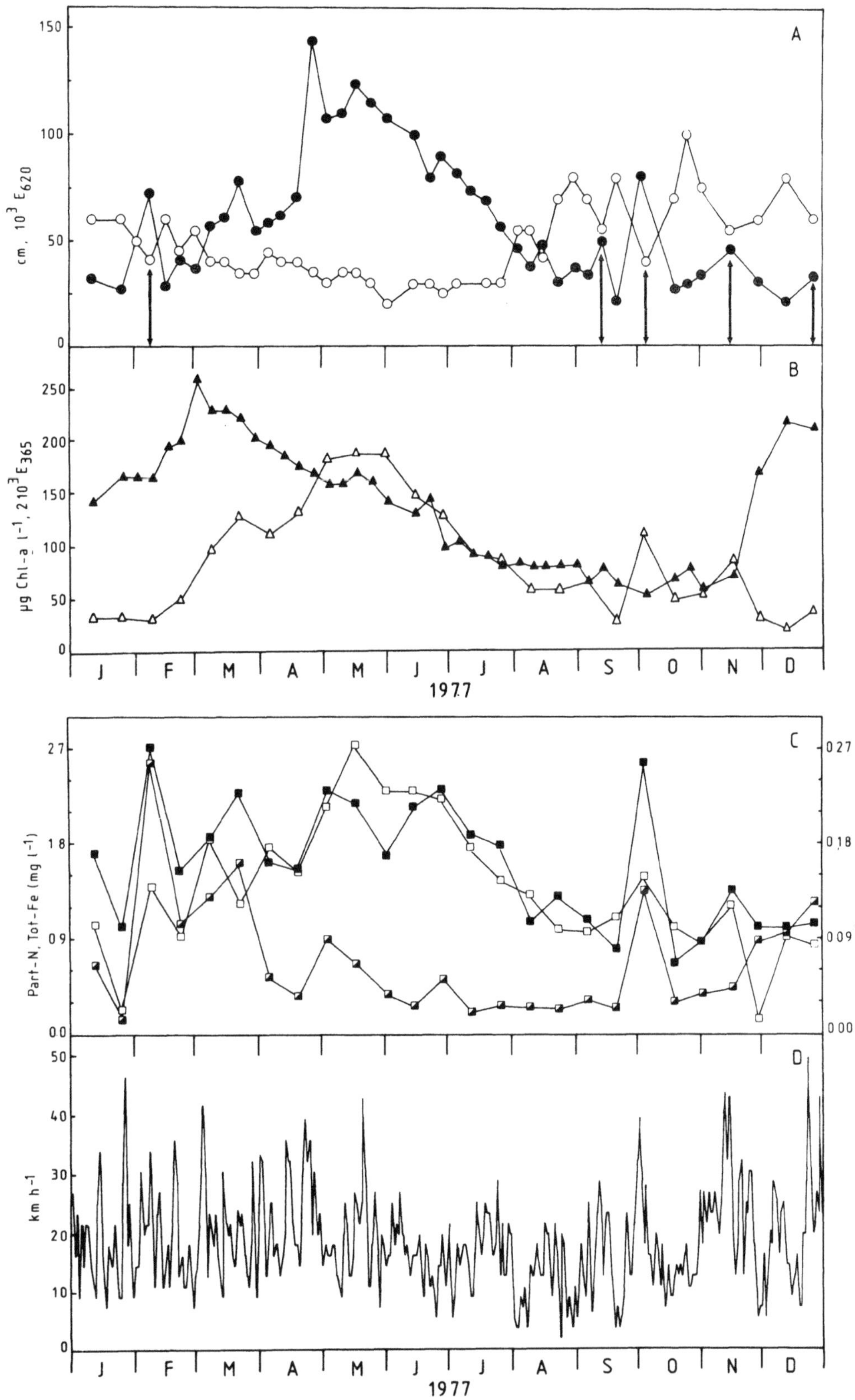

Fig. 2. Secchi-disc transparency (○) and turbidity (●) measured at 620 nm (E_{620}); for meaning of the arrows see text. B: Chlorophyll *a* concentration (△) and fulvic acid concentration (▲) measured as optical density at 365 nm (E_{365}). C: Concentration of Part-P (■), Part-N (□) and Tot-Fe (◪). D: Mean daily wind speed during 1977 as recorded at Leeuwarden (see Fig. 1B). All data concern samples taken during 1977 from section I of Tjeukemeer.

transparency varies between 20 and 100 cm. Remarkably though, a significant negative correlation between the Secchi-disc transparency of humus-rich Tjeukemeer water and the fulvic acid concentration is lacking but does exist with the turbidity ($P < 0.0005$). The turbidity is significantly correlated ($P < 0.0005$) with the chlorophyll *a* concentration (Fig. 2B), indicating the dominant role of algae in the turbidity and transparency of the humic Tjeukemeer water.

A comparison between turbidity, Secchi-disc transparency and the mean daily wind speed (Figs. 2A and 2D) shows that not only algae but also wind-caused resuspension of sedimented particles decreases the transparency. The dates when transparency minima and turbidity maxima are connected with wind speeds exceeding 30 km h^{-1} are marked with arrows in Fig. 2A. The transparency minima in early October and mid November coincide with chlorophyll *a* peaks, indicating the resuspension of sedimented algae. The wind-caused turbidity in early February is also reflected in high concentrations of Tot-Fe, Part-P and Part-N (Fig. 2C). During this period, no chlorophyll *a* peak was recorded, implying that the turbidity was caused by detritus. The high Tot-Fe peak in early October also indicates the involvement of detritus.

The wind-dependent resuspension of the top layer of the sediments in Tjeukemeer may of course have important impacts on the speciation and availability of algal nutrients. Thus the resuspended particles may absorb dissolved or colloidal nutrients, rendering them less available or, on the contrary, release organic or inorganic compounds stimulating algal growth. To date however, not much is known of these phenomena in Tjeukemeer.

Colour and Cl^- concentration

The Cl^- concentration of Tjeukemeer may exceed 200 mg l^{-1} (Golterman 1972; Golterman & Soekarjo 1973, 1974). This gives the lake an oligohaline character. The colour of the lake water is due to the presence of fulvic acids. Based on the colour standard of Hazen (1892), Tjeukemeer is humus-rich (20–100 mg l^{-1} of Pt). From these values De Haan (1975) calculated fulvic acid concentrations ranging between 3.4 and 13.7 mg l^{-1}. However, C.O.D. and colour measurements give much higher concentrations, i.e., between 20 and 60 mg l^{-1} (De Haan *et al.* 1982b). We consider the fulvic acid concentrations based on C.O.D. more reliable than those based on the Pt standard.

The colour and the Cl^- concentration of Tjeukemeer water vary strongly from year to year. The climate-dependent influx and outflux of water in the Friesian boezem was e.g. clearly demonstrated in the summers of 1972 and 1974 (Golterman 1972; Golterman & Soekarjo 1974). The influx of IJsselmeer water in the relatively wet summer of 1972 was about twice as low as that in the summer of 1974 (Figs. 3B and 3C). Consequently, Tjeukemeer received more IJsselmeer water in 1974 than in 1972. This is reflected in the quite different Cl^- concentrations of Tjeukemeer water during these years (Fig. 3A).

The months of January and February in 1972 were dry compared with those in 1974 (Fig. 3C). This is reflected in the rather small outlet of coloured polder water from the boezem during early 1972 (Fig. 3B). Therefore, the colour of Tjeukemeer was low and the Cl^- concentrations high during early 1972 (Fig. 3A). The inverse relationship between colour and Cl^- concentration in Tjeukemeer may be considered as a climatological indicator. The years with normal rainfall and evaporation would show the course of changes in colour and Cl^- as in 1974 (Fig. 3A). On the other hand, unusual courses of changes as in 1972 reflect abnormally wet summers and dry winters.

The usual colour and Cl^- courses are not inversely related in early spring (Fig. 3A). During this period the colour decreases while the Cl^- concentration remains constant. This phenomenon cannot therefore be explained by water movements, though influx of relatively less coloured polder water cannot entirely be excluded. It is likely that the decrease in colour in this period is caused by the pH increase due to photosynthesis and co-precipitation of $CaCO_3$ and fulvic acids (De Haan 1975). Besides fulvic acids, macro, minor and micro nutrients also may be co-precipitated with $CaCO_3$ due to interactions between $CaCO_3$ particles and dissolved and colloidal nutrient species.

Since Cl^- is not involved in chemical reactions or biological processes, its concentration is preferably used as an hydrological tracer in Tjeukemeer. Golterman *et al.* (1980) applied the Cl^- concentration to put up a water balance from which the P load of the lake was estimated. This method only gave an order

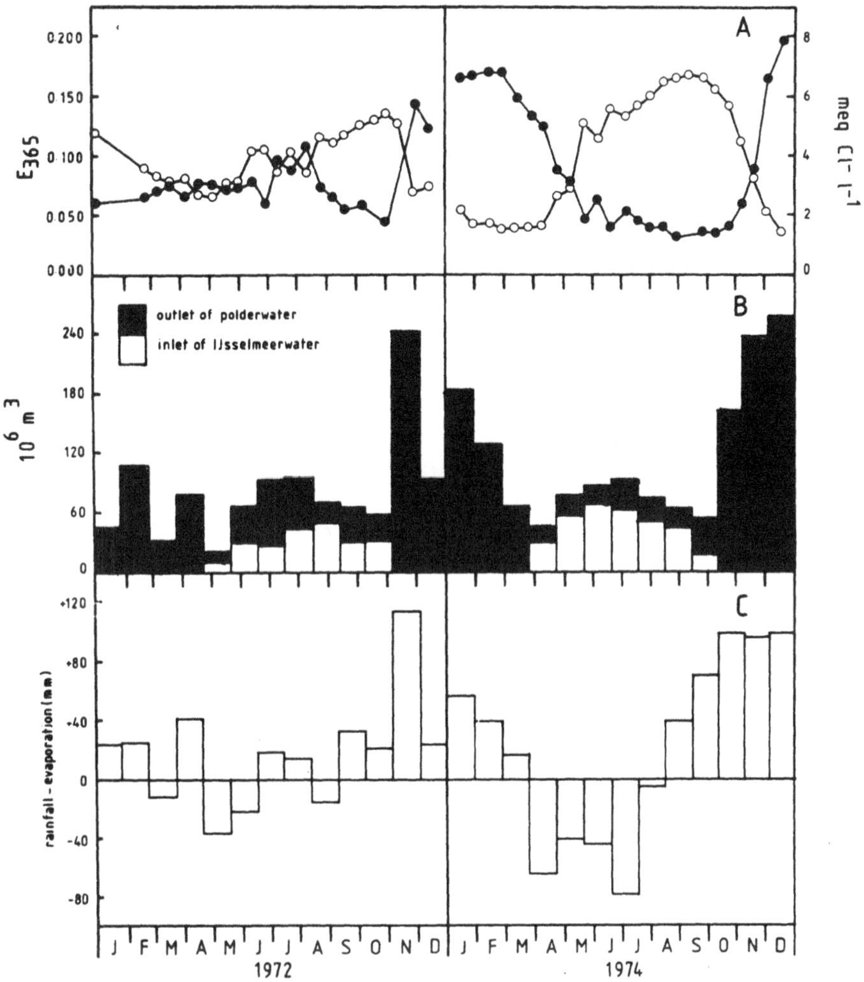

Fig. 3. A: Colour measured as optical density at 365 nm (●) and Cl⁻ concentration (○) of Tjeukemeer water at station 1 during 1972 and 1974. B: Inlet of IJsselmeer water (open blocks) and outlet of polder water (shaded blocks), at 4 weekly intervals, of the Friesian boezem during 1972 and 1974. Data from the Dutch Water Board, Department of Ways and Waterworks in the Province of Friesland. C: Rainfall – evaporation balance at 4 weekly intervals of the Friesian boezem during 1972 and 1974. Data from the Dutch Water Board, Department of Ways and Waterworks in the Province of Friesland.

of magnitude of the volume of water transported through the lake. Under normal climatic conditions, as in 1974, the lake is flushed once with IJsselmeer water in summer, and twice with polder water in winter. The corresponding Cl^- and fulvic acid loads are approximately 6 500 and 5 000 tons. After the outburst of botulism in 1975, the water authorities decided to flush the Friesian boezem more intensively in summer; this has led to even higher Cl^- loads.

From differences in the Cl^- concentration in West and East Tjeukemeer, Leenen (1982) derived a mixing rate of about 1 km^2 day^{-1}. This means that the incoming water is mixed through the lake in around 20 days. This rapid mixing time depends on the wind speed and may partly control the horizontal distributions of algae and (re)suspended material.

pH and alkalinity

The pH of humus-rich Tjeukemeer water always exceeds 7.5 (see e.g. Golterman 1968, 1969, 1971; Beattie *et al.* 1978). This unusually high pH value is caused by high alkalinity (Table 1; Fig. 4). Thus Tjeukemeer is an alkaline humus-rich lake, a prop-

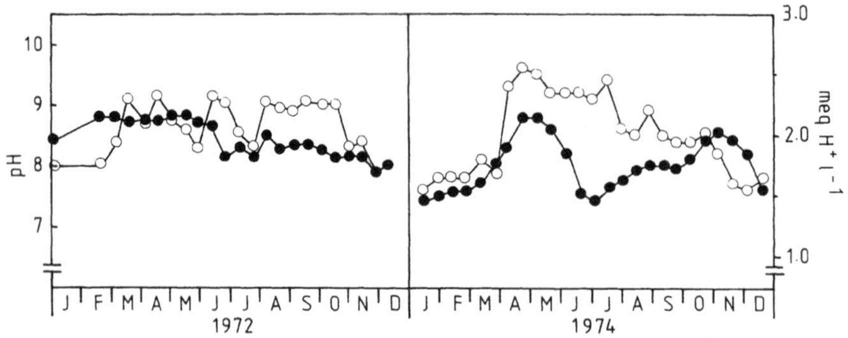

Fig. 4. pH (○) and alkalinity (meq H^+ l^{-1}) (●) of Tjeukemeer water at station 1 during 1972 and 1974.

erty not often encountered.

In years with normal climatic conditions, like 1974, the pH increases rapidly in March and April due to photosynthesis (Fig. 4 right). Generally the pH can rise about 2 units during spring, reaching values of 9.5 and higher (see e.g. Golterman & Soekarjo 1974). During the influx of IJsselmeer water in July and August, the pH drops about 1 unit when the gross primary production also decreases (Golterman 1969, 1970). Years with unusually dry winters and wet summers, like 1972, show less seasonally dependent pH fluctuations (Fig. 4 left).

Since pH and alkalinity are interrelated, the latter is also dependent on the influx of water from IJsselmeer and the polders (Figs. 3A and 4). However, the changes in alkalinity and Cl^- during 1974 do not correspond. This means that the alkalinity in the incoming waters is not constant. Probably the rates of photosynthesis and CO_2 diffusion from the air are involved in these changes in the alkalinity. It is noteworthy in this respect that the alkalinity minimum, usually occurring in June, coincides with the gross primary production maximum (Golterman 1969, 1970). The decrease in the alkalinity may have been partly caused by formation of CO_3^{2-} at pH 8.5 and precipitation of $CaCO_3$.

The high pH and alkalinity of Tjeukemeer water give rise to expectations that (bi)carbonato and hydroxy complexes are abundant metal species in the lake (Stumm & Morgan 1970). The solubility of these species determines mostly the availability of e.g. Fe, Mn, Cu and Zn. Yet almost no quantitative data are available relating the solubility of these metal species and their relative bio-availability at different pH and alkalinity.

As discussed before, the high pH may cause co-precipitation of $CaCO_3$, fulvic acids and nutrients, making the latter compounds less available.

Salinity, specific conductance, ionic composition and ionic balance

Speciation and availability of micro nutrients, i.e., trace metals, to algae is controlled mostly by the chemistry of the growth medium (Morel *et al.* 1979). In this respect knowledge of the concentration and composition of major ions is imperative. The concentration of major ions in Tjeukemeer, as reflected in the salinity, is high compared with the world mean of inland waters and varies by more than a factor of 2 (Table 2). It is evident that NaCl contributes considerably to the high salinity and determines the oligohaline character of the lake. The high NaCl load is due to the water from IJsselmeer which is fed by a branch of the river Rhine. The Cl^- concentration in this river may be as high as 400 mg l^{-1}, mainly of anthropogenic origin (Golterman 1975). In autumn and winter, when polder water is pumped into Tjeukemeer, the salinity decreases rapidly. However, the NaCl concentration of polder water is rather high due to upwelling haline water.

The dependence of the major ion concentrations on season is reflected in the specific conductance. Since Cl^- is among the ions which contribute mainly to salinity, it has a linear correlation with specific conductance. Influx of coloured polder water with relatively low salinity in winter coincides with decreasing specific conductance. However, since the colour does not contribute to the conductance, the colour and specific conductance are not linearly

Table 2. Composition of major ions in Tjeukemeer under normal and abnormal climatic conditions (unpublished results of Golterman).

	Wet summer (1972)			Normal summer (1974)			Normal winter (1974)			World mean	
	meq	meq %	mg l^{-1}	meq	meq %	mg l^{-1}	meq	meq %	mg l^{-1}	meq %[a]	mg l^{-1}[b]
Ca^{2+}	2.94	38	58.9	3.20	30	64.1	2.18	44	43.7	64	
Mg^{2+}	1.04	14	12.6	1.40	13	34.1	0.96	19	23.3	17	
Na^{+}	3.50	45	80.5	5.83	54	134.0	1.48	30	34.0	16	
K^{+}	0.24	3	9.4	0.27	3	10.6	0.33	7	12.9	3	
Cations	7.72		161.4	10.70		242.8	4.95		113.9		
HCO_3^-	2.07	27	126.3	1.75	17	106.8	1.53	38	93.4	74	
SO_4^{2-}	1.24	17	59.6	1.68	17	80.7	1.07	26	51.4	16	
Cl^-	4.15	56	147.1	6.72	66	238.2	1.46	36	51.8	10	
Anions	7.46		333.0	10.15		425.7	4.06		196.6		
Salinity	15.18		494.4	20.95		668.5	9.01		310.5		120

[a] According to Rodhe (1949); [b] According to Wetzel (1975).

correlated but inversely exponential (Fig. 5), as expected during a mixing of independent parameters. These interrelationships of colour, Cl^- concentration and specific conductance are highly significant ($P < 0.0005$) so that all the three parameters are suitable hydrological tracers. Since the colour, i.e., fulvic acids, are involved in co-precipitation with $CaCO_3$ and microbial degradation, Cl^- is the preferable hydrological tracer. In summer, when Tjeukemeer is flushed with IJsselmeer water, Na^+ and Cl^- are the most abundant ions, while in winter in the presence of polder water, Ca^{2+} and HCO_3^- dominate. The fluctuating concentration and composition of major ions will of course affect the speciation of micro nutrients. It is therefore reasonable to suppose that the speciation and availability of micro nutrients in Tjeukemeer change seasonally. This might influence the competition of different algal species for trace metals and find expression in the algal periodicity.

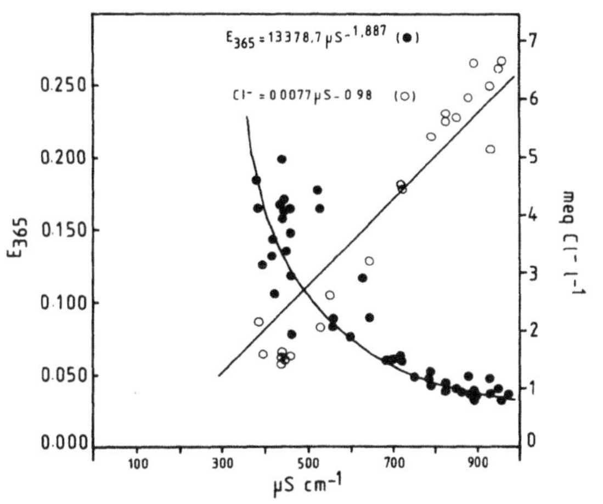

Fig. 5. Correlations between colour (E_{365}), Cl^- concentration and specific conductance ($\mu s\ cm^{-1}$) at 21 °C of Tjeukemeer water from section I during 1975.

Despite the dominance of Na^+ and Cl^- in summer, the Ca^{2+} and HCO_3^- concentrations (Table 1) give the lake water a permanently hard character. This hardness (6–9 German degrees) causes supersaturation of $CaCO_3$ when the pH exceeds 8.5. Because Ca^{2+} and HCO_3^- are determined in unfiltered water, one wonders to what extent colloidal and particulate $CaCO_3$ overestimate the true salinity and interfere in the excess of cations found in Tjeukemeer water. Golterman (1970) and Golterman & Soekarjo (1974) reported that this excess of cations was more pronounced in winter without speculating on the reasons for this finding. Their results suggest that the discrepancy in the ionic balance is connected with the type of water present in the lake (see e.g. Table 2). Indeed, more than 100 comparisons of the ionic balance between 1968 and 1980 reveal a significant correlation ($P < 0.005$, $r = 0.347$, $n > 100$) between this descrepancy and the fulvic acid concentration. This indicates that cations bound to fulvic acids are determined too during analysis, thus overestimating their ionic concentration. However, the low correlation coefficient (r) suggests that other factors not yet known may be involved.

Organic matter

Though we know that matter passing through a 0.2 μm filter is partly colloidal, in both this and the subsequent sections, matter $<0.2\ \mu m$ is defined as dissolved. The main constituent of the dissolved organic matter, i.e., fulvic acids, colour the lake water yellow-brown. The intensity of this colour, measured as optical density at 365 nm, is strongly correlated with the C.O.D. (De Haan *et al.* 1982b). Based on this correlation, the organic matter concentration is estimated to vary between 20 and 60 mg l^{-1}. This variation is controlled by the influx of water from the polders and IJsselmeer (Figs. 3A and 3B).

Besides fulvic acids, amino acids form an important constituent of the dissolved organic matter. Their mean concentration of 2.9 mg l^{-1} is 6% of the dissolved organic matter. As high as 13% may occur as amino acids simultaneously with both low concentrations of chlorophyll *a* and phaeopigments (Fig. 6). This suggests that the high amino acid concentration is related to dying algae, which is supported by algal counts (De Haan & De Boer 1979). These rather high amino acid concentrations reflect the eutrophic status of the lake. Unfortunately, the classic colorimetric sugar analysis using anthrone or orcinol sulfuric acid mixtures, appears to be unsuitable for humic solutions, and our knowledge about carbohydrates in Tjeukemeer is still fragmentary.

The fractionation and composition of fulvic acids have been intensively studied by gel filtration (De Haan 1972a, b; De Haan & De Boer 1978, 1979) and by Curie-point pyrolysis-mass spectrometry (De Haan *et al.* 1979, 1981). According to Sephadex G-25 gel filtration at pH 7.0, fulvic acids consist of three main fractions with different molecular weights (Fig. 7). Chemically, these fractions I, II and III show great similarity and may therefore be regarded as aggregates of different sizes consisting of phenols, carbohydrates and proteinaceous material. However, Sephadex gel filtration at pH 3.0 demonstrated that more than half of the at pH 7.0 simultaneously eluted carbohydrates, amino acids and fulvic acids are associated by acid-labile

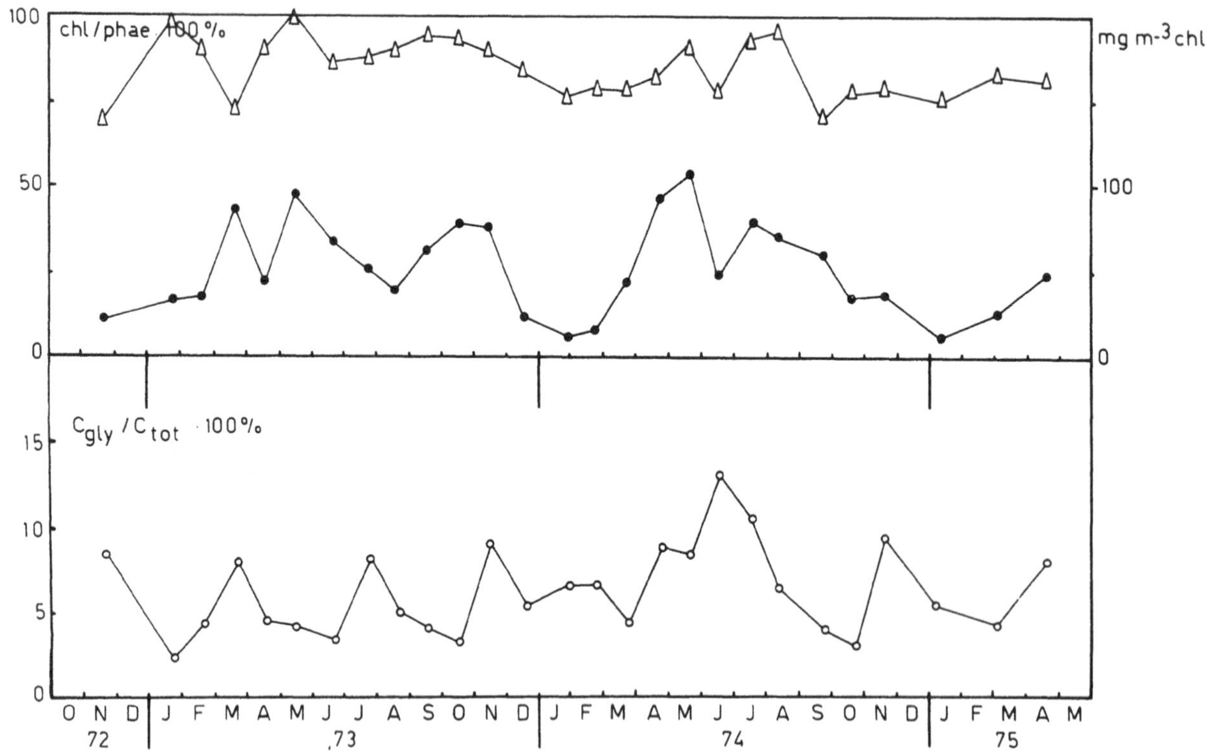

Fig. 6. Percentage of C in dissolved amino acids of the total dissolved organic C (○), chlorophyll *a* concentration (●) and the ratio of chlorophyll *a* to phaeopigment (△) in Tjeukemeer water at station 1 between October 1972 and May 1975. Chl = chlorophyll *a*, phae = phaeopigment, gly = glycine. (From De Haan & De Boer, 1979. Reproduced with permission of Schweizerbart'sche Verlagsbuchhandlung, Stuttgart, FRG.)

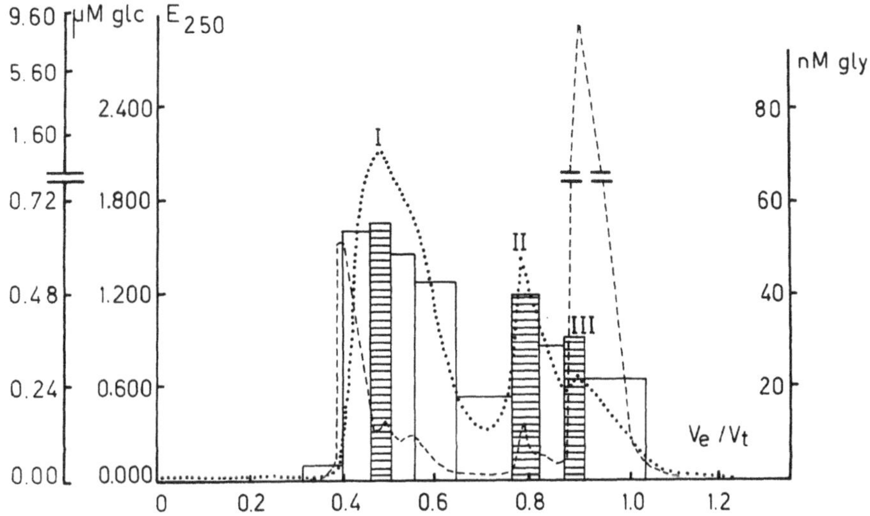

Fig. 7. Elution patterns of carbohydrates (----), amino acids (histogram bars) and fulvic acids (....) after Sephadex G-25 gel filtration at pH 7.0. Fulvic acids were estimated by light absorption at 250 nm. Shaded bars represent fractions analysed for carbohydrates and amino acids. Glc = glucose, gly = glycine, V_e = elution volume, V_t = total gel volume. (From De Haan & De Boer, 1978. Reproduced with permission of Pergamon Press Ltd., Oxford, England.)

hydrogen bonds (De Haan & De Boer 1978). The other half of the carbohydrates and amino acids are bound strongly to fulvic acids.

The metal binding properties of fulvic acids are well known (Schnitzer & Khan 1972) and are clearly demonstrated by the Cu binding capacity of the fulvic acid fractions from Tjeukemeer obtained after Sephadex gel filtration at pH 3.0 and 7.0 (Fig. 8). Given the high fulvic acid concentration in the lake, chelated metals may be an important factor in the availability of some of the trace metals essential for algal growth. On the other hand, fulvic acids may mask the toxicity of some heavy metals by chelation. Generally, as derived from their Cu binding capacity, fulvic acids are expected to play an important role in the distribution of metals over, and the functioning in, the different compartments of the Tjeukemeer ecosystem (Schnitzer & Khan 1972; Povoledo & Golterman 1975). The relative contribution of fulvic acids to the metal binding capacity of Tjeukemeer water is discussed in the next section.

Combining the results shown in Figs. 7 and 8, fulvic acids of relatively low molecular weight can bind Cu more effectively than the high molecular weight ones. Since the ratio of fulvic acids of high and low molecular weight changes seasonally (De Haan 1972b, 1975), the metal binding efficiency of fulvic acids in Tjeukemeer is not constant. This is clearly demonstrated by the Cu binding capacity per unit of optical density at 365 nm of Tjeukemeer and polder water (Fig. 9B). Usually, the former type of water contains more fulvic acids of low molecular weight, as predicted by the ratio of the optical densities at 250 and 365 nm (De Haan 1972b).

Metal binding capacity

As expected from the high concentration and the composition of dissolved organic matter in Tjeukemeer, the lake has a high metal binding capacity, e.g., as Cu the lake water can bind 250–1000 $\mu g\ l^{-1}$ (Fig. 9A). This is controlled by, among other factors, complexation and peptization of metals by fulvic acids. Complexed and peptized metal species will have different availabilities. That the algal growth inhibiting effect of Cu and the metal binding capacity in Tjeukemeer were inversely correlated in 1979, is noteworthy in this respect (De Haan *et al.* 1982a).

During abnormally wet summers, such as in 1979, when the lake contains relatively more polder water and has relatively higher E_{365} values (Fig. 9D), the metal binding capacity is apparently con-

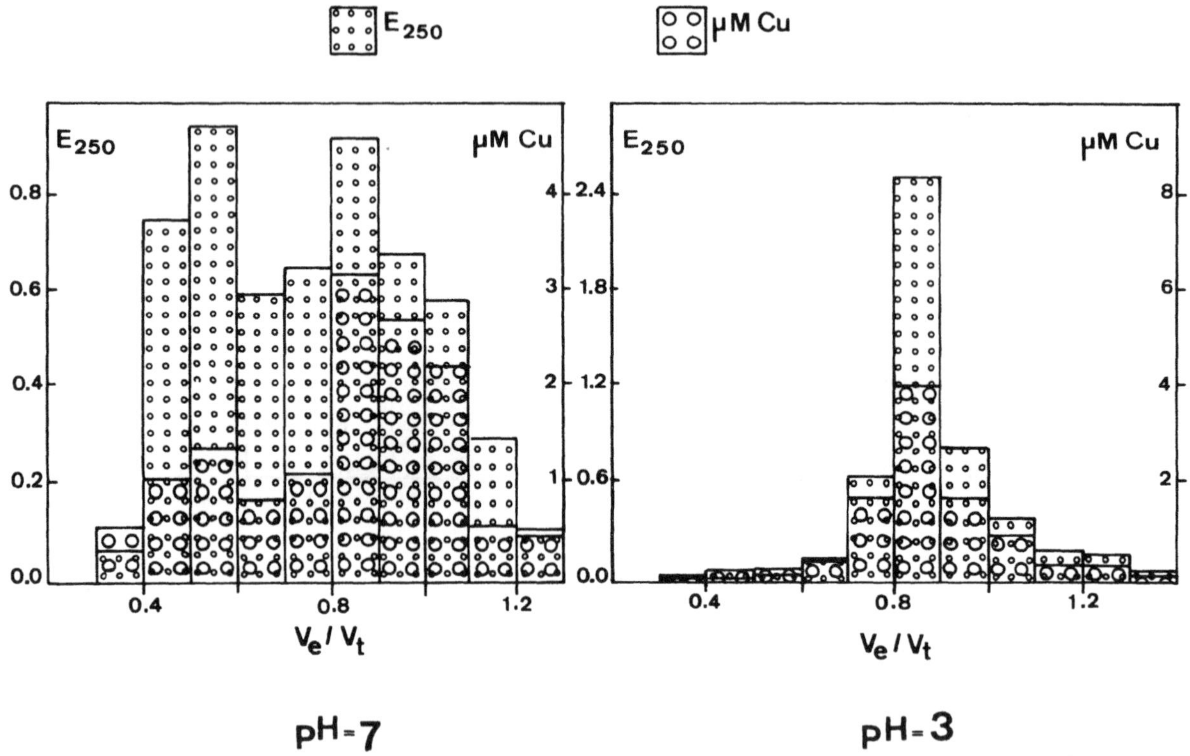

Fig. 8. Fulvic acid concentration and Cu binding capacity in pooled fractions after Sephadex G-25 gel filtration of fulvic acids at pH 3.0 and 7.0. One pool consists of 10 subsequent fractions eluted from the Sephadex column. Fulvic acids were estimated by light absorption at 250 nm, Cu binding capacity is expressed in μM of Cu. V_e = elution volume, V_t = total gel volume.

trolled by the concentration of allochthonous fulvic acids. In contrast, during summers with normal rainfall, when the lake is flushed with IJsselmeer water, the metal binding fluctuates quite independently of the water colour (Fig. 9B). Though the polder water contains 2–3 times higher concentrations of dissolved organic C than Tjeukemeer, the Cu binding capacity per unit organic C is consistently higher and more variable in the lake.

The variations in metal binding are similar to those of dissolved amino acids which are connected with waning algal populations (Figs. 6 and 9) (De Haan & De Boer 1979). So it is an obvious assumption that the metal binding capacity of Tjeukemeer water is also connected with the organic matter from algal cells. This idea is supported by the highly circumstantial occurrence of maximum numbers of *Aphanizomenon* and *Oscillatoria* spp. and the peak values of Cu binding capacity (De Haan *et al.* 1982b). Because polder water contains less algal biomass than that in Tjeukemeer, as reflected by chlorophyll *a* concentrations (Fig. 9C), and the organic matter is dominated by fulvic acids, the variations in metal binding do not appear to be induced by algae in this type of water.

Though more work is needed to demonstrate conclusively that the metal binding properties are greatly influenced by growth and death of abundant blue-green algae, evidence is accumulating that in eutrophic lakes like Tjeukemeer, speciation, availability and toxicity of metals are phycologically regulated.

Speciation of major, minor and micro nutrients as studied by ultra filtration

Major cations

The occurrence of colloidal $CaCO_3$ in Tjeukemeer has been discussed in connection with photosynthesis, pH and ionic balance. The high fulvic

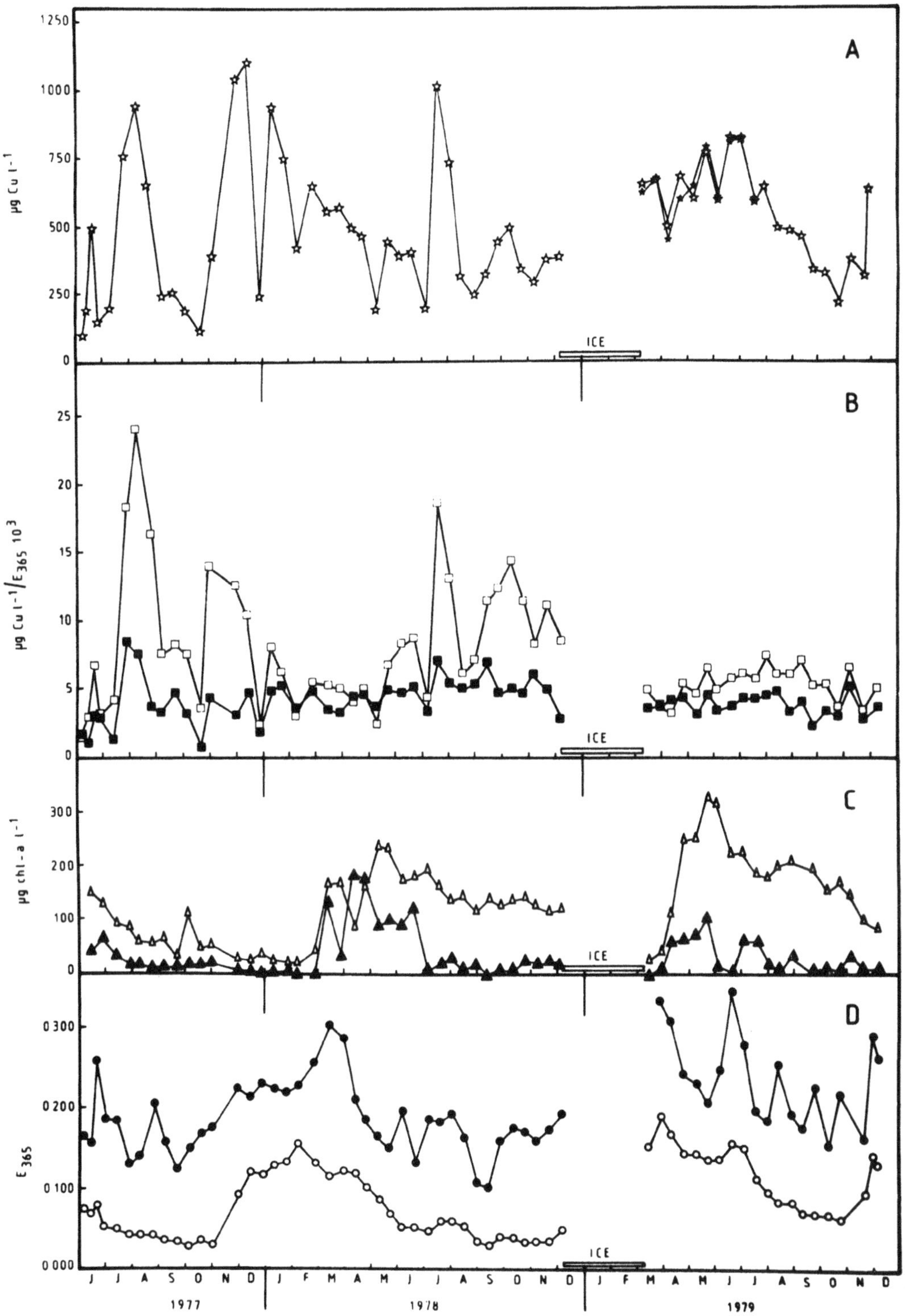

Fig. 9. A: Cu binding capacity (Cu-BC) as $\mu g\,l^{-1}$ of Cu measured with cuprizone (☆) and with dibenzyldithiocarbamate (★). B: Cu-BC per unit of optical density at 365 nm in Tjeukemeer (□) and polder water (■). C: Concentration of chlorophyll *a* in Tjeukemeer water (△) and polder water (▲). D: Concentration of fulvic acids in Tjeukemeer water (○) and polder water (●). All data shown were measured in water from section 1 of Tjeukemeer sampled between June 1977 and December 1979. (From De Haan *et al.*, 1982b. Reproduced with permission of Schweizerbart'sche Verlagbuchhandlung, Stuttgart, FRG.)

Table 3. Size fractionation of some major and minor elements in 0.2 μm filtered Tjeukemeer water from section I. Amounts are in %.

Element	<12.5 Å	<15 Å	<18 Å	<22 Å	<140 Å	<2.000 Å
Ca	61	82	95	97	97	100
Mg	66	84	98	100	100	100
Na	89	96	96	99	100	100
K	89	95	96	99	99	100
P	13	14	22	45	45	100
N	82	87	89	97	99	100
Fe	1	2	9	9	12	100
Si	98	98	98	99	99	100
org C	9	19	52	86	88	100

acid concentration may give rise also to chelated Ca and Mg. These different species of Ca and Mg may be involved in the discrepancy found in the ionic balance. To obtain more insight into the speciation of major cations, ultra filtration through pore sizes between 12.5 and 140 Å was applied. About 35% of Ca and Mg in Tjeukemeer was found to exist in particles with sizes between 12.5 and 18 Å (Table 3). About 70% of the fulvic acids occur in particles ranging from 15 to 22 Å. The partial overlap of these size ranges suggests that Ca and Mg are partly chelated by fulvic acids and partly a constituent of inorganic colloids. As expected, K and Na occur mainly in dissolved forms since 90% of these elements pass through filters with a pore size of 12.5 Å.

More work is needed to confirm whether the concentrations of colloidal Ca and Mg correspond to the found overestimation of major cations in the ionic balance. The preliminary results of ultra filtration do show that this technique may help gain insight into the speciation of major cations. Particularly in highly saline, alkaline lakes like Tjeukemeer, ultra filtration seems useful to determine true salinities and ionic balance, otherwise the cationic distribution may easily be overestimated by 25%.

Minor and micro elements

Until a few years ago, little was known about the occurrence of micro elements in Tjeukemeer. Recently, preliminary studies on some of the essential metals have indicated Cu concentrations between 10 and 60 $\mu g\ l^{-1}$ and Zn concentrations between 5 and 70 $\mu g\ l^{-1}$. Such high levels, particularly for Cu, point to an anthropogenic origin. More work should confirm the origin and levels of these metals and their impact on the phytoplankton of the lake.

For the concentration of minor nutrients with respect to the hydrology of Tjeukemeer, one is referred to Beattie *et al.* (1970, 1978) and Golterman (1974). Here, only some recently developed insights into the speciation of P, N, Fe and Si are given.

Arbitrary limits (0.2 or 0.45 μm) are still widely used to define dissolved aquatic matter. Particles > 0.2 μm can be roughly divided into micro-organisms, silt and detritus. Nutrients in these components are available for algae only if they are in equilibrium with the smaller, i.e., colloidal and dissolved, forms. The particulate nutrients are, therefore, of relatively minor importance in the study of relationships between speciation and bio-availability of nutrients, and are not further considered.

The size fractionation of nutrient forms < 0.2 μm is shown in Table 3. It is evident that the so-called dissolved P and Fe in Tjeukemeer water are not dissolved at all, but are more or less constituents of colloidals. More than half of the P and almost all the Fe is larger than 140 Å, indicating the occurrence of $FePO_4$ and $Fe(OH)_3$ colloids. On the other hand, 12% of the organic C is also larger than 140 Å. Remarkably, no Fe is found between 15 and 22 Å, in which range about 70% of the metal binding fulvic acids occur. This does not support the assumption of Golterman (1973, 1975) that fulvic – iron – phosphate complexes exist in Tjeukemeer. Apparently, the speciation of Fe is controlled more by the high pH than by the humic nature of the lake.

The truly dissolved PO_4–P fraction in Tjeukemeer is very small since about 80% of the P passing through 0.2 μm filters is larger than 12.5 Å (Table 3). However, roughly half of the P particles > 12.5 Å reacts as dissolved reactive P (DRP) during the molybdenum blue method due to hydrolysis at the ambient acidic pH. This means that this method applied to 0.2 μm filtrates of Tjeukemeer water seriously overestimates the concentration of PO_4–P. The occurrence of 2 DRP fractions, as demonstrated by ultra filtration of Tjeukemeer water, was reported earlier by Downes & Pearl (1978) and White & Payne (1980). These authors employed Sephadex gel filtration and demonstrated that the lakes and streams on North Island, New Zealand, in addition to PO_4–P, contain a reactive high molecular weight P fraction (RHMW-P). Both the P fractions were found to be available to algae, but the former gave faster growth rates. The relative availability of the 2 DRP fractions in Tjeukemeer is not yet known.

All the Si and almost all N in 0.2 μm filtered lake water is smaller than 12.5 Å (Table 3), and thus in truly dissolved inorganic form. The small percentage of N between 18 and 22 Å may be an organic N fraction. Because N appeared to be a major limiting nutrient in bioassays of Tjeukemeer water (De Haan *et al.* 1982a), we are interested in the chemistry and availability of organic N which is never depleted in the lake (see e.g. De Haan & De Boer 1979).

Summary

Man-made hydrology, agriculture and waste water disposal are the main influences on quality of the water in Tjeukemeer. Some striking anthropogenic reflections in the physico-chemical properties are the colour, and the eutrophic and oligohaline character of the lake.

An important consequence of the man-made hydrology is that the lake receives water of different origin and composition in different seasons. The amount of water from either source is determined by evaporation and rainfall. Therefore, salinity, pH, alkalinity and fulvic acid concentration – factors controlling the physico-chemical speciation of algal nutrients – fluctuate considerably. For instance, the Cl^- concentration in normal summers is about 250 mg l^{-1} and in winter only 50 mg l^{-1}. Correspondingly, the fulvic acid concentration, being inversely correlated with that of Cl^-, varies between 20 and 60 mg l^-. Annually, the pH ranges between 7.5 and 9.5.

The morphometry, the flat surroundings, the prevailing wind speed and the shallow ($\bar{z} = 1.5$ m) Tjeukemeer often cause complete mixing during prolonged periods. The resulting resuspension of peat and detritus reduces the transparency (20–100 cm). Remarkably, the brown colour of the humus-rich water hardly contributes to the transparency; light attenuation in summer is due mainly to algae. The wind-caused mixing of the water column and the top layer of the sediment may influence the speciation and availability of nutrients, the study of which is in progress.

The dissolved organic C pool of 10–30 mg l^{-1} is high and consists mainly of allochthonous fulvic acids giving the water its yellow-brown colour, the intensity of which is strongly correlated with the concentration of the former. The concentration and composition of organic matter appear to be controlled by hydrology and by microbial activity in Tjeukemeer. The concentration of amino acids reflects the eutrophic status of the lake. Since amino acids and fulvic acids have metal binding properties, it is not surprising that the metal binding capacity of Tjeukemeer water is relatively high (250–1000 μg l^{-1} of Cu). In periods when the lake contains polder water, fulvic acids contribute mainly to the metal binding. However, in the presence of IJsselmeer water, highly circumstantial evidence indicates that the metal binding capacity is controlled by organic matter from decaying algal populations. Thus metal speciation and algal periodicity might be connected in Tjeukemeer.

Tjeukemeer is supersaturated with $CaCO_3$. Particularly in spring, when the increase in photosynthesis results in a rise of pH, $CaCO_3$ and fulvic acids co-precipitate. Together with $CaCO_3$ and fulvic acids, algal nutrients may be precipitated, making them less available. On the other hand, fulvic acids may form colloidal species with Ca by peptization, which may also involve Mg and Fe. This phenomenon may contribute to the discrepancy found in the ionic balance – still a matter of study.

By using ultra filtration, the occurrence of colloidal Ca and Mg was clearly demonstrated. Also, most of the organic C (fulvic acids) occurs in colloidal structures. However, most of the Ca and Mg colloids and that of the fulvic acid colloids have different sizes, suggesting that colloidal Ca and Mg are only partly chelated by fulvic acids. Analogous findings for Fe, P and organic C cast doubt on the existence of the fulvic-iron-phosphate complex postulated by Golterman (1973, 1975) but support the existence of colloidal $FePO_4$ and $Fe(OH)_3$ peptized by fulvic acids. Apparently, the high pH controls the speciation of Fe to a larger extent than the high metal binding fulvic acid concentration.

A most interesting result of ultra filtration is that almost all the so-called dissolved P and Fe occur in colloidal structures. Probably due to the acidic pH in the molybdenum blue method, this colloidal P fraction is hydrolysed and determined as PO_4–P, thus overestimating the PO_4–P concentration. These results emphasize the usefulness of the ultra filtration technique in the study of the physico-chemistry of alkaline, saline and humic lakes like Tjeukemeer.

Studies comparing the bio-availability of the found colloidal and dissolved P and Fe species are underway.

Acknowledgements

It is impossible to thank all those who contributed to the present knowledge of the physico-chemistry of Tjeukemeer. However, I would like to mention those who played a major role in the investigations during the last decade: I am indebted very much to Dr. H. L. Golterman, whose scientific insight and stimulating ideas contributed greatly to this work. I also thank Dr. J. R. Moed for valuable discussions. Messrs. T. De Boer, H. A. Kramer and J. Voerman are thanked for their large share in the analyses of hundreds of samples which were collected by Mr. L. Lemsma and his crew. Miss Marian McLaughlin kindly helped in typing the manuscript.

References

Beattie, M., Bromley, H. J., Chambers, M., Goldspink, R., Vijverberg, J., van Zalinge, P. & Golterman, H. L., 1970. In: Kajak, Z. & Hillbricht-Ilkowska, A. (Eds.) Proc. IBP-UNESCO Symp., Kazimierz Dolny, Poland, pp. 421–446.

Beattie, D. M., Golterman, H. L. & Vijverberg, J., 1978. An introduction to the limnology of the Friesian lakes. Hydrobiologia 58: 49–64.

Beattie, D. M., 1978. Chironomid populations in the Tjeukemeer. Ph. D. thesis, Rijksuniversiteit Leiden. Drukkerij Zuid-Friesland, The Netherlands.

De Haan, H., 1972a. Some structural and ecological studies on soluble humic compounds from Tjeukemeer. Verh. int. Ver. theor. angew. Limnol. 18: 685–695.

De Haan, H., 1972b. Molecular-size distribution of soluble humic compounds from different natural waters. Freshwat. Biol 2: 235–241.

De Haan, H., 1975. Limnologische aspecten van humusverbindingen in het Tjeukemeer. Ph. D. thesis, Rijksuniversiteit Groningen. Boek- en offsetdrukkerij Zuid-Friesland, The Netherlands.

De Haan, H., 1977. Effect of benzoate on microbial decomposition of fulvic acids in Tjeukemeer (The Netherlands). Limnol. Oceanogr. 22: 38–44.

De Haan, H. & De Boer, T., 1978. A study of the possible interactions between fulvic acids, amino acids and carbohydrates from Tjeukemeer, based on gel filtration at pH 7.0. Water Res. 12: 1035–1040.

De Haan, H. & De Boer, T., 1979. Seasonal variations of fulvic acids, amino acids and sugars in Tjeukemeer, The Netherlands. Arch. Hydrobiol. 85: 30–40.

De Haan, H., De Boer, T. & Halma, G., 1979. Curie point pyrolysis mass-spectrometry of fulvic acids from Tjeukemeer, The Netherlands. Freshwat. Biol. 9: 315–317.

De Haan, H., Halma, G., De Boer, T. & Haverkamp, J., 1981. Seasonal variations in the composition of fulvic acids in Tjeukemeer, The Netherlands, as studied by Curie-point pyrolysis-mass spectrometry. Hydrobiologia 78: 87–95.

De Haan, H., Wanders, J. B. W. & Moed, J. R., 1982a. Multiple addition bioassay of Tjeukemeer water. Hydrobiologia 88: 233–244.

De Haan, H., De Boer, T. & Hoogveld, H. L., 1982b. Metal binding capacity in relation to hydrology and algal periodicity in Tjeukemeer, The Netherlands. Archiv. Hydrobiol. 92: 11–23.

Downes, M. T. & Pearl, H. W., 1978. Separation of two dissolved reactive phosphorus fractions in lakewater. J. Fish. Res. Bd Can. 35: 1636–1639.

Florence, T. M., 1977. Trace metal species in fresh waters. Water Res. 11: 681–687.

Frey, B. E. & Small, L. F., 1980. Effects of micro-nutrients and major nutrients on natural phytoplankton populations. J. Plankt. Res. 2: 1–22.

Golterman, H. L., 1968. Progress Report 1968, Limnological Institute, Nieuwersluis, The Netherlands.

Golterman, H. L., 1969. Progress Report 1969, Limnological Institute, Nieuwersluis, The Netherlands.

Golterman, H. L., 1970. Progress Report 1970, Limnological Institute, Nieuwersluis, The Netherlands.

Golterman, H. L., 1971. Progress Report 1971, Limnological Institute, Nieuwersluis, The Netherlands.

Golterman, H. L., 1972. Progress Report 1972, Limnological Institute, Nieuwersluis, The Netherlands.

Golterman, H. L., 1973. Deposition of river silts in the Rhine and Meuse Delta. Freshwat. Biol. 3: 267–281.

Golterman, H. L. & Soekarjo, R., 1973. Progress Report 1973. Limnological Institute, Nieuwersluis, The Netherlands.

Golterman, H. L. & Soekarjo, R., 1974. Progress Report 1974, Limnological Institute, Nieuwersluis, The Netherlands.

Golterman, H. L., 1974. De biologische productie in het Tjeukemeer. Vakbl. Biol. 7: 112–115.

Golterman, H. L., 1975. Physiological Limnology. Elsevier, Amsterdam.

Golterman, H. L., Voerman, J. & De Nie, H. W., 1980. Fosfaatbelasting van het Tjeukemeer. H_2O 13: 116–121.

Hazen, A., 1892. A new color standard for natural waters. Am. chem. J. 14: 300–310.

Leenen, J. D., 1980. Mengingsverschijnselen in het Tjeukemeer (Friesland). Laboratorium voor Hydraulica en Afvoerhydrologie, Agricultural University, Wageningen, The Netherlands. Nota 46.

Leenen, J. D., 1982. Hydrology of Tjeukemeer. Hydrobiologia 95: 199–203.

Mantoura, R. F. C., Dickson, A. & Riley, J. P., 1978. The complexation of metals with humic materials in natural waters. Estuar. Coast. Mar. Sci. 6: 387–408.

Moed, J. R., Hoogveld, H. L. & Apeldoorn, W., 1976. Dominant diatoms in Tjeukemeer (The Netherlands) II. Silicon depletion. Freshwat. Biol. 6: 355–362.

Moed, J. R. & Hoogveld, H. L., 1982. The algal periodicity in Tjeukemeer during 1968–1978. Hydrobiologia 95: 223–234.

Morel, F. M. M., Rueter, J. G., Anderson, D. M. & Guillard, R. R. L., 1979. Aquil: a chemical defined phytoplankton culture medium for trace metal studies. J. Phycol. 15: 135–142.

Povoledo, D. & Golterman, H. L. (Eds.), 1975. Humic Substances: Their Structure and Function in the Biosphere. Pudoc, Wageningen, The Netherlands.

Rodhe, W., 1949. The ionic composition of lake waters. Verh. int. Ver. theor. angew. Limnol. 10: 377–386.

Schnitzer, M. & Khan, S. U., 1972. Humic Substances in the Environment. M. Dekker, New York.

Stumm, W. & Morgan, J. J., 1970. Aquatic Chemistry. John Wiley, New York.

Vijverberg, J., 1976. The effect of food quantity and quality on the growth, birth-rate and longevity of Daphnia hyalina Leydig. Hydrobiologia 51: 99–108.

Wetzel, R. G., 1975. Limnology. W. B. Saunders, Philadelphia.

White, E. & Payne, G., 1980. Distribution and biological availability of reactive high molecular weight phosphorus in natural waters in New Zealand. Can. J. Fish. aquat. Sci. 37: 664–669.

The algal periodicity in Tjeukemeer during 1968–1978

J. R. Moed & H. L. Hoogveld
Limnological Institute, Tjeukemeer Laboratory, De Akkers 47, 8536 VD Oosterzee, The Netherlands

Keywords: algal periodicity, Tjeukemeer, diatoms, blue-green algae, green algae, biovolume

Abstract

The seasonal periodicity of algal numbers and biomass and of chlorophyll-*a* was investigated in the shallow, eutrophic, alkaline and peaty lake, Tjeukemeer, The Netherlands, during 1968–1978. Diatoms and *Scenedesmus* spp. were abundant throughout the whole period.

Several blue-green algal species, occurring since 1971, usually dominated for most of the year. Relative abundance of algal species in Tjeukemeer during the study period were compared with that in 1955–1960.

Introduction

Tjeukemeer is a shallow, eutrophic, alkaline and peaty lake with a complex hydrology. This lake is part of the Frisian boezem system, which is used to regulate the water table in polders, i.e. low-lying pasture grounds. Thus, Tjeukemeer is filled with superfluous drainage water during the autumn and winter, but with replenishment water from IJsselmeer during the spring and summer. Beattie *et al.* (1972) and Golterman (1974) have described the morphometry, hydrology and hydrochemistry of Tjeukemeer, whereas Beattie (1978) and Beattie *et al.* (1978) have reported similar aspects of the Frisian lakes as a whole. The eutrophic status of Tjeukemeer is confirmed by the maximal concentrations of Tjeukemeer is confirmed by the maximal concentrations of Tot-P_{diss} (0.3 mg ml^{-1}), (NO_3 + NH_3)-N (4–6 mg ml^{-1}) and of SiO_2-Si (5 mg ml^{-1}), measured in 0.2 μm filtered lake water at the onset of the algal growth (Moed 1979). Detailed physico-chemical data are given by De Haan (1982).

Since 1968 a regular counting of the planktonic algae in Tjeukemeer has been carried out, except during May–October 1971; this was initiated within the framework of the International Biological Programme (Golterman 1968). The published data are scanty but information can be found in the annual Progress Reports, on the algal development in 1970 (Golterman 1974), the appearance of the diatom *Diatoma elongatum* (Moed & Hoogveld 1975) and on the wax and wane of diatoms in relation to silicon depletion (Moed *et al.* 1976). Golterman (1974) concluded that no clear patterns in the occurrence of the blue-green algae could be recognized. This paper contains a detailed account of the periodicity of the planktonic algae in Tjeukemeer for the period 1968–1978, in order to test, amongst other things, the validity of Golterman's conclusion for later years.

Materials and methods

In the period 1968–1973, fortnightly samples were taken from the upper 20 cm of three stations (Beattie *et al.* 1972). From 1974 the upper metre of the water column was sampled, while from November 1975 onwards the number of sampling stations was extended to ten (Moed *et al.* 1976). Within two hours of sampling the samples were mixed and treated with Lugol reagent (Schwoerbel 1966).

Hydrobiologia 95, 223–234 (1982). 0018-8158/82/0952-0223/$01.20.

The algae were counted following the method of Utermöhl (1958). For studies on the algal periodicity attention was paid to the abundant species. The term is used arbitrarily depending on whether a density of 1 000 or more individuals per ml was reached at least once in a year. Both cells, threads and colonies are used as counting-units. The data are presented in numbers of individuals demonstrating an algal periodicity.

For a calculation of the algal volume in 1977, the dimensions of 100 individuals of *Oscillatoria agardhii* were measured on 24 dates. The other filamentous algae were measured, similarly, only if their numbers amounted to at least 20% or more of the total numbers of the threads: this was the case for *Oscillatoria redekei* and *Lyngbya* spp. 7 times, *Aphanizomenon flos-aquae* 5 times and for *Planctonema lauterbornii* 11 times. Occasionally algal volumes of the remaining species, if important, were measured. The algal volumes agree reasonably with those of Nauwerck (1963).

It is doubted whether *O. redekei* (Geitler 1932), *Oscillatoria limnetica* (Prescott 1970) and *Lyngbya* spp. (Prescott 1970) are different species. Studies on the taxonomical problems regarding this are in progress.

Chlorophyll-*a* was extracted using 80% ethanol, and heating for 5 min at 75 °C in glass tubes (Moed 1973). The concentration was calculated after Nusch & Palme (1975).

Results

Previous algal data

Preceding the studies at our Institute, Leentvaar (1955a, b, 1956, 1957, 1958, 1963) examined the plankton of Tjeukemeer as a part of his plankton investigations of Dutch lakes. He listed the relative abundance in 3–5 categories (Table 1). We consider the species in category 3 and above as being domi-

Table 1. Plankton studies in Tjeukemeer 1955–1960 (Leentvaar 1955a, b, 1956, 1957, 1958, 1963).

	1955 Sp	1955 Su	1956 Sp	1957 Su	1958 Sp	1958 Su	1958 A	1960 Sp	1960 Su	1960 A
Diatoms										
Melosira spp.	3	1	5	1	1	.	3	.	3	3
Unicell. centr. diat.	.	.	1	.	.	3	3	.	3	3
Asterionella formosa	4	1	5	1	1	3	3	1	.	.
Other Pennatae	.	.	1	1	.	3	3	3	3	3
Green algae										
Scenedesmus spp.	1	1	1	3	3	3	3	3	3	3
Pediastrum spp.	1	1	4	3	3	3	3	3	3	3
Dictyosphaerium sp.	.	.	.	.	.	3	.	3	.	3
Tetraëdron sp.	.	.	.	1	.	.	.	.	3	3
Green filaments	.	.	.	4	.	.	.	3	3	3
Other green algae	.	.	.	.	.	3	.	3	3	3
Blue-green algae										
Microcystis aeruginosa	1	1	2	1	.	3	3	3	3	3
Anabaena sp.	.	1	2	.	.	3	.	3	.	.
Oscillatoria agardhii + *O. redekei*	.	.	.	.	.	.	.	1	.	.
Aphanizomenon flos-aquae	.	.	.	.	.	1	3	3	.	3
Blue-green filaments (incl. *Lyngbya*)	.	4	.	.	.	1	3	1	1	1
Other *blue-green algae* (mainly *Merismopedium* and *Chroococcus*)	1	1	1	1	.	.	3	3	3	3

1 = present
2 = few
3 = many
4 = very many
5 = mass

. = not observed
Sp = spring
Su = summer
A = autumn

N.B.: For 1958 and 1960 only 1, 3 and 5 are used!

nant. Species dominant in at least two years were: the blue-greens, *Microcystis aeruginosa* (1958, 1960), *Anabaena* sp. (1958, 1960) and *Aphanizomenon flos-aquae* (1958, 1960); the green algae, *Scenedesmus* spp. (1957, 1958, 1960), *Pediastrum* spp. (1956–1958, 1960) and *Dictyosphaerium* sp. (1958, 1960); and the diatoms, *Melosira* spp. (1955, 1956, 1958, 1960) and *Asterionella formosa* (1955, 1956, 1958).

For comparison with the present algal densities, Leentvaar supplied us – in 1980 – with samples taken in the summer and autumn of 1960. The density of *Scenedesmus* spp., 200 colonies ml^{-1}, was the highest. The near-absence of *M. aeruginosa* and *A. flos-aquae* disagrees with Leentvaar's data (Table 1). Furthermore, our observations suggest that in 1980 algal densities in Tjeukemeer were around two orders of magnitude higher than they were in 1960. However, in 1960 water was sampled with a bucket at the shore and filtered through a 26 μm net (Leentvaar, pers. comm.). Therefore a quantitative comparison of data of 1960 with the recent ones is not possible.

Abundant species in the period 1968–1978

To establish algal periodicity in Tjeukemeer, the species having 'abundant density' (see Materials and methods) in three or more years in the period 1968–1978 were selected; their courses and also that of *Asterionella formosa*, one of the main dominant diatoms about 15 years earlier (Table 1), are plotted in Figs. 1 and 2. To complete the picture of the abundant algae, the unspecified species are enumerated in Table 2, together with those abundant for less than 3 years.

Among the diatoms, *A. formosa* occurred in low densities until 1977 and 1978, when this species became abundant. *Diatoma elongatum* was abundant from the autumn of 1971 onwards (Moed & Hoogveld 1975), while *Melosira* spp. and unicellular centric diatoms were abundant throughout the whole period.

Several species contributed to the annual main maximum of *Melosira*. Boesewinkel-De Bruyn (1971) reported that during 1968–1971 *M. italica* was the main species, whereas *M. granulata* was

Table 2. Unspecified groups of algae exceeding a density of 1 000 ind. ml^{-1} (abundant) in Tjeukemeer in the period 1968–1978, dates in parentheses. Species abundant for less than 3 years are also listed.

	1969	1970	1971	1972	1973	1974	1975	1976	1977	1978
Unspecified pennate diatoms	2 650 (16.04)	3 800 (22.04)	4 800 (31.03)	1 380 (15.03)	2 200 (28.03)	2 100 (22.03)	1 640 (07.04)			
Unspecified green flagellates	1 080 (21.05)			3 500 (12.07)	2 900 (28.03)			3 800 (13.04)	1 360 (03.05)	7 400 (06.06)
Green uni-cells				4 650 (13.06)		1 480 (07.05)	3 220 (29.07)	3 900 (20.03)		1 700 (05.07)
Unspecified blue-green filaments				3 045 (14.11)	3 620 (09.10)	30 360 (05.06)		5 920 (25.05)		6 520 (02.05)
Cryptomonas spp.					1 450 (30.01)					
Ankistrodesmus sp.										5 900 (06.06)
Lagerheimia spp.		2 620 (10.06)					1 120 (15.07)			
Oocystis spp.		1 920 (03.06)					1 600 (15.07)			
Tetraëdron spp.		1 000 (10.06)								
Tetrastrum staurogeniaeforme			1 720 (14.04)	1 120 (17.04)						
Anabaena flos-aquae							1 980 (26.08)			1 150 (18.07)
Microcystis aeruginosa							1 080 (12.08)			

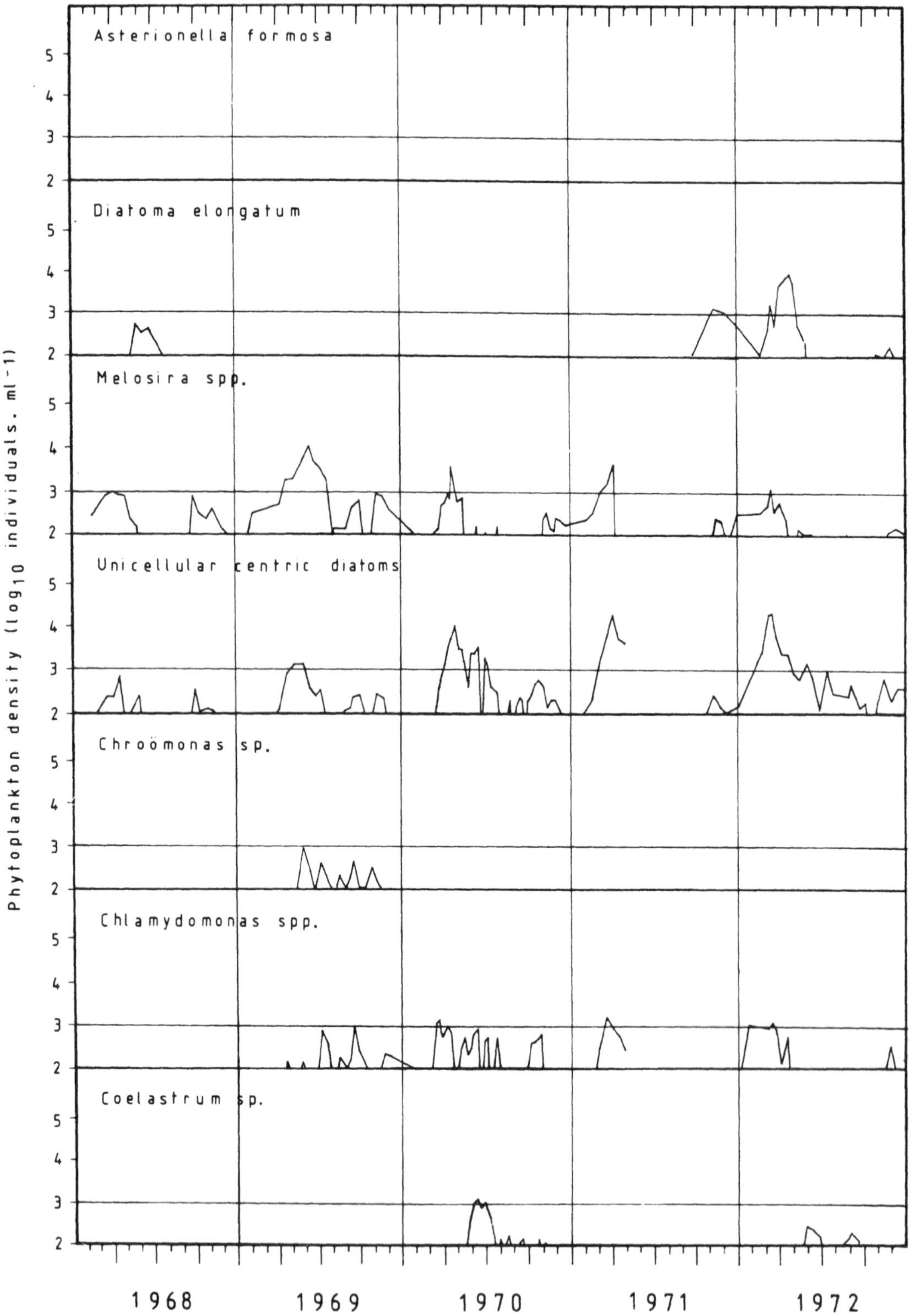

Fig. 1. Seasonal changes of the diatoms, the Cryptophyta and some green algae, abundant in at least three years in Tjeukemeer

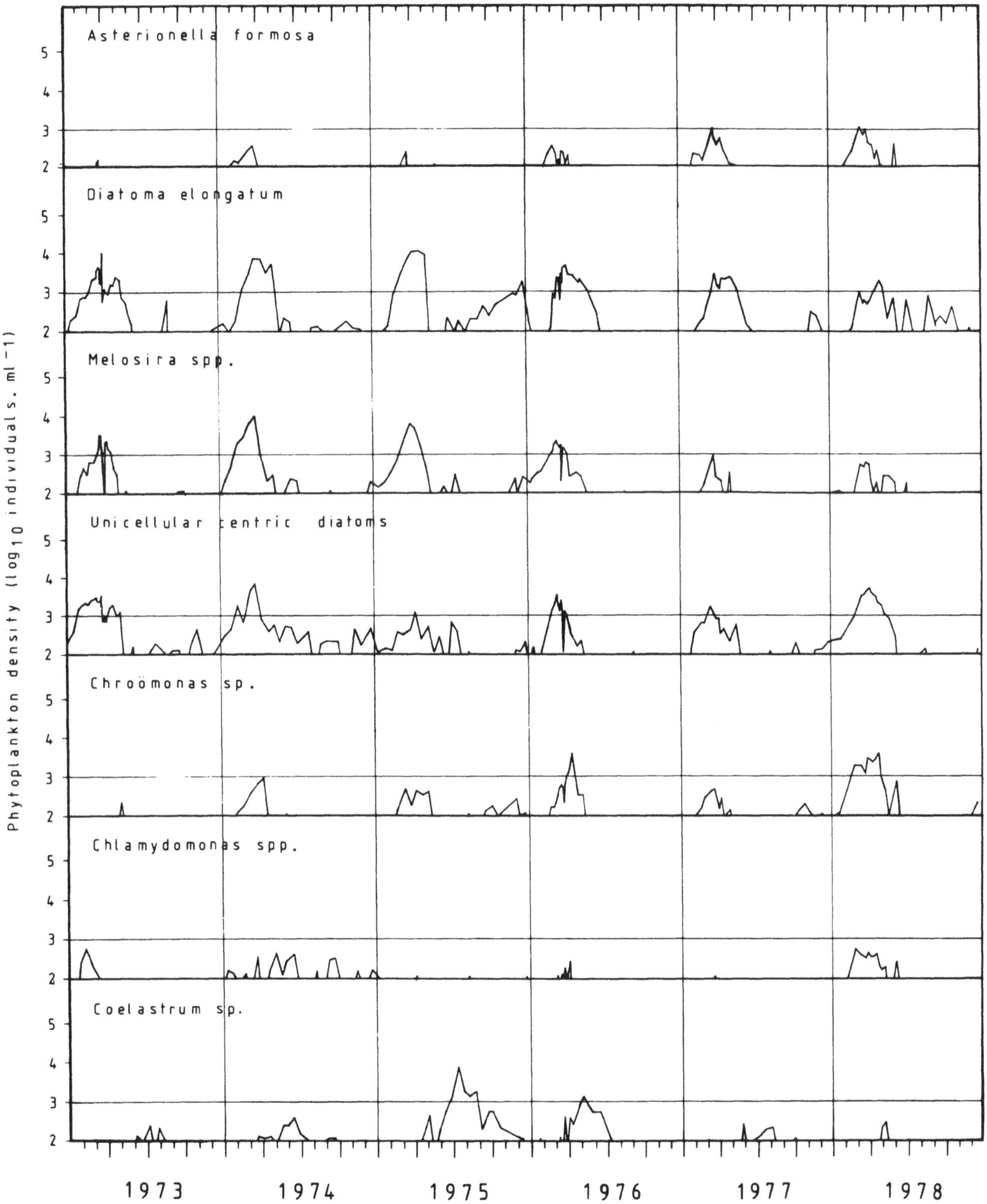

in the period 1968–1978; *Asterionella formosa* is also included.

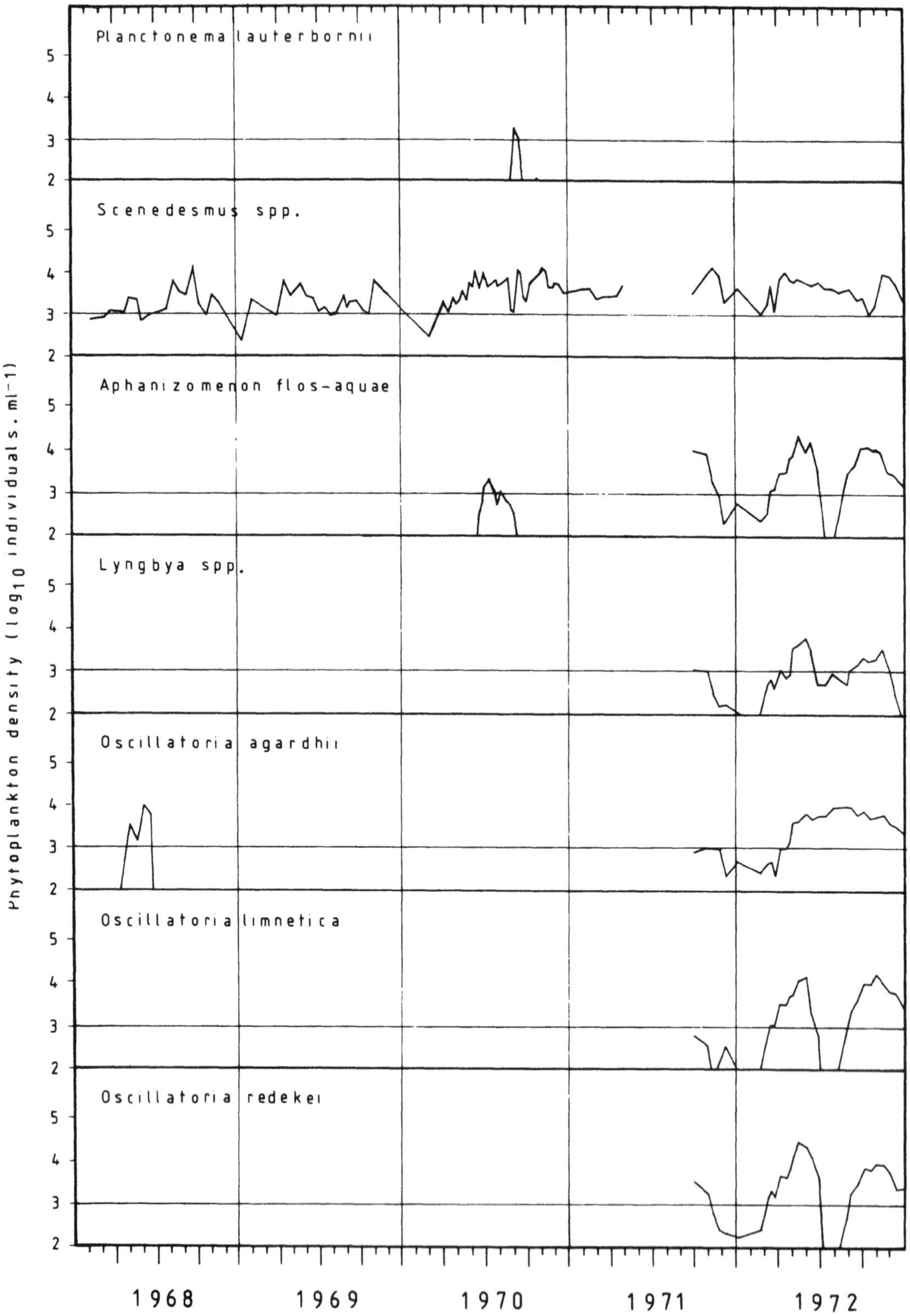

Fig. 2. Seasonal changes of the blue-greens and rest of the green algal species (see Fig. 1), abundant in a least three years

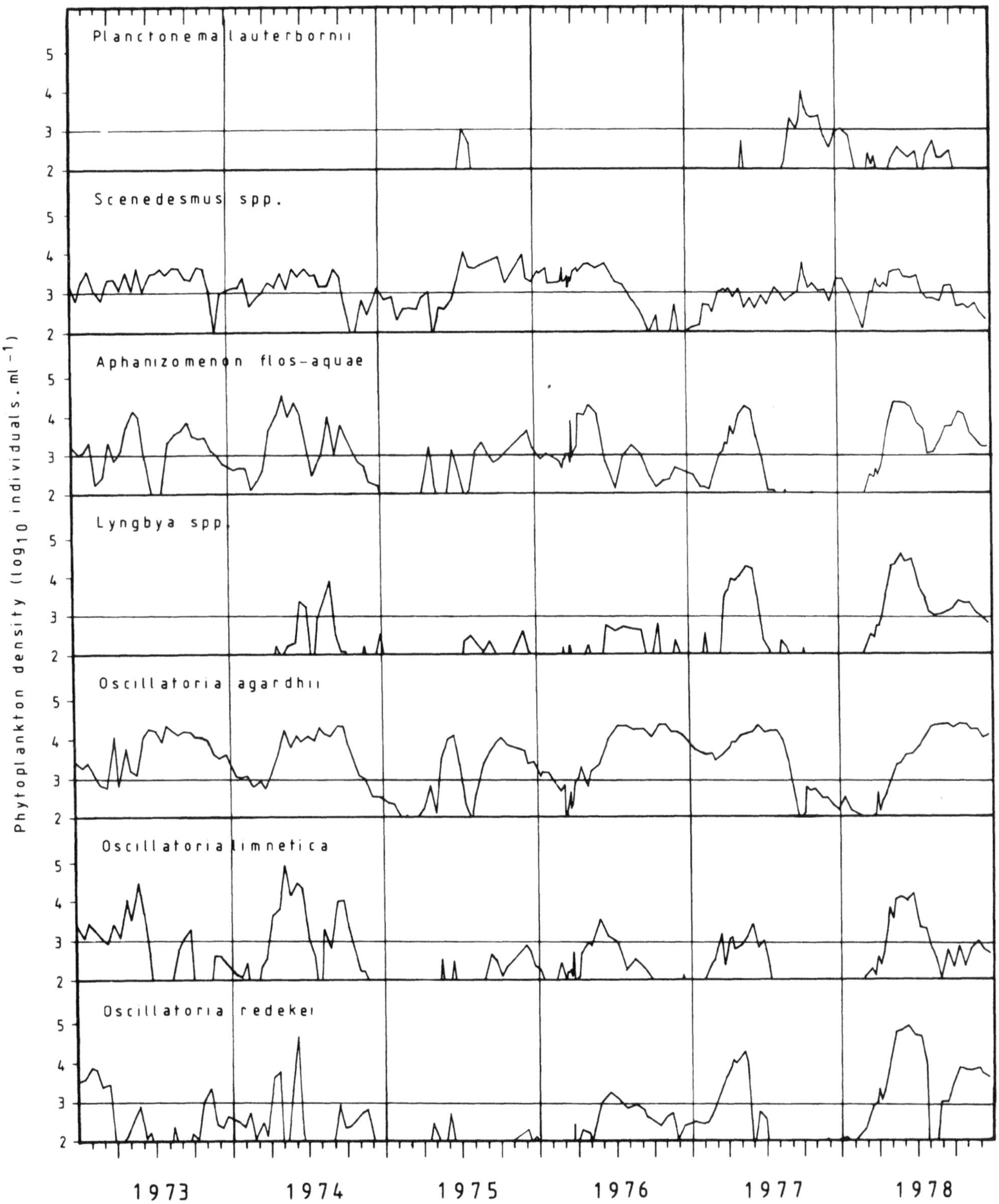

in Tjeukemeer in the period 1968–1978.

observed quite often. During 1972–1975 *M. italica* was also the main contributor, while *M. islandica* is a good second now (maximally 25%). The situation was the opposite in 1976 and 1977, but the contribution of both species was roughly similar again in 1978. Before 1972 the annual main maximum of the centric diatoms was comprised of *Actinocyclus normanni, Stephanodiscus astraea* and *S. astraea* var. *minutula* (Boesewinkel-De Bruyn 1971). Since 1972 the latter species have contributed 75% to the total numbers of the unicellular diatoms.

The Cryptophyceae consisted mainly of *Chroomonas* sp. and several species of *Cryptomonas*. In the period 1970–1972 these species were not distinguished from one another. However, recent re-investigations indicate abundance of *Chroomonas* sp. (3 800 cells ml^{-1}) in April 1972, while this species also reached abundant numbers in 1974, 1976 and 1978. *Cryptomonas* spp. were abundant in 1973 only (Table 1).

Chlamydomonas spp., *Coelastrum* sp. and a green filamentous alga, presumably *Planctonema lauterbornii*, became abundant in 1970, 1975 and 1977. Regarding the dominance of *Pediastrum* spp. in 1956–1960 (Table 1), the annual maxima during 1968–1978 varied between 300 and 900 colonies ml^{-1}. *Scenedesmus* spp. were the most numerous green algal group with abundant numbers throughout the whole 1968–1978 period. Also, the unspecified green flagellates and green unicells were abundant for several years (Table 2).

In the period 1971–1978, the blue-green algae *Aphanizomenon flos-aquae, Lyngbya* spp., *Oscillatoria agardhii, O. limnetica* and *O. redekei* were more or less abundant. Data on densities of blue-green algae in 1971 are partly missing due to sampling errors. In previous years *A. flos-aquae* was dominant in 1958 and 1960 (Table 1), whereas we observed abundance in 1970. In the period before 1970 *O. agardhii* was only abundant for a short period in 1968. The numbers of unspecified blue-green filaments were surprisingly high in 1974 (Table 2), due to identification problems. *Anabaena flos-aquae* and *Microcystis aeruginosa* were dominant in 1958 and 1960 (Table 1) but in the period 1968–1978 abundance was reached not earlier than 1975 when the usual blue-greens developed badly.

Algal periodicity

The lake has shown a pronounced algal periodicity since 1972 because most of the abundant species from year to year are the same, and their periods of maxima correspond from year to year. The diatoms developed mainly in February–March, although in a few cases the numbers were also relatively high in autumn (Fig. 1). The maxima of abundant diatoms were in March or April. This also holds good for the unspecified diatoms (Table 2). The main maximum of *Diatoma elongatum* is observed most often a few weeks later than that of *Melosira* spp. and the unicellular centric diatoms.

Chroomonas sp., the main representative of the Cryptophyta, also had maximal numbers in April (Fig. 1). In contrast, the maximal numbers of the various green algal species did not occur in the same short period. For *Chlamydomonas* spp. and *Coelastrum* sp. it was in March and May–June, respectively (Fig. 1). In 1970, 1975 and 1977 a green filamentous alga, probably *Planctonema lauterbornii*, exceeded a density of 1 000 filaments ml^{-1} in July–August when the usual abundant blue-greens were absent (Fig. 2). *Scenedesmus* spp. behaved differently in that they had several maxima in various periods (Fig. 1). This may be caused by grazing by zooplankton.

Since 1972, the abundant blue-greens have developed certain patterns. The onset of the blue-green algal growth was observed generally in March, often a few weeks later than that of the diatoms. This difference is conceivably owing to temperature or light or both since during this period nutrient concentrations in the lake are high. The maxima are observed in May–June, whereas in July the number of filaments are very low (Fig. 2). Thereafter, a second phase can often be observed with maxima in September–November and a relatively low number during winter. The periodicity of *Oscillatoria agardhii* differs from that of the other blue-greens. The onset of growth is often observed somewhat later. Also, this alga is capable of maintaining high numbers during the summer. Only in July 1975 and August 1977 was hardly any filament present. It was in these periods that considerable amounts of the green filamentous *P. lauterbornii* were observed.

Considering the period 1968–1978, *Melosira* spp., unicellular centric diatoms and *Scenedesmus* spp. were abundant almost every year, but for the diatoms only, the maximal numbers were restricted to a relatively short period (March–April). In contrast, *Scenedesmus* spp. showed maximal numbers in one or more seasons. Therefore the algal periodicity during 1968–1978 was restricted to *Melosira* spp. and the unicellular centric diatoms. For the period 1972–1978, *Diatoma elongatum, Oscillatoria agardhii, O. redekei, O. limnetica, Aphanizomenon flos-aquae* and *Lyngbya* spp. should also be included. In this period *Chroomonas* sp. became abundant in alternate years and exhibited algal periodicity as well.

Algal volume and chlorophyll

Since 1971 concentrations of blue-green algae have increased greatly with main maxima usually in May–June, except for *O. agardhii,* which has a maximum a few months later. The diatoms had maxima in the March–April period. These conclusions are derived from data on numbers per ml, ignoring the volume differences between both species as well as between individuals of the same species. It was treated this way since data on biovolume or dry weight are lacking. The volume contribution (%) of diatoms, Cryptophyta, blue-greens and green algae in different seasons can be given for 1977 only (Fig. 3). During February–March 1977 diatoms increasingly contribute, followed by the blue-greens and green algae. Such a sequence agrees with the quantitative indications obtained from the numbers present in 1977. However it should be emphasized that Fig. 3 does not reveal the usual dominant position of the blue-greens during August–December, because in this period in 1977 a bloom of a green filamentous algae, probably *P. lauterbornii,* developed.

Also it was checked whether the course of changes in the diatoms and blue-greens agreed with that of the chlorophyll-*a* concentration (Fig. 4), which is often used as a biomass parameter. Data for 1968 are scanty. In 1969 and 1970 chl-a concentration reached 100 mg m^{-3} maximally except for higher value measured in April 1970. This high value probably relates to a diatom maximum. In the second half of 1971 the values were now and then higher than 100 mg m^{-3}. In the following years, ignoring the March–April period with its relatively high numbers of diatoms, maximal chl-a concentrations were 175–275 mg m^{-3} (May–June), except for 1975. Thus the changes in the blue-greens (Fig. 2), namely low numbers in 1968–1970, increasing numbers in 1971 and relatively high numbers dur-

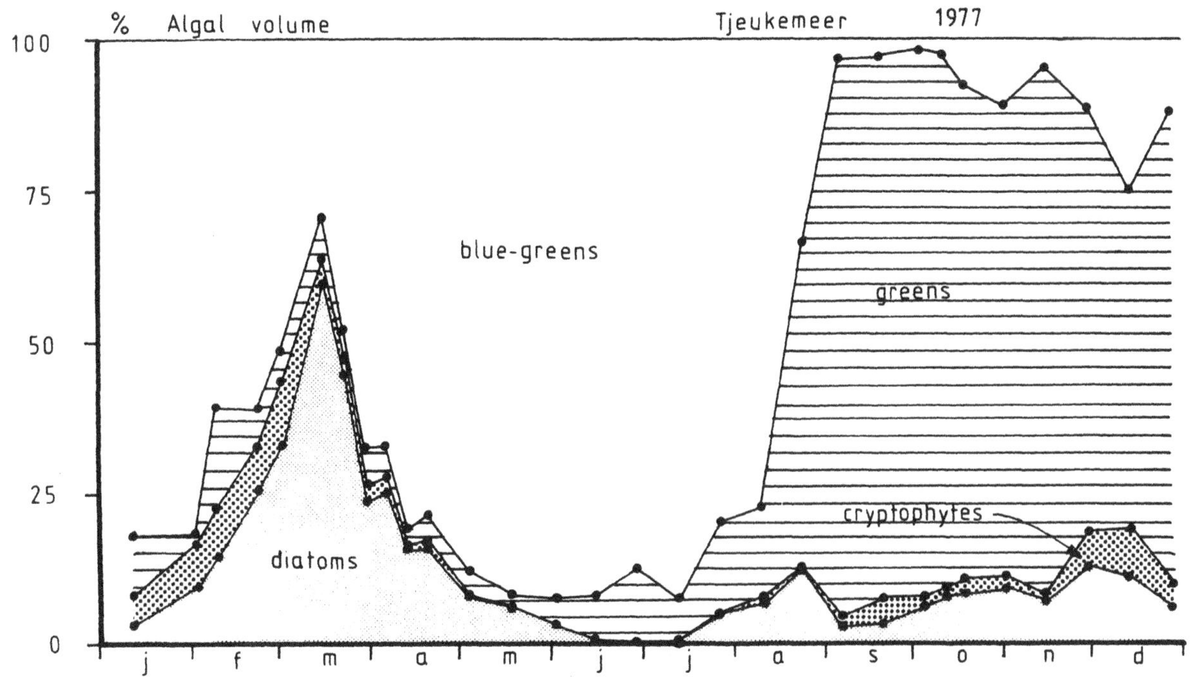

Fig. 3. Per cent contribution of the various algal groups to the total algal volume in Tjeukemeer in 1977.

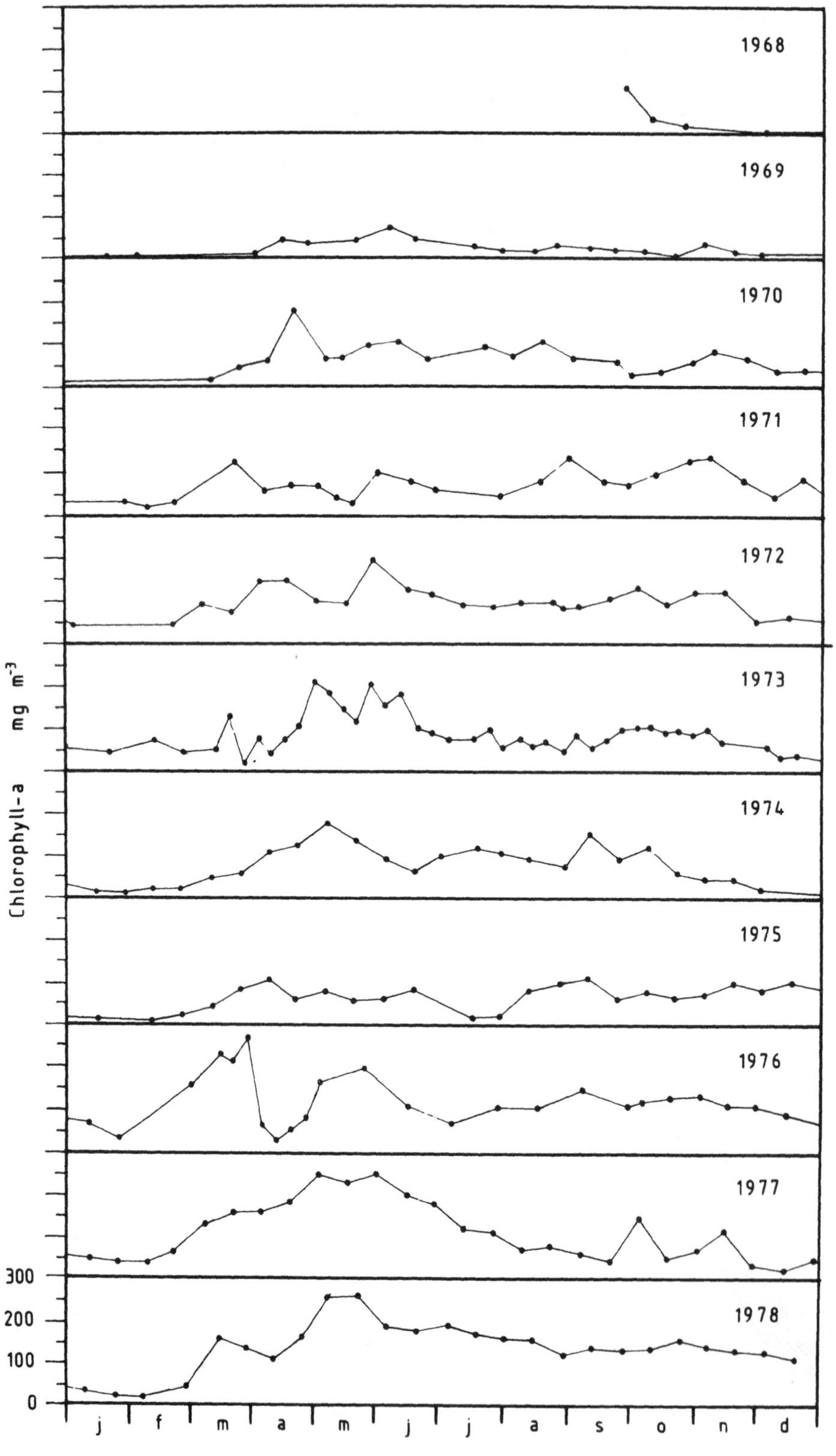

Fig. 4. Changes in chlorophyll-*a* concentrations in Tjeukemeer in the period 1968–1978.

ing 1972–1978 (except for 1975), correspond to the general course of chl-a in those years. In short, the main chl-a maximum is observed around the annual maximum of blue-green filaments in May–June. The diatoms contribute significantly to the total algal numbers, with maxima in the March–April period, and as expected, a maximum or plateau of chl-a concentration generally occurred.

Discussion

Summarizing the data on algal abundance the most striking change in algal composition concerns the blue-greens. A striking increase in algal filaments was observed from autumn 1971 onwards. A similar increase in number of cells showed the diatom *Diatoma elongatum*, previously only listed as 'many' in 1958 (Table 1). *Asterionella formosa* on the other hand reached large numbers in 1956 and 1958, but seemed to show a decrease in abundance in later years with only a reasonable numbers of cells in 1977 and 1978. An explanation for these phenomena is difficult. P- and N-concentrations did not change much throughout the observation period. In the winter of 1968 the concentrations of NH_3-N, PO_4-P and Tot-P already amounted to 2, 0.3 and 0.75 mg l^{-1}, respectively (Golterman 1969). A major factor for low numbers of blue-green algae might have been sand suction in 1968 and 1969 in the western part of the lake. This caused a tremendous turbidity and consequently reduction of light penetration in the lake (Bromley 1970). Also, Berger (1975) mentions sand suction as a possible reason for the disappearance of *Oscillatoria agardhii* from the Eemmeer between July 1972 and May 1974. Another factor might have been an increase in inflow of IJsselmeer water during 1971 compared with previous years (Moed & Hoogveld 1975). However, we do not know which factor in this IJsselmeer water has eventually formed growth conditions for blue-greens and *D. elongatum*. More study is needed to evaluate the role of the specific hydrology of the lake. In this respect the presence of a deep depression in the lake of about 5–10 m (area 20 ha, i.e. around 1% of the lake's surface area) may play an important role for some species to escape from washing-out during periods with heavy rainfall in autumn and winter, when the growth is greatly reduced.

Analyses of algal numbers, algal volumes in 1977 and chlorophyll-*a* suggest that since 1972 blue-green algae have dominated in both number and volume for by far the greatest part of the year. This is in agreement with later data (Hoogveld 1980, 1981).

Golterman's conclusion (1974) that no clear pattern in the occurrence of the blue-green algae could be recognized need to be revised. Our results showed an onset of the blue-green algal growth generally in March with maxima in May–June. A second maximum often occurred in September–November. The periodicity of *O. agardhii* differs distinctly from this pattern.

The mass-development of diatoms and blue-greens in the first part of the year is conceivable in relation to the nutrient loading of the lake. During the previous wet season the lake is filled with nutrient rich water, mainly originating in agricultural areas (De Haan 1982). In agreement with this, maxima of blue-green filaments and chlorophyll-*a* concentrations in May–June coincide with relatively low nutrient concentrations in the lake. During summer mainly *O. agardhii* and *Scenedesmus* spp. can maintain relatively high numbers, perhaps related to increase in mineralization rates.

The explanation of algal periodicity in Tjeukemeer is the objective of a broad study, which includes the role of nutrient availability and limitation, temperature, light, and the lake hydrology (Moed 1979; De Haan *et al.* 1982).

Summary

The seasonal periodicity of algal numbers and biomass and of chlorophyll-*a* was investigated in the shallow, eutrophic, alkaline and peaty lake Tjeukemeer, during 1968–1978.

With regard to species composition and algal numbers, two periods can be distinguished, viz. 1968–1970 and 1972–1978. The diatom *Melosira* spp., unicellular centric diatoms and green algae (*Scenedesmus* spp.) were abundant during the whole period, while the diatoms showed a distinct periodicity. This holds good for the diatom *Diatoma elongatum*, the cryptophyte *Chroomonas* sp. and the blue-greens *Oscillatoria agardhii*, *O. redekei*, *O. limnetica*, *Lyngbya* spp. and *Aphanizomenon flos-aquae* in the latter period (Figs. 1 and 2). Possible explanations for the increase in abundant species from 1972 onwards are given. In the years

1955–1960 (Table 1), *Melosira* spp., *Scenedesmus* spp. and *A. flos-aquae* were dominant in at least some of the years.

The determinations of algal numbers, algal volumes and chlorophyll-*a* indicate that since 1972 the diatoms have contributed more to the total algal volume in February and March, and the blue-greens in the rest of the year. A striking exception was observed in 1977 (Fig. 3), when *Planctonema lauterbornii* occurred in mass numbers during a period in which the blue-greens failed to grow.

Acknowledgements

We are grateful to Dr. H. L. Golterman, who started the routine algal studies and to Mrs. P. J. Boesewinkel-De Bruyn, Mr. T. H. Frank and Drs. J. H. Coers, who counted the samples in some years. Special thanks are due to Mr. L. Lemsma for sampling in the field. We are also indebted to Mr. N. J. van Benthem, for photographic work, and to Dr. H. de Haan, Dr. R. D. Gulati and Dr. S. Parma for their comments on the manuscript.

References

Beattie, D. M., 1978. Chironomid populations in the Tjeukemeer. Ph. D. thesis, University of Leiden. 150 pp.

Beattie, D. M., Bromley, H. J., Chambers, M. R., Goldspink, R., Vijverberg, J., Zalinge, N. P. van & Golterman, H. L., 1972. Limnological studies on Tjeukemeer – a typical Dutch polder reservoir. In: Kajak, Z. & Hillbricht-Illkowska, A. (Eds.) Proc. I.B.P.-Unesco Symposium on Productivity Problems of Freshwaters, pp. 421–447. Kazimierz Dolny, Poland.

Beattie, D. M., Golterman, H. L. & Vijverberg, J., 1978. An introduction to the limnology of the Friesian lakes. Hydrobiologia 58: 49–64.

Berger, C., 1975. De eutrofiëring en het voorkomen van Oscillatoria agardhii Gom. in de randmeren van Flevoland. H_2O 17: 340–350.

Boesewinkel-De Bruyn, P. J., 1971. Verslag van het phytoplanktononderzoek op het Tjeukemeer in de jaren 1968–1970. Internal Report No. 1971–1 (in Dutch), Limnological Institute, Tjeukemeer Laboratory, Oosterzee (The Netherlands).

Bromley, H. J., 1970. A study of the zooplankton of Tjeukemeer. Student report 1970–4, Limnological Institute, Tjeukemeer Laboratory, Oosterzee (The Netherlands).

De Haan, H., 1982. Physico-chemical environment in Tjeukemeer with special reference to speciation of algal nutrients. Hydrobiologia 95: 205–221.

De Haan, H., Wanders, J. B. W. & Moed, J. R., 1982. Multiple addition bioassay of Tjeukemeer water. Hydrobiologia 88: 233–244.

Geitler, L., 1932. Cyanophyceae. In: Rabenhorst's Kryptogamen-Flora, 14.Bd. Akademische Verlagsgesellschaft, Leipzig. 1196 pp.

Golterman, H. L., 1968. In: Netherland's participation in the International Biological Programme, Progress Report 1966–1967, pp. 29–31. Netherland's Committee for the I.B.P., North-Holland Publishing Company, Amsterdam.

Golterman, H. L., 1969. Annual report of the Limnological Institute, Nieuwersluis, for the year 1968. Verh. k. nederl. Akad. Wet. Afd. Natuurk. rks. 2, 58: 33.

Golterman, H. L., 1974. De biologische produktie van het Tjeukemeer. Vakbl. Biol. 54: 112–116.

Hoogveld, H. L., 1980. Fytoplankton van het Tjeukemeer, 1979. Internal report No. 1980–5 (in Dutch, with English summary), Limnological Institute, Tjeukemeer Laboratory, Oosterzee (The Netherlands).

Hoogveld, H. L., 1981. Fytoplankton van het Tjeukemeer, 1980. Internal report 1981–11 (in Dutch, with English summary), Limnological Institute, Tjeukemeer Laboratory, Oosterzee (The Netherlands).

Leentvaar, P., 1955a. Verslag van twee bezoeken in april en juni aan het Tjeukemeer. Internal Report (in Dutch), RIVON, Leersum, 4 pp.

Leentvaar, P., 1955b. Hydrobiologische waarnemingen in Zuid-Friesland. Levende Nat. 58: 180–184.

Leentvaar, P., 1956. Voorjaarsplankton van enige Nederlandse meren. Levende Nat. 59: 132–137.

Leentvaar, P., 1957. Plankton van de Nederlandse Meren, Zomer 1957. Internal report (in Dutch), RIVON, Leersum.

Leentvaar, P., 1958. Plassenplankton 1958. Internal report (in Dutch), RIVON, Leersum.

Leentvaar, P., 1963. Resultaten van het hydrobiologisch onderzoek van oppervlaktewater in 1960. Water 47: 203–207.

Moed, J. R., 1973. Effect of combined action of light and silicon depletion on Asterionella formosa Hass. Verh. int. Verein. Limnol. 18: 1367–1374.

Moed, J. R. & Hoogveld, H. L., 1975. Dominant diatoms in Tjeukemeer. The appearance of Diatoma elongatum (Lyngb.) Agardh. Freshwat. Biol. 5: 159–165.

Moed, J. R., Hoogveld, H. L. & Apeldoorn, W., 1976. Dominant diatoms in Tjeukemeer (The Netherlands). II. Silicon depletion. Freshwat. Biol. 6: 355–362.

Moed, J. R., 1979. Onderzoekprogramma (1979–1983) van de Werkgroep Algologie (in Dutch). Nota 1979–12, Limnological Institute, Tjeukemeer Laboratory, Oosterzee (The Netherlands).

Nauwerck, A., 1963. Die Beziehungen zwischen Zooplankton und Phytoplankton im See Erken. Symb. bot. ups. 17: 5. 163 pp.

Nush, E. A. & Palme, G., 1975. Biologische Methoden für die Praxis der Gewässeruntersuchung. Gwf-Wasser/Abwasser 116, H. 12: 562–565.

Prescott, G. W., 1970. Algae of the Western Great Lakes Area. W. M. C. Brown, Dubuque, Iowa. 977 pp.

Schwoerbel, J., 1966. Methoden der Hydrobiologie. Kosmos Gesellschaft der Naturfreunde. Franckh'sche Verlagshandlung, Stuttgart. 207 pp.

Utermöhl, H., 1958. Zur Vervollkommnung der quantitativen Phytoplankton Methodik. Mitt. int. Verein. Limnol. 9: 1–38.

Population dynamics and production of *Daphnia hyalina* Leydig and *Daphnia cucullata* Sars in Tjeukemeer

J. Vijverberg and A. F. Richter
Limnological Institute, Tjeukemeer Laboratory, De Akkers 47, 8536 VD Oosterzee, The Netherlands

Keywords: population dynamics, secondary production, Cladocera, *Daphnia,* simulation

Abstract

As part of a research programme on the food chains in Tjeukemeer, the *Daphnia hyalina* and *Daphnia cucullata* populations were studied for three successive years. To analyse the factors regulating the production of these two species, their population parameters (density, size distribution, fecundity) and population dynamics (birth rate, mortality rate) were studied and related to environmental factors. Since *Daphnia* in Tjeukemeer shows continuous recruitment, the population dynamics model 'INSTAR' was developed and used to integrate field data with laboratory data on development rates and length–weight relationships. The dynamics of the *Daphnia* species are mainly regulated by temperature and fish predation, the latter affects both birth rate and mortality. Total annual *Daphnia* production was 3.1–6.9 g org. dry wt m^{-2}, and annual P/B ratio ranged from 25 to 40 for *D. cucullata* and from 45 to 49 for *D. hyalina.*

Introduction

This article presents part of a study on the production of copepods and cladocerans in Tjeukemeer, a shallow eutrophic polder reservoir in the north of the Netherlands. Earlier articles have dealt with the chemical composition and energy contents of copepods and cladocerans (Vijverberg & Frank 1976), the effect of food quantity and quality on a laboratory population of *Daphnia hyalina* (Vijverberg 1976), the population structure, life histories and abundance of copepods (Vijverberg 1977) and the effect of temperature in laboratory studies on development and growth of copepods and cladocerans (Vijverberg 1980).

The Cladocera represent 80–85% of the mean annual zooplankton biomass. Of the eight common cladoceran species, *Daphnia hyalina* is the most important in terms of biomass. This species is, after *Leptodora kindtii* whose densities are generally low, the largest zooplankton species present. Tjeukemeer is inhabited by a large population of planktivorous fish, consisting of both young (0+) and older fish. The most important zooplankton predator among the 0+ fish is the smelt *Osmerus eperlanus* and among the older fish is the bream *Abramis brama* (Hazelzet 1972; van Densen 1978; Goldspink 1978). As these fish feed positively size selectively *Daphnia hyalina* is the most important fish-food species of the zooplankton community. Compared with *D. hyalina, D. cucullata* is a smaller and generally less abundant species.

In order to analyse the factors regulating the production of these two species, both their population parameters (fecundity, size distribution) and population dynamics were studied and related to environmental factors. Since cladocerans in Tjeukemeer show continuous recruitment, a population dynamics model is used to integrate field data with laboratory data on development rates and length–weight relationships.

Hydrobiologia 95, 235–259 (1982). 0018-8158/82/0952-0235/$05.00.

Study area

Tjeukemeer is a shallow (mean depth = 1.5 m), eutrophic freshwater lake with a surface area of about 21 km^2. The littoral zone is poorly developed; the index of shoreline development (D_l) is 1.25 (Hutchinson 1957). Tjeukemeer is part of an interconnected system of water bodies (the Frisian 'boezem') which act as water reservoirs for the surrounding polders (Beattie *et al.* 1978). The replacement time of the water in the lake is 3-6 months. For a more detailed description of the lake and the position of the sampling stations, see de Nie *et al.* (1980).

Materials and methods

Field data

For three years (1969–1971), water samples were taken from ten fixed stations at weekly intervals, except during ice cover. This resulted in 130 collecting dates. Samples were taken with a 5 l cylindrical Friedinger closing sampler. At each station two samples were taken, one just below the surface and the other just above the bottom. The samples were concentrated by filtration through 120 μm mesh plankton gauze and preserved in 4–5% formalin. All samples collected on one date were combined and sub-sampled with the whirling apparatus of Kott (1953). On each collecting date one horizontal tow net sample was collected. This material was preserved in 95% ethanol to prevent loss of eggs (Haney & Hall 1973) and was only used for determination of fecundity.

The samples, collected on each date, were processed in the following ways:

1. To identify species, and for counting, 1/10 of the original mixed sample, representing 10 l of lake water, was completely examined. Densities for each sequential day were calculated by linear interpolation between sampling dates.

2. To identify the size distribution, from a mixed sample usually 100 individuals of each species were measured. Length measurements were made under the microscope with a micrometer eyepiece to the nearest 0.03 mm. Individuals were measured from the top of the head or, if a 'helmet' was present, from the base of the helmet to the base of the tail spine.

3. To determine fecundity the population was divided into arbitrarily defined size classes: 0.77–0.94, 0.95–1.12. 1.13–1.30, 1.31–1.48, 1.49–1.66, 1.67–1.84, 1.85–2.02 and 2.03–2.20 mm. Generally 10 but at least 4 ovigerous females were examined from each size class containing adult females. We did not differentiate between eggs and embryos, both being counted as eggs. The relation between the number of eggs per ovigerous female (F) and the length (L, mm) of the female is described by:

$$F = a\,L^b \quad (1)$$

or

$$\ln F = \ln a + b \ln L \quad (2)$$

where ln a is the intercept and b the regression coefficient. Eq. (2) is an ln–ln transformation of the power function Eq. (1), and is used to facilitate regression calculations. In the calculations we used the mean egg number per ovigerous female per size class and the mean length of each size class.

Only parthenogenetic egg counts were used for these calculations. In the model for each sequential day the numbers of eggs for each size of female were calculated by linear interpolation between sampling dates. The percentage of ovigerous females in relation to total adult females was calculated by examination of at least 100 adult females from each species. Parthenogenetic and resting (sexual) eggs were distinguished.

Water temperature was measured daily. Phytoplankton densities, chlorophyll-*a* content and primary production were determined bi-weekly. Chlorophyll-*a* concentrations were measured in ethanol according to procedure 7.8.2. IBP manual nr. 8 (Golterman 1970). Gross primary production was measured using the ^{14}C method (Beattie *et al.* 1972).

Laboratory estimates

The mean length increment per week per size class in relation to temperature and the duration of egg development in relation to temperature were assessed under semi-natural conditions in the laboratory. The results are given in Vijverberg (1980). Length increments per week were obtained at temperatures of 2.5, 5, 10, 15, 20 and 25 °C. Growth at intermediate temperatures was calculated by linear interpolation.

During the development of the population dynamics model it became evident that we needed additional experimental information to enable more realistic simulations. Firstly we needed to know the duration time of juvenile and adult instars in relation to temperature. Since the duration time of the adult instars at a given temperature is constant and approximately equal to the duration of egg development (Vijverberg 1980), it was sufficient to assess the duration of the juvenile instar as a percentage of the adult instar at the same temperature (17.5 °C). To identify instars, their length was observed every 12 h; apart from this, methods and culture conditions were the same as those described by Vijverberg (1980). Secondly we needed more precise information about the size of the new born. At 17.5 °C we cultured *D. hyalina* and *D. cucullata* under standard conditions and related the length of new born to the length of the mother.

Length–weight relationships were assessed (see Vijverberg & Frank 1976). Weight per individual is expressed as μg org. dry wt, this value being exclusive of both ash weight and chitine weight. During the filtering procedure, which preceded the chemical determinations, essentially all eggs were lost from the broodchambers, so the weight (W) in the equation:

$$W = c\, L^{d} \tag{3}$$

represents somatic weight only.

Population dynamics model

The technical aspects of the model 'INSTAR' were developed by the second author (A.F.R.) in cooperation with Dr. P. Hogeweg (Department of Bio-Informatics, University of Utrecht, Utrecht, The Netherlands). The details of the structure of the computer programme together with a discussion on the value of this type of approach compared with other models already in use is given by Hogeweg & Richter (1982). Consequently we will confine ourselves here to outlining the structure of the model and to summarizing its most important properties. The model is designed to simulate:

1. reproduction,
2. individual growth (length increment per moult),
3. density changes per size class owing to birth, death or growth.

The model used here differs from the more commonly used continuous simulation models where all variables of the system are transformed simultaneously at each timestep (Argentesi *et al.* 1974; de Wit & Goudriaan 1974). It is a discrete event oriented simulation model written in SIMULA-67 (Dahl *et al.* 1970). In this type of simulation an event is the transformation of some of the variables of the model system (Hogeweg 1974). Which parameters are transformed differs per event. Between events nothing happens in the model system; the time elapse between events need not be constant but can itself be a variable of the system. Since growth by moulting is a discontinuous process, event-orientated simulation is especially suited for the simulation of growth in cladocerans. The same applies for life cycle phenomena such as egg production, egg development and hatching of new born, which are closely related to moulting in cladocerans. The model is based on the life cycle events of an individual *Daphnia* (Fig. 1). Output was organized per size class, but individuals in the model system kept acting as independent entities. The *D. hyalina* population was divided into four, the *D. cucullata* population into three size classes (see Tables 6 and 7). The duration of an instar in the field (D) is calculated stepwise in the following way:

$$D = f_i \frac{D(T(t_1)) + D(T(t_2))}{2} \tag{4a}$$

$$t_2 = t_1 + f_i \times D\,(T(t_1)) \tag{4b}$$

where:

- f_i = juvenile instar duration expressed as a fraction of the adult instar duration at a constant temperature T,
- $D(T)$ = the adult instar duration at a constant temperature T,
- $T(t)$ = the water temperature as a function of time t,
- t_1 = the moment at which the present instar developed by moulting from preceding instar.

When a model *Daphnia* is juvenile, length increment per moult is calculated as follows: At birth the combined durations of the successive juvenile in-

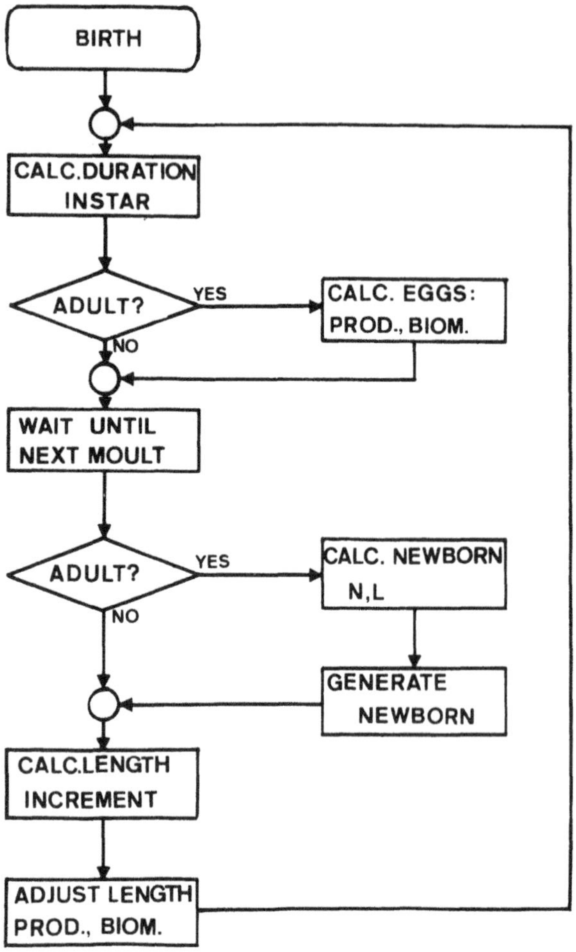

Fig. 1. Simplified flow diagram of the life cycle of an individual *Daphnia* in the simulation model 'INSTAR'. For explanation see text.

stars are calculated according to Eq. (4), the time at which the animal becomes adult can now be assessed. The length of the first adult instar ($L♀_{ad}$) at a particular time was calculated from the field observations. The observed 'minimum length of an adult female' is certainly an underestimation of the mean length of the first adult instars since only a few of the smallest females in a sample were measured. For this reason the minimum length was slightly increased, taking into account the probable variance of the length of the first adult instar as judged from field samples. Minimum length values were increased by 0.09 mm for *D. hyalina* and by 0.06 mm for *D. cucullata*. The length of a new born (L_{nb}) is related to the length of the mother and is calculated using Eq. (6) (see Results: duration and growth of juvenile instars). Thus, the length increment per moult of the ith juvenile instar (dL_i) is:

$$dL_i = \Delta L_i \, (L♀_{ad} - L_{nb}) \tag{5}$$

where ΔL_i is the length increment per moult of the ith juvenile instar expressed as a fraction of the total juvenile length increment from new born to first adult instar.

When a model *Daphnia* is adult, length increment at moulting is calculated as the product of the duration of the preceding instar and the mean length increment per unit of time. The length increment depends on both water temperature during the instar duration and the size of the female; the relation between length increment, female size and water temperature is given by Vijverberg (1980).

Somatic weight is calculated using Eq. (3); somatic weight increments are assessed after each moult. This information is used for the calculation of the standing crop biomass (B) and the biomass production per time unit (P).

When the animal is adult, eggs appear in the broodchamber a short time after each moult. The brood size is mainly derived from field observations: it is the number of eggs per ovigerous female of a particular size multiplied by the percentage of reproductively active adult females. The number of reproductively active females is derived from the percentage of adult females but corrected for the time just after moulting when eggs are still in the ovaries and not yet in the broodchamber.

From our laboratory observations it seemed a likely assumption that 100% of the *D. hyalina* adult population was reproductively active, as the percentage of females with eggs was about 85%, and for *D. cucullata* this was about 80%. Consequently, to assess the percentage of reproductively active females in the populations of *D. hyalina* and *D. cucullata*, the percentages of adult females with eggs, according to field observations, were increased by 15 and 20% respectively. The egg weight is considered to be equal to the weight of the resulting newborn and is calculated using Eq. (6) (see p. 248). This information is used for the calculation of the standing crop biomass (B) and the gonad production per time unit (P_g). Gonad production added to somatic production gives the total biomass production (P).

In the simulation, mortality occurs twice a day. For each size class the number of animals in the simulation is compared with those in the field at the same time, the difference being mortality. This approach to estimating mortality is very similar to that of Elster (1954, 1955), Edmondson (1960, 1968) and Bernardi (1974), but their models do not take into account the egg mortality caused by the death of adult females.

Results and discussion

Physical conditions

Due to the shallow, wind-exposed nature of the lake there is a high degree of water mixing. The oxygen concentration in the lake water is similar over the whole water column and very near to saturation value. Probably as a result of the almost continuous water mixing, *Daphnia* does not show a distinct vertical distribution pattern nor vertical diurnal migration (de Nie *et al.* 1980).

The water is very turbid, Secchi-disc readings varying between 20 and 40 cm. The low light transparency is only partly caused by the high algal standing crops. An important feature is the high incidence of peat in the lake substrates. Even at moderate wind speeds peat particles whirl up from the bottom into the water column. During winter, spring and autumn the transparency of the water is further decreased by large amounts of polder water supplied to the lake by canals from the surrounding polders. This water contains high concentrations of brown humic compounds (de Haan 1972).

Thermal stratification develops only on exceptionally calm days in summer, but is only 50 cm deep and never persists for more than a few days. Diurnal temperature variation usually showed a variation from 1 to 1.5 °C; only occasionally were higher variations of up to 3 °C recorded. Ice cover was present during winter; it varied from 1 month in 1970/71 to 2 and 3 months in 1968/69 and 1969/70 respectively. There was a marked seasonal difference in water temperature. The highest water temperatures, 23–24 °C, were recorded during July and August. In Table 1 water temperature is expressed as day-degrees above 10 °C. A value of 10 °C was chosen because we did not observe any noticeable population increase of *Daphnia* below this temperature. There are also differences in water temperature between years. In 1971 spring water temperatures were relatively high in comparison with both other years; 1970 showed somewhat higher temperatures during May and June, compared with 1969.

Table 1. Annual and seasonal water temperature changes in Tjeukemeer in day-degrees above 10 °C.

	1969	1970	1971
January	-227	-276	-250
February	-238	-233	-165
March	-258	-202	-188
April	-72	-77	-23
May	+89	+123	+151
June	+227	+269	+203
July	+271	+228	+301
August	+293	+282	+277
September	+181	+160	+184
October	+64	+25	+64
November	-90	-80	-113
December	-259	-179	-140
Annual	-9	+40	+301

Algal succession, algal biomass and primary production

There is a marked algal succession (Moed & Hoogveld 1982). During early-mid spring diatoms (*Diatoma elongatum* (Lyngb.) Agardh., *Melosira* spp., unicellular diatoms) and flagellates (*Chroömonas* spp. and *Cryptomonas* spp.) dominate the flora. In May diatoms and flagellates are replaced by green algae (*Scenedesmus* spp.) or by filamentous blue-green algae (*Aphanizomenon flos-aquae* (L) Ralfs. and *Oscillatoria* spp.). During winter diatoms reappear but green algae remain dominant while blue-green algae are absent or only occasionally present. The filamentous blue-green algae were practically absent during 1969, sub-dominant in 1970 and dominant over green algae in 1971.

The phytoplankton standing crop biomass was expressed as μg chlorophyll-*a* l^{-1}; the seasonal variation is given in Fig. 2 (broken line). Mean annual values for the growing season (March–September) ranged from 35 in 1969 to 72 and 73 $\mu g\ l^{-1}$ in 1970 and 1971. The chlorophyll content was markedly lower during winter; the values during the growing season showed large variations.

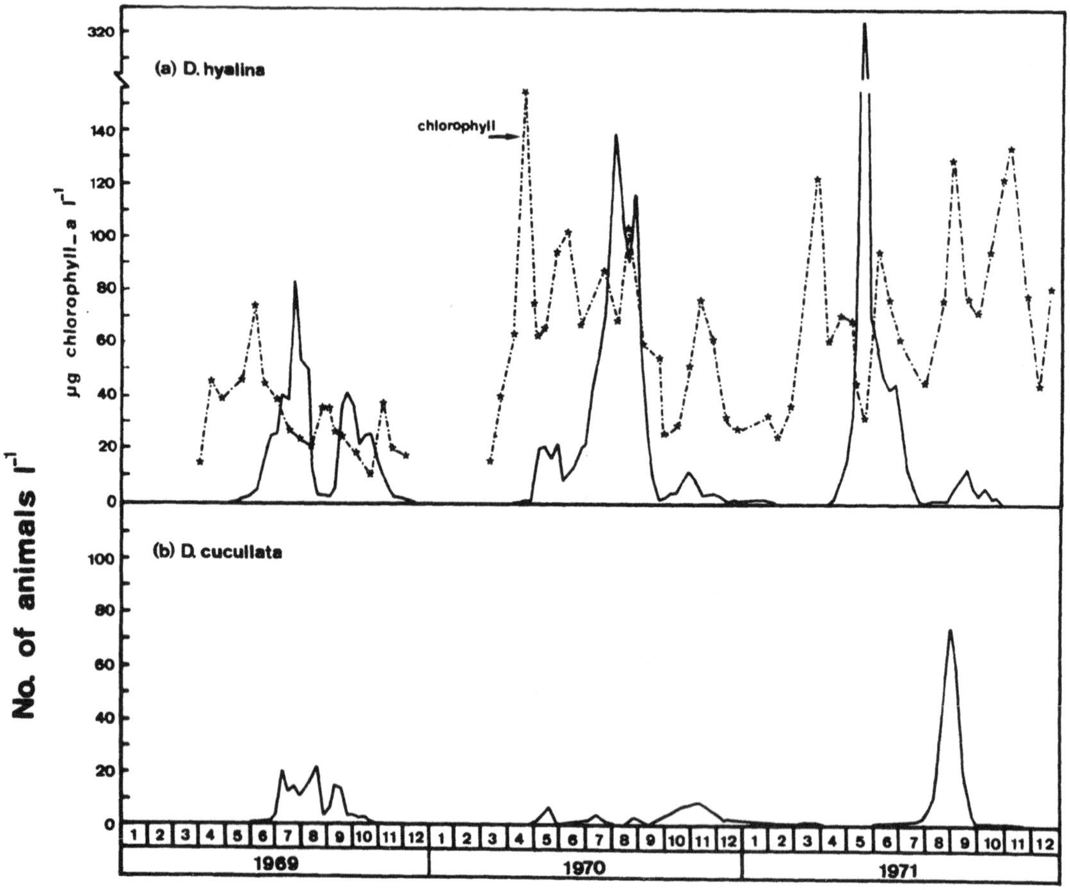

Fig. 2. Chlorophyll-*a* concentration in $\mu g\ l^{-1}$ (broken line) and population densities in numbers per litre (solid line) of (a) *D. hyalina* and (b) *D. cucullata.*

The mean daily gross primary production during the growing season was 1.1 g C per m^2 in 1969, 3.2 g C per m^2 in 1970 and 2.2 g C per m^2 in 1971. Only 1970 showed a distinct seasonal variation in production rates, low values in early spring (0.2–1.9 g C m^{-2} day^{-1}) and high values during June and July (7.6–8.8 g C m^{-2} day^{-1}). The total gross annual primary production varied from 0.3 kg C m^{-2} in 1969 to 0.8 kg C m^{-2} and 0.6 kg C m^{-2} in 1970 and 1971.

Stocks of planktivorous fish

The bream (*Abramis brama*) dominates the fish biomass. Its population consisted of about 15 different year classes, two of them (1959 and 1963) being particularly strong. Because of this population structure, annual variation in standing stock biomass was rather small during the research period.

In 1969 the bream population represented a mean standing crop biomass of 87 kg fr wt ha^{-1} and showed an annual P/B of 0.4 (Goldspink 1978). Assuming a growth efficiency of 0.1 (Humphreys 1979) and an assimilation efficiency of 0.8 (Winberg 1956) the total annual consumption of the bream and smelt populations can be estimated. The growth efficiency will be dependent on the size class structure of the population, this being higher for the smaller animals. As the value given by Humphreys is a mean value based on a literature review, food consumption calculated in this way will underestimate the bream and overestimate the smelt consumption to some extent. The total annual consumption of the bream population amounted to $87 \times 0.4 \times (\frac{100}{10}) \times (\frac{10}{8}) \approx 450$ kg fr wt ha^{-1}. Nearly all this consumption took place between May and September. During this period the bream population consumed the equivalent of 3.4% of its own standing stock biomass per day. On an annual basis the diet of the Tjeukemeer bream, expressed in biomass units, consists for one half of zooplankton and for the other half of chironomids. When *Daphnia* is abundant, which is usually between May and July, their diet contains mainly *Daphnia* (Hazelzet 1972). During this period the numbers of *Daphnia* consumed are very high and equivalent to about 3 kg fr wt ha^{-1} day^{-1}.

Among the young fish the smelt (*Osmerus eperlanus*) is generally the most important predator on zooplankton. The smelt consists for approx. 99% of one year class (0+). They are able to spawn as 1+. Annual variations in standing stock are large, sometimes annual mean standing stocks differing from each other by a factor of 50. It is a pity that no detailed information is available about the smelt in Tjeukemeer before 1971. A survey during the summer and autumn of 1970 indicated low standing stocks of smelt. In contrast, 1971 showed a much higher mean standing stock of 15–20 kg fr wt ha^{-1} and a production of 60 kg fr wt ha^{-1} yr^{-1} (van Densen 1973). The total annual consumption of the smelt population during that year will have been 60 $\times (\frac{100}{10}) \times (\frac{10}{8}) \approx 750$ kg fr wt ha^{-1}. This planktivorous fish, which feeds positively size selectively, eats predominantly *Daphnia* when available (van Densen 1978). Predation pressure of smelt on *Daphnia* was high from the end of June to the beginning of September. As during 1971 about one-third of their energy intake was from *Daphnia*, the predation pressure exercised by the smelt population was equivalent to 250 kg fr wt ha^{-1} yr^{-1}.

Daphnia population densities

Changes in the population densities of *D. hyalina* and *D. cucullata* during 1969, 1970 and 1971 are given in Fig. 2. Generally *D. hyalina* showed two population maxima a year, the first and highest maximum occurring in spring or summer, and the second maximum in late summer or early autumn. During mid and late summer, densities declined to low values. In 1971 *D. hyalina* densities started to increase a month earlier than during both previous years; the first maximum was reached in mid May, approximately 2 months earlier than in 1969 and 1970. These time differences were almost certainly caused by the high water temperatures prevailing in March, April and May 1971 (Table 1). The seasonal periodicity in *D. hyalina* densities in 1969 and 1970 were similar. The sudden decrease in densities at the end of May 1970 cannot be explained by a decrease in water temperature during this period.

There is some indication that *D. hyalina* during its first population maximum was overgrazing the algal crops, causing a sudden drop in the chlorophyll content. This could have happened in 1969 and 1971 when chlorophyll-*a* reached minimum values of 25–30 μg l^{-1} during the period of maximum population. However, this was not the case during 1970 when the chlorophyll content was high during the first population maximum and both chlorophyll and population minima were reached simultaneously.

Population densities of *D. cucullata* were lower than those of *D. hyalina*. *D. cucullata* reached its highest densities usually during mid and late summer. Its densities were usually low during periods when *D. hyalina* occurred at high densities.

Since samples taken at different stations were pooled, it was not possible to directly calculate 95% CL of the mean population densities. During 1968 the same sampling scheme was employed as during 1969–71, but samples were not pooled or sub-sampled. Sample variances were calculated for the different species at each sampling date, and they correlated strongly with population densities (de Nie *et al.* 1980). These regressions were used to calculate the 95% CL of the mean *D. hyalina* and *D. cucullata* densities between 1969 and 1971, the variance introduced by sub-sampling also being taken into account (Fig. 7).

Size distribution

The size distribution of a cladoceran population at a certain moment is an important population characteristic, since it is not possible to directly assess age or instar number from morphological characteristics. Both mortality and birth rate are strongly dependent on the size distribution of the population: mortality, because predation as a rule is size selective (Dodson 1974; Zaret 1978; Lynch 1979), and birth rate, because fecundity (F) is a power function of the female length (L.).

The size distribution of *Daphnia* showed a marked seasonal variation (Fig. 3).

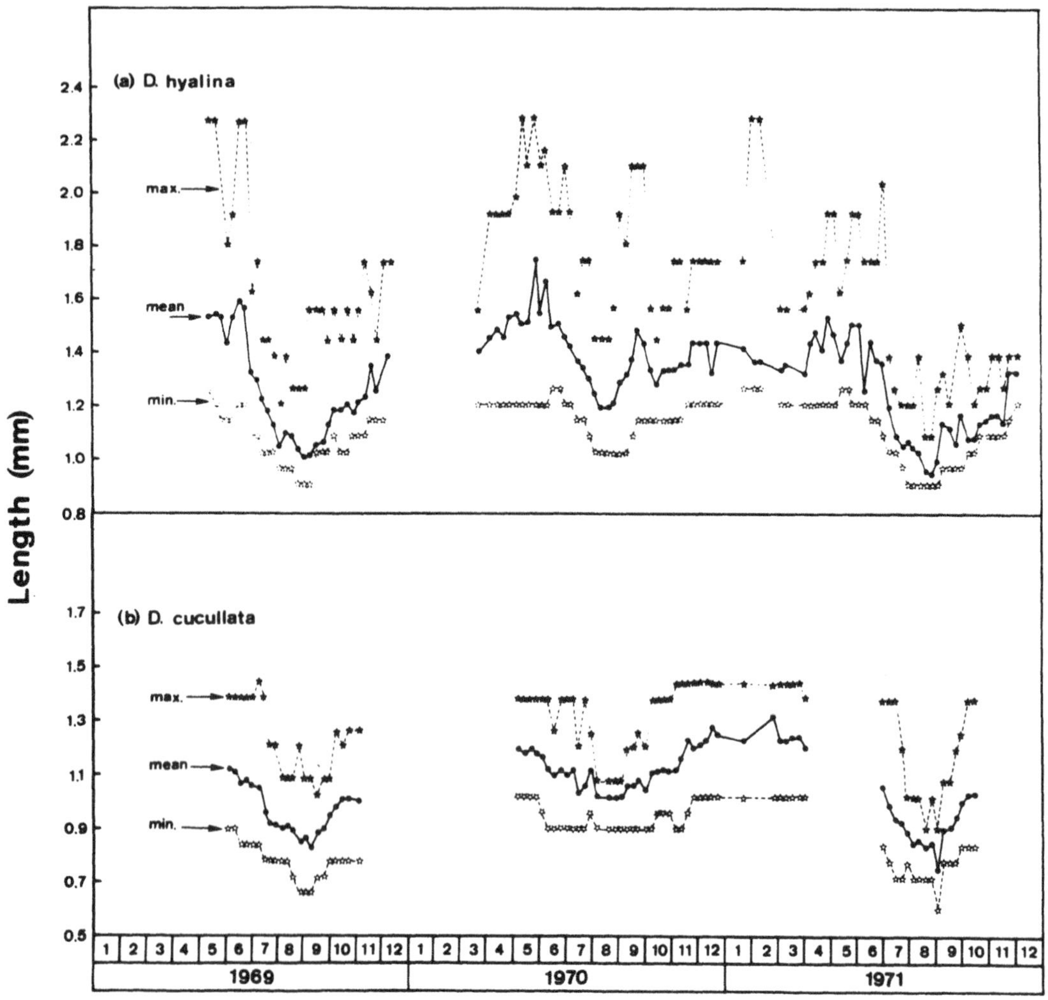

Fig. 3. Minimum, mean and maximum lengths of adult females of (a) *D. hyalina* and (b) *D. cucullata*.

The mean lengths of adult females were greatest during spring and smallest during mid and late summer. During autumn and winter intermediate values were reached. The minimum adult length, which is the length at which the animal becomes adult, changed in a similar way. In *D. hyalina* the difference in size between minimum and mean length was largest during spring, intermediate during autumn and winter and smallest during summer and early autumn. *D. cucullata* showed no such distinct seasonal variation in size difference, the absence of which is most probably an indication of the effect of positive size selective fish predation, which is most severe during summer and early autumn and acts most heavily on the larger of the two *Daphnia* spp.: *D. hyalina*. The maximum length, being the length of the largest female encountered in a 10 l sub-sample, followed basically the same pattern as the minimum and mean length. The largest females observed measured 2.3 (*D. hyalina*) and 1.5 mm (*D. cucullata*) in length, which corresponds with the maximum length we observed in the laboratory cultures. It takes a new born 7–10 weeks at 15 °C to reach such a length. Thus, such peaks in maximum length must coincide with periods of low mortality. The variations in mean and minimum length correlate well with the seasonal variation in water temperature (Table 1), the smallest individuals being found during periods of high water temperature. Culture results with *D. hyalina* (Vijverberg 1976, 1980) showed that the observed variations in minimum length were at least partly due to the combined effect of food quality, food quantity and temperature. At intermediate temperatures (10, 15 °C) the first adult instar was longer than at high (20, 25 °C) and low (2.5, 5 °C) temperatures. At a constant temperature, high food levels resulted in larger growth increments per moult, and thus in larger first adult instars. This may explain the changes in minimum length observed during winter. The difference in size between minimum length and mean length was largest during 1970 for both *Daphnia* spp. This is an indication of a lower mortality in the larger size classes during that particular year. There was also a marked annual difference in the minimum and mean size of both *Daphnia* spp. The lowest minimum and mean length values where reached during the mid and late summer in 1969 and 1971, whereas these values were much higher in the summer of 1970. This cannot be explained by differences in temperature regime. Temperatures were quite similar during this period in the three successive years, except for August 1971 which showed somewhat lower temperatures than August 1970 (Table 1). Thus a factor other than temperature must be responsible for this difference. The annual variation in positively size selective predation pressure exerted by fish is most probably responsible. The smelt, the most dominant 0+ fish in the lake, shows marked annual differences in population densities.

These densities were at least an order of magnitude higher during 1971 than in 1970 (van Densen, pers. comm.). Smelt feed preferentially on *Daphnia* even when *Daphnia* densities are rather low and other zooplankton species are much more abundant (van Densen 1973). The larger animals are selected and, because most of these larger animals are reproductive females, this selection should lead to a gradual reduction in the mean size of the first adult instar. This interpretation is corroborated by the observations of Hebert (1974) and Hrbáček & Hrbáčkova-Esslová (1960). Hebert showed that *Daphnia* populations in natural habitats generally consist of a number of ecologically differentiating clones. Hrbáček & Hrbáčkova-Esslová studied the growth rate of different strains of *D. hyalina* and *D. cucullata* under standardized laboratory conditions. They observed that *Daphnia* strains from waterbodies with high stocks of planktivorous fish showed lower growth increments per moult, resulting in smaller first adult instars. It is possible that the *Daphnia* population in Tjeukemeer consists of slow and fast growing clones, whose relative frequencies are changed by seasonal and annual differences in fish predation.

Reproductive cycle and fecundity

Reproductive cycle. As a rule, most eggs produced by *Daphnia* females are parthenogenetic eggs which develop into free living juveniles without fertilization; these eggs usually develop into female offspring, but when a female is crowded and food supply is rapidly reduced, she produces male progeny (Banta & Brown 1929; Slobodkin 1954; Young 1978). Shortly after males are produced, adult females in the population produce sexual eggs; these eggs must be fertilized in order to develop. The carapace surrounding the broodchamber thickens and encloses one or two eggs, forming an ephippium. In *D. hyalina* each ephippium always contained two eggs, while this varied from one to two in *D. cucullata*. Sexual eggs normally diapause and hatch during the subsequent year in late winter or early spring (Smyly 1979). In 1969 and 1971 *D. hyalina* showed two sexual periods a year (Fig 4), the first between May and August being associated with the first population maximum, and the second period during September and October with the second population maximum. During the second sexual period resting eggs were produced even when the population densities of *D. hyalina* were below 10 individuals l^{-1}. It seems unlikely that during this period crowding was an important stimulus for the production of males and resting eggs. In 1969 and 1971 *D. cucullata* showed one sexual period a year during late summer or early autumn. Here too resting egg production was not necessarily associated with high population densities. It is probable that during autumn the photoperiod is of more importance than crowding. Stross & Hill (1965) showed experimentally that short day periods (12 h light, 12 h dark) increased the resting egg production in *D. pulex*. The intensity of resting egg production by *D. hyalina* and *D. cucullata*, though sometimes different in particular years, was similar. Because of the much higher population densities of *D. hyalina* that species produced annually many more resting eggs than *D. cucullata*. *D. hyalina*

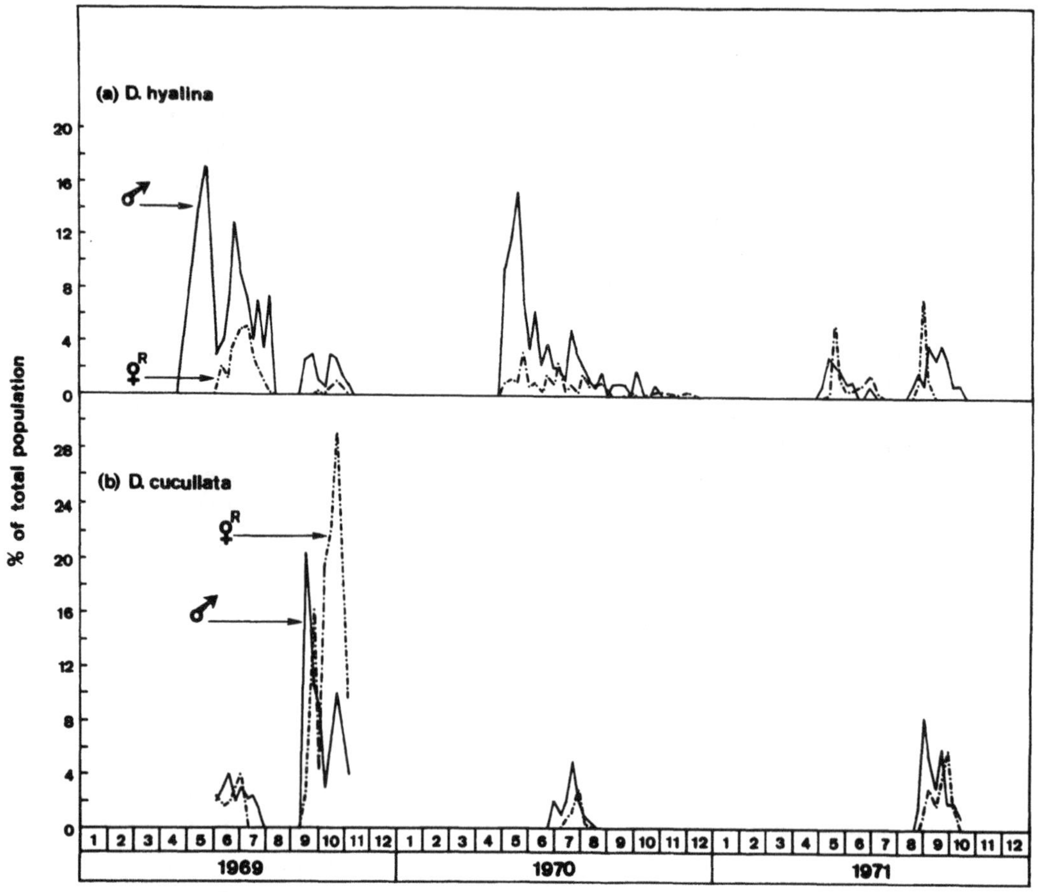

Fig. 4. Relative densities of males and females with resting eggs expressed as a percentage of the total population of (a) *D. hyalina* and (b) *D. cucullata.*

produced per annum 25–58 resting eggs l^{-1} yr^{-1}, while *D. cucullata* produced only 0.25–8.35 resting eggs l^{-1} yr^{-1}. As will be shown later this is more than enough to explain the population increase in the spring of the subsequent year.

Fecundity. Fecundity is defined as the mean number of parthenogenetic eggs per adult female. This figure contains two components: the reproductive ratio, defined as the percentage of ovigerous females in relation to the total number of adult females, and the number of eggs per ovigerous female (F).

The reproductive ratio changes seasonally (Fig. 5). In *D. hyalina* low values are associated with high population densities, high summer temperatures and the presence of females, with resting eggs. Minima of about 20% are reached during mid summer. Higher values of around 85% were observed during autumn, winter and spring. In 1971 the reproductive ratio of *D. hyalina* remained high, although population densities reached exceptionally high values. In 1971 the population maximum was reached in May, about 2 months earlier than in 1969 and 1970. Water temperatures remained low and food quality relatively high. In *D. cucullata* the lowest ratios were usually observed during mid and late summer and during autumn. Minima ranged from 20 to 40%. As in *D. hyalina* these minima were associated with the presence of females with resting eggs, but they were not necessarily correlated with high population densities or high water temperatures.

In 1969 also a spring minimum was observed which was associated with females with resting eggs; during the rest of the year the ratio was approx. 80%.

For each sampling the number of eggs per ovig-

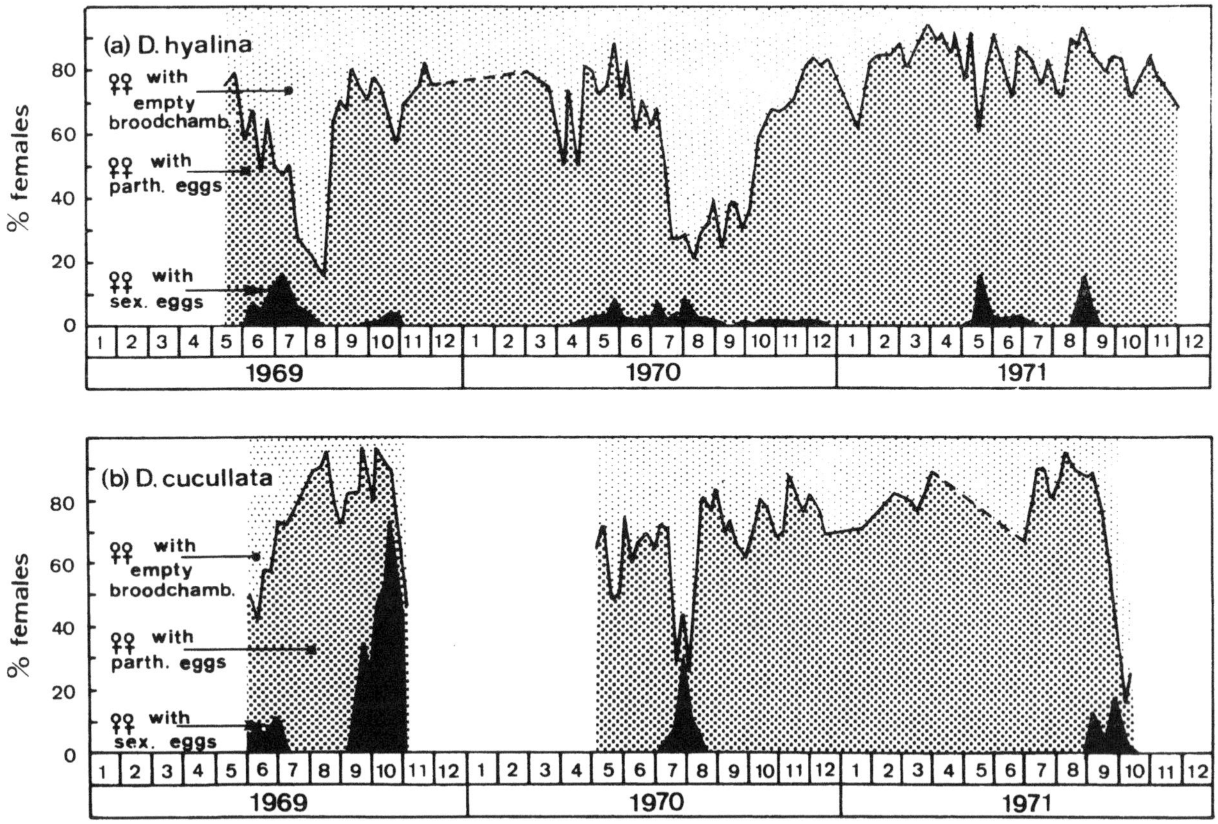

Fig. 5. Reproductive ratio of (a) *D. hyalina* and (b) *D. cucullata*.

erous female (F) was related to female length (L) using Eq. (1). These results, together with the observations on the reproductive ratio, were used in the 'INSTAR' model. The F-values showed a marked seasonal variation, which was positively correlated with the reproductive ratio. Using the reproductive ratios of *D. hyalina* (Fig. 5a) we divided 1969, 1970 and 1971 into four seasonal periods: spring, summer, autumn and winter. Due to annual differences in temperature regime these periods were not the same each year, and they are given in Table 2. The seasonal and annual variations in number of eggs per ovigerous female are given in Fig. 6 and Table 3a, b. The differences between regression lines were tested 'pairs wise' with the F-test ($P < 0.05$). We first compared the residual variances with the two-tailed F-test; if these showed no significant differences, then the slopes and finally the elevations of the lines were tested with a one-tailed F-test (Snedecor & Cochran 1967, pp. 432–436). Regression lines were assumed to be different when either their slopes or their elevations were significantly different. The regression lines were compared for seasonal effects; the results of the significance tests are summarized in Table 3b. The largest differences were observed between spring and summer values, the latter being significantly lower than the former. Autumn values were usually in between spring and summer values. Both species showed relatively high winter values, which were significantly different from summer values. The annual variation for the same species in the same season was generally small and only signif-

Table 2. Seasonal periods chosen for the calculation of the number of eggs per ovigerous female (F) in 1969, 1970 and 1971.

Season	1969	1970	1971
Spring	20/4–15/7	24/3 –14/7	2/3–11/5
Summer	22/7–26/8	21/7 –15/9	18/5–22/6
Autumn	2/9–25/11	22/9 –24/11	29/6–30/11
Winter		1/12–22/12	22/1–23/2

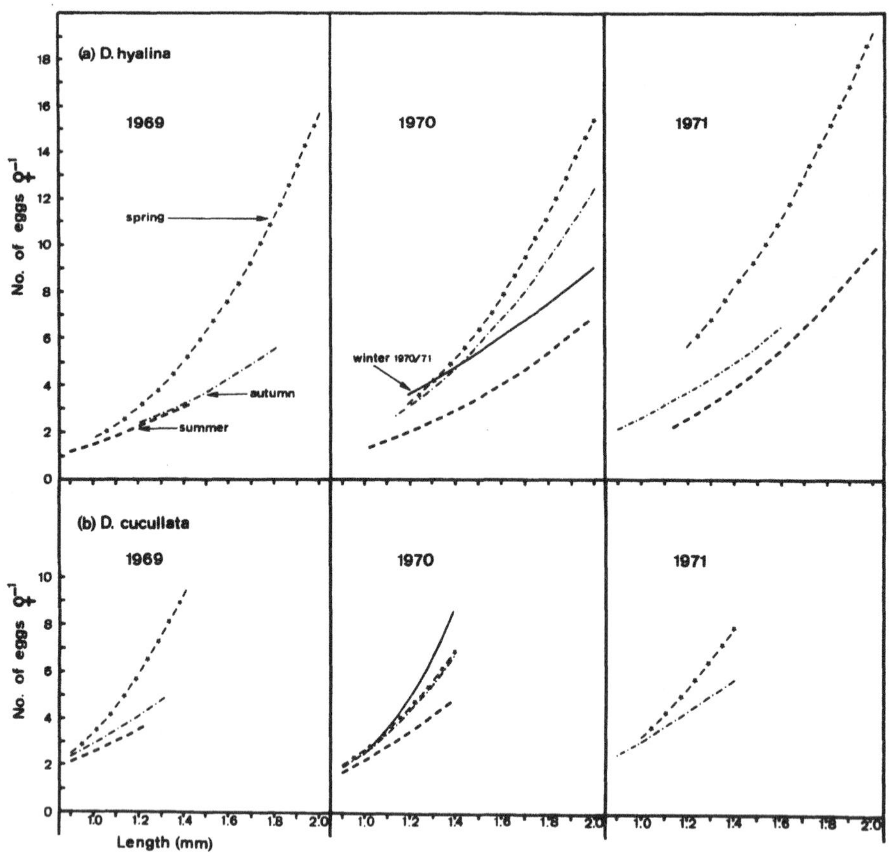

Fig. 6. Seasonal and annual variation in the number of eggs per ovigerous female (F) for (a) *D. hyalina* and (b) *D. cucullata*.

icant in a few cases. During 1970 both species showed rather high autumn values, while *D. hyalina* showed relatively high values during spring and summer 1971. There was much less reduction in egg number during summer for *D. cucullata* than for *D. hyalina*.

In order to facilitate comparison of fecundity for different sampling dates the egg number was calculated for a *D. hyalina* female of standard size (L = 1.3 mm) using Eq. (1), and multiplied by the reproductive ratio. The so-calculated standard egg production (SEP) (Hebert 1977) was compared with the seasonal variation in chlorophyll-*a* content (Fig. 2). During the period of the first population maximum we found a strong negative correlation between fecundity (SEP) and chlorophyll-*a* content in 1969 and 1971. However, there was no relationship in 1970, when fecundity (SEP) dropped to low values during the first population maximum while the chlorophyll-*a* concentration was still high.

In both species the mean number of eggs per adult female was influenced more by seasonal variations in the reproductive ratio and the size distribution of the adults than by variations in the relation between egg number and female length. A sudden reduction in fecundity of *D. hyalina* during summer, generally at the time of the first population maximum, was also reported by Nauwerck (1963), George & Edwards (1974), Bernardi & Canali (1975), Gliwicz (1977), Smyly (1979), Jones *et al.* (1979), Stebler (1979) and Seitz (1980). This seems to be a common feature of all large *Daphnia* spp. Small *Daphnia* spp. like *D. cucullata* seem much less affected by adverse summer conditions (Stebler 1979; Seitz 1980).

The negative correlation which we found between the chlorophyll-*a* content and the population density of *D. hyalina* (1969, 1971) and between chlorophyll-*a* content and fecundity (SEP) (1969, 1971) for this species during its first population

Table 3a. Values for Eq. (2) relating number of eggs per ovigerous female (F) to female length (L, mm); n is the number of observations.

Species	Year	Season	ln a ± 95% CL	b ± 95% CL	n	r
D. hyalina	1969	Spring	0.55 ± 0.22	3.22 ± 0.56	45	0.87
		Summer	0.39 ± 0.06	2.12 ± 0.29	19	0.92
		Autumn	0.44 ± 0.15	2.20 ± 0.61	41	0.75
	1970	Spring	0.60 ± 0.19	3.12 ± 0.44	71	0.87
		Summer	0.22 ± 0.15	2.41 ± 0.44	37	0.89
		Autumn	0.65 ± 0.23	2.67 ± 0.65	32	0.87
		Winter	0.94 ± 0.19	1.77 ± 0.45	40	0.79
	1971	Spring	1.29 ± 0.24	2.44 ± 0.54	43	0.83
		Summer	0.50 ± 0.27	2.60 ± 0.63	26	0.84
		Autumn	0.97 ± 0.10	1.89 ± 0.46	58	0.75
D. cucullata	1969	Spring	1.23 ± 0.12	2.98 ± 0.65	24	0.91
		Summer	0.94 ± 0.21	1.73 ± 1.07	14	0.72
		Autumn	1.05 ± 0.16	1.91 ± 0.71	22	0.80
	1970	Spring	1.01 ± 0.10	2.76 ± 0.49	35	0.91
		Summer	0.77 ± 0.16	2.41 ± 1.19	18	0.74
		Autumn	1.01 ± 0.09	2.80 ± 0.47	30	0.94
		Winter	0.92 ± 0.15	3.68 ± 1.18	18	0.83
	1971	Spring	1.21 ± 0.08	2.60 ± 0.42	14	0.97
		Autumn	1.15 ± 0.09	1.84 ± 0.46	47	0.77

Table 3b. Comparison of regression lines (Eq. 2) describing the seasonal variation in fecundity of *D. hyalina* and *D. cucullata*.

Species	Season	1969		1970		1971	
		Spring	Summer	Spring	Summer	Spring	Summer
D. hyalina	Spring		O		+		+
	Summer	O		+		+	
	Autumn	+	–	–	+	+	–
	Winter			O	+		
D. cucullata	Spring		+		+		
	Summer	+		+			
	Autumn	+	–	–	+	O	
	Winter			–	+		

O = residual variances are significantly different.
– = regression lines are not significantly different.
+ = regression lines are significantly different (F-test, $P < 0.05$).

maximum suggests that an overgrazing of the algal crops caused a reduction in fecundity. However, this was not substantiated by the observations from 1970, when such relationships were absent. This indicates that more factors than food quantity alone influence fecundity. Fecundity depends not only on the amount of food intake but also on assimilation efficiency and the level of metabolic demands. Both assimilation efficiency and metabolic demands are strongly temperature dependent (Slobodkin 1954; Lynch 1977). This means that for each species there is a certain temperature above which the utilization of food for reproduction will decrease. Our results suggest that *D. hyalina* reaches this optimum temperature earlier than *D. cucullata*. The assimilation efficiency is also greatly influenced by food quality (Schindler 1970; Arnold 1971; Porter 1973; Gulati 1975). Food quality in Tjeukemeer was much better during spring than summer (Vijverberg 1976; Moed & Hoogveld, 1982). During spring, diatoms and flagellates were present in high densities while colonial green algae and/or filamentous blue-green algae dominated during summer. Thus the seasonal variation in reproductive ratio and F-value is probably more influenced by changes in food quality and high summer temperatures than by changes in the total algal biomass. Fecundity as a whole was primarily influenced by the size distribution of the adult population.

Duration and growth of juvenile instars

There exists a positive correlation between the length of the new born (L_{nb}, mm) and the length of their mother (L_m, mm). This relationship can be described by a linear regression:

$$L_{nb} = p + q\, L_m \qquad (6)$$

where p is a constant and q the regression coefficient. For *D. hyalina* this relationship is: $L_{nb} = 0.293 + 0.183\, L_m$ ($r = 0.97$; $n = 167$) and for *D. cucullata* $L_{nb} = 0.272 + 0.168\, L_m$ ($r = 0.96$; $n = 203$). A similar relationship between size of the mother and her offspring for *Daphnia* was reported by Green (1954) and Hrbáčkova (1974).

The duration time of juvenile and first adult instars of *D. hyalina* and *D. cucullata* is given in Table 4. *D. hyalina* showed 4 and *D. cucullata* 3 juvenile instars. A constant number of 4 juvenile instars for *D. hyalina* was also reported by Green (1956). Navaneethakrishnan & Michael (1971) reported 5 and Hrbáčkova-Esslová (1971) found a more variable number of 4–7 juvenile instars for this species. The duration of the first three juvenile instars of *D. hyalina* were very similar, but the duration of the fourth juvenile instar was significantly longer (t-test, $P \leqslant 0.05$). The duration of each of the four juvenile instars was significantly shorter than those of the first adult instar (t-test, $P \leqslant 0.05$). In *D. cucullata* the development time of each juvenile instar was very similar, each being significantly shorter than the development time of the first adult instar (t-test, $P \leqslant 0.05$). For both species the duration of the juvenile instars was approximately half the duration of the adult instars. This agrees well with the observations of Murugan & Sivaramakrishnan (1973) on *Simocephalus* and of Bottrell (1975) on eight littoral species of Cladocera. The duration of the adult instar at a given temperature is constant and independent of instar number (Hall 1964; Bottrell 1975; Vijverberg 1980).

The juvenile growth, expressed as length increment per moult (mm), is given in Table 5. The length increment per moult is very similar for the four different juvenile instars of *D. hyalina*. The length increment of the first juvenile instar of *D. cucullata* is essentially less than that of the second and third instar. Green (1956) measured juvenile growth of eight different *Daphnia* spp. Generally, he found the smallest length increments in the first juvenile instar and the largest in the last or last-but-one pre-adult instar. This agrees well with out results on *D. cucullata*, but is slightly different from what we found for *D. hyalina*.

Table 4. Duration time of juvenile and first adult instars of *D. hyalina* and *D. cucullata* at 17.5 °C, in days and as a percentage of the duration of the adult instar. Mean ± 95% CL. Number of observations in parentheses.

Species	Instar no.	Duration time (days)	(%)
D. hyalina	1	1.7 ± 0.09 (15)	49
	2	1.5 ± 0.11 (15)	43
	3	1.7 ± 0.09 (15)	49
	4	2.1 ± 0.15 (15)	60
First adult:	5	3.5 ± 0.09 (15)	100
D. cucullata	1	1.8 ± 0.15 (16)	58
	2	1.9 ± 0.10 (16)	61
	3	1.7 ± 0.15 (16)	55
First adult:	4	3.1 ± 0.06 (40)	100

Table 5. Mean length ± 95% CL and length increment per moult of juvenile instars of *D. hyalina* and *D. cucullata* at 17.5 °C. Length increment absolute (mm) and as percentage of the total length increment from first juvenile instar to first adult. Number of observations in parentheses.

Species	Instar no.	Mean length (mm)	Length Increment (mm)	(%)
D. hyalina	1	0.61 ± 0.02 (20)	0.17	22
	2	0.78 ± 0.02 (20)	0.23	29
	3	1.01 ± 0.02 (20)	0.19	25
	4	1.20 ± 0.03 (20)	0.18	23
D. cucullata	1	0.47 ± 0.01 (23)	0.10	20
	2	0.57 ± 0.01 (21)	0.19	37
	3	0.76 ± 0.02 (24)	0.22	43

Population dynamics

Four simulations were carried out to simulate seasonal changes in population densities:

Simulation 1. The model used is primarily based on density, population structure and fecundity of the field population, the development rate of the parthenogenetic eggs, the growth rate of juveniles and adults in relation to size and temperature, and the water temperature in the lake. Simulations start in an early spring situation when the field densities are $\geqslant 2$ individuals l^{-1}. Mortality is assumed to be always $\geqslant 0$.

Simulation 2. The simulation differs from sim. 1 only by the fact that the population is initiated by the hatching of resting eggs. It is based on the number of resting eggs produced during the previous year. Since no observations on resting egg production were available for 1968, we assumed that these were the same number as were produced during the subsequent year. We assumed a mortality rate for resting eggs during diapause of 50% (Smyly 1979; Stross *et al.* 1979). In the model system simulations were started when the water temperature in the lake increased above 5 °C, but not earlier than 1 March, since both temperature and day length are involved in the hatching of resting

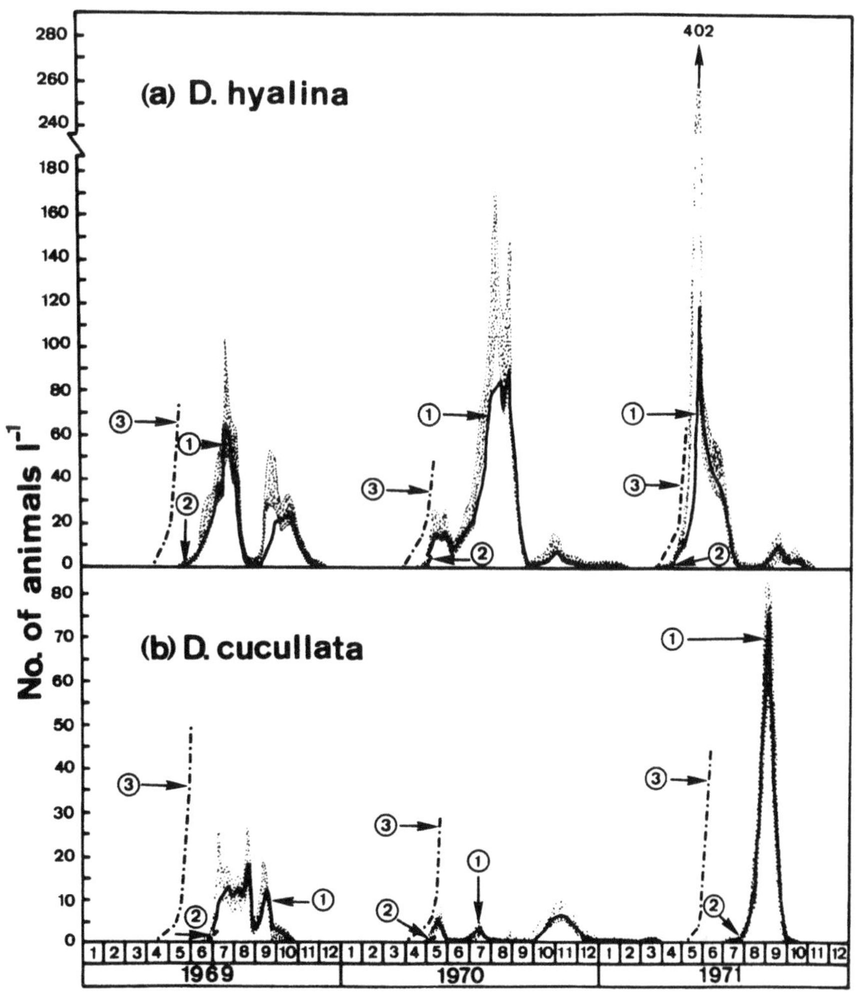

Fig. 7. Observed and simulated population densities of (a) *D. hyalina* and (b) *D. cucullata*. Shaded area represents 95% CL of observed population densities. For explanation see text.

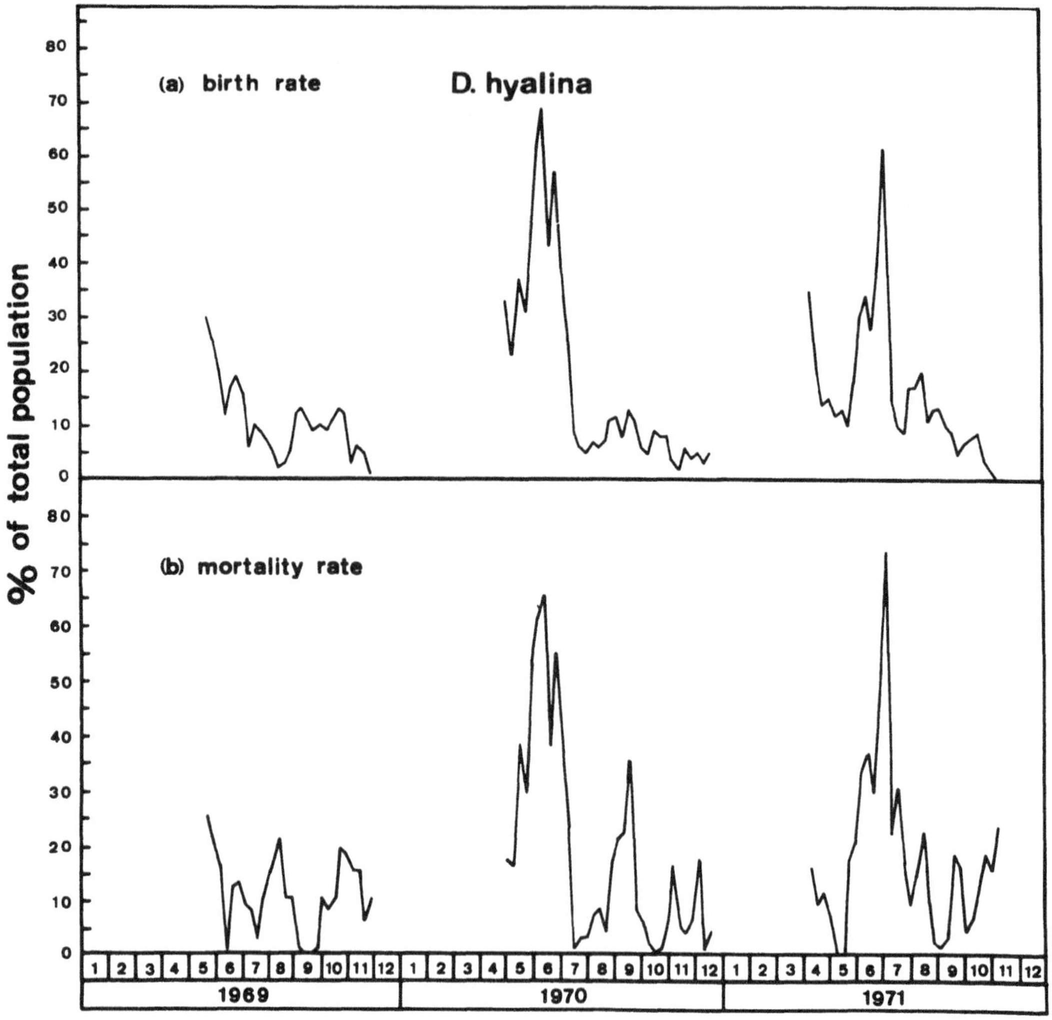

Fig. 8. Seasonal and annual variations in mean daily birth rates (b, %) per week an mean daily mortality rates (d, %) per week of *D. hyalina* (sim. 1).

eggs (Smyly 1979). In the model system 4% of the surviving resting eggs hatched each day over 25 days. Mortality is always assumed to be ⩾0.

Simulation 3. The same simulation as sim. 2, but mortality was assumed to be 0.

Simulation 4. The same simulation as sim. 1, but in this model system mortality could also be <0. Mortality is computed by comparing simulated densities with observed field densities. Since these figures are both estimates, they are subject to error and the mortality calculated in this way may become negative.

The results of the simulations are shown in Fig. 7. The sim. 1 predicted densities are generally a little below the observed densities (Fig. 1) but are usually well within the 95% CL of the density estimates. Only occasionally do the predicted densities fall outside these limits; this is generally the case during periods of steep exponential population increase.

The increase in population densities out of resting eggs (sim. 2) is usually somewhat too low, although potentially there are enough eggs to produce a strong population increase in early spring (sim. 3). This discrepancy is most probably caused by the behaviour of the resting eggs, which float on the surface after being produced and presumably

hatch in the littoral zone during the subsequent year (Hebert 1978). Since our fixed sampling stations are all in the open water zone of the lake this could possibly explain the underestimation of the population densities during spring. In the model system an underestimation of the field densities leads via an overestimation of the mortality to a too low simulation density.
density.

For the calculation of the birth and mortality rates we used sim. 1. The results are given in Figs. 8 and 9 and in Tables 6 and 7.

The birth rate of *D. hyalina* shows a distinct seasonal pattern. In spring and early summer, when population densities are still medium-low, birth rates are high. This is primarily an effect of the size structure of the population, containing many large females, the high egg number per ovigerous female and the high reproductive ratio. Temperature regime is also favourable, usually above 10 °C but well below 20 °C. When during the first population maximum the density increases, the birth rate decreases.

In 1971, when the first population maximum was reached early in the growing season, we observed a second birth rate peak during June and early July. This did not result in a second population maximum. A mid-summer minimum which coincided with a low fecundity and a small mean adult size was observed during all three years, but was most pronounced in 1969 and 1970. During early autumn the birth rate increased somewhat, mainly due to an increased fecundity and a larger mean adult size. From November onwards, when temperature dropped below 10 °C, the birth rate decreased to 0–5% day^{-1}.

The birth rate of *D. cucullata*, unlike *D. hyalina*, shows no distinct seasonal pattern. Throughout

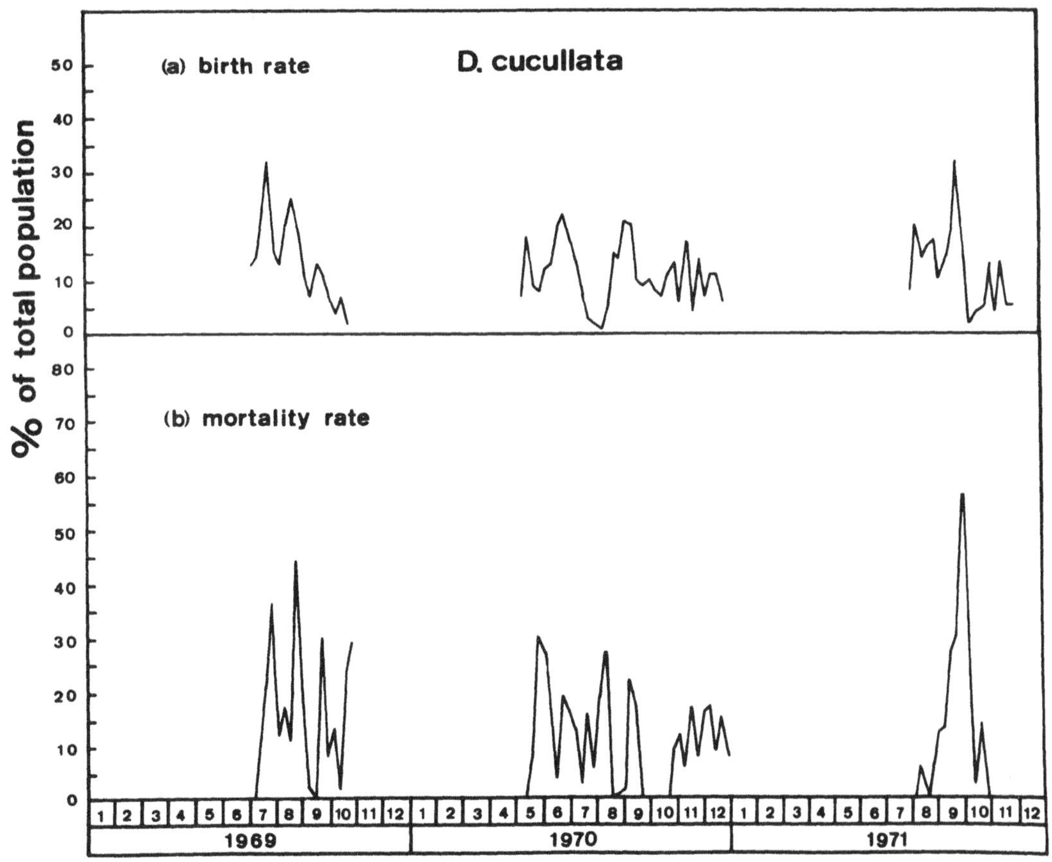

Fig. 9. Seasonal and annual variations in mean daily birth rates (b, %) per week and mean daily mortality rates (d, %) per week of *D. cucullata* (sim. 1).

Table 6. Mean daily size specific mortality rates (d, %) of *D. hyalina* in Tjeukemeer per month and per year (sim. 1). Size classes: 1 = 0.42–0.76; 2 = 0.77–1.12; 3 = 1.13–1.48; 4 = ⩾1.49 mm.

	1969				1970				1971			
	1	2	3	4	1	2	3	4	1	2	3	4
April	–	–	–	–	–	–	–	–	15	7	0	2
May	53	1	0	9	47	11	14	0	3	11	6	1
June	18	15	2	3	130	17	16	2	75	4	11	19
July	8	4	15	29	40	0	1	35	43	29	76	44
August	0	3	164	–	0	3	4	341	9	5	145	–
September	0	0	10	29	35	10	13	42	1	12	85	–
October	16	8	3	299	0	0	4	51	0	10	114	–
November	12	23	6	220	6	10	11	26	0	2	60	–
December	12	9	6	20	4	9	10	2	3	1	69	–

Table 7. Mean population densities and mean mortality rates per size class of *D. hyalina* and *D. cucullata* between April and December. Size classes, *D. hyalina* as in Table 6, for *D. cucullata* 1 = 0.42–0.76; 2 = 0.77–1.12; 3 = ⩾1.13 mm. Mortality (sim. 1), biomass 'lost' (sim 4).

	Year	*D. hyalina*				*D. cucullata*		
		1	2	3	4	1	2	3
Mean density	1969	5 685	6 528	2 873	400	3 253	1 752	55
(Nos m^{-2})	1970	9 188	10 832	8 153	1 523	1 823	998	467
	1971	8 423	6 030	4 948	1 723	6 605	2 677	10
Mean relative	1969	37	42	19	3	64	35	1
densities	1970	31	36	27	5	55	30	14
(%)	1971	40	29	23	8	71	29	0.1
Mean mortality	1969	636	354	383	156	731	137	33
(Nos m^{-2} day^{-1})	1970	2 690	463	371	520	241	43	30
	1971	2 347	576	476	180	1 406	162	40
Mean relative	1969	42	23	25	10	81	15	4
mortality	1970	67	11	9	13	77	14	10
(%)	1971	66	16	13	5	87	10	3
Mean mortality	1969	15	8	26	87	21	9	51
rate	1970	33	8	9	62	12	7	22
(% day^{-1})	1971	17	9	63	17	25	4	274
Mean mortality	1969	12	15	40	33	60	25	15
in org. dry wt	1970	22	9	18	51	44	22	33
(%)	1971	27	16	33	25	71	18	12
Mean biomass 'lost'	1969	1.3	1.6	4.5	3.6	0.9	0.4	0.3
(mg org. dry wt m^{-2} day^{-1})	1970	5.0	2.0	4.2	11.7	0.4	0.1	0.3
	1971	6.3	3.8	7.7	5.8	1.5	0.4	0.3

summer and early autumn maxima may be reached from 20 to 30% day^{-1}. The negative effect of population density on birth rate is rather weak. Low values are usually associated with the presence of a high proportion of females with resting eggs in the population. During winter the birth rate is approx. 10% day^{-1}, which is high considering the prevailing low temperatures.

The maximum birth rates recorded here for *D. hyalina* were approximately twice as high as those observed for *D. cucullata*. Thus, under favourable conditions, *D. hyalina* has a higher potential for population increase than *D. cucullata*. To judge from the fecundity and population structure data, *D. hyalina* meets these favourable conditions earlier in the year than *D. cucullata*. During summer *D. cucullata* often shows higher birth rates than *D. hyalina*, but during spring *D. hyalina* generally has a much higher birth rate. We did not find any relation between birth rate and primary production or between birth rate and chlorophyll-*a* concentration.

In general, seasonal changes in mortality rate correspond with changes in birth rate (Figs. 8 and 9). For *D. hyalina*, in spring birth rates are generally higher than mortality rates. Both rates are rather high. The steep decrease in population densities after the first population maximum is reached is the combined effect of high mortality and low birth rates. During early autumn birth rates are usually somewhat higher than the mortality rates, giving rise to a second population maximum. At the end of autumn birth rates decline below the mortality rate, causing the decrease in the population densities to very low values.

For *D. cucullata* a clear pattern could only be observed during 1971. The decline in population densities after the first population maximum is reached is caused mainly by a sudden increase in the mortality rate and much less by a reduction in the birth rate.

Our values on birth and mortality rates are not directly comparable with most results reported in the literature. Nearly all previous authors have used Edmondson's formula (Edmondson 1960, 1972, 1974) to calculate birth rates. Only recently did Seitz (1979) and Threlkeld (1979) recognize the fact that the use of this method introduces a serious bias in the birth rate estimates, and thus also in the mortality rate estimates. This error is made because of the egg mortality caused by the relatively high mortality of large ovigerous females. When the mortality of adult ovigerous females is relatively high, which will be the case when positively size selective predation by planktivorous fish is an important cause of death, this can lead to an overestimation of the birth rate by about 50% (Threlkeld 1979).

Since the 'INSTAR' model used in the present study accounts for egg mortality caused by the mortality of ovigerous females, it would be expected that our values for birth and death rates would be generally lower than those reported in the literature, but this is not the case. Our mean annual values of 10–20% day^{-1} for *D. hyalina* and 5–10% day^{-1} for *D. cucullata* agree well with most figures reported in the literature for *Daphnia* spp. of comparable size inhabiting productive lakes in the temperate climate zone (Hall 1964; Tappa 1965; Wright 1965; Dodson 1972; Bernardi 1974; Seitz 1980). Somewhat lower values of 5–10% day^{-1} in productive lakes were reported for *D. hyalina* by George & Edwards (1974) and Smyly (1979).

The mortality rates per size group are given in Tables 6 and 7. Between April and June the first size class of *D. hyalina* showed relatively high mortality rates. This may be the result of over-simplifications in the model (sim. 1) or may have had natural causes. In the first case, the hatching of resting eggs or the growth rate of the first generation born from resting eggs was most probably involved. In the second case, it seems likely that the invertebrate predator *Cyclops vicinus*, which reached its highest densities during this period, was responsible. However, for the rest of the year both species generally exhibited a strong size dependent mortality, the largest individuals showing the highest mortality rate. The rate values did not depend on the absolute *Daphnia* size but on their relative size within the population. When the largest size class dies out, which was observed for *D. hyalina* during 1969 and 1971, the highest mortality rates shifted from size class 4 to size class 3. Size dependent mortality, as observed here, strongly affected the rate of population decrease, because not only were the adult females killed but also numerous eggs or embryos which these large females carry with them. It is not likely that this is the effect of an increased physiological mortality in old age. Only a few individuals during the beginning of the growing season reached a size corresponding with old age; during the rest of the season *Daphnia* never reached old age. The physiological mortality, as we observed for these size classes in the laboratory under semi-natural conditions, was only 1–2% day^{-1}. Similar low physiological mortality rates for juveniles and adults of *Daphnia* below the maximum size were reported by Hall (1964), Kozlowski (1978) and Threlkeld (1979a). Thus it seems most likely that the observed size dependent mortality is caused by the positive size selective predation by planktivorous fish. Although the mortality rate is a convenient way of expressing mortality, especially for showing the effect of predation pressure, it also has its restrictions. For instance the very high mean monthly mortality rates of >100% day^{-1} which we found occasionally for the largest size classes of both *Daphnia* spp. are only possible when the population densities in these size classes are relatively low. Thus high mortality rates do not necessarily indicate high absolute mortality. To facilitate the discussion on size dependent mortality, some relevant information is summarized in Table 7. The data are presented on an annual basis (April–December). Relative density gives the percentage distribution of the densities over the size classes; relative mortality gives the percentage distribution of the mortality over the size classes. The absolute mortality is not only expressed as numbers per size class per m^2 per day, but also as 'loss' of org. dry wt per size class per m^2 per day. Two important conclusions may be drawn. Firstly, mean relative mortality among the smallest individuals (size class 1) is very high: about 40–70% for *D. hyalina* and about 80–90% for *D. cucullata*. Secondly, the 'loss' of biomass as a result of mortality is highest in the larger individual s (classses 3 and 4) of *D. hyalina*. In *D. hyalina* about 70% and in *D. cucullata* about 20% of the net production is 'lost' in these size classes.

The high absolute mortality in the first class may be due to starvation (Hebert 1978; Gliwicz & Siedlar 1980), to an increased mortality during the time of hatching (Prepas & Rigler 1978) or to invertebrate predation (Anderson 1970; Brandl & Fernando 1975; Kerfoot 1977). However, since mortality during hatching and the juvenile stage was rather low in our culture under semi-natural conditions using lake water as culture medium (Vijverberg 1980), this is most probably the effect of invertebrate predation. Several carnivorous species were present: *Leptodora kindtii* reached densities of 0.1–0.6 ind. l^{-1} during summer, *Acanthocyclops robustus* and *Mesocyclops leuckarti* reached densities of 30–100 and 10–30 ind. l^{-1} during the growing season, and the opossum shrimp *Neomysis integer* often reached densities of 10–20 ind. m^{-2} in August–September. Predacious water mites are also present but were not studied.

The large proportion of biomass 'lost' as animals larger than 1.1 mm is most probably the effect of positively size selective fish predation. The amount of biomass presumably 'lost' to the planktivorous fish is 210 kg fr wt ha^{-1} in 1969 and approx. 450 kg fr wt ha^{-1} in 1970 and 1971. On account of the production of the most abundant planktivorous fish (bream and smelt), their growth efficiency (Humphreys 1979), their assimilation efficiency (Winberg 1956) and the proportion of *Daphnia* consumed by bream and smelt, we estimated the amount of *Daphnia* consumed by these planktivores. In 1971 when the smelt was abundant the fish consumed 470 kg fr wt of *Daphnia* ha^{-1} yr^{-1}, while in 1970 when the smelt was less abundant fish consumption was about 200 kg fr wt ha^{-1} yr^{-1}. The values for biomass 'lost' and biomass consumed are in good agreement with each other.

Standing crop biomass and production

For the calculation of birth and mortality rates we used sim. 1. Frequently during periods of strong population increase in that simulation, a time lag appeared between the observed field densities and the simulated densities (Fig. 7). This did not seriously bias our results since these time lags were not large and because results were analysed predominantly on a relative time scale. Distinctly, for the estimation of biomass and production it is essential to calculate values on an absolute time scale. For that reason we used here sim. 4 instead of sim. 1. This simulation allows for negative mortality to ensure that simulated densities are equal to the observed field densities at all times. The results are given in Tables 8 and 9.

There is a marked seasonal variation in standing crop biomass, production and productivity. Peaks of standing crop biomass and production coincide with population maxima. In *D. hyalina* generally half of the annual production is realized during the 1–2 months of the first population maximum. An-

Table 8. Monthly and annual biomass, production and productivity data for *D. hyalina* in Tjeukemeer. Biomass is expressed as mg org. dry wt m^{-2} (sim. 4).

	Mean biomass (B)			Production (P)			P/B ratio		
	1969	1970	1971	1969	1970	1971	1969	1970	1971
January	–	–	13.2	–	–	11.9	–	–	0.9
February	–	1.3	3.1	–	1.1	4.3	–	0.8	1.4
March	1.4	0.8	1.4	1.3	2.2	2.9	0.9	1.4	2.1
April	1.2	7.3	86.8	1.6	10.7	255.9	1.3	1.5	2.9
May	10.7	249.7	990.2	34.3	882.3	3 682.7	3.2	3.1	3.7
June	142.2	190.3	398.6	573.8	778.1	1 728.7	3.8	4.8	4.3
July	235.6	403.9	47.9	1 027.1	1 516.4	264.8	4.4	3.8	5.5
August	54.8	569.4	10.0	405.7	2 420.2	67.2	7.4	4.3	6.7
September	124.2	129.6	33.3	466.7	339.1	197.1	3.8	2.6	5.9
October	137.3	36.2	12.4	457.7	105.1	55.3	3.3	2.9	4.5
November	25.2	29.8	1.9	52.7	65.5	4.0	2.1	2.2	2.1
December	2.5	11.8	0.6	2.1	16.6	0.6	0.8	1.4	1.0
Annual	61.3	135.8	133.3	2 987.0	6 136.2	6 275.4	48.7	45.2	47.1

Table 9. Monthly and annual biomass, production and productivity data for *D. cucullata* in Tjeukemeer. Biomass is expressed as mg org. dry wt m^{-2} (sim. 4).

	Mean biomass (B)			Production (P)			P/B ratio		
	1969	1970	1971	1969	1970	1971	1969	1970	1971
January	–	–	6.5	–	–	7.1	–	–	1.1
February	–	–	1.7	–	–	0.9	–	–	0.5
March	–	–	4.3	–	–	4.0	–	–	0.9
April	–	–	–	–	–	–	–	–	–
May	–	9.7	–	–	38.9	–	–	4.0	–
June	3.9	3.5	1.0	10.3	9.1	3.1	2.6	2.6	3.1
July	49.7	8.4	2.7	153.1	26.6	13.3	3.1	3.2	4.9
August	34.6	3.1	79.3	131.7	21.0	281.6	3.8	6.8	3.6
September	23.8	5.7	89.8	84.1	19.1	257.5	3.5	3.4	2.9
October	4.9	25.9	1.3	9.7	60.4	2.9	2.0	2.3	2.2
November	0.3	36.5	–	0.4	36.4	–	1.3	1.0	–
December	–	12.2	–	–	10.7	–	–	0.9	–
Annual	9.8	8.8	15.6	389.3	222.2	570.4	39.7	25.3	36.6

nual P/B ratios of *D. hyalina* are 10–50% higher than those of *D. cucullata*. The highest monthly P/B ratios are observed during June–August, the lowest during winter. Maximum ratios differ from minimum ratios by an order of magnitude. There is a marked difference in the P/B ratio of the different size classes: the larger the size the lower the ratio (Table 10). Usually more than half of the total annual production was gonad production (Table 11). This is to be expected from a population which contains for a large part parthenogenetically reproducing females. The total annual *Daphnia* production represented 1–2% of the total annual gross primary production. A similar value was found by George & Edwards (1974) for *D. hyalina* in a eutrophic reservoir in Wales. However, most authors report essentially higher values in the range of 6–13% (Wright 1965; Jones *et al.* 1979; Lei & Armitage 1980). The absolute production values given here for *D. hyalina* in Tjeukemeer are 3–7 times lower than those reported by Bernardi & Canali (1976) for Lago Maggiore and 5–12 times lower that the values given by George & Edwards (1974) and Jones *et al.* (1979) for the same species. This

Table 10. Annual P/B for *D. hyalina* and *D. cucullata* (sim. 4), given in size classes as shown in Tables 6 and 7.

	D. hyalina			*D. cucullata*		
	1969	1970	1971	1969	1970	1971
Size class 1	90.7	88.4	74.3	65.8	59.2	58.3
Size class 2	55.2	56.9	78.5	33.8	26.5	28.5
Size class 3	30.8	33.8	43.4	11.8	8.0	7.6
Size class 4	33.1	34.9	31.2			
Population mean	48.7	45.2	47.1	39.7	25.3	36.6

Table 11. Annual gonad production (P_g) of *D. hyalina* and *D. cucullata* expressed as mg org. dry wt m^{-2} and as a percentage of total annual production (sim. 4).

	D. hyalina			*D. cucullata*		
P_g	1969	1970	1971	1969	1970	1971
Org. dry wt	1 284	3 804	3 138	272	140	376
%	43	62	50	70	63	66

was not caused by a lower productivity but by a 5–30 times lower standing crop biomass. The annual P/B values given here for *D. hyalina* and *D. cucullata* are relatively high. Most other authors report values of about 20 for copepods and cladocerans (Waters 1977). The values given by Wright (1965) for *D. schødleri*, Winberg (1972) for *D. cucullata* in Lake Ilmen (USSR) and Lei & Armitage (1980) for *D. ambigua* in a pond in Kansas, are higher and more comparable with our values.

Conclusions

Although changes in food quantity and quality (probably also temperature) had a marked effect on the fecundity of *Daphnia* in the course of the year, quantitatively the size distribution of adults was a factor of much more importance.

Most probably the size structure of the population is primarily influenced by predation, both by fish and invertebrates, and only secondarily by temperature and food conditions. Fish predation, being positively size selective, strongly affects the mature population and thus the population fecundity. That is why fish predation is much more effective in influencing *Daphnia* population densities than invertebrate predation, which acts predominantly on the immature stages.

Temperature is important mainly because it influences the egg duration time. Thus the dynamics of the two *Daphnia* populations in Tjeukemeer are mainly regulated by fish predation and temperature. As a result, mean population densities, mean standing crop biomass and biomass production were relatively low. These conclusions are in agreement with Zaret (1978) who hypothesized that predation is a primary factor in structuring freshwater systems.

The differences in the observed seasonal variation in densities between *D. hyalina* and *D. cucullata* are most likely caused by seasonal variations in predation pressure, food quality and temperature. *D. hyalina* is more sensitive than *D. cucullata* to the adverse effects of fish predation, low food quality and high summer temperatures. Consequently spring and early summer circumstances are more favourable for *D. hyalina*, while *D. cucullata* is better adapted to the summer situation.

The productivity of the *Daphnia* populations was high when compared with most results given in the literature. Since productivity within the population is closely related to size this is evidently caused by the population size structure which contained predominantly juveniles and small females for most of the year. The observed low gross ecological efficiency between the algal community and the *Daphnia* populations is predominantly an effect of intensive *Daphnia* predation by planktivorous fish, food quantity and quality being of only secondary importance.

Summary

The *Daphnia hyalina* and *Daphnia cucullata* populations in Tjeukemeer were studied over three successive years. To analyse the factors regulating the production of these two species, their population parameters and population dynamics were studied and related to environmental factors. Since the populations show continuous recruitment a model was developed to integrate field data with laboratory data.

Both species showed a marked seasonal variation in their size distribution, mean adult size being highest during spring and winter and lowest during mid-summer. The relation between egg number per ovigerous female with season varied markedly in *D. hyalina* but less in *D. cucullata*. *D. hyalina* showed its highest fecundity during spring and its lowest during mid-summer. Since brood size is exponentially related to female length, the total number of eggs in the population was generally influenced more by variations in the size distribution of the population than by variations in the fecundity.

D. hyalina showed its highest birth rates (30–70% day^{-1}) during spring and early summer. *D. cucullata* reached several maxima during summer and early autumn of 20–30% day^{-1}. The relative mortality (Nos. $Daphnia^{-1}$ day^{-1}) was size dependent, and it acted most severely on the largest females. The absolute mortality was expressed as numbers (Nos. m^{-2} day^{-1}) and as 'biomass-lost' (g org. dry wt m^{-2} day^{-1}). In *D. hyalina* absolute mortality expressed as numbers was highest in the smallest size class, but when expressed as 'biomass-lost' it was highest in the larger size classes (length $>$1.1 mm). A good agreement was found between the amount of 'biomass-lost' as *Daphnia* $>$1.1 mm and the amount of *Daphnia* biomass consumed by the two most important planktivorous fish spp. in the lake.

Although changes in food quality have a significant effect on the fecundity of *Daphnia* it is concluded that the dynamics of the two *Daphnia* populations in Tjeukemeer are mainly regulated by water temperature and fish predation. Fish predation affects both mortality and birth rate. The birth rate is negatively influenced by selective fish predation in two different ways: firstly by the high predation pressure on large ovigerous females, and secondly by the reduction of the population to small females with a low fecundity.

Probably as a result of the high predation pressure by the fish, mean population densities and mean standing crop biomass of *Daphnia* are relatively low and productivity is rather high. Total annual *Daphnia* production was 3.1–6.9 g org. dry wt m^{-2}, and annual P/B ratio ranged from 25 to 40 for *D. cucullata* and from 45 to 49 for *D. hyalina.*

Acknowledgements

Throughout this work we have been indebted to the members of the Tjeukemeer Foodchains and Production Research Group of the Limnological Institute in Oosterzee, we would especially like to thank Mr. W. L. T. van Densen and Mr. H. W. de Nie for much useful advice and discussion. We are also indebted to Professor K. Bakker and Dr. H. L. Golterman, both from the University of Leiden, and also to Dr. P. Hogeweg from the University of Utrecht, for their interest and highly valued criticism during the research period and the preparation of the manuscript. Finally we would like to thank Miss M. E. McLaughlin who made the English text more readable.

References

Anderson, R. S., 1970. Predator–prey relationships and predation rates for crustacean zooplankters from some lakes in western Canada. Can. J. Zool. 48: 1229–1240.

Argentesi, F. R., Bernardi, R. de and Di Cola, G., 1974. Mathematical models for the analysis of population dynamics in species with continuous recruitement. Mem. Ist. Ital. Idrobiol. 31: 245–275.

Arnold, D. E., 1971. Ingestion, assimilation, survival and reproduction of Daphnia pulex fed seven species of blue-green algae. Limnol. Oceanogr. 16: 906–920.

Beattie, D. M., Bromley, H. J., Chambers, M., Goldspink, R., Vijverberg, J., Zalinge, N. P. van and Golterman, H. L., 1972. Limnological studies on Tjeukemeer, a typical Dutch 'polder reservoir'. In: Kaják, Z. and Hillbricht-Illkowska, A. (Eds.) Productivity Problems of Freshwaters. Proceedings of an IBP–UNESCO Symposium on Productivity Problems of Freshwaters, pp. 421–447. Kazimierz Dolny, Poland.

Beattie, D. M., Golterman, H. L. and Vijverberg, J., 1978. An introduction to the Limnology of the Friesian Lakes. Hydrobiologia 58: 49–64.

Banta, A. M. and Brown, L. A., 1929. Control of sex in Cladocera. 1. Crowding the mother as means of controlling male production. Physiol. Zool. 2: 80–92.

Bernardi, R. de, 1974. The dynamics of a population of Daphnia hyalina Leydig in Lago Maggiore, Nothern Italy. Mem. Ist. Ital. Idrobiol. 31: 221–243.

Bernardi, R. de and Canali, S., 1975. Population dynamics of pelagic cladocerans in Lago Maggiore. Mem. Ist. Ital. Idrobiol. 32: 365–392.

Bernardi, R. de and Di Cola, G., 1976. Instantaneous growth rates and production estimation in natural zooplankton populations. Mem. Ist. Ital. Idrobiol. 33: 105–123.

Bottrell, H. H., 1975. Generation time, length of life, instar duration and frequency of moulting, and their relationship to temperature in eight species of Cladocera from the River Thames, Reading. Oecologia (Berl.) 19: 129–140.

Brandl, Z. and Fernando, C. H., 1975. Investigations of the feeding of carnivorous cyclopoids. Verh. int. Verein. Limnol. 19: 2959–2965.

Densen, W. L. T. van, 1973. Populatiedynamika en produktie van spiering, Osmerus eperlanus eperlanus, in het Friese merengebied. Student Report 1973-6, Limnological Institute, Nieuwersluis/Oosterzee, The Netherlands. 41 pp.

Densen, W. L. T. van, 1978. Food of young-of-the-year fish in Tjeukemeer, The Netherlands. In: Proceedings of the Zodiac Symposium on Adaptation, pp. 131-132. PUDOC, Wageningen, The Netherlands.

Dahl, O. J., Myhrhaug, B. and Nygaard, K., 1970. Simula information, common base language. Publication no. S-22, Norwegian Computer Centre, Oslo, Norway. 145 pp.

Dodson, S. I., 1972. Mortality in a population of Daphnia rosea. Ecology 53: 1011-1024.

Dodson, S. I., 1974. Zooplankton competition and predation: an experimental test of the size-efficiency hypothesis. Ecology 55: 605-614.

Edmondson, W. T., 1960. Reproductive rates of rotifers in natural populations. Mem. Ist. Ital. Idrobiol. 12: 21-77.

Edmondson, W. T., 1968. A graphical model for evaluating the use of the egg ratio for measuring birth and death rates. Oecologia (Berl.) 1: 1-37.

Edmondson, W. T., 1972. Instantaneous birth rates of zooplankton (comment). Limnol. Oceanogr. 17: 792-795.

Edmondson, W. T., 1974. Secondary production. Mitt. int. Verein. Limnol. 20: 229-272.

Elster, H. J., 1954. Über die Populationsdynamik von Eudiaptomus gracilis Sars und Heterocope borealis Fischer im Boden-Obersee. Arch. Hydrobiol. Suppl. 20: 546-614.

Elster, H. J., 1955. Ein Beitrag zum Produktionsbiologie des Zooplankton. Verh. int. Verein. Limnol. 12: 404-411.

George, D. G. and Edwards, R. W., 1974. Population dynamics and production of Daphnia hyalina in a eutrophic reservoir. Freshwat. Biol. 4: 445-465.

Gliwicz, Z. M., 1977. Food size selection and seasonal succession of filter feeding zooplankton in a eutrophic lake. Ekol. Polska 25: 179-225.

Gliwicz, Z. M. and Siedlar, E., 1980. Food size limitation and algae interfering with food collection in Daphnia. Arch. Hydrobiol. 88: 155-177.

Goldspink, C.R., 1978. The population density, growth rate and production of bream, Abramis brama, in Tjeukemeer, The Netherlands. J. Fish. Biol. 13: 499-517.

Golterman, H. L., 1970. Methods for chemical analysis of freshwaters. IBP Handbook no. 8, revised 2nd printing. Blackwell, Oxford. 166 pp.

Green, J., 1954. Size and reproduction in Daphnia magna (Crustacea: Cladocera). Proc. Zool. Soc. London. 124: 535-545.

Green, J., 1956. Growth, size and reproduction in Daphnia (Crustacea: Cladocera). Proc. Zool. Soc. London. 126: 173-204.

Gulati, R. D., 1975. A study on the role of herbivorous zooplankton community as primary consumers of phytoplankton in Dutch lakes. Verh. int. Verein. Limnol. 19: 1202-1210.

Haan, H. de, 1972. Some structural and ecological studies on soluble humic compounds from Tjeukemeer. Verh. int. Verein. Limnol. 18: 685-695.

Hall, D. J., 1964. An experimental approach to the dynamics of a natural population of Daphnia galeata mendotae. Ecology 45: 94-112.

Haney, J. F. and Hall, D. J., 1973. Sugar-coated Daphnia: a preservation technique for Cladocera. Limnol. Oceanogr. 18: 331-333.

Hazelzet, J. C. A. M., 1972. The bream population of the Tjeukemeer. Student Report 1972-3, Limnological Institute, Nieuwersluis/Oosterzee, The Netherlands. 96 pp.

Hebert, P. D. N., 1974. Ecological differences between genotypes in a natural population of Daphnia magna. Heredity 33: 327-337.

Hebert, P. D. N., 1977. Niche overlap among species in the Daphnia carinata complex. J. Anim. Ecol. 46: 399-409.

Hebert, P. D. N., 1978. The adaptive significance of cyclomorphosis in Daphnia: more possibilities. Freshwat. Biol. 8: 313-320.

Hogeweg, P., 1974. Concepts for simulation, a comparison based on simulation languages. In: Buffart, H. F. J. M. and Oud, J. H. L. (Eds.) Developments in Systems Theory I, pp. 119-135. Stenfert Kroese, Leiden, The Netherlands.

Hogeweg, P. and Richter, A. F., 1982. INSTAR, a discrete event model for simulating zooplankton population dynamics. Hydrobiologia 95: 275-285.

Hrbáčkova-Esslová, M., 1971. The size distribution of neonates and growth of Daphnia hyalina Leydig (Crustacea, Cladocera) from Lake Maggiore under laboratory conditions. Mem. Ist. Ital. Idrobiol. 27: 357-367.

Hrbáčkova, M., 1974. The size of primiparae and neonates of Daphnia hyalina Leydig (Crustacea: Cladocera) under natural and enriched food conditions. Vest. Cs. Spol. Zool. 38: 98-105.

Hrbáček, J. and Hrbáčkova-Esslová, M., 1960. Fish stock as a protective agent in the occurrence of slow developing dwarf species and strains of the genus, Daphnia. Int. Revue ges. Hydrobiol. 45: 355-358.

Humphreys, W. F., 1979. Production and respiration in animal populations. J. Anim. Ecol. 48: 427-453.

Hutchinson, G. E., 1957. A Treatise on Limnology, Vol. I. Wiley, New York. 1015 pp.

Jones, H. R., Lack, T. J. and Jones, C. S., 1979. Population dynamics of Daphnia hyalina var. lacustris in Farmoor I, a shallow eutrophic reservoir. J. Plankton. Res. 1: 45-65.

Kerfoot, W. C., 1977. Implications of copepod predation. Limnol. Oceanogr. 22: 316-326.

Kott, P., 1953. Modified whirling apparatus for subsampling of plankton. Austr. J. Mar. Freshwat. Res. 4: 387-393.

Kozlowski, J., 1978. Efficiency of exploited populations of Daphnia magna. Experimental study and computer simulation model. Pol. Arch. Hydrobiol. 25: 803-831.

Lei, C. and Armitage, K. B., 1980. Ecological energetics of a Daphnia ambigua population. Hydrobiologia 70: 133-143.

Lynch, M., 1977. Fitness and optimal body size in zooplankton populations. Ecology 58: 763-774.

Lynch, M., 1979. Predation, competition and zooplankton community structure: an experimental study. Limnol. Oceanogr. 24: 253-272.

Moed, J. R. and Hoogveld, H. L., 1982. The algal periodicity in Tjeukemeer during 1968-1978. Hydrobiologia 95: 223-234.

Murugan, N. and Sivaramakrishnan, K. G., 1973. The biology of Simocephalus acutirostratus King (Cladocera: Daphniidae)-laboratory studies of life span, instar duration, egg production, growth and stages in embryonic development. Freshwat. Biol. 3: 77–83.

Nauwerck, A., 1963. Die Beziehungen zwischen Zooplankton und Phytoplankton im See Erken. Sym. Bot. Upsal. 17: 1–163.

Navaneethakrishnan, P. and Michael, R. G., 1971. Egg production and growth in Daphnia carinata King. Proc. Indian Acad. Sci. 73: 117–123.

Nie, H. W. de, Bromley, H. J. and Vijverberg, J., 1980. Distribution patterns of zooplankton in Tjeukemeer, The Netherlands. J. Plankton Res. 2: 317–334.

Porter, K. G., 1973. Selective grazing and differential digestion of algae by zooplankton. Nature. London 244: 179–180.

Prepas, E. and Rigler, F. H., 1978. The enigma of Daphnia death rates. Limnol. Oceanogr. 23: 970–988.

Schindler, J. E., 1970. Food quality and zooplankton nutrition. J. Anim. Ecol. 40: 589–595.

Seitz, A., 1979. On the calculation of birth rates and death rates in fluctuating populations with continuous recruitment. Oecologia (Berl.) 41: 343–360.

Seitz, A., 1980. The coexistence of three species of Daphnia in the Klosterzee. I. Field studies on the dynamics of reproduction. Oecologia (Berl.) 45: 117–130.

Slobodkin, L. B., 1954. Population dynamics in Daphnia obtusa Kurz. Ecol. Monogr. 24: 69–88.

Smyly, W. J. P., 1979. Population dynamics of Daphnia hyalina Leydig (Crustacea: Cladocera) in a productive and unproductive lake in the English Lake District. Hydrobiologia 64: 269–278.

Snedecor, G. W. and Cochran, W. G., 1967. Statistical Methods, 6th edn. The Iowa State University Press, Iowa 493 pp.

Stebler, R., 1979. Das pelagische Crustaceenplankton des Bielersees: Abundanzdynamik, Produktion und Sukzession. Schweiz. Z. Hydrol. 41: 2–34.

Stross, R. G. and Hill, J. C., 1965. Diapause induction in Daphnia requires two stimuli. Science 150: 1462–1464.

Stross, R. G., Nobbs, P. A. and Chisholm, S. W., 1979. SUNDAY, a simulation model of an arctic Daphnia population. Oikos 32: 349–362.

Tappa, D. W., 1965. The dynamics of the association of six limnetic species of Daphnia in Aziscoos Lake, Maine, Ecol. Monogr. 35: 395–423.

Threlkeld, S. T., 1979. Estimating cladoceran birth-rates: the importance of egg mortality and the egg age-distribution. Limnol. Oceanogr. 24: 601–612.

Threlkeld, S. T., 1979a. The midsummer dynamics of two Daphnia species in Wintergreen Lake, Michigan. Ecology 60: 165–179.

Vijverberg, J., 1976. The effect of food quantity and quality on the growth, birth-rate and longevity of Daphnia hyalina Leydig. Hydrobiologia 51: 99–108.

Vijverberg, J., 1977. Population structure, life histories and abundance of copepods in Tjeukemeer, The Netherlands. Freshwat. Biol. 7: 579–597.

Vijverberg, J., 1980. Effect of temperature in laboratory studies on development and growth of Cladocera and Copepoda from Tjeukemeer, The Netherlands. Freshwat. Biol. 10: 317–340.

Vijverberg, J. and Frank, Th. H., 1976. The chemical composition and energy contents of freshwater copepods and cladocerans in relation to their size. Freshwat. Biol. 6: 333–345.

Waters, T. F., 1977. Secondary production in inland waters. Adv. Ecol. Res. 10: 91–164.

Winberg, G. G., 1956. Rate of metabolism and food requirements of fish. Fish. res. Bd Can Transl. Ser. 194.

Winberg, G. G., 1972. Etudes sur le bilan biologique énergetique et la productivité des lacs en Union Sovétique. Verh. int. Verein. Limnol. 18: 39–64.

Wit, C. T. de and Goudriaan, J., 1974. Simulation of ecological processes. Centre for Agricultural Publishing and Documentation, Wageningen, The Netherlands. 167 pp.

Wright, J. C., 1965. Population dynamics and production of Daphnia in Canyon Ferry Reservoir, Montana. Limnol. Oceanogr. 10: 583–590.

Young, J. P. W., 1978. Sexual swarms in Daphnia magna, a cyclic parthenogen. Freshwat. Biol. 8: 279–281.

Zaret, T. M., 1978. A predation model of zooplankton community structure. Verh. int. Verein. Limnol. 20: 2496–2500.

Population dynamics and production of *Acanthocyclops robustus* (Sars) and *Mesocyclops leuckarti* (Claus) in Tjeukemeer

J. Vijverberg & A. F. Richter
Limnological Institute, Tjeukemeer Laboratory, De Akkers 47, 8536 VD Oosterzee, The Netherlands

Keywords: population dynamics, secondary production, cyclopoid copepods, simulation

Abstract

The population dynamics and production ecology of the two dominant copepod species, *Acanthocyclops robustus* and *Mesocyclops leuckarti,* in Tjeukemeer (The Netherlands) were studied for three successive years. Since copopods in Tjeukemeer show continuous recruitment, a population dynamics model 'INSTAR' was developed and used to integrate field data on population density, population structure and fecundity and laboratory data on development rates and length–weight relationships. The pattern of size specific mortality indicated that both invertebrate and vertebrate predation were important in the regulation of population numbers.

Introduction

This paper presents part of the results of a study on the population dynamics and production of copepods and cladocerans in Tjeukemeer, a shallow, eutrophic reservoir in the north of the Netherlands. Previous papers were concerned with length–weight relationships of copepods and cladocerans (Vijverberg & Frank 1976), the population structure and population densities of copepods (Vijverberg 1977), the effect of temperature in laboratory studies on growth and development of copepods and cladocerans (Vijverberg 1980) and the population dynamics and production of *Daphnia* spp. (Vijverberg & Richter 1982). The work was carried out within the framework of an IBP project which also included studies on higher and lower trophic levels (Beattie *et al.* 1971).

The copepods represent about 20% of the mean annual zooplankton biomass. They dominate the cladocerans in winter and early spring and often during autumn too, although they reach their highest densities during late spring and summer. The lake is inhabited by a large population of planktivorous fish, mainly by the young (0+) of various species, and older bream (van Densen 1978; Goldspink 1978; Lammens 1982).

The study deals with two copepod species which are abundant particularly during the growing season (May–September): *Acanthocyclops robustus* (Sars), the dominant copepod species in the lake, and *Mesocyclops leuckarti* (Claus) (subdominant). The studies on two other copepod species, *Eurytemora affinis* (Poppe) and *Cyclops vicinus* (Uljanin), both of which are abundant during spring only, will be presented in a separate paper.

In order to analyse the factors regulating the production of the two former species, their population dynamics were studied and, whenever possible, related to the environmental factors. Since copepods in Tjeukemeer show continuous recruitment, just as cladocerans do, a population dynamics model was used to integrate field data on population densities, population structure and fecundity and laboratory data on development rates and length–weight relationships.

Hydrobiologia 95, 261–274 (1982). 0018-8158/82/0953-0274/\$02.80.

Study area

Tjeukemeer is a shallow (mean depth = 1.5 m), eutrophic, freshwater lake with a surface area of about 21 km². The littoral zone is poorly developed; the index of shoreline development (D_l), calculated according to Hutchinson (1957) is 1.25. For a detailed description of the lake, its reservoir function and the position of the sampling stations see Vijverberg (1977) and de Nie *et al.* (1980).

Material and methods

Field data

For three years (1969, 1970 and 1971), the lake was sampled using a 5 l Friedinger sampler at fixed stations at weekly intervals, except during ice cover. This resulted in 130 collecting dates. At each station two samples were taken, one just below the surface and the other just above the bottom, and these were concentrated using a 80 μm mesh plankton gauze.

Population densities, relative abundance of instars, mean size of instars, sex ratio and fecundity were measured (Vijverberg 1977). The copepodites were measured under the microscope from the anterior end of the cephalothorax to the tip of the furcal ramus. We did not discriminate between the different naupliar instars. Initially the nauplii of the four copepod species were counted together; this gave a total naupliar density. Later, the naupliar density of each species was estimated from the relative abundance of the first copepodite instar of that particular species. Density for each day was computed by linear interpolation between sampling dates.

Water temperatures measured daily showed a pronounced seasonality. The highest temperatures, 23–24 °C, were recorded during July and August. There was an ice cover during winter lasting 1–3 months. The monthly water temperatures, expressed as day-degrees above 10 °C, are given by Vijverberg & Richter (1982) in Table 1.

Laboratory estimates

The durations of egg development, and combined naupliar and combined copepodite instars in relation to temperature were measured in the laboratory under semi-natural conditions. The methods, culture conditions and the results are given by Vijverberg (1980). Additional culture experiments were carried out at 17.5 °C to assess the relative duration of the separate copepodite instars. We took readings every 12 h and the culture conditions were the same as in the previous experiments.

For calculating individual biomass, the length–weight relationship for the cyclopoid copepods of Tjeukemeer (Vijverberg & Frank 1976) was used.

$$W = 1.74\ L^{1.68} \qquad (1)$$

where W = weight as μg ind.$^{-1}$ (organic matter minus chitine)
L = body length (mm).

Population dynamics model

The technical aspects of the 'INSTAR' model were developed by one of us (A.F.R.) in co-operation with Dr. P. Hogeweg (The Netherlands). It is a discrete event-oriented model designed to stimulate reproduction, individual growth by moulting and density changes per instar owing to birth, death or growth. The structure of the model together with some of its most important properties are summarized by Vijverberg & Richter (1982). A more detailed account is given by Hogeweg & Richter (1982).

The model used here for the calculation of copepod birth rate, mortality rate, production and productivity is essentially the one used to study the population dynamics and production of *Daphnia* spp. (Vijverberg & Richter 1982). Some minor changes were introduced to adapt the model to the life cycle events of an individual copepod (Fig. 1). Output was organized per instar or size class (Table 1), but individuals in the model system kept acting as independent entities.

The durations of the separate copepodite instars were computed from the lake temperatures, the relation between temperature and the duration of the combined copepodite instars, and from the relative duration of each specific copepodite instar at 17.5 °C. Growth in length per moult was estimated from the mean length of the instars in the field population at that time.

Somatic weight of copepodite instars is calculated using Eq. (1). On the basis of the results of Lair (1978), we estimated the weight per egg to be 16% of

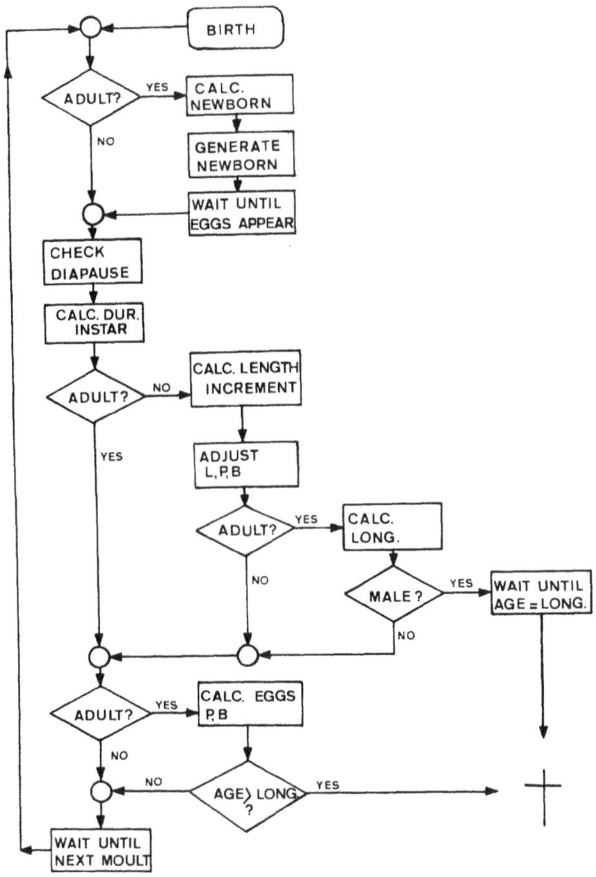

Fig. 1. Simplified flow diagram of the life cycle of an individual copepod in the simulation model 'INSTAR'. For explanations see text.

the weight of the CI; the mean weight of the naupliar instars I–IV was assumed to be equal to the egg weight. The weight of the 5th and last naupliar instar was taken as 50% of the weight of the first copepodite instar (Burgis 1975), and its duration as 20% of the duration of the combined naupliar instars (Miller *et al.* 1977). Weight increments were assessed after each moult and were used for the calculation of the standing crop biomass (B) and the biomass production per unit of time (P).

The animal becomes sub-adult in the 5th copepodite instar (CV). Fifty per cent of the CIV instars develop into females. The number of females, percentage of females with brood and the brood size were derived from field observations (Vijverberg 1977). Egg durations in relation to temperature and the time lapse between broods were derived from laboratory data (Vijverberg 1980). This information is used for the estimation of standing crop biomass (B) and gonad production (P_g). Gonad production is added to somatic production to give the total biomass production (P).

In the simulation model, mortality occurs twice a day. For each instar class the number of copepods in the simulation is compared with those in the field at the same time, the difference being mortality. This approach is similar to that of Argentesi *et al.* (1974) and Bernardi & Di Cola (1976), but their models are not based on the individual life cycle of zooplankton species and do not take into account the egg mortality caused by the death of ovigerous females.

Results and discussion

Duration of copepodite instars

The instar durations of *A. robustus* were shorter than those of *M. leuckarti* (Table 2). Males developed faster than females, although the time difference was only small in the case of *M. leuckarti*. The duration of the *A. robustus* CV (♀s) instar was significantly longer, compared with CV (♂s) and younger copepodite instars. Such a long duration

Table 1. Mean length (L, mm) with S.D. of naupliar and copepodite size classes. Number of observations in parentheses.

	Size class no. Instar	1 Nauplia	2 CI	3 CII & CIII	4 CIV & CV	5 ♂ (CVI)	6 ♀ (CVI)
A. robustus	1969	0.2	0.39 ± 0.05 (12)	0.52 ± 0.08 (24)	0.77 ± 0.13 (36)	0.82 ± 0.10 (12)	1.12 ± 0.11 (12)
	1970	0.2	0.42 ± 0.04 (14)	0.55 ± 0.06 (28)	0.80 ± 0.13 (42)	0.86 ± 0.08 (14)	1.22 ± 0.12 (14)
	1971	0.2	0.44 ± 0.04 (13)	0.59 ± 0.08 (26)	0.80 ± 0.13 (39)	0.87 ± 0.04 (13)	1.24 ± 0.08 (13)
M. leuckarti	1969	0.2	0.41 ± 0.01 (12)	0.50 ± 0.04 (24)	0.71 ± 0.08 (36)	0.76 ± 0.02 (12)	0.92 ± 0.02 (12)
	1970	0.2	0.41 ± 0.02 (13)	0.53 ± 0.05 (26)	0.72 ± 0.07 (39)	0.76 ± 0.03 (13)	0.94 ± 0.02 (13)
	1971	0.2	0.41 ± 0.01 (14)	0.50 ± 0.04 (28)	0.70 ± 0.07 (42)	0.72 ± 0.04 (14)	0.92 ± 0.04 (14)

Table 2. Duration in days of copepodite instars of *A. robustus* and *M. leuckarti* at 17.5 °C, as a percentage of the duration of the combined copepodite instars (CI–CV). Mean ± 95% CL. Number of observations in parentheses.

Species	Instar no.	Duration (days)	(%) ♂	(%) ♀
A. robustus	CI	1.2 ± 0.13 (16)	23	22
	CII	1.0 ± 0.06 (16)	20	18
	CIII	0.9 ± 0.17 (15)	18	16
	CIV	0.8 ± 0.12 (16)	16	15
	CV-♂s	1.2 ± 0.20 (7)	23	
	CV-♀s	1.6 ± 0.20 (9)		29
M. leuckarti	CI	1.6 ± 0.11 (15)	22	22
	CII	1.2 ± 0.13 (15)	17	16
	CIII	1.5 ± 0.25 (14)	21	20
	CIV	1.5 ± 0.25 (10)	21	20
	CV-♂s	1.4 ± 0.30 (9)	19	
	CV-♀s	1.6 ± 0.23 (10)		22

of CV and/or CIV has often been observed in copepods (Eckstein 1964; Ivanova 1973; Munro 1974; Jacobs & Bouwhuis 1979; Jamieson 1980; Vidal 1980). In *M. leuckarti* the copepodite instar durations are nearly constant. An exception is perhaps the relatively short duration of the CII, although differences were only significant when compared with the durations of the CI and CV (♀s). A moulting pattern in which each moult to moult phase is completed in a nearly constant time is called isochronal development (Miller *et al.* 1977). Isochronal development as observed here for *M. leuckarti* is exceptional for cyclopoid copepods. To date it had only been known for a few calanoid copepod species, mainly from marine habitats (Katona 1971; Johnson & Miller 1974; Corkett & McLaren 1978).

Population dynamics

The changes in the population densities and population structure were simulated in three different ways:

Simulation 1. Each simulation of the population development covered 8–9 months of field data. Thus, three simulations were needed for the three sampling years. The simulation of the *A. robustus* population development was started as soon as the field population reached a density of 2.0 copepodites l^{-1}. The population growth of *M. leuckarti* was initiated as soon as the diapause of the resting stage CV was broken, which was usually in March. Mortality is always assumed to be ≥0.

Simulation 2. The same as Sim. 1, but mortality of the nauplii could also be <0. This seems to be a paradox but mortality is computed by comparing simulated densities with those observed in the field and since these figures are both subject to error, the mortality calculated in this way may be negative.

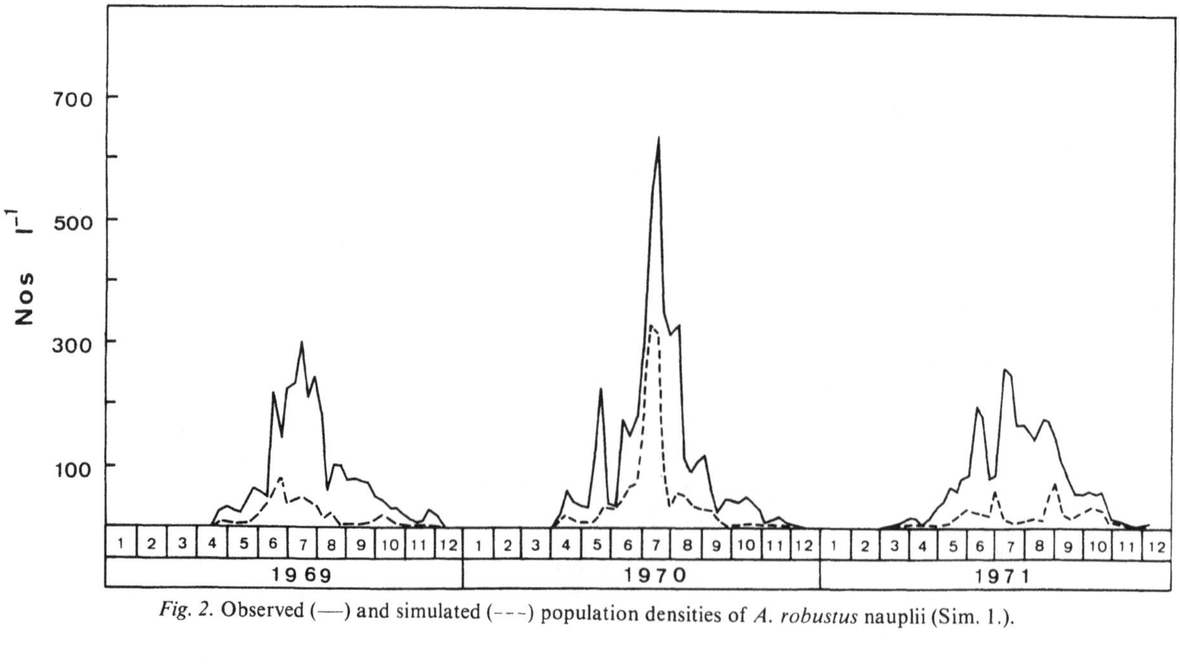

Fig. 2. Observed (—) and simulated (---) population densities of *A. robustus* nauplii (Sim. 1.).

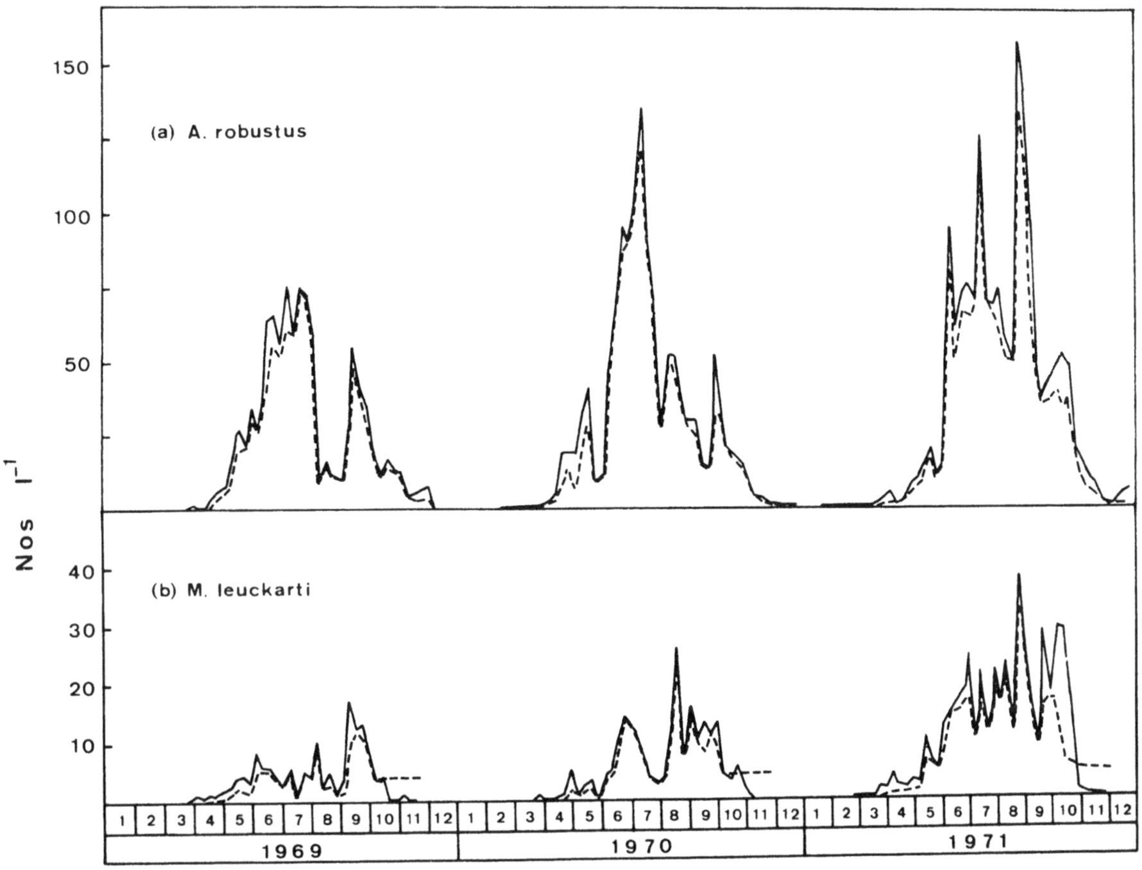

Fig. 3. Observed (—) and simulated (---) population densities of total copepodites (Sim. 2). (a) *A. robustus*, (b) *M. leuckarti*. Note the different scales.

Simulation 3. The same as Sim. 2, but mortality of the copepodite instars could also be <0.

Simulation 1 was used for the nauplii densities of *A. robustus* (Fig. 2). For all three years the simulated densities are markedly lower than those observed in the field. A similar discrepancy was observed for *M. leuckarti*. Apparently the recruitment of nauplii from eggs was underestimated. That this inconsistency in our data is due to an error in the durations of eggs and/or combined naupliar instars is unlikely since temperature is the main regulating factor here. Most probably we underestimated the number of eggs in the population due to the vertical distribution of the ovigerous females. In an earlier study we found no significant difference between the densities in the near-surface and near-bottom samples for *A. robustus* and only a slight difference for *M. leuckarti* (de Nie *et al.* 1980). It is possible however, that a substantial part of the adult population was concentrated in or just above the bottom substrate. This layer could not be sampled with the Friedinger type of volume sampler which we used, consequently resulting in an underestimation of the egg stock. This hypothesis is corroborated by the composition of the bream's diet. This fish is the most abundant benthivore in the lake. The large numbers of adult cyclopoid copepods encountered in their stomachs are usually associated with micro- and macro-benthic species and not with pelagic zooplankton species (Lammens, pers. comm.). This indicates that adults of cyclopoid copepods live as micro-benthos in or on the bottom substrate. This aspect needs more attention in the near future.

Due to the inaccuracy of egg-stock estimates we had to use Sim. 2 instead of Sim. 1 to simulate the population dynamics of the copepodites. For both species the densities of copepodites obtained by simulation were close to those observed in the field

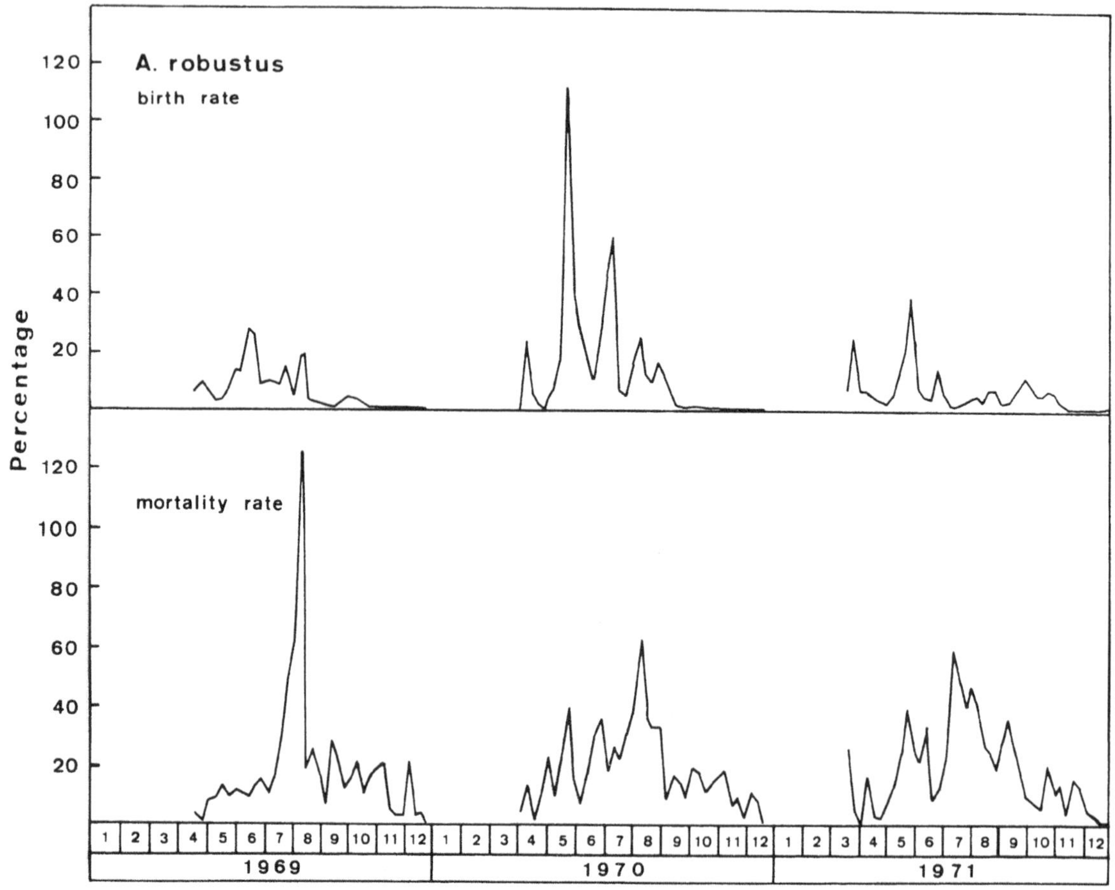

Fig. 4. Seasonal and annual variations in mean daily birth rates (b, %) per week and mean daily mortality rates (d, %) per week of *A. robustus* (Sim. 2).

(Fig. 3). Even the strong population increases were usually well simulated. The winter diapause of *M. leuckarti* was also included in the simulation. During the second half of September the first CV instars went into diapause and by late October most individuals were in their resting stage CV. In March of the following year diapause was broken and the first CV instars developed into adults. In the lake nearly all diapausing CV instars sink to the bottom and thus are not encountered in the collected samples. This explains why the simulated densities in October–November are higher than those observed in the field.

The birth rate is defined as the percentage of nauplii born per copepod per day. Mean rates were computed for weekly intervals. The peaks in the birth rates of *A. robustus* were generally observed during May–August (Fig. 4a) when the fecundity was still relatively high (Vijverberg 1977); the high water temperatures ensured a short egg duration of 1–2.5 days. The high birth rates, i.e. >50% copepod^{-1} day^{-1}, occurred during the summer of 1970, due mainly to a relatively high percentage of ovigerous females in the population (Vijverberg 1977).

The seasonal pattern in the birth rates of *M. leuckarti* differed from that of *A. robustus* (Fig. 5a). The highest rates were generally observed during July–August, except in 1971, when slightly higher rates were observed in April. The variations in birth rate reflect the seasonal changes in fecundity and water temperature. Fecundity showed two annual maxima, in April and during July–September, the first being somewhat higher than the second (Vijverberg 1977). As water temperatures were highest during June–August, the birth rate maxima were reached usually during July–August. A somewhat different pattern observed in 1971 can be explained by the higher water temperatures in April 1971 than in April 1970.

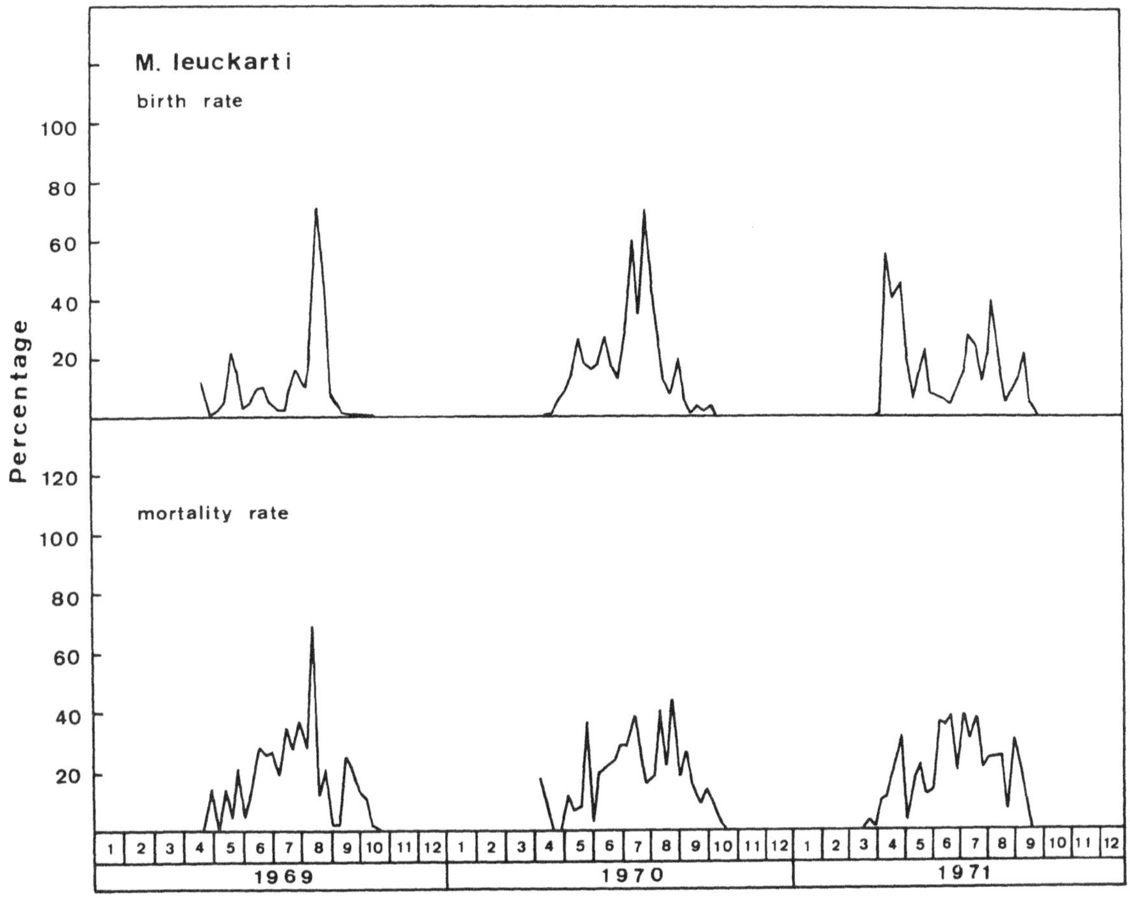

Fig. 5. Seasonal and annual variations in mean daily birth rates (b, %) per week and mean daily mortality rates (d, %) per week of *M. leuckarti* (Sim. 2).

The mortality rate is defined here as the percentage of copepodites (CI–CVI) dying per copepodite per day. The seasonal pattern was very similar for both species during the three years, although *A. robustus* reached noticeably higher maximum rates (60–120%) than *M. leuckarti* (40–70%). The highest rates were usually observed during July–August, when predation pressure exercised by fish was most severe. The relatively low mortality rate in the case of *M. leuckarti* compared with those of *A. robustus* during autumn and winter is an artefact. Since in September *M. leuckarti* went into diapause in the bottom substrate, and thus the data on the densities of the resting stages were lacking, we assumed zero mortality for diapausing individuals. However, in this phase they were predated by the benthivorous bream, although the predation pressure was relatively mild (Lammens, pers. comm.).

Our results on birth and mortality rates are not fully comparable with most of those reported in the literature. Firstly, we underestimated the number of females with eggs and thus the birth rate too, due to sampling errors; secondly nearly all previous authors used Edmondson's formula (Edmondson 1960, 1974) to compute the birth and death rates, which according to Seitz (1979) and Threlkeld (1979) introduces a serious bias in the birth rate estimates, and thus also in the mortality rate estimates. This error is caused because of egg mortality as a result of mortality of adult ovigerous females. This error is particularly significant when positive size-selective predation by planktivorous fish is an important cause of death, and can therefore, lead to a considerable overestimation of the birth rate. According to Threlkeld (1979), this may cause an overestimation of the birth rate by about 50% for *Daphnia*.

The death and birth rates found by us are very

Table 3. Daily means of size-specific mortality rates (d, %) of *A. robustus* in Tjeukemeer in different months during 1969, 1970 and 1971 (Sim. 2). Size classes as in Table 1.

	1969				1970				1971			
	3	4	5	6	3	4	5	6	3	4	5	6
March	–	–	–	–	–	–	–	–	–	0	31	20
April	0	0	26	0	15	0	14	12	6	4	2	9
May	8	13	0	9	18	30	4	13	29	8	10	28
June	9	2	4	33	19	7	5	64	12	3	9	51
July	17	27	6	32	20	24	7	29	44	9	11	117
August	33	14	23	105	38	15	19	60	35	1	9	81
September	3	3	11	75	5	4	11	93	39	3	15	48
October	4	23	14	39	6	29	12	15	4	18	10	18
November	13	9	10	30	5	8	25	103	7	14	13	28
December	12	8	22	48	0	0	15	25	0	3	1	0

Table 4. Daily means of size-specific mortality rates (d, %) of *M. leuckarti* in Tjeukemeer during different months in the years 1969, 1970 and 1971 (Sim. 2). Size classes as in Table 1.

	1969				1970				1971			
	3	4	5	6	3	4	5	6	3	4	5	6
March	–	–	–	–	–	–	–	–	–	0	64	230
April	0	0	19	100	0	0	6	14	0	0	1	0
May	2	3	7	21	10	11	12	13	11	12	15	10
June	11	4	28	20	15	6	13	6	6	3	9	19
July	20	14	16	15	53	6	13	10	38	5	14	30
August	40	1	23	26	10	25	49	24	30	6	18	18
September	9	7	69	12	14	1	82	39	21	9	42	62
October	27	1	34	7	0	1	140	40	13	1	–	–
November	–	0	–	–	–	0	–	–	–	0	–	–
December	–	0	–	–	–	0	–	–	–	0	–	–

similar to the those reported by Cummins *et al.* (1969) for *Diaptomus siciloides* and *Cyclops vernalis* in the eutrophic Sanctuary Lake (Penn., U.S.A.). They also compare well with the rates given by Burgis (1971) for *Thermocyclops hyalinus* in the tropical and eutrophic Lake George (Uganda). The mortality rates are also similar to those of *Acartia clausii* in a temperate lagoon in Wasington D.C., U.S.A. (Landry 1978). However, the rates observed by us are much higher than those reported for calanoid copepods in lakes in New Zealand (Chapman 1973; Green 1976; Burns 1979).

The mortality rates per size group are given in Tables 3–6. Because of our underestimation of the egg stock, an accurate estimate of the nauplii mortality was not possible. Initial simulations (Sim. 2) to estimate the CI mortality rate showed enormously high mortality rates, often about 1 000% copepodite^{-1} day^{-1}. This seemed unrealistic when compared with the mortality rates of usually <40% for the CII and CIII instars.

Although the recruitment from eggs to naupliar instars in our simulation proved to be too low in comparison with the observed field densities, the recruitment from the naupliar instar (NV) to the first copepodite instar (CI) appeared to be much too high. This can be explained by very high mortality in the naupliar phase, causing a progressive decrease in the densities of the naupliar instars (NI–NV). Since we had no detailed knowledge of the age distribution within the naupliar phase, the age-specific mortality of nauplii could not be properly simulated. The problem was provisionally solved by creating a second naupliar instar class (NV) and by assuming the mortality rate of CI to be equal to that of size class 3 (CII & CIII). The density of the NV instar was computed by simulation, using the NV duration and the CI mortality as input. Because we underestimated the density of the adult population, the mortality rates of males and females were overestimated to some extent. The magnitude of error depends mainly on the fraction of the population missed with the employed sampling procedure.

Neither species showed a distinct seasonal pattern in its size specific mortality (Tables 3 and 4). For the size class 6 of *A. robustus* the mortality rate was generally higher in the females than in the males. The mortality rates for the size classes 3 to 5 were similar to each other. *M. leuckarti* males and females showed the highest mortality rates, while CII, III exhibited a higher rate than CIV, V. The field mortality rates were much higher than the physiological mortality rates of about 2% observed for these instars in the laboratory cultures. Thus it is very likely that the mortality in the field populations is due mainly to predation. High mortality rates do not necessarily mean a high absolute mortality, e.g. mortality rates as high as 100% ind.$^{-1}$ day^{-1} were observed only in size classes with very low densities. The data on size dependent mortality, on an annual basis, are summarized in Tables 5 and 6. The relative density gives the percentage distribution of the density over the size classes. The absolute mortality is expressed both as numbers per size class m^{-2} day^{-1}, and as 'loss' of organic matter per size class day^{-1}.

Table 5. Mean population densities and mean mortality rates per size class of *A. robustus*. Size classes as in Table 1. Mortality (Sim. 2), biomass 'lost' (Sim. 3).

	Year	Size classes 2	3	4	5	6
Mean density	1969	6 478	8 272	5 945	6 996	1 086
(nos. m^{-2})	1970	10 120	11 229	6 748	7 569	1 234
	1971	14 029	12 699	10 699	9 577	1 509
Mean relative	1969	22	29	21	24	4
densities (%)	1970	27	30	18	21	3
	1971	29	26	22	20	3
Mean mortality	1969	–	1 258	944	759	594
(nos. m^{-2} day^{-1})	1970	–	2 622	1 327	827	663
	1971	–	4 251	642	1 201	972
Mean relative	1969	–	35	27	21	17
mortality (%)	1970	–	48	24	15	12
	1971	–	60	9	17	14
Mean mortality	1969	–	11	10	10	43
rate (% day^{-1})	1970	–	14	12	12	49
	1971	–	21	7	10	43
Mean mortality	1969	–	13	28	20	39
in org. dry wt.	1970	–	21	26	16	37
(%)	1971	–	26	3	24	47
Mean biomass 'lost'	1969	–	0.4	0.9	0.6	1.2
(mg org. dry wt	1970	–	1.1	1.3	0.8	1.9
m^{-2} day^{-1})	1971	–	1.5	0.2	1.4	2.8

Table 6. Mean population densities and mean mortality rates per size class of *M. leuckarti*. Size classes as in Table 1. Mortality (Sim. 2), biomass 'lost' (Sim. 3).

	Year	Size classes 2	3	4	5	6
Mean density	1969	896	1 201	2 772	369	327
(nos. m^{-2})	1970	1 482	1 749	3 351	487	644
	1971	3 258	3 391	6 684	1 551	933
Mean relative	1969	16	22	50	7	6
densities (%)	1970	19	23	43	6	8
	1971	21	21	42	10	6
Mean mortality	1969	–	194	70	109	68
(nos. m^{-2} day^{-1})	1970	–	338	148	164	105
	1971	–	861	141	346	256
Mean relative	1969	–	44	16	25	15
mortality (%)	1970	–	45	20	22	14
	1971	–	54	9	22	16
Mean mortality	1969	–	14	6	29	15
rate (% day^{-1})	1970	–	17	5	33	16
	1971	–	19	4	19	29
Mean mortality	1969	–	24	7	30	39
in org. dry wt	1970	–	20	17	28	35
(%)	1971	–	22	5	44	29
Mean biomass 'lost'	1969	–	0.10	0.03	0.12	0.16
(mg org. dry wt	1970	–	0.12	0.10	0.16	0.21
m^{-2} day^{-1})	1971	–	0.27	0.06	0.54	0.35

Two important conclusions are drawn: first, the relative mortality among the smallest individuals (size class 3) is rather high, namely 35–60% for *A. robustus* and 45–55% for *M. leuckarti;* secondly, the 'loss' of biomass as a result of mortality is the highest in the adults (size classes 5 and 6): 50–70% for *A. robustus* and 60–70% for *M. leuckarti* of the net production is lost in the adult stage.

The relatively high mortality in the nauplii and early copepodite instars may be due to starvation, to an increased mortality during moulting, or to predation. However, since mortality during the juvenile phase was rather low in our cultures, under semi-natural conditions using lake water as culture medium, high mortality is most likely caused by predation. As fish, except for their larvae, feed predominantly on the advanced copepodite instars (Nilssen 1977; van Densen, in prep.; Lammens, pers. comm.), invertebrate predators are most likely to be responsible for the naupliar mortality. Advanced stages of cyclopoid copepods, the raptorial cladoceran *Leptodora kindtii* and the opossum shrimp *Neomysis integer* were all present during most of the year and are known to feed on nauplii and small copepodites (Cummins *et al.* 1969; McQueen 1969; Monakov 1972; Brandle & Fernando 1975; Bremer & Vijverberg 1982).

The large proportion of biomass 'lost' in the adult phase is most probably the effect of positive size-selective predation by fish. In Tjeukemeer the bream (*Abramis brama*) dominates the fish biomass. Its population consists of around 15 different year classes; the classes 1959 and 1963 are particularly strong. Because of this, the annual variation in standing stock biomass was rather small during the research period. The annual production of bream in 1969 was about 34 kg fr wt ha^{-1} (Goldspink 1978). Among the young fish, the smelt (*Osmerus eperlanus*) is generally the most important zooplankton predator. The smelt stock shows large variations from year to year (van Densen, in prep.). Detailed information about the smelt populations in 1969 and 1970 is lacking. A survey during summer and autumn of 1970 indicated low standing stocks, as against a much higher mean annual stock of 15–20 kg fr wt ha^{-1} and a production of about 60 kg fr wt ha^{-1} yr^{-1} in 1971 (van Densen 1973). Bream and smelt fed mainly on copepods during periods when *Daphnia* was not available. On an annual basis, copepods comprised about 10% (dry wt) of the diet of the bream, and about 25% of that of the smelt (Lammens 1982; van Densen, in prep.). Assuming a growth efficiency of 0.1 (Humphreys 1979) and an assimilation efficiency of 0.8 (Winberg 1956), the consumption of copepods by fish can be calculated. It must be noted that the growth efficiency depends upon the size class structure of the population and is higher for smaller animals. Using the value given by Humphreys, which is a mean value based on a literature review, the calculated food consumption of the bream is underestimated and that of the smelt overestimated to some extent.

Table 7. Monthly and annual biomass, production and productivity data for *A. robustus* in Tjeukemeer. Biomass is expressed as mg organic dry wt m^{-2} (Sim. 3).

	Mean biomass (B)			Production (P)			P/B ratio		
	1969	1970	1971	1969	1970	1971	1969	1970	1971
January	–	–	–	–	–	–	–	–	–
February	–	–	–	–	–	–	–	–	–
March	–	–	2.2	–	–	4.8	–	–	2.2
April	5.2	17.1	7.1	13.5	62.5	28.2	2.6	3.5	4.0
May	39.0	49.7	26.8	155.9	299.7	201.3	4.0	6.0	7.5
June	91.3	113.1	106.0	517.7	857.3	706.4	5.7	7.6	6.7
July	107.9	180.4	119.0	1 038.4	1 456.8	1 238.3	9.6	8.1	10.4
August	39.6	71.1	132.2	642.2	853.8	962.4	16.2	12.0	7.3
September	50.7	42.5	95.0	390.3	266.0	476.9	7.7	6.2	5.0
October	24.3	34.1	62.8	100.9	134.3	245.8	4.1	3.9	3.9
November	10.8	8.7	12.4	28.4	31.4	30.8	2.6	3.6	2.5
December	2.8	2.7	10.7	5.5	5.3	16.5	2.0	1.9	1.5
Annual	31.0	43.6	48.0	2 892.5	3 967.0	3 911.6	93.2	91.1	81.4

Table 8. Monthly and annual biomass, production and productivity data for *M. leuckarti* in Tjeukemeer. Biomass is expressed as mg organic dry wt m^{-2} (Sim. 3).

	Mean biomass (B)			Production (P)			P/B ratio		
	1969	1970	1971	1969	1970	1971	1969	1970	1971
January	–	–	–	–	–	–	–	–	–
February	–	–	–	–	–	–	–	–	–
March	–	–	1.3	–	–	4.7	–	–	3.6
April	2.4	2.6	6.4	5.7	6.4	14.0	2.4	2.4	2.2
May	4.4	4.4	13.0	16.7	21.7	94.3	3.8	4.9	7.3
June	8.7	16.4	26.3	54.2	93.2	163.6	6.3	5.7	6.2
July	6.3	13.8	24.7	61.0	91.9	185.7	9.7	6.7	7.5
August	8.6	19.3	33.2	105.3	172.7	217.8	12.2	9.0	6.6
September	17.2	19.6	25.6	126.5	116.2	170.0	7.4	5.9	6.6
October	10.4	10.2	28.1	10.9	4.0	26.6	1.0	0.4	0.9
November	9.3	9.0	18.6	0	0	0	0	0	0
December	9.3	9.0	18.6	0	0	0	0	0	0
Annual	6.4	8.7	16.4	380.3	506.0	876.3	59.5	58.1	53.5

The bream consumed annually:

$34 \times [\frac{100}{10}] \times [\frac{10}{8}] \times [\frac{10}{100}] = 43$ kg fr wt of copepods ha^{-1}.

For the smelt during 1971 this was:

$60 \times [\frac{100}{10}] \times [\frac{10}{8}] \times [\frac{25}{100}] = 188$ kg fr wt of copepods ha^{-1} yr^{-1}.

Thus in 1971 when the smelt was abundant, the fish in Tjeukemeer consumed about 230 kg fr wt of copepods ha^{-1}, while in 1970 this ranged from 50 to 100 kg fr wt ha^{-1}.

This total amount of copepod biomass 'lost' in the adult phase during each year was computed from the data in Tables 6 and 7. To compare the results with the consumption figures, biomass was expressed in fresh weight; organic matter (dry wt) was assumed to be 10% of fresh weight (Winberg *et al.* 1971). Annually, the copepod biomass 'lost' in the adult phase was approximately 80 in 1969, 110 in 1970 and 180 kg fr wt ha^{-1} in 1971. These values compare fairly well with the previously computed fish consumption estimates.

Biomass and production

For the calculation of birth and mortality rates we used Sim. 2. During periods of strong population increase, a time lag regularly appeared between the observed field densities and the simulated ones (Fig. 3). This did not seriously bias our results since these time lags were not great and because results were analysed predominantly on a relative time scale. However, for the estimation of standing crop biomass, production and productivity it is essential to calculate values on an absolute time scale. Thus we used here Sim. 3 instead of Sim. 2. This simulation allows for negative mortality in naupliar and copepodite instars, to ensure that simulated densities match the observed field densities.

Both species exhibit a pronounced seasonality in their standing crop biomass, production and productivity (Tables 7 and 8). *A. robustus* showed its highest production during July; 75–80% of the annual production occurred from June to August. In contrast, *M. leuckarti* had its highest production during August and September; 85–95% of its annual production took place during the 4 months from June to September. The annual P/B ratio of *A. robustus* was about 35% higher than that of *M. leuckarti*. The highest monthly P/B ratio was observed during July–August, coinciding with the water temperature maximum.

The P/B ratio of the different size classes differed markedly; the nauplii (size class 1) ratio was comparatively high (Table 9). The relatively low productivity of size class 4 of *M. leuckarti* is due to the diapause of their CIV instars, these resting stages show an individual growth rate of zero. When the adult instar is reached the somatic growth stops. In

Table 9. Annual P/B for *A. robustus* and *M. leuckarti* (Sim. 3), given per size class as shown in Table 1.

	A. robustus			*M. leuckarti*		
	1969	1970	1971	1969	1970	1971
Size class 1	282.1	233.7	239.6	264.2	213.9	205.9
Size class 2	62.0	55.2	59.5	40.5	46.1	38.0
Size class 3	72.9	59.3	67.3	58.8	53.7	56.3
Size class 4	59.9	89.0	60.1	12.9	15.5	18.3
Size class 5	0	0	0	0	0	0
Size class 6	15.4	31.4	9.3	11.4	18.9	17.6
Population mean	93.2	91.1	81.4	59.5	58.1	53.5

Table 10. Total annual production (P, mg org. dry wt m^{-2}) and annual gonad production (P_g, %) and nauplii production (P_n, %) expressed as a percentage of the total production (Sim. 3).

	A. robustus			*M. leuckarti*		
	1969	1970	1971	1969	1970	1971
P	2 893	3 967	3 912	380	506	876
P_g	4	8	5	4	7	8
P_n	70	64	61	73	68	64

the males P/B equals zero; in females egg production results in a ratio of around 10 to 30. The gonads contributed 4–8% to the total copepod production, half of which was nauplii production (Table 10). The latter is a crude estimate since we lacked detailed information about the age dependent mortality within the naupliar phase, but undoubtedly it points to the importance of the nauplii production.

Most production values reported for individual copepod species range from 0.2 to 10 g dry wt m^{-2} (see Waters 1977 for review). Our values for *A. robustus* and *M. leuckarti* are well within this range. However, annual P/B ratios for these species are markedly higher than the most reported for copepod species in both temperate and tropical water bodies, namely, from 5 to 30 (Kibby 1971; Lévêque *et al.* 1972; Burgis 1974; Bosselman 1975; George 1976; Green 1976; Burns 1979). The annual P/B ratio of 58 given by Landry (1978) for *Acartia clausii* in a temperate lagoon in Washington D.C. (U.S.A.) is more comparable with our results.

The two main reasons for the high ratio found by us are: first, the nauplii contribution to the total copepod stock is proportionally large and since the nauplii are several times more productive than copepodites, a high P/B ratio for the population as a whole is the result. Secondly, the durations of the combined copepodite instars assessed here and in a previous study (Vijverberg 1980) are relatively short when compared with those cited in the literature. The relevant literature concerns only *M. leuckarti* (Monakov 1958; Smyly 1961; Gophen 1976; Jamieson 1980); no data were available for *A. robustus.* For corresponding temperatures (range, 15–25 °C) most authors report durations 1.3–4.5 times longer than ours. As these authors used artificial culture media, comprising only one or two food species as against natural food used by us, it is difficult to interpret these differences. Both food quantity and quality, and in some cases also hereditary factors, are involved (for a detailed discussion see Vijverberg 1980). It is to be expected, however, that the use of natural food and food concentrations in laboratory cultures, as in the present study, will give more realistic estimates of the development rates in the field.

In conclusion, the simulations revealed unexpected gaps in our knowledge about the population dynamics and production ecology of *A. robustus* and *M. leuckarti* in the Tjeukemeer. Future research will be directed at examining the abundance of the different naupliar instars in the field, their durations and the micro-distributions of adult copepods in and just above the substrate.

Our results show that cyclopoid copepods play a significant role in the economy of the Tjeukemeer ecosystem. They are important food organisms for most 0+ fish, 1+ and older bream and probably also for other invertebrates, as advanced cyclopoid instars, *Leptodora* and *Neomysis.*

As a secondary producer *A. robustus* is only of subsidiary importance to *Daphnia hyalina,* which with a production of 3.1–6.9 g org. dry wt m^{-2} yr^{-1} is the most important secondary producer of the zooplankton community in the Tjeukemeer.

Summary

The population dynamics and production ecology of *Acanthocyclops robustus* and *Mesocyclops leuckarti* in Tjeukemeer, a shallow lake in the Netherlands, were studied during three successive years. Since the populations show continuous recruitment a discrete event-oriented simulation model, based on the life cycle of an individual copepod, was used to integrate the field data with the laboratory data.

The simulations showed several gaps in our studies, the most important of these being lack of data on the micro-distribution of the females in and just above the bottom, abundance and durations of different naupliar instars. Because we overlooked a part of the female population, both the egg stock and birth rates were underestimated and the mortality rates of the adults overestimated.

Birth rates varied usually between 5 and 20% of the total population per day. Mortality rates of combined copepodite instars usually ranged from 5 to 30% day^{-1}. Nauplii and adults showed markedly higher mortality rates than those of the juvenile copepodite instars.

The annual P/B ratio of the total copepod population was relatively high: 81–93 for *A. robustus* and 53–59 for *M. leuckarti*. This was due to the high proportion of nauplii in the population and to the relatively high individual growth rate of copepodites. The total annual production of *A. robustus* and *M. leuckarti* amounted to 2.9–3.9 and 0.4–0.9 g org. dry wt m^{-2}, respectively. More than half of this was nauplii production; egg production contributed only 4–8%.

Acknowledgements

We are indebted to the members of the Food-chains and Production Research Group of the Tjeukemeer laboratory, Limnological Institute, Oosterzee. We are also grateful to Dr. P. Hogeweg of the University of Utrecht for her guidance in the modelling and to Miss M. E. McLaughlin for improving the English text.

References

Argentesi, F. R., Bernardi, R. de & Di Cola, G., 1974. Mathematical models for the analysis of population dynamics in species with continuous recruitment. Mem. Ist. Ital. Idrobiol. 31: 245–275.

Beattie, D. M., Bromley, H. J., Chambers, M., Goldspink, C. R., Vijverberg, J., Zalingen, N. P. van & Golterman, H. L., 1972. Limnoligical studies on Tjeukemeer, a typical Dutch 'polder reservoir'. In: Kajak, Z. & Hillbricht-Illkowska, A. (Eds.) Productivity Problems of Freshwaters, pp. 421–447. Proceedings of an IBP-UNESCO Symposium on Productivity Problems of Freshwaters Kazimierz Dolny, Poland.

Bernardi, R. de & Di Cola, G., 1976. Instantaneous growth rates and production estimation in natural zooplankton populations. Mem. Ist. Ital. Idrobiol. 33: 105–123.

Bosselman, S., 1975. Population dynamics of Eudiaptomus graciloides in Lake Esrom. Arch. Hydrobiol. 75: 329–346.

Brandl, Z. B. & Fernando, C. H., 1975. Food consumption and utilization in two freshwater cyclopoid copepods (Mesocyclops edax and Cyclops vicinus). Int. Revue Ges. Hydrobiol. 60: 471–491.

Bremer, P. & Vijverberg, J., 1982. Production, population biology and diet of Neomysis integer (Leach) in a shallow Frisian Lake (The Netherlands). Hydrobiologia 93: 41–51.

Burgis, M. J., 1971. The ecology and production of copepods, particularly Thermocyclops hyalinus, in the tropical Lake George, Uganda. Freshwat. Biol. 1: 169–192.

Burgis, M. J., 1974. Revised estimates for the biomass and production of zooplankton in Lake George, Uganda. Freshwat. Biol. 4: 535–541.

Burgis, M. J., 1975. The biomass of cyclopoids in two Thames Valley reservoirs. Verh. int. Verein. Limnol. 19: 1834–1840.

Burns, C. W., 1979. Population dynamics and production of Boeckella dilatata (Copepoda, Calanoida) in Lake Hayes, New Zealand. Arch. Hydrobiol. Suppl. 54: 409–465.

Chapman, A., 1973. Calamoecia lucasi (Copepoda, Calanoida) and other zooplankters in two Rotorua, New Zealand, lakes. Int. Revue Ges. Hydrobiol. 58: 79–104.

Corkett, C. D. & McLaren, I. A., 1978. The biology of Pseudocalanus. Adv. mar. Biol. 15: 1–231.

Cummins, K. W., Costa, R. R., Rowe, R. E., Moshiri, G. A., Scanlon, R. M. & Zajdel, R. K., 1969. Ecological energetics of a natural population of the predaceous zooplankter Leptodora kindtii Focke (Cladocera). Oikos 20: 189–223.

Densen, W. L. T. van, 1973. Populatiedynamika en produktie van de spiering, Osmerus eperlanus eperlanus, in het Friese merengebied. Student Report 1973-6, Limnological Institute, Nieuwersluis/Oosterzee, The Netherlands, 41 pp.

Densen, W. L. T. van, 1978. Food of young-of-the-year fish in Tjeukemeer, The Netherlands. In: Proceedings of the Zodiac Symposium on Adaptation, pp. 131–132. PUDOC, Wageningen, The Netherlands.

Densen, W. L. T. van. The ecology of 0^+ fish in Tjeukemeer, The Netherlands. II. Feeding behaviour. In prep.

Densen, W. L. T. van. The ecology of 0^+ fish in Tjeukemeer, The Netherlands. I. Population densities, growth and production. In prep.

Eckstein, H., 1964. Untersuchungen über den einflusz des Rheinwassers auf die Limnologie des Schluchsees. IV. Untersuchungen über die Einbürgerung des Eudiaptomus gracilis Sars in den Schluchsee. Arch. Hydrobiol. Suppl. 28: 119–182.

Edmondson, W. T., 1960. Reproductive rates of rotifers in natural populations. Mem. Ist. Ital. Idrobiol. 12: 21–77.

Edmondson, W. T., 1974. Secondary production. Mitt. int. Verein Limnol. 20: 229–272.

George, D. G., 1976. Life-cycle and production of Cyclops vicinus in a shallow eutrophic reservoir. Oikos 27: 101–110.

Goldspink, C. R., 1978. The population density, growth rate and production of bream, Abramis brama, in Tjeukemeer, the Netherlands. J. Fish Biol. 13: 499–517.

Gophen, M., 1976. Temperature effect on lifespan, metabolism, and development time of Mesocyclops leuckarti (Claus). Oecologia 25: 271–277.

Green, J. D., 1976. Population dynamics and production of the calanoid copepod Calamoecia lucasi in a northern New Zealand lake. Arch. Hydrobiol. Suppl. 50: 313–400.

Hogeweg, P. & Richter, A. F., 1982. INSTAR, a discrete event-oriented model for simulating zooplankton population dynamics. Hydrobiologia 95: 275–285.

Humphreys, W. F., 1979. Production and respiration in animal populations. J. Anim. Ecol. 48: 427–453.

Hutchinson, G. E., 1957. A Treatise on Limnology, Vol. I. Wiley, New York. 1015 pp.

Ivanova, M. V., 1973. Growth patterns of copepod crustaceans (in Russian). Hydrobiol. J. 9: 15–21.

Jacobs, R. P. W. M. & Bouwhuis, A. M. J., 1979. The year cycle of Eudiaptomus vulgaris (Schmeil, 1896) (Copepoda, Calanoida) in a small, acid water body during 1973. Development in the natural habitat and relationships between temperature and duration of developmental stages. Hydrobiologia 64: 17–36.

Jamieson, C. D., 1980. Observations on the effect of diet and temperature on rate of development of Mesocyclops leuckarti (Claus) (Copepoda, Cyclopoida). Crustaceana 38: 145–154.

Johnson, J. K. & Miller, C. B., 1974. Dynamics of isolated plankton populations in Yaquina Bay, Oregon. Oregon State Univ. Eng. Exp. Sta. Circ. 46: 27–35.

Katona, S. K., 1971. Ecological studies on some planctonic marine copepods. Ph.D. thesis, Harvard University. 145 pp.

Kibby, H. V., 1971. Energetics and population dynamics of Diaptomus gracilis. Ecol. Monogr. 41: 311–327.

Lair, N., 1978. Repartition spatio-temporelle, biomasse et production des population zooplanctoniques du lac d'Aydat en periode estivale. Hydrobiologia 61: 237–256.

Lammens, E. H. R. R., 1982. Growth and diet of the bream, (Abramis brama) in relation to its feeding conditions in Tjeukemeer. Hydrobiologia 95: 311–320.

Landry, M. R., 1978. Population dynamics of production of a planktonic marine copepod, Acartia clausii, in a small temperate lagoon on San Juan Island, Washington. Int. Revue Ges. Hydrobiol. 63: 77–119.

Lévêque, C., Carmouze, J. P., Dejoux, C., Durand, J. R., Gras, R., Iltis, A., Lemoalle, J., Loubens, G., Lauzanne, L., Saint-Jean, L., 1972. Recherches sur les biomasses et la productivité du lac Tchad. In: Kajak, Z. & Hillbricht-Illkowska, A. (Eds.) Productivity Problems of Freshwaters, pp. 421–447. Proceedings of an IPB-UNESCO Symposium on Productivity Problems of Freshwaters. Kazimierz Dolny, Poland.

McQueen, D. J., 1969. Reduction of zooplankton standing stocks by predaceous Cyclops bicuspidatus thomasi in Marion Lake, British Columbia. J. Fish. Res. Bd Can. 26: 1605–1619.

Miller, C. B., Johnson, J. K. & Heinle, D. R., 1977. Growth rules in the marine copepod genus Acartia. Limnol. Oceanogr. 22: 326–335.

Monakov, A. V., 1958. The life cycle of Mesocyclops leuckarti Claus (Copepoda, Cyclopoidea) (in Russian). Dokl. Akad. Nauk. SSSR 120: 419–421.

Monakov, A. V., 1972. Review of studies on feeding of aquatic invertebrates conducted at the Institute of Biology of Inland Waters, Academy of Sciences, USSR. J. Fish. Res. Bd Can. 29: 363–383.

Munro, I. G., 1974. The effect of temperature on the development of egg, naupliar and copepodite stages of two species of copepods, Cyclops vicinus Uljanin and Eudiaptomus gracilis Sars. Oecologia 16: 355–367.

Nie, H. W. de, Bromley, H. J. & Vijverberg, J., 1980. Distribution patterns of zooplankton in Tjeukemeer, The Netherlands. J. Plankton Res. 2: 317–334.

Nilssen, J. P., 1977. Cryptic predation and the demographic strategy of two limnetic cyclopoid copepods. Mem. Ist. Ital. Idrobiol. 34: 187–196.

Seitz, A., 1979. On the calculation of birth-rates and death-rates in fluctuating populations with continuous recruitment. Oecologia 41: 343–360.

Smyly, W. J. P., 1961. The life-cycle of the freshwater copepod Cyclops leuckarti Claus in Esthwaite Water. J. Anim. Ecol. 30: 153–159.

Threlkeld, S. T., 1979. Estimating cladoceran birth-rates: the importance of egg mortality and the egg age distribution. Limnol. Oceanogr. 24: 601–612.

Vidal, J., 1980. Physioecology of zooplankton. II. Effects of phytoplankton concentration, temperature, and body size on the development and moulting rates of Calanus pacificus and Pseudocalanus sp. Mar. Biol. 56: 135–146.

Vijverberg, J., 1977. Population structure, life histories and abundance of copepods in Tjeukemeer, The Netherlands. Freshwat. Biol. 7: 579–597.

Vijverberg, J., 1980. Effect of temperature in laboratory studies on development and growth of Cladocera and Copepoda from Tjeukemeer, The Netherlands. Freshwat. Biol. 10: 317–340.

Vijverberg, J. & Frank, Th. H., 1976. The chemical composition and energy contents of freshwater copepods and cladocerans in relation to their size. Freshwat. Biol. 6: 333–345.

Vijverberg, J. & Richter, A. F., 1982. Population dynamics and production of Daphnia hyalina Leydig and Daphnia cucullata Sars in Tjeukemeer. Hydrobiologia 95: 235–259.

Waters, T. F., 1977. Secondary production in inland waters. Adv. Ecol. Res. 10: 91–163.

Winberg, G. G., 1956. Rate of metabolism and food requirements of fish. Fish. Res. Bd Can., Transl. Ser., 194.

Winberg, G. G. (Ed.) & collaborators, 1971. Symbols, units and conversion factors in studies of freshwater productivity. IBP Central Office, London. 23 pp.

INSTAR, a discrete event model for simulating zooplankton population dynamics

P. Hogeweg[1] & A. F. Richter[2]
[1] *Bioinformatica, University of Utrecht, Padualaan 8, 3584 CH Utrecht, The Netherlands*
[2] *Limnological Institute, Tjeukemeer Laboratory, De Akkers 47, 8536 VD Oosterzee, The Netherlands*

Keywords: simulation model, zooplankton population dynamics, discrete event formalism, individual-oriented modelling

Abstract

In this paper we discuss the basic principles of discrete event, individual oriented, data based modelling in ecology, and we present an application of this modelling strategy. The strategy is contrasted with some more conventional modelling strategies with respect to its purpose, its basic units and its heuristic properties.

INSTAR applies this modelling strategy to the simulation of the fluctuations of the population structure and density of microcrustaceans through the year. The model encompasses one microcrustacean species at a time, and its interface with the rest of the ecosystem; it has been applied to several Cladocera and Copepoda species in a shallow eutrophic lake in the Netherlands (Vijverberg & Richter 1982a, b). Possibilities for extending the model are discussed.

Introduction

Formal modelling serves a number of very different purposes in ecology. Failure to recognise that there are different types of modelling formalisms may easily lead to the adoption of a formalism less suitable for the purpose intended. For instance, the purpose of classical population models is to provide a unifying baseline for thinking about ecosystems. Therefore classical population models contain as little detail as possible about the relations studied; they can therefore be termed 'minimal models' or 'minimodels'. As 'paradigm systems' (Hogeweg & Hesper 1981a) they play an indispensable role in ecology. The minimodels have shown that mini-formulations are sufficient to provide a rich set of behaviour patterns, including many relations which seem at first sight counter-intuitive but which are in fact common to both models and ecosystems. However, attempts to fit actual data to such mini-models do not really increase our ecologically knowledge, whether such a fit is obtained or not: the minimodels typically imply assumptions about the populations studied, which are known to be false, and they ignore details in which we are interested; in addition, the values of the parameters obtained by fitting data to the models are often not interesting in themselves (e.g. because they are not independently measurable).

On the other hand, the purpose of many contemporary ecological models is to predict the behaviour of a particular ecosystem under the influence of changing parameters or driving variables. Patten (1974) pointed out that very simple linear models often provide better predictions for the global variables typically asked for in this context than do more elaborate population type models. This, we think, is not due to the 'inherent linearity of ecosystems' as Patten assumed but is because the behaviour of linear models is known well enough to force them into the correct (or desired) behaviour when enough free (uninterpreted) parameters are available, as is always the case in such models. Non-linearities on the other hand usually occur in more detailed population oriented models, and therefore such models cannot be so easily forced into 'repres-

Hydrobiologia 95, 275-285 (1982). 0018-8158/82/0953-0275/$02.20.

enting the world view of the user' (the goal of many simulation models according to Palme (1979)). Obviously such practices do not provide much new insight into ecosystems.

In contrast to the two purposes of ecological modelling mentioned above, the purpose of much ecological research is to obtain a more detailed view of the behaviour of the various species in particular ecosystems. In this context the purpose of formal models is to represent current knowledge as faithfully as possible so that this knowledge can be tested for sufficiency in generating the overall behaviour of the ecosystems, and thus reveal lacunae in our current knowledge. The model presented in this paper is designed exclusively with this purpose in mind: to 'debug' current knowledge about populations by assessing, by means of formal modelling, the implications of this knowledge in the ecosystem under consideration.

Formalisms for (simulation) models

Almost all ecological models are formulated in terms of differential or difference equations. This holds for all the types of ecological models mentioned above: the classical population models were originally formulated in one or other of these formalisms (Lotka (1925) and Volterra (1926) used differential equations, Nicholson (1933) first used difference equations), and minimodels still use these formalisms almost exclusively. Prediction-oriented models using linear or non-linear systems theory also fall into this category, and knowledge-oriented models, which are usually developed as an elaboration of the models mentioned earlier, also employ these formalisms. Originally such models could be analysed exclusively by analytical means but nowadays they are most often analysed by computer simulation; this is the only mode for analysing the more elaborate prediction- and knowledge-oriented models, whereas in minimodels simulation is almost always used alongside analytical results (Hogeweg & Hesper 1978).

However, in the context of digital simulation, differential and difference equation models and their elaborations containing switches etc. can be seen as special cases of the discrete event formalism (Hogeweg & Hesper 1979; Zeigler 1976). According to this view the selection of 'interesting' variables and events is at the base of every modelling effort. Interesting variables are those variables which change the course of behaviour of other variables in an unexpected way (Hogeweg & Hesper 1980, 1981a, c). Different modelling formalisms can be seen as different strategies for the choice of interesting events. In continuous (i.e. differential equation) models the underlying conceptualisation is that everything is changing all the time, and that all these changes are interesting. Nevertheless, in digital simulation, changes of course occur discretely: the choice of the step-size (fixed or variable) is based on the calculability of the next state or on output requirements (i.e. 'interestingness' for other variables or the user). Fixed time-step models (i.e. difference equation models) assume that changes in all variables occur simultaneously at discrete points in time and that only then do they influence other variables; thus events and interesting events are assumed to be equivalent. Contrariwise, the underlying conceptualisation of discrete event formalism is that 'once in a while something interesting is happening somewhere in the system' (Hogeweg 1974), and therefore this type of formalism does not enforce a uniform choice of interesting events as do the other two formalisms, but stimulates the researcher to recognise which events are crucial in the system and which are not.

In our view minimodels as well as knowledge-oriented models can profit from the greater flexibility of the discrete event formalism, and from the emphasis it lays on the selection of interesting event. Moreover, discrete event formalism makes it possible to use individuals and their information processing as basic units.

In the next section we describe the heuristically useful (i.e. useful for helping to solve ecosystem problems but not absolutely necessary) properties of discrete event models. In the subsequent section we present the model INSTAR which applies the discrete event formalism to a model simulating the growth of microcrustacean populations. In other words we use the formalism in an application for which the more conventional formalisms could be used as well but which may, nevertheless, profit from our approach. Finally, in the last section, we show how the zooplankton model could be extended so as to incorporate a larger part of an ecosystem as well as a larger part of the modelling formalism we have in mind.

Discrete event, individual-oriented modelling

Timing regime

Our primary conception of ecosystems includes processes which are viewed as happening once in a while (e.g. births, deaths, sometimes growth of individuals, predation) and processes viewed as being continuous (e.g., photosynthesis, respiration, sometimes growth of individuals, etc.). The latter group of processes often contains on/off switches, i.e. events occurring once in a while and redirecting the course of behaviour of some variables, whereas the former processes can be (and often are) viewed as continuous due to averaging over many individuals. The given categorisation of processes therefore actually represents different time-scales and/or different averagings over, say, organisms or molecules.

Applying the discrete event formalism we can use different time-scales for different processes in the same model. A special case is to view some processes as continuous.

If (as is the case for the discrete event processes in one and the same model) averagings over large non-homogeneous populations are not enforced on cases where more detailed knowledge is available, and if unwarranted detail is not enforced in cases where only global information is available, then new knowledge can be incorporated in the model as it becomes available. Moreover, within the context of discrete event, individual-oriented models the continuous processes often turn out to be simple processes containing very few variables, which are isolated from other variables in between events which might be interesting just because they change the *de facto* connectivity of the continuous variables. Therefore the continuous processes often do not have to be solved by integration procedures although this is quite feasible if necessary, because they can be solved explicitly for the intervals between events, or because they are known to behave in a certain way from previous simulations or from previous experience in the current simulation. Thus, not all intermediate values of the variables have to be computed, because the values can be obtained by symbolic integration or looking up in a table whenever some part of the simulation needs this information (Hogeweg & Hesper 1979, 1981a; Zeigler 1978).

Individual-oriented simulation

Recently the obvious point has been made (MacMahon et al. 1981) that ecosystems consist of interacting organisms being physical units in space/-time rather than of populations which are not physically connected and cannot directly/causally affect each other. In other words, in the dynamic energy and information network of an ecosystem, organisms (individuals) form tighter clusters than populations (Hogeweg & Hesper 1979). However, conventional modelling approaches employ almost fixed interaction structures (networks) between the chosen variables. Individuals, however, do not have such a fixed interaction network with other individuals all the time and therefore cannot be used as units in such simulations.

The fixed interaction structure enforces a top-down approach to modelling: the choice of variables (i.e. which of the disconnected units are lumped and which strongly connected units (e.g. individuals) are split up into several compartments (e.g. phosphate, nitrate, etc.)) depends on other variables because each combination of interactions has to be lumped in a separate compartment. The incorporation of new variables in the model yields new combinations of interactions and therefore requires re-lumping and re-splitting of variables. In knowledge-oriented models a bottom-up approach is required so that new knowledge can be incorporated incrementally into the model in a modular fashion.

An individual seems to be the best module in knowledge-based ecosystems models. An individual organism is the basic unit of the autecological research on which ecosystems research should depend. Moreover an individual is important as an information processing unit. When we use an individual as the information processing unit in our models, we are forced to make it explicit which information is supposed to be available to the individual and thus see whether such assumptions are reasonable or not (Hogeweg & Hesper 1979). The more flexible interaction structure and the more flexible timing regime of the discrete event formalism make it feasible to use individuals as units. Such a unit describes (our knowledge about) the potential behaviour of an individual of a certain type (species) throughout its life and the dependence of its behaviour on the environment and other organisms it encounters. Ideally such an organism (module) can actively extract all the information it needs from its environment (the simulation model) whether this information is explicitly available or not: only in this way can new knowledge about the information used by that organism be incorporated into the model without it being necessary to change other parts of the model. Techniques to enable model parts to achieve such self-sufficiency are being developed (Hogeweg & Hesper 1981a), but in the model described below we use standard discrete event simulation techniques of the general purpose simulation language SIMULA/67 (Dahl et al. 1970).

INSTAR, a discrete event model for simulating the growth of populations of microcrustaceans

Outline of the model

The purpose of the INSTAR model is to integrate laboratory data and field data on the growth of microcrustacean species in order to display the combined effect of all non-included factors. It incorporates laboratory data on length increment per unit of time of individuals in relation to their current length, egg development time and instar duration in relation to stage and to water temperature.

The field data that are included concern the density of size classes (Cladocera) or stages (Copepoda) of the species under consideration and the fecundity of the adult animals in terms of percentage of egg-bearing females and number of eggs per female per size class. (For a detailed description of the sort of data required by the model, see the appendix on the input and output of the model, and Vijverberg & Richter 1982a, b.) These data are made to fit the model (if possible) by means of the 'black-box-fish' which consumes the required quantity of zooplankton to bring its density down to the observed density. Thus the model contains (only) one free variable (all others are obtained from measurements), but this variable can be adjusted as needed; its behaviour is not restricted by hypotheses as it is in models with badly known parts (i.e. freely adjusted parts), except that the black-box-fish is not allowed to eat a negative amount. The generated behaviour of the black-box-fish shows us the combined effect of predation and mortality as well as discrepancies between laboratory and field conditions. When data on predation become available the black-box-fish can be gradually replaced by 'real'-fish; finally the black-box-fish is reduced to representing only natural mortality and 'wrong assumptions'. The less the black-box-fish represents the easier it becomes to evaluate the acceptability of its behaviour, but even before that, its behaviour can be evaluated to such an extent that it can suggest new directions of research (see below and Vijverberg & Richter 1982a, b).

ZOOPL, the basic unit of the model

The basic unit of the model is ZOOPL, which represents a zooplankton individual or, for practical reasons, a small group of individuals born almost simultaneously and having similar mothers (similar with respect to size, see below).

ZOOPL is implemented in SIMULA/67 as a CLASS; different species, in particular Cladocera and Copepoda, are implemented as separate subclasses of ZOOPL (i.e. ZOOPL *CLASS* CLAD and ZOOPL *CLASS* COP). These subclasses contain different procedures for length increase, fertility and descent mechanism. Other subclasses, for other microcrustaceans, can be easily added to the model.

The interesting events in the life of a cladoceran or copepod seem to be birth, death, moultings and diapause, at least as far as a model on population growth is concerned. At the time of these the individuals change their interrelationships with their environment, e.g. at the time of moulting they become larger and or change shape and therefore may change their food habits; moreover their size increase increases their visibility and palatability for their predators.

ZOOPL decribes the life history of the zooplankter under consideration in terms of the interesting events which are scheduled at the appropriate times.

*ZOOPL CLASS CLAD: the life history of Cladocera**

The birth of a zooplankter which appears at the

* The life history described here is according to the current version of the model and differs slightly from that used in Vijverberg & Richter (1982a).

time of moulting of the mother is represented in the model by the creation of a ZOOPL: the fate of the new borns becomes independent of the fate of the mother (before that time, death of the mother implies death of the young; this factor has been omitted from some other zooplankton growth models based on compartments for different size classes and eggs - Argentesi *et al.* (1974), Berdordi (1974), Berdordi & di Cola (1974). Formally the new ZOOPL is created by the mother, although in the model the procedure GATHER interferes to gather the similar simultaneously-born young of several similar mothers; our description, however, will be in terms of a single individual (a case which can be used in the model). Newly created ZOOPLs are either 'newborns', starting their development immediately, or resting eggs. In the latter case the ZOOPLs are quiescent until activated in the next 'spring', and then develop in the same way as newborns.

The new ZOOPL is assigned an initial size which depends on the size of the mother:

$$L_{nb} = p + q\,L_m \qquad (1)$$

where L_{nb} = size of new-born,
L_m = size of mother,
p and q = regression parameters obtained from experiments.

In initial versions of the model the dependence of the size of the newborns on the size of the mother was not included. This led to negative estimates of infant instar durations in some periods of the year; hence we did experiments which disclosed size relationship between mother and young.

The duration of adult instars at a given water temperature is constant and approximately equal to the egg development time (Bottrell 1975b; Vijverberg 1980). The duration of juvenile instars on the other hand is not constant but is dependent on the instar stage they are in. ZOOPL determines the duration of an instar at the moment ZOOPL enters that instar on the basis of current and future temperature and, in the case of juvenile animals, ZOOPL determines which instar it is in.

The duration of an instar is calculated by using the following equation:

$$D_i = \alpha_i\,(D_{Temp_{(t_1)}} + D_{(Temp_{(t_2)})}) \,/\, 2$$
$$t_2 = t_1 + D_{Temp_{(t_1)}} \qquad (2)$$

where t_1 = the time when the ZOOPL enters the instar,
D_t = the duration of an instar at temperature t_1
Temp subscript (t) = the temperature at time,
α_i is a correction factor,
D_i is known from laboratory experiments on egg turnover time: see Vijverberg (1980).

The durations of juvenile instars are assumed to be fixed fractions of the adult instars. Thus at moulting time ZOOPL schedules its next active phase at the time calculated for the next moulting. Moreover it adjust its properties: length and biomass. Length increment in fact occurs only at that time. For adults length increment is calculated from the duration of the last instar and from the laboratory data on length increment per unit time at different temperatures and for different instars:

$$\delta L_{\,t,s} = D_k\,\frac{G_{(temp_{t_0},\,s)} + G_{(temp_{t_1},\,s)}}{2} \qquad (3)$$

where L_{ts} = the length increment of an animal of size s at temperature t,
G = the observed (interpolated) growth rate, evaluated for the temperature at t_0, the beginning of the previous instar, and t_1, the current time,
D_k = the duration of the previous instar.

Since the laboratory data on growth rates proved to be unsatisfactory for simulating juvenile growth, field data were used instead.

The length increment per juvenile instar was calculated taking into account the length of the newborn (equation 1) and the length of the first adult stage observed in the field, i.e. the length of the smallest egg-bearing females, which, in one case, varied strongly during the year (Vijverberg & Richter 1982).

$$\delta L = \alpha_i\,(L_A - L_{NB}) \qquad (4)$$

where L_A = minimal length of egg-bearing adults,
L_{NB} = minimal length of juveniles at birth,
α_i = fraction for the ith instar)

Whereas length increment occurs at the time of moulting only, biomass increase is supposed to be a continuous process. Nevertheless biomass increase is calculated only at the time of moulting, except when needed at intermediate times (e.g. because it is eaten), in which case it is calculated for that time (see 'Timing regime'). Mean biomass of an instar is supposed to be directly proportional to the length of the animal in that instar:

$$B_i = cL_i^d \quad (5)$$

where c and d are regression coefficients.
However, while the length of an individual remains constant the biomass is supposed to increase gradually during an instar. Because the size of the animal is constant between moultings we assume a linear growth of biomass in contrast to the frequent assumptions of exponential growth, where increase in biomass increases feeding capacity. Therefore biomass at time t (in between moultings) is calculated as

$$B_t = \frac{(t - t_1)}{(t_1 - t_0)} (B_i - B_{i-1}) \text{ when } t < t_1 + \tfrac{1}{2}(t_2 - t_1)$$

and (6)

$$B_t = \frac{(t_2 - t)}{t_3 - t_2} (B_{i+1} - B_i) \text{ when } t > t_1 + \tfrac{1}{2}(t_2 - t_1)$$

where t_0, t_1, t_2, t_3 refer to the time of entry into the previous star, the current instar, the following instar and departure from the following instar respectively. In addition the animal contains biomass of its young. For simplicity and because of the lack of data, linear growth is again assumed.

$$BY_t = \frac{t - t_0}{t_1 + \tfrac{1}{2} D_1 - t_0} B'_{(t_1 + \frac{1}{2} D_1)} \quad (7)$$

where BY_t = the biomass of the young at time t_j, t_0 and t_1 = time of entry into and departure from the current instar,
$B'_{t_1 + \frac{1}{2} D)}$ = the biomass of the newborns at $t_1 + \frac{1}{2} D_1$ calculated by equation (5),
D_1 = the duration of the first instar as calculated from equation (2).

In the present model intermediate values are calculated whenever the 'black-box-fish' eats. Output requirements could be another reason for updating biomass at intermediate times but at present output is always preceded by the consumption of food by the black-box-fish so that no separate updating is required.

ZOOPL CLASS COP: the life history of Copepoda

The differences between Copepoda and Cladocera are: Because the juvenile instars of Copepoda are morphologically different, the laboratory data relate to instar duration times, rather than to size increase per unit of time. Instar duration times are available for all juvenile stages, but are lumped for nauplia.
For the same reason instars of Copepoda are recognisable in the field, whereas instars of Cladocera are not. Therefore in the Copepoda model field data on size of the instars are used instead of laboratory data on growth rates. The size variations through the season may be caused by food availability etc. as well as by temperature (compare next section).

The size of the copepod instars do not affect the behaviour of ZOOPL *CLASS* COP directly, whereas for ZOOPL *CLASS* CLAD they do. The development times of copepods are unaffected and the size of the next instar is again obtained from the field as is the fertility of the adult copepods: size differences do not influence the fertility, whereas they do in the case of cladocerans. However, the sizes are used for biomass estimates to assess the amount of food transferred to the next trophic level.

Other differences are that in the copepod a diapause may occur in the last stages before the adult stages, whereas in the case of Cladocera a diapause only occurs in the form of winter eggs, and that the characteristics of nauplia are not dependent on their mother (all mothers being equal as far as the model is concerned).

Summarising, ZOOPL and its specialisations CLAD and COP describe the life history of the modelled species in terms of birth, moulting, diapause, death and the production of new ZOOPLs as important events, and in terms of the increase of biomass of animals and eggs/larvae as continuous variables which are updated at the time of the moultings and when death occurs.

Embedding of ZOOPL in the ecosystems

Interface with lower trophic levels. The hypothesis that the available data may be sufficient to simulate the fluctuations of the zooplankton species in the field involves far-reaching assumptions about the interface of the zooplankton with its food species. These assumptions are:

1. The duration of instar is independent of food supply.
2. Fecundity depends on the availability of food.
3. The length increment per moult does not depend on food supply.

The first assumption will be generally true for adult cladocerans living under eutrophic conditions (Bottrell 1975b; Vijverberg 1980, Vijverberg & Richter 1982a). Only when food is extremely scarce will the instar duration increase noticeably (Vijverberg 1976). It is not known, however, whether the duration of juvenile instars is also independent of food supply.

The duration of the copepod instars is also assumed to be dependent only on temperature and not on environmental conditions including food supply; we think that this assumption is reasonably acceptable in the conditions studied by us (Vijverberg & Richter 1982b) although under different conditions an effect of food supply was demonstrated (Miller *et al.,* 1977).

The second assumption is clearly true; it has been shown both experimentally and from field observations that food quantity and food quality substantially influence the fecundity of *Daphnia hyalina* (Vijverberg 1976; Vijverberg & Richter 1982a).

The last assumption is probably false. In Cladocera the length increment per moult, like the fecundity, most probably is influenced by the food supply. However, since this effect is difficult to measure, and we do not have sufficient data available, we have not taken this effect into account so far. In the future, however, we hope to establish a link between length increment per moult and fecundity (Vijverberg, personal communication), and thus indirectly link length increment and food availability.

Thus we need at least an interface of ZOOPL linking it and the lower trophic levels through the fecundity. This interface is accomplished by incorporating in the simulation field data on brood size per size class throughout the year. For Cladocera a second interface with the lower trophic levels consists of field data on the size of the first adult instar, which varies greatly through the year and influences the fecundity because of the strong relation between size and fecundity.

For Copepoda we use field data for fecundity as well as length increment per instar (and thus assumption 3 is not used, although it is probably valid).

Thus we use observed properties of the zooplankton to interface ZOOPL with the lower trophic levels of the ecosystem. In so doing we lump all factors which influence these properties, including non-food factors. The price paid for not explicitly modelling the influence of the lower trophic levels on these properties is that the model, as it stands, can be used only when these properties are known throughout the simulation period; therefore, the model cannot be used for prediction in years not observed in such detail.

Interface with higher trophic levels, the 'black-box-fish'. The embedding of ZOOPL in higher trophic levels is achieved through the 'black-box-fish'. The black-box-fish BITE stands in place of 'real' fish, i.e. modules representing individual fish of various species described in terms of their feeding habits and other sources of mortality.

The feeding habits of the 'black-box-fish' are such that they meet the mortality requirements that emerge by fitting simulated densities to observed densities. Feeding is supposed to be a continuous process which is approximated by a stepwise Euler integration with a step size of 12 hours. Thus, every 12 hours BITE reduces the zooplankton population by an amount determined by the data concerning the population density of the various size classes/stages. Whenever the population density of ZOOPLs in a size class/stage exceeds the density observed at that time in the field (or obtained by interpolation between field observations), BITE eats the required amount to reduce the density to the observed density, otherwise no food is consumed (normally, negative amounts of food are not accepted, i.e. too low densities are not adjusted to the observed densities, whereas too high densities are, although there is a key which does allow for negative mortality, see Vijverberg & Richter (1982a)). Note that in the case where size classes are recognised (cladocerans), ZOOPLs of different ages can be lumped in one size class (or ZOOPLs of the same age can be split up into different size classes) because of differences in growth conditions. Food selection by BITE is, within size class/-stage, random, i.e. when ZOOPLs represent one animal, the chance that the animal will be killed will equal the fraction of the population decrease required in the animal's size class/stage. When ZOOPLs represent several animals the number of animals per ZOOPL is decreased according to this fraction. Thus BITE supplies us with a mortality curve through the year, given that the assumptions in the simulation are correct.

(In)validation of the model. The free adjustment of the mortality may seem to render the model insensitive to falsification. This is not entirely true: the postulated growth rates may be incapable of producing the observed population increase, and the observed mortality curve may be 'unexpected'. In a certain sense both these types of invalidation of the model did in fact occur:

In *Daphnia hyalina,* the observed population increase in spring, especially the sudden occurrence of large quantities of quite large animals shortly after very low densities, cannot be simulated by the model (Vijverberg & Richter 1982a). The time-lags, i.e. the time needed to grow to such sizes simply is not available. In continuous, compartment models this effect is less conspicuous because a compartment starts to 'leak' to the next compartment as soon as it is filled, since the population within a compartment is supposed to be homogeneous (i.e. composed of all ages) and a fixed fraction (possibly variable with temperature) is supposed to be transferred to the next compartment per unit time. In the case of zooplankton species the homogeneity assumptions seem to be invalid in spring: growth rates are very strongly dependent on temperatures of around 10–15 °C and the instar duration times are of the same order of magnitude as the rate of temperature increase. Therefore this discrepancy between the model and the data seems to point to sources of population increase other than the animals observed: e.g. influx from the poorly sampled inshore zone, where resting eggs may accumulate and develop into large animals and subsequently enter the open water. Further research is needed.

Likewise, the Copepoda model could not produce the observed increase in densities and so has revealed gaps in our knowledge about, for example, nauplia recruitment (see Vijverberg & Richter 1982b).

2. High mortality after the initial spring *Daphnia* bloom occurs early in the season before the 0^+ fish can be held responsible. This was an unexpected result of the first simulation runs. Further research disclosed the important role of bream as a predator of larger daphnias (Vijverberg & Richter 1982a). The influence of invertebrates on the smaller animals, which is high at times is at present under investigation.

Thus we see that discrepancies can occur between the model and the data. Such discrepancies can and should lead to further research rather than be seen as falsifications of the model only.

Possible ways in which the model can be extended

As mentioned above, the interface of the simulated microcrustacean population with the remainder of the ecosystem is, in the present version of the model, achieved entirely by including field data on the (supposedly) relevant properties of the zooplankters, e.g. fertility and (in the case of copepods) size increase. Therefore the zooplankton populations can be considered in isolation. The price paid is that the model cannot be used unless these data ara available. Moreover the observed mortality is lumped over a variety of causes, which makes the model less amenable to evaluation (see previous section). However, it is possible to extend the present model incrementally by including in it other populations and their hypothesised or observed interactions with the zooplankters. For example, data on the standing crop of various fish populations and the amount of food they eat could gradually replace the 'black-box-fish'.

Again, an entirely data based mode of operation is feasible: data on stomach contents of a predator and standing crop, together with estimates on food turnover time (a difficult matter), readily yield the amount of zooplankton mortality for which this predator is responsible and narrow down the 'rest' mortality of the 'black-box-fish'. Such a data based fish in fact does not have to 'know' the density and size distribution of the zooplankters: 'they eat what they eat'.

Alternatively, the amount eaten by the fish could

be made not to depend on field data when the amount eaten is related to the population structure of the zooplankton populations, e.g. in terms of density (in)dependence, size dependence and preference index. However one defines such an interrelationship, it seems again heuristically useful to use the individual fish as the basic unit and its birth and death as interesting events while updating size increase of the individuals and a change of feeding habits if and only if the food conditions change significantly. The time scale for the various events of the fish can differ greatly. For example, when the death rate is constant (i.e. for periods in which death rates may be assumed constant) the time of death can be scheduled for each individual by making a random drawing from a negative exponential distribution (the result may fall outside the simulated period, of course): the chances of death do not have to be evaluated at every step. On the other hand, feeding habits may change whenever the structure of the zooplankton populations changes: in between, the feeding habits can be assumed to be constant, as the size increase of fishes occurs on a larger timescale than changes in the zooplankton population. Thus, whenever relevant changes occur in the zooplankton population, the fish can adjust its parameters accordingly. For intermediate times the amounts eaten by the fish can be solved explicitly from the prevailing parameters, e.g. the time of an interesting event for a zooplankton entity, the amount eaten or the chance of having been eaten) is estimated in this way.

Alternatively, feeding can occur at (small) fixed time-steps (i.e. Euler integration) in a similar fashion as is used in the present model for the 'black-box-fish'.

Note, however, that in the case of this feeding habit mode of operation, data (simulated or from the field) should be available for all food species, whereas for the data mode of operation the influence of one predator on one of its preys can be assessed. Note, moreover, that the relation between zooplankton and fish is in principle symmetric: the interface of a data oriented fish model with 'black-box-food' can be extended by incorporating the zooplankton data.

Another useful way of extending the model is to modify the feeding habits of the 'black-box-fish' without violating its definition as adjusting the simulated densities of the various size- classes/-stages to the observed densities. For example it could be made to prefer certain food items to other ones by allowing it to distinguish among animals within the groups recognised in the field observations. An interesting example along this line would be to make it prefer egg-bearing females to barren ones. This distinction is very likely to be made by some real fish: the visibility as well as the food value increases during egg-bearing. Moreover such a preference could influence the simulation results strongly as it can cause a strong reduction in birth rate and transfer to the next size-class/stage, whereas the mortality remains the same within the size-class under consideration. For copepods data are available about the duration of the barren period in between egg-bearing periods. For cladocerans it is not known whether the reduction in the percentage of egg-bearing females is due to an increase in the barren period or due to the fact that individuals complete an instar without producing eggs. In both cases it is interesting to study the effect in relation to the habits of the black-box-fish. To implement this extension we could include the initiation of egg-bearing as an interesting event in the life-cycle of the zooplankters (i.e. in ZOOPL), and model the growth by some continuous function which again is updated only when needed (i.e. when the black-box-fish has to decide that it prefers a particular animal rather than only to assess the amount eaten), and to include the appropriate choice mechanism is the black-box-fish. If egg development were the same for all animals within a size-class/stage (i.e. no barren instars occurred), the adjustments of the model could be limited to the black-box-fish, without changing ZOOPL: it could deduce the egg-bearing status of the ZOOPLs from the time they entered their present instar and eat accordingly. In either case, however, the distinction does not have to be made by all fish present in the model. An important property of ecosystems is, we think, that each of its members has its own 'world view', i.e. entities which are indistinguishable for some may serve as 'predator or prey' for others. The individual-oriented approach to simulation make it feasible to implement these different world views without just performing the same operation on a number of distinguished compartments, as would be the case in compartment models.

APPENDIX

Input and output of INSTAR.

INPUT

Field data (as function of time)
1. Water temperature
2. Density of each size class or stage
3. Mean egg number per ovigerous female (for cladocerans for each adult size class)
4. Percentage ovigerous females
5. Mean length of individuals of each stage (for copepods only)
6. Minimal length of ovigerous females (for cladocerans only)

Laboratory data
1. Instar duration time as function of temperature and size class/stage
2. Length increment per unit of time as function of temperature and size class (for cladocerans only)
3. Length of newborn as function of maternal length (for cladocerans only)
4. Weight as function of length

OUTPUT (for each size class/stage, as function of time)
1. Simulated density
2. Simulated biomass
3. Number of births per unit of time
4. Number of deaths per unit of time
5. Production of biomass by length growth or fecundity per unit of time
6. Loss of biomass by mortality per unit of time

Discussions and conclusions

The basic features of the model presented in this paper are as follows:

1. It is a knowledge-oriented model: its purpose is to display gaps in our knowledge rather than to produce the best possible fits disregarding the inadequacies of the formulation.
2. It is a data-based model: it uses data on the relevant properties of a population to embed it into its ecosystem rather than known or postulated relations to other members of the ecosystem. The structure of the model is such, however, that the latter can gradually replace the former.
3. It is an individual-oriented model: the basic unit is an individual, which is modelled in terms of its changing interactions with other individuals rather than by splitting it up over several compartments in order to make it possible to use a fixed interaction structure.
4. It is a discrete event model: only interesting events occur in the model, i.e. not all variables are updated at fixed time steps.

We have discussed at length in the preceding sections why these features may be advantageous even in the case of zooplankton populations, which we regard as being less individualised than some other parts of ecosystems.

However, all the advantages are of a heuristic rather than a practical nature. The more flexible formalism does not enforce a simple model structure which makes model development relatively harder and program runs more costly, without necessarily producing a better fit with the simulated data; these properties may, if you are so inclined, be listed as disadvantages of our approach.

Acknowledgements

We thank Dr. J. Vijverberg for supplying us with the problem and the data for which the model was made. We thank him and Dr B. Hesper for useful discussions and for reviewing the manuscript. We thank Mrs. S. M. McNab for corrections in the English text.

References

Argentesi, F. R., Bernardi, R. de & Di Cola, G., 1974. Mathematical models for the analysis of population dynamics in species with continuous recruitment. Mem. Ist. ital. Idrobiol. 31: 245–275.

Bernardi, R. de, 1974. The dynamics of a population of Daphnia hyalina Leydig in Lago Maggiore, Northern Italy. Mem. Ist. ital. Idrobiol. 31: 221–243.

Bernardi, R. de & Di Cola, G., 1976. Instantaneous growth rates and production in natural zooplankton populations. Mem. Ist. ital. Idrobiol. 33: 105–123.

Bottrell, H. H., 1975a. Generation time, length of life, instar duration and frequency of moulting, and their relation to temperature in eight species of Cladocera from the River Thames, Reading. Oecologia 19: 129–140.

Bottrell, H. H., 1975b. The relationship between temperature and duration of egg-development in some epiphytic Cladocera and Copepoda from the River Thames, Reading, with a discussion of temperature functions. Oecologia 18: 63–84.

Dahl, O. J., Myhrhaug, B. & Nygaard, K., 1970. Simula information common base language. Norwegian Computing Center, Publ. No. S.22

Dussart, B., 1969. Les Copépodes des eaux continentales d'Europe occidentale, Tome 2: Cyclopoides et biologie. Boubée, Paris, 292 pp.

Hogeweg, P., 1974. Concept for simulation, a comparison based on simulation languages. In: Buffort, H. F. J. M. & J. H. L. (Eds.) Developments in Systems Theory, Vol. I, pp. 119–135. Stenfert Kroese, Leiden.

Hogeweg, P. & Hesper, B., 1978. Interactive instruction on population interactions. Comput. Biol. Med. 8: 319–327.

Hogeweg, P. & Hesper, B., 1979. Heterarchical selfstructuring simulation systems, concept and applications in biology. In: Zeigler, B. P. (Ed.) Methodologies of Modelling & Simulation, pp. 221–231. North-Holland.

Hogeweg, P. & Hesper, B., 1981a. Two predators and a prey in a patchy environment; an application of MICMAC modelling. J. theor. Biol. 93: 311–332.

Hogeweg, P. & Hesper, B., 1981b. On the role of OBSERVERS in large scale systems. In: Proc. of the UKSC/81 Conf. on Computer Simulation, Harrogate, England, pp. 420–425. Westbury House.

Hogeweg, P. & Hesper, B., 1981c. Selfstructuring simulation systems. In: Ören, T. (Ed.) Proc. Cybersoff Symposium, Namur (Belgium).

Lotka, A. J., 1925. Elements of Physical Biology. Williams & Wilkins, Baltimore.

Miller, C. B., Johnson, J. K. & Heinle, D. R., 1977. Growth rules in the marine copepod genus Acartia. Limnol. Oceanogr. 22: 326–335.

Mac Mahon, J. A., Schimpf, D. J., Andersen, D. C., Smith, K. G. & Bayn, R. L., Jr., 1981. An organism centered approach to some community and ecosystem concepts. J. theor. Biol. 88: 287–308.

Nicholson, A. J., 1933. The balance of animal population. J. Anim. Ecol. 2: 132–178.

Volterra, V., 1926. Variazione e fluttuazioni del numero d'individui in specie animali canviventi Mem. Accad. Nazl. Lincci 2: 31–113 (transl. in Chapman, R. N., 1931, Animal Ecology, McGraw Hill, New York).

Vijverberg, J., 1976. The effect of food quantity and quality on the growth, birth rate and longevity of Daphnia hyalina Leydig. Hydrobiologia 51: 99–109.

Vijverberg, J., 1980. Effect of temperature on laboratory studies on development and growth of Cladocera and Copepoda from Tjeukemeer, The Netherlands. Freshwat. Biol. 10: 317–340.

Vijverberg, J. & Richter, A. F., 1982a. Population dynamics and production of Daphnia hyalina Leydig and Daphnia cucullata Sars in Tjeukemeer. Hydrobiologia 95: 235–259.

Vijverberg, J. & Richter, A. F., 1982b. Population dynamics and production of Acanthocyclops robustus (Sars) and Mesocyclops leuckarti (Claus) in Tjeukemeer. Hydrobiologia 95: 261–274.

Zeigler, B. P., 1976. Theory of Modelling & Simulation. John Wiley, New York.

Zeigler, B. P., 1978. Multi-model, multi-formalism modelling; an ecosystem example. In: Halfon, E. (Ed,) Theoretical Ecosystems. Academic Press.

Distribution and production of the larval chironomid populations in Tjeukemeer

D. M. Beattie
Limnological Institute, Tjeukemeer Laboratory, De Akkers 47, 8536 VD Oosterzee, The Netherlands

Keywords: chironomids, larval populations, distribution, production, shallow lakes

Abstract

The distribution and production of larval chironomids in Tjeukemeer, a shallow, moderately large and wind exposed lake, were estimated between 1969 and 1972. The following seven species together contributed more than 95% of all the individuals collected: *Chironomus anthracinus, Chironomus plumosus, Cladotanytarsus mancus, Glyptotendipes pallens, Polypedilum nubeculosum, Polypedilum scalaenum* and *Stictochironomus histrio.* Density and biomass distribution depended generally on the substrate type and its stability. The standing crop biomass was highest in the reed-beds and in a narrow 100 m wide inshore zone and low in the large offshore zone. Wind induced wave action, causing substrate instability, is likely to be the main factor responsible for these low standing crops. The mean annual larval production in the lake was: 1969, 1.8; 1970, 0.4; 1971, 1.7 and 1972, 3.7 g dry wt m^{-2}.

Introduction

Chironomid larvae are an important component of the food chain in Tjeukemeer. During the study period (1969–1972) the benthivorous fish stock amounted to about 100 kg fr wt ha^{-1}, of which around 87 kg was bream *(Abramis brama)* and about 10 kg eel *(Anguilla anguilla)* (Beattie *et al.* 1972; Goldspink 1978). The share of larval chironomid biomass in the diet of the 1+ and older bream was roughly 50%, while in eel it was about 60% (Hazelzet 1972; de Nie 1982).

The purpose of this study was twofold: firstly to study the larval chironomid density and biomass distribution in time and space and to relate this to environmental factors, and secondly to assess the chironomid larval production and relate this to the utilization by benthivorous fish in the lake.

Study area

The Tjeukemeer is a shallow (mean depth 1.5 m) and moderately large (21 km^2) freshwater lake. Being surrounded by a low dike and flat land, with almost no trees, the lake is greatly exposed to wind. Its bottom consists of a mosaic pattern of approximately 60% peat, 35% sand and 5% mud. The macrophyte vegetation along the shore is constituted chiefly of dense *Phragmites australis* beds. The total area of these beds is small, i.e. about 1% or 20 ha of the lake area. The 'inshore zone', defined as a 100 m broad zone from the outer margin of the reed-beds towards the centre of the lake, runs parallel to the shore. This zone consists mainly of sand (about 70%) and mud (about 30%) and represents approximately 10% of the total lake area. The rest of the lake, the 'offshore zone', constitutes about 90% of the total lake area. Since the densities in the different substrates in this zone were comparable, only the mud was sampled. The positions of the sampling stations are indicated in Fig. 1.

Hydrobiologia 95, 287–306 (1982). 0018-8158/82/0953-0287/$04.00.

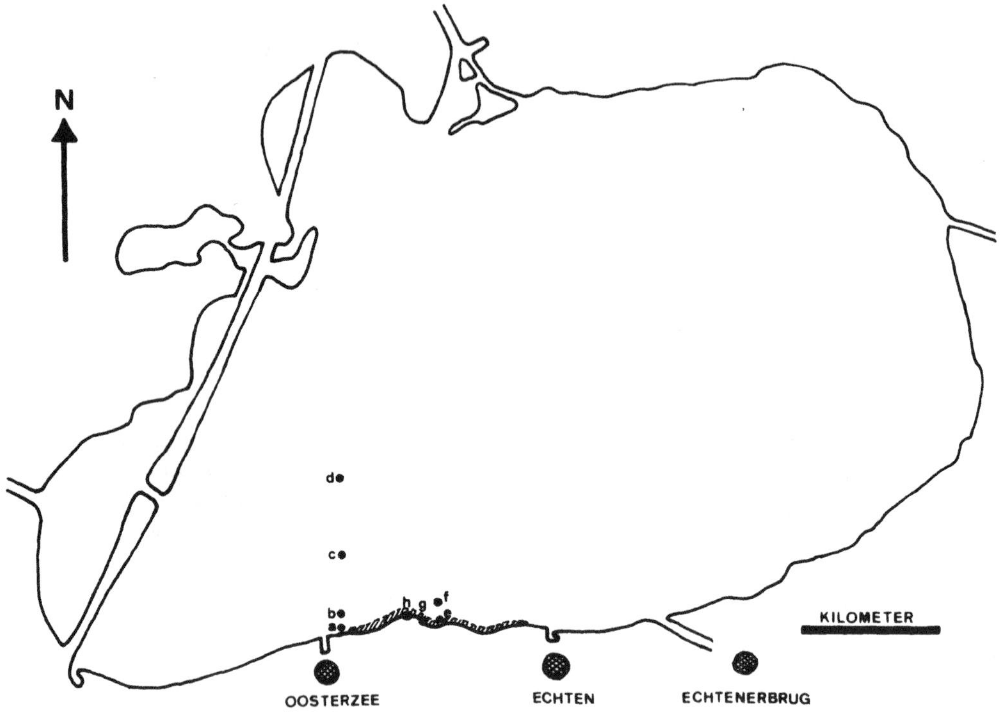

Fig. 1. Map of Tjeukemeer showing the position of sampling stations: a, b, c and d are the 5, 100, 500 and 1 000 m stations in mud substrate; e and f are the 5 and 100 m stations in sand substrate; g and h are the stations in the reed-bed; transect sampling was carried out along the a–d trajectory.

Methods

In the soft peat and mud substrates a modified Ekman-Birge grab (Beattie 1979), mounted on the end of a pole, was used to take bottom samples. Each sample had a surface area of 225 cm^2. In the mud it was possible to extract a 25 cm deep sample but in peat this was limited to 10 cm. However in both cases 95% of the total chiromonid larvae could be sampled.

In the harder sand substrate a Petersen grab was used in conjunction with a boat hook which allowed the sampler to be pushed 5 cm into the sand. Each sample had a surface area of 225 cm^2, but the mode of operation of the jaws of this apparatus unevenly enclosed the lowest layer of the sample. However, since the depth distribution of chironomid larvae was restricted to a few centimetres due to hardness of the substrate more than 90% of the larvae could be sampled. Thus the results from samples taken with different samplers and in different substrates could be compared without any correction.

Samples were brought back to the laboratory in buckets and processed immediately by sieving through a 0.3 mm mesh size sieve using a shower appliance. This dissociated the larger peat fibres from the chironomid larvae and also washed away smaller organic detritus and inorganic particles without damaging the larvae. The sample residue was transferred from the sieve to a small bucket, inundated with a sugar solution (specific gravity 1.07) and stirred vigorously (Anderson 1959). The larvae quickly floated to the surface and, depending upon the size, were either scooped off with a piece of plankton net or removed with the aid of pincers and transferred to a dish of water. This flotation process is very efficient and removes almost all of the larvae. The larvae were then sorted into species and counted by instar.

Sand samples were allowed to remain in the collection buckets and subjected to dislocation by a high pressure waterjet. This efficiently separated the chironomid larvae from the sand grains. As the larvae remained in suspension and the sand grains rapidly sedimented out, they could be decanted into a plankton net and subsequently transferred to a sorting dish. All the larvae from a particular set of samples were then mixed, a sub-sample taken from each species and preserved in 4% formalin for dry weight determination.

Preparations of head capsules were used to identify representatives of the different types of larvae found in the lake, using the identification keys of Johannsen (1937a, b), Chernovskii (1949), Bryce (1960) and Moller Pillot (1978). For species confirmation, larvae from a large autumn sample were cultured in the laboratory and the adult males were identified mainly using Lindner (1962) and Pinder (1978).

For dry weights (mg) the larvae were dried at about 75 °C for 24 hours and then weighed in groups of ten individuals or more on a micro-balance. Prior to this the larvae were kept for at least two weeks in formalin to reach a stable condition (Parma 1971). The formalin dry weight should be multiplied by 1.1 to get the actual dry weight, but unless otherwise stated, all the values given in this paper are uncorrected. The population dynamics of the larval populations were studied by following the pulses in the population development of the individual species. The time of emergence was estimated from the decrease in the numbers of 4th instar larvae and the appearance of pupae.

Results and discussion

Out of the recorded taxa (Table 1), seven together contributed more than 95% of all individuals collected. These are: *Chironomus* f.l. *anthracinus, Chironomus plumosus* (L), *Cladotanytarsus* f.l. *mancus, Glyptotendipes pallens* (Meigen), *Polypedilum nubeculosun* (Meigen), *Polypedilum scalaenum* (Schrank) and *Stictochironomus histrio* (Fabr.).

Table 1. List of chironomid taxa found in Tjeukemeer 1969–72.

Chironomus f.l. *anthracinus*
Chironomus plumosus (L)
Cladotanytarsus f.l. *mancus*
Cryptochironomus supplicans (Meigen)
Glyptotendipes pallens (Meigen)
Polypedilum nubeculosum (Meigen)
Polypedilum scalaenum (Schrank)
Procladius f.l.
Stictochironomus histrio (Fabr.)

The terms 'biomass' and 'numerical' dominants have been employed in the following account to indicate the two groupings of chironomid species in the Tjeukemeer. In the biomass dominant species the individual 4th instar larvae are relatively large (1–7 mg) but have in general low densities ($\leqslant 100$ m^{-2}); in contrast, the numerical dominant species have high population densities ($\geqslant 5000$ m^{-2}) but low individual biomass (about 1 mg).

Density and larval biomass gradients

The transect through the mud substrate was sampled in March when larvae of the overwintering generation had a stable population density and approached their maximum standing crop biomass (see Fig. 1). The bulk of the biomass dominants *(Chironomus plumosus, Glyptotendipes pallens, Polypedilum nubeculosum* and *Stictochironomus histrio)* were restricted to the inshore zone (Fig. 2). The numerical dominants *(Polypedilum scalaenum* and *Cladotanytarsus mancus)* occurred slightly further from the shore. The total larval standing crop biomass decreased sharply away from the reed-beds, corresponding to a change in larval density (Fig. 2B). The larval biomass was concentrated in the inshore zone and was per m^2 about 6 times higher than in the offshore zone. Although this is generally true for all species considered, the biomass dominants show a higher degree of concentration (6:1) than the numerical dominants (3:1).

Based on this information the four sampling stations chosen in the lake were 5, 100, 500 and 1 000 m from the edge of the reed-bed. The data from the former two stations are used to describe the population dynamics and mean standing crop biomass of chironomid species in the inshore zone, and those from the latter two stations that for the offshore zone. In order to examine the contrast in chironomid larval production within the inshore zone, two sampling stations at 5 and 100 m were chosen in a sand area.

Chironomus plumosus. Based on the larval density along the transect (Fig. 2), the distribution of *C. plumosus* larvae may be divided into three zones. This is true for both overwintering and summer generations (Fig. 3A) and all the zones run parallel

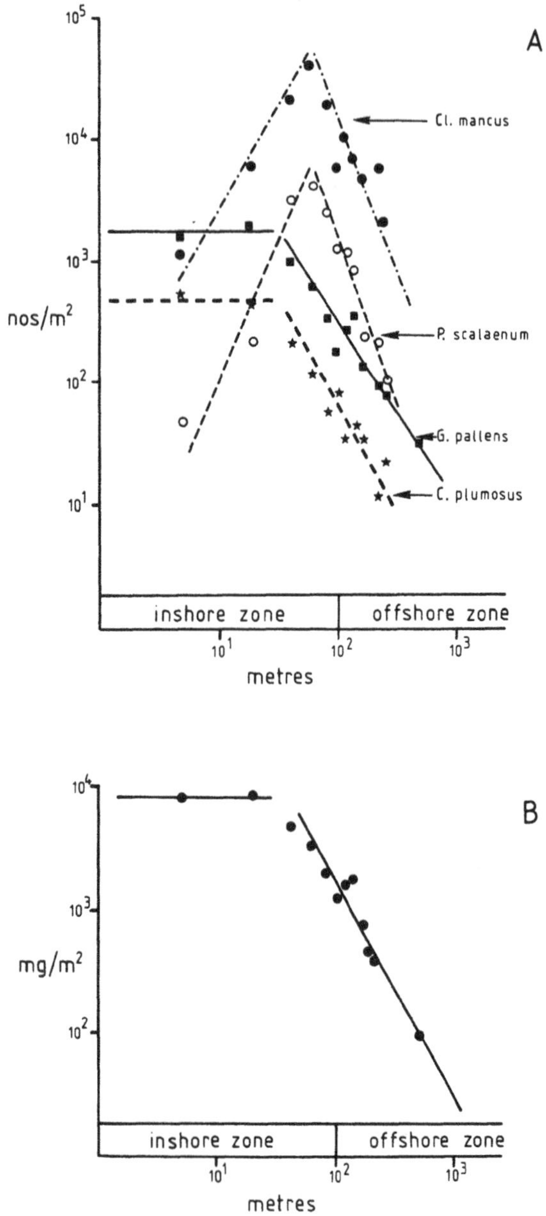

Fig. 2. The larval density gradients (numbers m^{-2}) of four chironomid species (A) and the gradients in total chironomid larval biomass (g dry wt m^{-2}) (B).

to the shore. The first and the most important zone is at its maximum 25 m wide and is immediately adjacent to the edge of the reed-bed having the highest larval densities of *C. plumosus* for the whole lake. The second zone, with a maximum width of 140 m, extends over the steepest part of the gradient slope and has low larval densities. The third zone is about 100 m wide and merges imperceptibly into the vast central region of the lake at approximately 200 m from the reed-bed. It has the lowest larval densities of the three zones. *C. plumosus* larvae were absent from the 1 000 m station at all times. Each zone has a contiguous larval distribution ($s^2 > \bar{x}$).

Glyptotendipes pallens. The distribution of *G. pallens* larvae may also be divided into three zones (Fig. 2) corresponding closely to those already described for the *C. plumosus* larvae. This zonation is most marked in the overwintering generation (Fig. 3B) and again all zones run parallel to the shore. The first and most important zone is at its maximum 30 m wide, adjacent to the reed-beds. This zone, except for the reed-beds, has the highest larval densities of *G. pallens* in the lake. The second zone has a maximum width of 150 m extending over the steepest part of the gradient slope and has moderate larval densities. The third zone has the lowest larval densities.

Polypedilun scalaenum. The bulk of *P. scalaenum* 2nd instar larvae occur in a zone running parallel to the shore and extending from 25 to 150 m (Fig. 2). Within this zone there is a larval density gradient from about 4 000 m^{-2} to about 100 m^{-2}. Continuous sampling during the autumn showed that initially this gradient extended right up to the reed-bed margin. However, whereas beyond 25 m the population development continued unhindered, in the zone adjacent to the reed-beds it was probably prevented by the simultaneous development of a high density of *C. plumosus.* This may have happened in the following way: initially the 4th instar larvae grazed down the eggs and very small larvae of *P. scalaenum* on the mud surface and later they prevented the establishment of larger *P. scalaenum* larvae by their foraging activity in the upper layers of the mud. A similar 'sweeping clean' activity was recorded for the *C. anthracinus* population in Lake Esrom (Jónasson 1972) but in that case the effect was intra-specific.

Cladotanytarsus mancus. The transect data (Fig. 2) show that *C. mancus* has a larval density distribution pattern similar to that of *P. scalaenum,* a density gradient from about 30 000 m^{-2} to about 2 000 m^{-2} with the bulk of the population in the

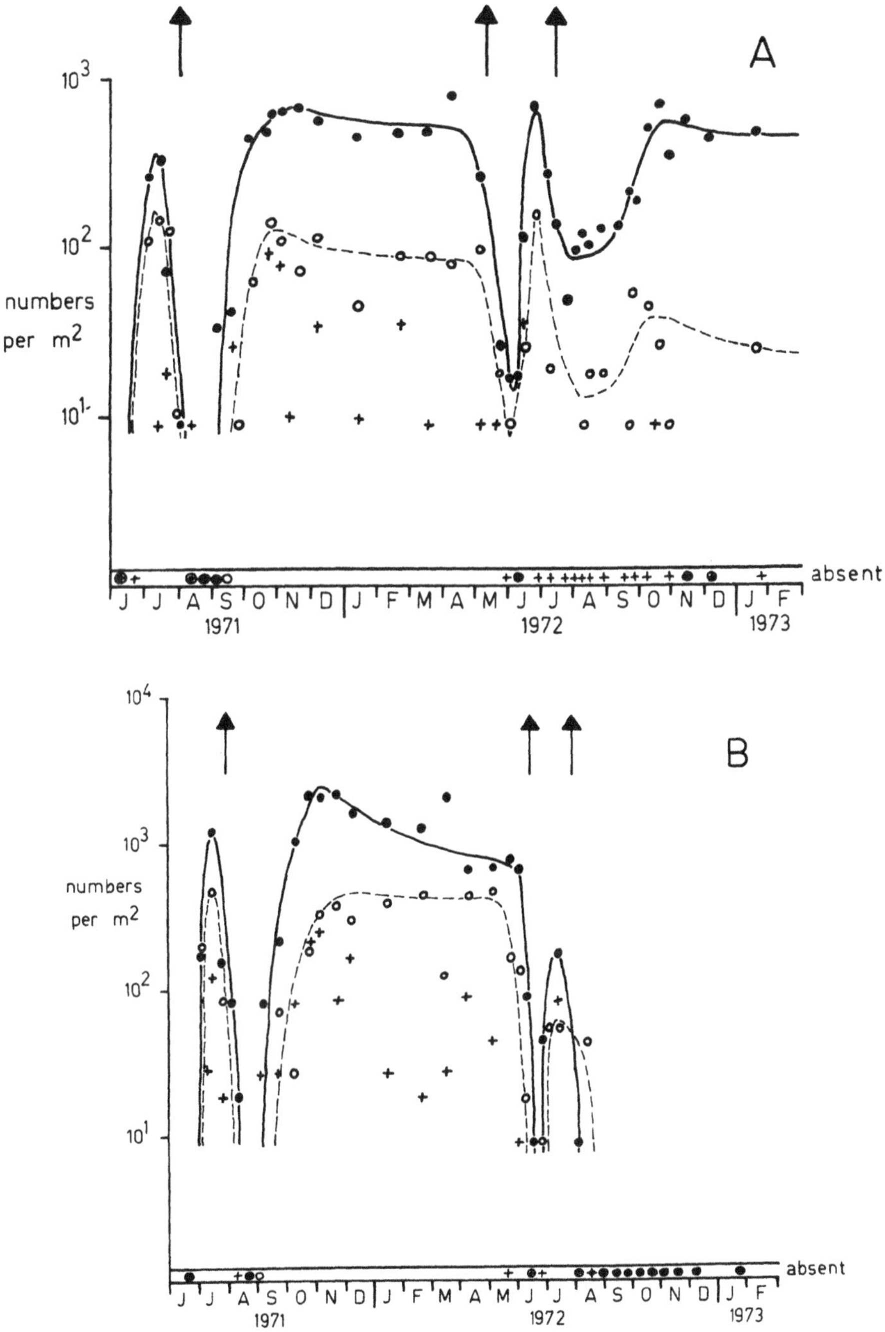

Fig. 3. Density (numbers m^{-2}) of (A) *C. plumosus* and (B) *G. pallens* 4th instar larvae at the 5 (●——●), 100 (○——○) and 500 m (+ +) stations in the mud with arrows indicating adult emergence.

inshore zone. Although densities at the 500 m station were quite high, very few larvae were recorded at the 1 000 m station.

In the most productive years, the size of the *C. mancus* population in Tjeukemeer was almost an order of magnitude greater than that of *P. scalaenum.* Adjacent to the reed-bed margin *C. mancus* probably suffers similarly due to interspecific competition with *C. plumosus,* as has already been described for *P. scalaenum.* As it has a 2nd instar diapause stage, there is a redistribution of larvae during the stormy weather in autumn.

A distinct density gradient as observed here for Tjeukemeer was also observed for other large shallow lakes which lack a macrophytic bottom vegetation in their offshore zone. Neusiedlersee (Schiemer 1979) is a large (300 km^2) shallow (mean depth 1 m) lake very much exposed to the wind, just like Tjeukemeer. The highest densities and biomass values were found in a sheltered zone close to the shore, while the values in the wind exposed central part of the lake were reduced respectively, to 27 and 10% of those observed in the inshore zone. Lake Arresø is a shallow (3.0 m), wind exposed lake with a surface area of 41 km^2. A particularly strong density gradient was observed here: larval densities decreased from 5 000 indiv. m^{-2} in the reed-beds to 20–40 indiv. m^{-2} in the offshore zone (Jónasson & Lindegaard 1979). Wolderwijd is a moderately large (18 km^2) and shallow (mean depth 2 m) lake and, like Tjeukemeer and Neusiedlersee, much exposed to the wind. The chironomid larval densities in this lake also decreased progressively from the shore to the centre of the lake (Beattie 1981). In these shallow lakes wave action stirs the bottom substrate, even at moderate windspeeds.

In loose, flocculating sediment without the stabilizing influence of macrophytic vegetation, chironomids have difficulty in building their tubes and surviving (Maitland *et al.* 1972; Jónasson & Lindegaard 1979). Thus it is likely that the lack of substrate stability is responsible for the observed low larval densities in the centre of large, shallow, wind exposed lakes.

Annual and seasonal variations of densities and biomass in the mud

Biomass dominant species

Throughout the study period the larvae of species in the genus *Chironomus* were never numerous and almost absent in some years. *G. pallens* is the biomass co-dominant with *Chironomus* in the inshore zone and although it contributes greatly to the standing crop biomass it presence is fortuitous and may be viewed as a side effect of the main *G. pallens* population in the adjacent reed-bed.

Chironomus spp. In the inshore zone *C. anthracinus* was present in sufficient numbers in only the first year (320 m^{-2}) of the study, but in the subsequent two years was very poorly represented (10 m^{-2}). In the fourth year it was replaced by *C. plumosus,* which occurred in reasonable numbers. The observed changeover of *Chironomus* species does not seem to be a permanent shift caused by a fundamental change in the lake environment, like the one observed by Sokolova (1961) in the Utscha reservoirs, but represents fluctuations of *Chironomus* species in the environs of the Tjeukemeer. The population dynamics of *C. plumosus* will be described here, but *C. anthracinus,* when present, has the same essential characteristics except that the lower individual biomass of the larvae gives a 55% lower production at comparable densities.

C. plumosus has at least two discrete larval generations per year in the Tjeukemeer (Fig. 3). The 1971–1972 overwintering generation began to develop in the second half of September and all the larvae were in the 4th instar stage by the end of October and the population was at its maximum density. These larvae then underwent three growth phases. The first, a period of 10 weeks, was from September to December, during which time the dry weight per larva increased by about 1 500% (Fig. 4).

The population density remained constant during the autumn; thus the standing drop biomass increased, due to the growth increment of the larvae, from 0.028 to 1.97 g m^{-2} (Fig. 5). During the second phase, a period of about 14 weeks from December to mid-March, in which the temperature of the lake water dropped below 5 °C, dry weight per larva decreased by 8%. This was probably the result of a temperature triggered non-obligatory diapause, as was observed by Fischer (1974) for the same species. During the third phase of about 7 weeks, from mid-March to May, the dry weight per larva increased by 28%. Since the population density remained constant during the spring, the standing crop biomass also increased.

In 1972 the adults emerged during the first two weeks of May and most of the *C. plumosus* pupated, leaving a residual population of only 30 m^{-2}. The precise time of emergence can vary from year to year but is always during May or early June and is determined by the completion of a temperature-controlled development cycle. For field studies it is convenient to express the length of the larval life-cycle as accumulated day-degrees of temperature (Miller 1941; Judd 1953; Konstantinov 1958). Thus,

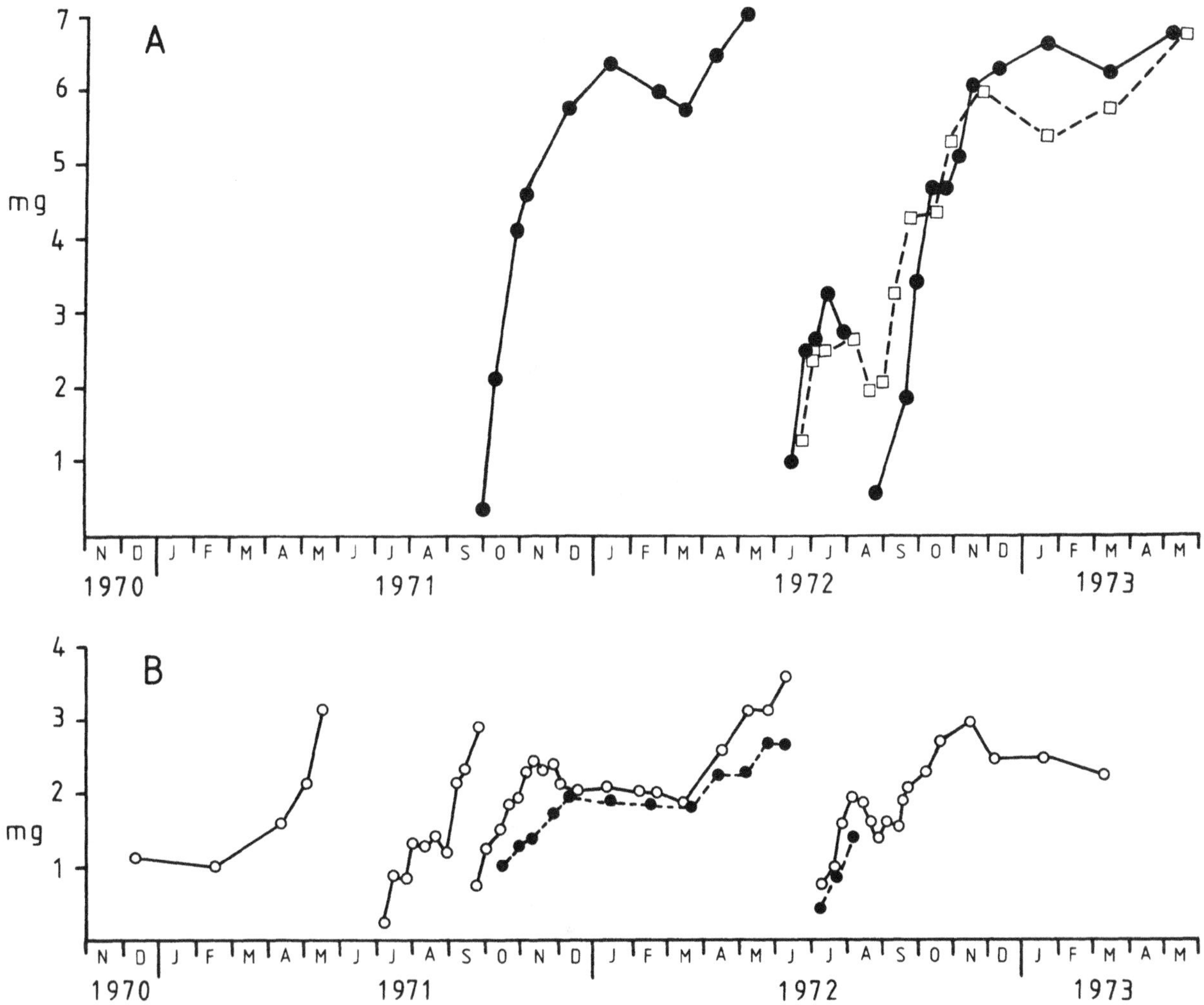

Fig. 4. Changes in individual 4th instar dry weight of (A) *C. plumosus* larvae in the mud of the inshore zone (●——●) and reed-beds (□----□), and (B) *G. pallens* larvae in the mud of the inshore zone (●----●) and the reed-beds (○——○).

in 1972 *C. plumosus* emerged during the second week of May after accumulating 2269 day-degrees, in 1971 in the first week of June after accumulating 2275 day-degrees.

A summer generation of *C. plumosus* followed the emergence but there was a delay of about four weeks before the development of the larval population could be detected in the bottom fauna. By mid-June, larvae were in the 4th instar stage; by mid-July, the larvae had increased to a dry weight of 3.5 mg per individual and pupated, giving rise to small imagos which contrasted with the large imagos of the overwintering generation. The maximum population density attained was similar to that of the overwintering generation. Although a residual population (100 m^{-2}) remained during the second half of July, emergence occurred throughout August. These larvae maintained a constant dry weight. At the end of August, growth recommenced and the life-cycle completed concurrently with the next overwintering generation.

The accumulated day-degrees for the summer generations of *C. plumosus* were 742 and 720 in 1971 and 1972 respectively and the generation times in both years were very similar. Thus, the annual variation in the emergence period of the summer generation seems to be dependent on the time of emergence and egg-laying of the previous overwin-

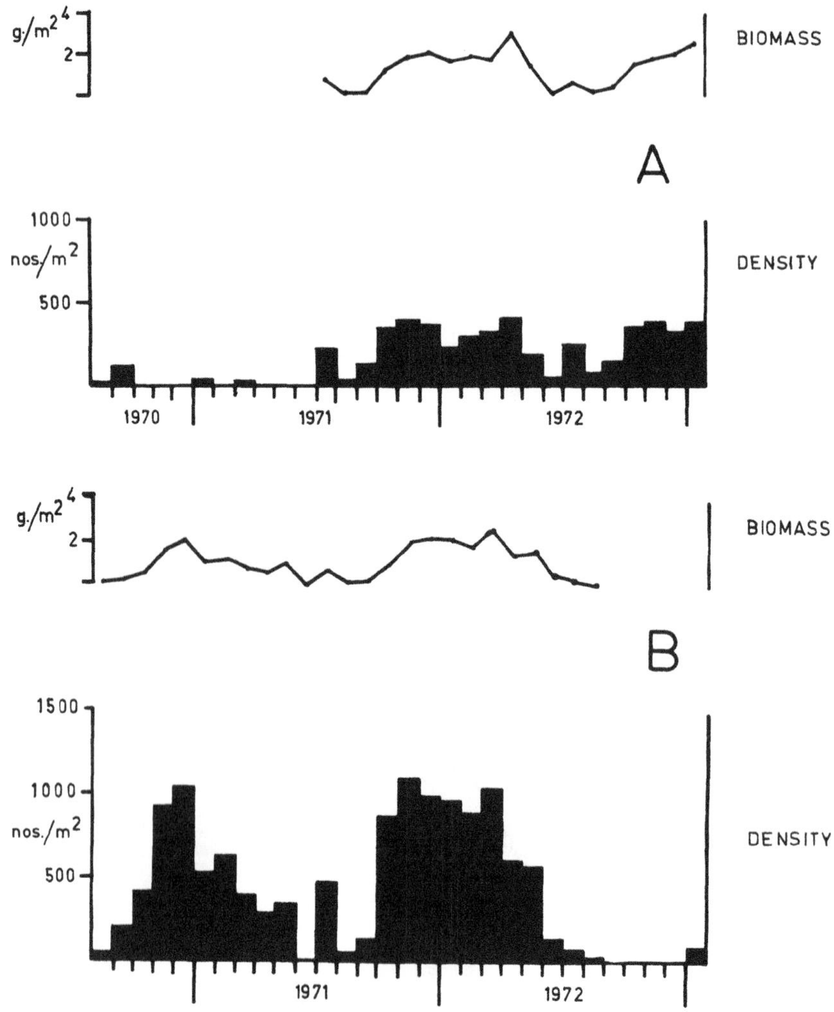

Fig. 5. Changes in mean monthly biomass (g dry wt m^{-2}) and density (numbers m^{-2}) of (A) *C. plumosus* and (B) *G. pallens* 4th instar larvae in the mud of the inshore zone.

tering generation, rather than on the larval development rate. Therefore, under favourable circumstances, a second summer generation may occur. The generation times of *C. plumosus* in the Tjeukemeer are in agreement with those determined experimentally (Tubb & Dorris 1965) at temperatures similar to those in Tjeukemeer.

The growth rate of *C. plumosus* larvae in Tjeukemeer may reach 0.54 mg per week in both the overwintering and summer generations; this is in close agreement with the experimentally determined maximum growth of 0.55 mg per week (Gizinskii & Wisniewski 1971).

When larval food is non-limiting, as is the case in Tjeukemeer, the growth rate is to a large degree temperature independent and occurs at temperature ranges of 5–15 °C in summer (Hilsenhoff 1966). Therefore the prevailing temperatures in summer and winter allow both generations larval growth to the critical pupation weight in a very short time (Jónasson 1965). In the summer generation this is reached after about five weeks and most of the larval population emerges. However, in the overwintering generation there is no emergence after five weeks, despite the larvae having reached the critical pupation weight; and the larvae continue to grow at the same rate during the autumn. After the winter growth pause, the larvae resume growth in the spring at a rate of 0.18 mg per week until they emerge.

This phenomenon can be accounted for by temperature, which independently affects both growth and development (Konstantinov 1958). During summer the continuous high temperature permits a constant and high development rate resulting in a short generation time. The larval phase is determined when development is completed and the critical pupation weight is achieved. Both these events occur within a few days of each other. Thus, despite the high larval growth rate, relatively small pupae and adult are produced. From autumn to spring, low temperatures give rise to a slow development rate resulting in a long generation time. Therefore a large pupa and imago is produced. Similar growth cycles are known for other *Chironomus* species (Sokolova 1961; Klötzli *et al.* 1974; Ringe 1974).

In very shallow lakes with no thermal stratification, such as those in Friesland, the temperature of the bottom substrate makes at least two, and sometimes three, generations of *Chironomus* larvae per year possible (Jancovič 1971; Zbarakh 1971; MacKay 1977a). In deeper lakes, such as Lago di Varese (Bonomi 1962) and the Utscha reservoirs (Sokolova 1961), temperature influences are related to both time and space. Two *Chironomus* generations per year are possible in the littoral zone, which reacts rapidly to temperature fluctuations, but only one generation is possible in the benthic zone, due to constantly low hypolimnetic temperatures.

Glyptotendipes pallens. In the first two years of study the *G. pallens* populations in the inshore zone were very poor and during the third and fourth years they were moderate. This variation from year to year is similar to that recorded for the *Chironomus* species, but is dependent upon the supply of migratory larvae from the reed-beds and not upon direct egg laying in the inshore zone.

G. pallens has only two generations per year in Tjeukemeer (Fig. 3). The 1971–72 overwintering generations built up in the inshore zone during September and October due to the migration of 3rd and 4th instar larvae from the adjacent reed-beds. The 5 m station was the first to be populated, followed by the 100 and 500 m stations. The rapid population growth of 3rd and 4th instar larvae in the reed-beds most probably produced intra-specific competition for space and thus larval migration towards the under-populated open water (Hruska 1974). The maximum 4th instar population density was reached by mid-November.

Over about 7 weeks in October and November, dry weight per larva increased by 74% (Fig. 4) and the standing crop biomass doubled (Fig. 5). During winter, over about 16 weeks from late November to early April, dry weight per larva decreased by 4%, resulting in a slight decrease in standing crop biomass. The larvae recommenced growth in spring and in 10 weeks from April to June, dry weight per larva increased by about 40%; however, since the larval density decreased strongly, the standing crop biomass decreased too.

The actual reduction in larval density was confined to the 5 m station (Fig. 3B) where fish predation was more effective on the high larval density next to the reed-bed. The decrease continued until the larval density was reduced to a level at which fish predation became apparently ineffective, and similar to that at the 100 m station. The population of the *G. pallens* winter generation emerged in June.

The summer generation of *G. pallens* followed the emergence and after two weeks the build-up of the larval population due also to larval migration could be detected in the bottom fauna. The larvae had only one growth period of around 4 weeks in July, the larva increasing in dry weight by about 250%. The summer generation of *G. pallens* varies in size from year to year, but it never attains such high population densities as the overwintering generation because the migration time is restricted by the short life-cycle. The maximum standing crop biomass of the summer generation was recorded in 1972. In late July the total population emerged; a second summer generation was recorded in neither the open water nor the reed-beds.

There is a delay of approximately two months between the emergence of the summer generation and the commencement of the succeeding overwintering generation which contrasts strongly with the almost immediate development of the summer generation. Thus the mechanism controlling the life cycle of *G. pallens* permits only two generations per year and restricts the potential biomass production. The growth rates of *G. pallens* larvae in both autumn and spring may be temperature dependent, but as in *C. plumosus,* does not solely depend upon attainment of a critical pupation weight.

Numerically dominant species

The magnitude of the populations of *P. scalaenum* and *C. mancus* varied from year to year (Table 2). Both species are mainly confined to the inshore zone of the Tjeukemeer. Although *C. mancus* contributes relatively little biomass to the chironomid community, it represents the most important chironomid food sources for the bottom feeding bream (Hazelzet 1972; Lammens 1982).

Polypedilum scalaenum. At the onset of larval development in September the density at the 5 m station was 37% of that at the 100 m station (Table 3). In early November this was reduced to 20% and later this month it was only 2%. During late autumn the wind-induced wave action brings both the uppermost layer of the bottom substrate and the small 2nd instar larvae of *P. scalaenum* into suspension. Thus in this period larvae are passively redistributed mainly in the inshore zone. However, only in December when the *C. plumosus* larvae cease their feeding activities, could this have a noticeable effect on larval distribution, namely the increase in larval density at the 5 m station to 23%. But in the spring the *C. plumosus* larvae began feeding again and the density was further reduced to 3%. However in some years, e.g. in 1969, there was no *in situ* development of a *P. scalaenum* population in September and the entire inshore zone was dependent upon redistributed larvae for a population build-up in November and December. Such a redistribution of larvae does not occur in the summer generations due to the absence of a prolonged 2nd instar stage and fewer storms over the lake. The distribution of 3rd and 4th instar larvae of *P. scalaenum* depends not only on the final 2nd instar distribution in late April, but also on the migration of 3rd and 4th instars from the high density areas. The offshore zone had a low density of randomly distributed larvae at all times.

There are usually two discrete generations of *P. scalaenum* larvae per year in the Tjeukemeer and overwintering generations have considerable annual variation (Table 3, Fig. 6). During September 1971 the *P. scalaenum* eggs laid by the females of the preceding summer generation hatched and the 2nd instar larvae rapidly built up a high population

Table 3. P. scalaenum larval densities (numbers m^{-2}) at the 5 and 500 m stations and at the 5 m station as a percentage of that at the 100 m station. P = pupae.

	5 m	200 m	%	Dominant instar stage
September	1449	3879	37	2
November (early)	657	3267	20	2
November (late)	54	3258	2	2
December	1143	4959	23	2
April	36	1188	3	2
May (early)	1263	2484	51	3
May (late)	783	1679	45	4
June (early)	18	378	5	P

Table 2. Maximum population densities (numbers m^{-2}) of *P. scalaenum* and *C. mancus* larvae and pupae in the mud of the inshore zone, as the mean of the 5 and 100 m stations.

Instar		2nd	2nd	2nd	3rd	4th	Pupae	4th
Month		Sept.	Dec.	Mar.	Apr.-May	May-June	June	July
1969-70	*P. scalaenum*	0	972	540	306	441	126	387
	C. mancus	0	2835	1931	320	297	126	554
1970-71	*P. scalaenum*	1944	1746	2979	2061	1962	171	59
	C. mancus	1800	1728	1107	432	1143	81	621
1971-72	*P. scalaenum*	3650	4959	1575	2484	1679	378	1395
	C. mancus	16394	17784	13437	5616	3141	135	603
Mean	*P. scalaenum*	1865	2559	1698	1617	1361	225	614
	C. mancus	6065	7449	5492	2123	1527	114	593

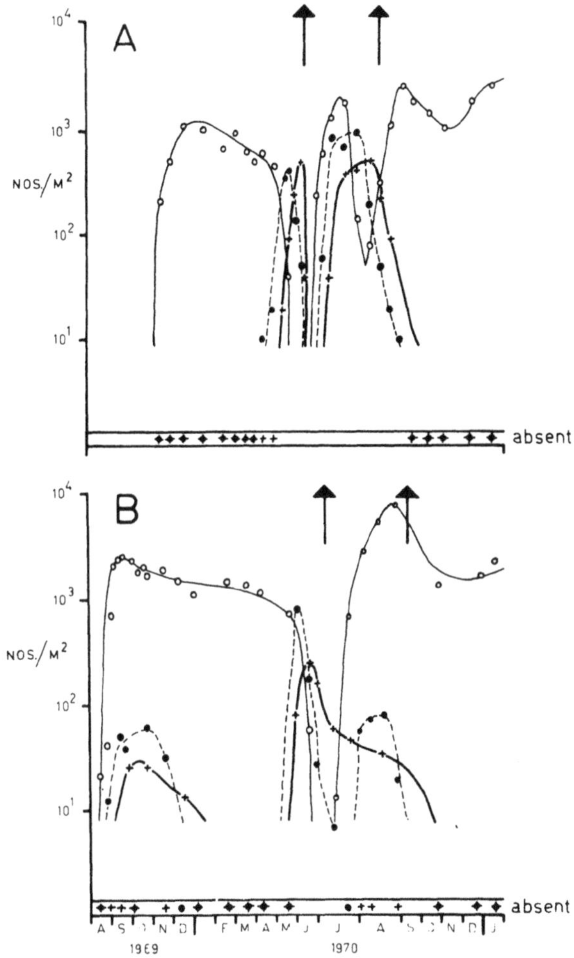

Fig. 6. Density (numbers m^{-2}) of *P. scalaenum* in (A) mud and (B) sand of the inshore zone, given as: 2nd (○——○), 3rd (●----●) and 4th (+—+) instars with arrows indicating adult emergence.

at the end of the month. The larvae remained in the 2nd instar throughout the autumn and winter due to diapause, despite the autumn temperatures favourable for development. During winter and early spring the larval density was reduced. By March the 2nd instar diapause was broken and the larvae had resumed growth.

The larval population had a maximum 3rd and 4th instar density in early and late May, respectively. Pupation and emergence occurred in June. Four weeks later, 2nd instar larvae were again present in the inshore zone, but developed without diapause, directly through to pupation in 4 to 5 weeks in mid-August. The maximum 4th instar density was relatively high in 1972. By late September all larvae of the summer generation had emerged.

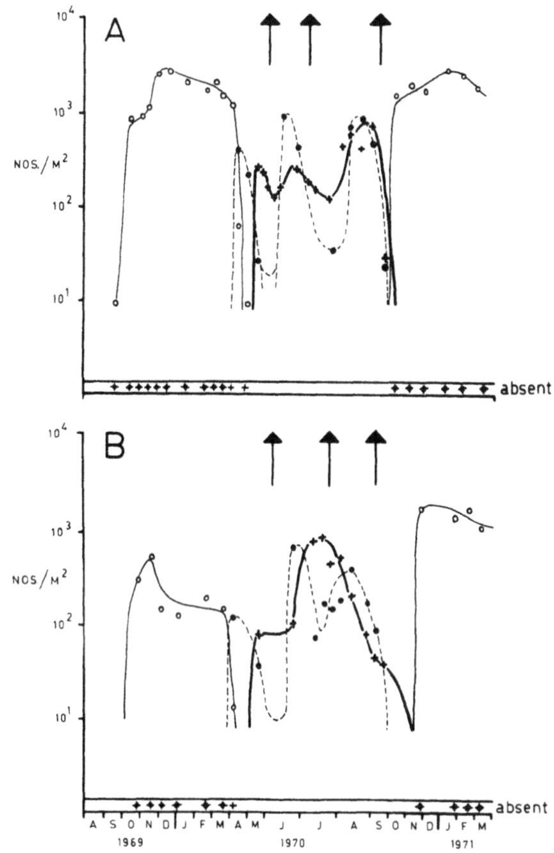

Fig. 7. Density (numbers m^{-2}) of *C. mancus* in (A) mud and (B) sand of the inshore zone, given as: 2nd (○——○), 3rd (●----●) and 4th (+—+) instars, with arrows indicating adult emergence.

During the diapause period, from September to March, there was no growth of 2nd instar larvae. The main larval growth phase occurred after diapause was broken and the larvae had developed through to 3rd and 4th instars.

The different length of life-cycle for the two generations of *P. scalaenum* is interesting but, unlike for *C. plumosus,* cannot be directly accounted for by temperature dependent development. In the *P. scalaenum* summer generation development is continuous and the 2nd instar larvae are present for only a few days. In contrast, in the overwintering generation they are present for six months, despite the temperature of the initial phases in each generation being the same.

Cladotanytarsus mancus. C. mancus has at least two, and usually three, larval generations per year in the Tjeukemeer mud (Fig. 7). There are large annual variations in population size of the overwintering generations, although the population size of the summer generations remains remarkably constant (Table 3). There is also a rapid build up of *C. mancus* 2nd instars in September. The 2nd instar larvae remained in diapause throughout the winter and resumed development and growth only in March after the diapause was broken. As with *P. scalaenum*, the *C. mancus* population can also be built up by redistribution of larvae in late autumn.

In early spring the larval density was still high, but as the larvae developed to pupae there was continuous mortality due to the onset of fish predation. Larval growth and development were more rapid than in *P. scalaenum*. The population reached maximum 3rd instar density in mid-April and maximum 4th instar density in early May. The larvae pupated in late May and the population emerged. Immediately thereafter, 2nd instars were again present in the inshore zone. In late June, i.e. 4 weeks later, they pupated without an intervening diapause.

The emerging adult females laid eggs which gave rise to the second summer generation, which also consisted of non-diapausing larvae. These larvae developed continuously for 4–5 weeks and pupated in mid-August. The maximum 4th instar densities of the first and second summer generations of 1972 were both very similar to each other and to those in other years.

The rapid rate of larval growth and development in all instars during the summer results in an almost continuous recruitment to the next phase. The summer generations overlap considerably since the population development is less synchronized than in winter. This leads to a more or less continuous presence of *C. mancus* larvae throughout the summer, in which period they are the most important chironomid larvae in the bream diet (Hazelzet 1972; Lammens 1982).

C. mancus has a pattern of alternating generations very similar to that of *P. scalaenum* except that the extra summer generation is more usual.

Other species

The other chironomid species present in the lake in some years (Table 1) contributed in a small way to the total biomass production of the chironomid community.

Polypedilum nubeculosum. This species was present in 1968 and 1972. The most important of the occasional species, the population is restricted to the inshore zone. In 1972 this species had three generations.

Stictochironomus histrio. This species prefers sand but also occurs in the mud in some years. It was also restricted to the inshore zone and, except for a few metres next to the reed-bed margin, had a low larval density of less than 100 m^{-2}.

Cryptochironomus supplicans. These larvae were occasionally found in the inshore zone. Its larval density was less than 10 m^{-2}.

Procladius spp. This predatory chironomid was rare and larval density did not exceed 5 m^{-2}.

Annual and seasonal variations of densities and biomass in the sand

Throughout the study period the sand area had poor chironomid larval populations, even of sand-preferring species such as *Stictochironomus histrio* and *Cryptochironomus supplicans*. Because of the low densities of *Chironomus* spp. and *G. pallens*, the term 'biomass' dominants is inappropriate for such species here. But the numerical dominants *P. scalaenum* and *C. mancus* were present each year in substantial numbers and consequently had a more important role in the standing crop biomass of the chironomid community than in the mud.

Numerically dominant species

The population dynamics of *P. scalaenum* and *C. mancus* in the sand are described using data from the 5 m station, which is representative of the whole inshore zone. For comparison with the population densities in the mud, data from the 100 m station are used.

Polypedilum scalaenum. This species is mainly confined to the inshore zone in the sand. The larval densities at the 5 and 100 m stations had a gradient effect similar to that seen in the mud, but the densities at the edge of the reed-beds were the highest. However, the final 4th instar densities are very similar at both stations (Table 4).

The overwintering generation of *P. scalaenum*

Table 4. The larval densities (numbers m^{-2}) of the different instars of *P. scalaenum* and *C. mancus* at the maximum periods of development in the mud and sand of the inshore zone.

	P. scalaenum					*C. mancus*					
	Mud			Sand		Mud			Sand		
	5	100	500	5	100	5	100	500	5	100	
2nd	527	3 650	41	3 067	1 971	7 799	16 394	6 813	3 330	1 418	Oct.
2nd	0	1 575	0	1 944	1 332	5 220	13 437	7 785	2 295	90	March
3rd	1 263	2 484	24	1 175	158	995	5 616	567	1 395	225	Apr.–May
4th	783	1 679	9	297	207	653	3 141	135	1 373	225	May–June
Pupae	18	378	18	23	27	27	108	0	90	0	June–July
4th	117	1 395	0	0	0	0	603	0	3 443	513	July

has a life-cycle similar to that in the mud but is a month longer in duration (Fig. 6). The summer generation was not distinct.

The development of eggs through to diapausing 2nd instar larvae occurred in September but also as early as mid-August. This is probably caused by differing emergence times in sand and mud. The initial density of the 2nd instar population can be quite large. However, in the years under consideration, by the end of September this was reduced to a stable overwintering density, a density similar to that recorded at the 100 m station in the mud for the 2nd instar of *P. scalaenum*. Although the lowering of the water temperature during winter breaks the larval diapause at the same time both in the mud and sand the subsequent rate of larval growth and development is lower in the latter case. Thus in 1970 the presence of 3rd instar larvae was delayed until the end of May.

The adults emerged from the sand at the beginning of July, about three weeks later than in the mud, but in some years it may still give rise to a summer generation. When 2nd instar larvae appeared by early August the 3rd and 4th instar larvae developed in late summer (Fig. 6), and some adults emerged. Despite the meagre standing crop biomass of *P. scalaenum* in the sand, its contribution to the chironomid community biomass could be important in some years.

The 2nd instar larval density in autumn at 100 m was lower in the sand than in the mud. However, by spring both 2nd instar densities were reduced to a similar level. Thereafter the sand population loses more larvae so that the pre-pupational 4th instar density is only 12% of that in the mud. The crude mortality of *P. scalaenum* from 2nd instar in autumn to pre-pupal 4th instar in the sand is 90%, and in the mud about 55%. The greater larval mortality and a three week longer life-cycle in the sand show that this substrate is inferior to the mud, probably lacking an adequate organic food source.

In summary, the *P. scalaenum* larval population in the sand, in the virtual absence of biomass dominant chironomid species, has a more significant share in the standing crop of the chironomid community than in the mud.

Cladotanytarsus mancus. From the field data it is not possible to conclude that *C. mancus* had a density gradient in the sand. However, marked differences between the 5 and 100 m stations in some years would lend support to the presence of a gradient (Table 4).

The life-cycle of *C. mancus* in the sand and the mud were similar: we usually observed three generations per year (Fig. 7). Due to considerable overlap between summer generations, so that it is difficult to discriminate between them, two summer generations may be an underestimation.

At the end of September eggs laid by females of the preceding summer generation hatched and developed through to give a maximum density of diapausing 2nd instar larvae during November. As with *P. scalaenum*, the 2nd instar build-up can occur in November and December by the redistribution of larvae from other high density areas. The size of the overwintering population can vary from year to year, its density in the sand being an order of magnitude less than in the mud. After diapause, the 2nd instar larvae resume growth in spring. In 1971 the population reached a maximum 3rd instar density in May and 4th instar density in early June. The

C. mancus larvae emerged from the sand in early June and their population was greatly reduced.

As with *P. scalaenum*, the standing crop biomass of *C. mancus* in the sand is relatively small, but in years without biomass dominant species it makes a similarly important contribution to the total biomass of the chironomid community in the sand. The 2nd instar larval density in autumn at 100 m is very much lower in the sand than in the mud, and in spring this difference is even greater. By the 4th instar stage the population density in the sand is only 7% of that in the mud. The crude mortality of *C. mancus* in the sand (about 85%) and mud (about 80%) from 2nd instars in autumn to prepupal 4th instars is comparable. Even though the *C. mancus* larval population in the sand is much smaller than in the mud, it is still important as a source of food for the bream (Hazelzet 1972; Lammens 1982).

Other species

Chironomid species other than *P. scalaenum* and *C. mancus* occurred sporadically in the sand, and the densities were low.

Glyptotendipes pallens. This species occurred in the sand every year as a concomitant effect of the high densities in the adjacent reed-beds. *G. pallens* 4th instar larvae appeared in the sand in September and October. In November a stable overwintering density was reached at the inshore stations. In most years the fortuitous presence of *G. pallens* provided the bulk of the chironomid larval biomass in the sand.

Stictochironomus histrio. This species prefers sand but contrary to expectations it is not the dominant chironomid in the sand areas. The maximum recorded density at the inshore stations was less than 150 m^{-2}. In some years this species was virtually absent, but when present its contribution to the biomass of the total chironomid community was similar to *G. pallens*.

Polypedilum nubeculosum. This species was an insignificant member of the chironomid community in the sand. In most years it was absent.

Cryptochironomus supplicans. This sand-preferring species was present in most years, but had larval population densities less than 10 m^{-2}.

Annual and seasonal variations of densities and biomass in the reed-beds

The lake has no typical littoral region but there is a zone around the shore approximating to this habitat. This zone is about 10 m wide, extends around 90% of the lake shore, and has a shoreline depth of a few centimetres, deepening to 50–100 cm at the boundary with the open water. The substrate is a 1–2 cm layer of sand covered sparsely by detritus. Underneath this is a firm, deep sand layer bound together by the roots mainly of dense *Phragmites* stands and *Typha* as sub-dominant in some areas. The reed-beds are broken down each year by the autumn gales and winter ice. The effects of this destruction are only felt above the substrate, although the ice can freeze solid down to the sand.

The greater part of the fauna in the reed-beds is made up of three species, namely *Gammarus tigrinus, Asellus aquaticus* and *Glyptotendipes pallens.* Other species do occur but are less significant with regard to productivity. *G. tigrinus* and *A. aquaticus* have been studied in detail by Chambers (1977a, b). The following account is mainly confined to a study of the life-cycle of *G. pallens* in the reed-beds and how it influences the open-water chrionomid fauna.

Glyptotendipes pallens. A preliminary survey of the chosen reed-bed indicated that for over 90% of the area covered by the *G. pallens* population (see Fig. 1), it had a contiguous distribution and a high 4th instar larval density. In 1970 two sets of samples taken towards each end of the reed-bed showed a higher concentration of larvae in the east than in the west. In the mid section of the reed-bed, random samples were taken. A reasonable representation of *G. pallens* population behaviour, applicable to the majority of the reed-bed, was obtained. *G. pallens,* having two generations per year, is a permanent member of the reed-bed fauna with little annual variation in the mean 4th instar density (Fig. 8).

The overwintering generation in 1971 began to develop in late September and the 3rd instars reached a maximum density in October. Throughout the rest of the autumn and winter it was reduced to a residual density of about 200 m^{-2}. In March, development was resumed and many 2nd instar larvae moulted to the 3rd instar. The similarly low density of 3rd instars in autumn and spring occurred because this was only a transitory stage in the life-cycle.

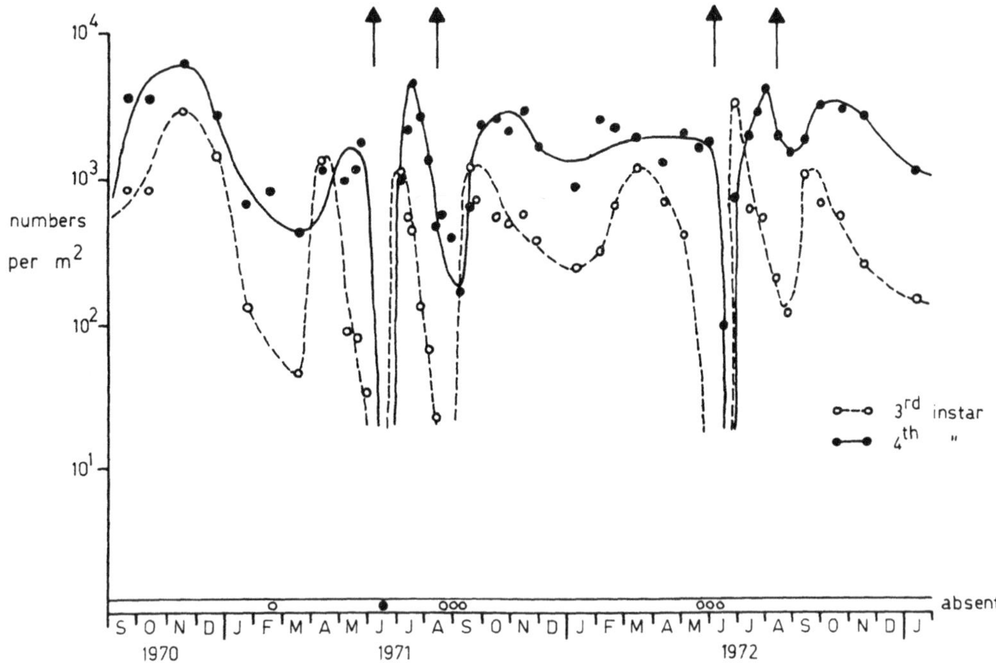

Fig. 8. Density (numbers m^{-2}) of *G. pallens* 3rd (O----O) and 4th (●——●) instars in the reed beds.

The accumulative 4th instar stage, after a long period of population build-up during October and November, reached a higher maximum larval density in late autumn. Throughout the winter there was an overall decrease of 25% in the 4th instar density. When the maximum 4th instar densities were reached, larval migration to the inshore zone took place. In spring the lake began to warm up and the further larval development increased the 4th instar density slightly during May 1972, compared with a much greater increase in the spring of 1971. Unlike the autumn population, this 'boosted' spring population suffered almost no reduction in density until the entire population emerged. The second period of development during spring was not detectable in the inshore zone and there was no egg laying in this area by *G. pallens*. Pupation of the individual larvae lasted only a few days but the emergence period of the whole population can extend up to three weeks, with its maximum in the second week of June.

The eggs laid by females emerging from the overwintering generation developed almost immediately and rapidly reached their 3rd and 4th instar maxima in early and late July, respectively. The peak emergence period of the *G. pallens* summer generation can vary by up to three weeks, i.e. somewhere between the last week of July (as in 1971) and mid-August (as in 1972). By November 1970 the larvae showed a maximum standing crop biomass of about 8 g m^2. After the larval development in spring, standing crop biomass decreased to around 3 g m^2 (Fig. 9). The trend for the 1971–72 overwintering generation was similar, except for the higher standing crop biomass in late spring.

From late September until the end of November, the 4th instar larvae grew rapidly (Fig. 4B). The length of this autumn growth period is variable and depends upon the water temperature. By late November growth stopped and from mid-December to late March the larvae lost 18% of their weight. In March the larvae began to grow again and over a period of about 12 weeks dry weight per larva increased by 85%.

The summer population generally emerged in early August, but in 1972 substantial numbers of larvae did not emerge but continued to grow. These larvae formed a significant proportion (60%) of the succeeding overwintering generation.

The life-cycle characteristics of *G. pallens* indicate a temperature dependent development mechanism similar to that already described for *C. plumo-*

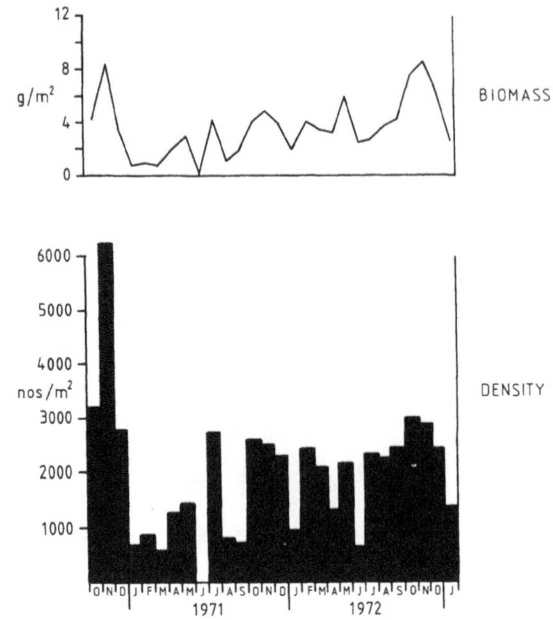

Fig. 9. Changes in mean monthly densities (numbers m^{-2}) and biomass (g dry wt m^{-2}) of *G. pallens* 4th instar larvae in the reed-beds.

sus. The more rapid growth of *G. pallens* in the Tjeukemeer reed-beds than that of the migrant larvae in the inshore zone resulted in a 15% higher pre-pupation weight. The lower growth rates of *G. pallens* larvae in both autumn and spring compared with summer growth rates were due to temperature differences.

Chironomus plumosus. This was the only chironomid species which had a large larval population in the reed-beds. This was only during the summer of 1972. In other years the larvae occurred only sporadically. Following the emergence of *C. plumosus* in May 1972 from the inshore zone, the 4th instar larvae developed substantially in the reed-beds. The predominantly NW wind, with a speed of 9–12 m per second, blew the emerging adult insects into the reed-beds, where egg-laying was confined to the sheltered stems. The 4th instar density reached a maximum of more than 5 000 m^{-2} within a month, the dry weight per larva increasing to 2.5 mg. By late June more than 90% of the population had emerged but no subsequent overwintering population developed.

Production

Production in the reed-beds and the inshore zone was calculated per cohort of each species by multiplying the mean standing crop per cohort with a P/B ratio of 5. According to Waters (1979), the cohort P/B ratio of benthic invertebrates is fairly constant (range 4–6). Thus, the use here of a ratio of 5 seems reasonable. Annual production of species was calculated by adding together the cohort production of the different generations for each year. For convenience the whole production of the overwintering generation was included in the year of emergence. The production in the off-shore zone was computed by mulitplying the mean annual standing crop biomass in this zone, expressed as a fraction of the standing crop biomass in the inshore zone, with the production in the inshore zone.

The annual variation in production of each species and in the total larval production of chironomids in the inshore zone is considerable (Table 5).

Table 5. Annual production (g dry wt m^{-2}) of the main chironomid species in the inshore zone.

	Mud				Sand				Whole zone			
	1969	1970	1971	1972	1969	1970	1971	1972	1969	1970	1971	1972
C. anthracinus												
	5.8	0.40	7.8	20.0	2.8	0.00	0.37	2.8	3.7	0.12	2.6	8.0
C. plumosus												
G. pallens	0.26	0.66	4.5	5.0	0.26	0.38	0.18	0.05	0.26	0.47	1.48	1.5
P. scalaenum	1.4	0.25	1.4	2.7	0.21	0.54	0.11	0.05	0.57	0.46	0.50	0.84
C. mancus	7.8	0.78	0.98	3.9	0.02	0.54	0.78	0.15	2.35	0.61	0.84	1.3
S. histrio	0.00	0.00	0.00	0.22	0.03	0.05	0.78	0.29	0.02	0.04	0.55	0.27
P. nubeculosum	0.07	0.00	0.00	1.4	0.07	0.00	0.00	0.27	0.07	0.00	0.00	0.61
Total	15.4	2.1	14.7	33.2	3.4	1.5	2.2	3.6	7.0	1.7	6.0	12.5

Usually *Chironomus* realized the highest production, followed by *G. pallens* and *C. mancus*. Generally *Chironomus* contributed about 50% and the latter two species 15–20% each to the total annual production. The production in the reed-beds depends mainly on *G. pallens* but *G. plumosus*, when present, also contributes substantially to the total larval production. The total production per m^{-2} is approximately 4–10 times higher here than in the inshore zone (Table 6). Due especially to the low production in the large offshore zone, the mean annual production calculated over the whole lake area is low compared with the values reported for most other waterbodies (Table 7). For most lakes and reservoirs in the temperate zone, larval chironomid production ranges from 10 to 30 g dry wt m^{-2} yr^{-1} (see for review MacKay 1977b; Waters 1977; Schiemer 1979). The few estimates for eutrophic shallow lakes usually also fall within this range. Potter & Learner (1974) recorded 20 g for a eutrophic reservoir in S. Wales and Kajak *et al.* (in press) reported 26 g for Lake Mikolaişkie and 8–22 g dry wt m^{-2} yr^{-1} for Lake Warniak in Poland. The total annual chironomid production in Loch Leven (Scotland) in 1971 also fell within this range, but during 1970 the production was much higher (Charles *et al.* 1974; Maitland & Hudspith 1974). The

Table 6. Annual production (g dry wt m^{-2}) of the main chironomid species in the reed-beds.

	Production	
	1971	1972
G. pallens	24.0	28.5
C. plumosus	0.0	32.5
Total	24.0	61.0

Table 7. Annual production (g dry wt m^{-2}) of larval chironomids in the reed-bed, the inshore zone and the offshore zone and the mean annual production for the whole lake area. NS = not sampled.

	Surface area %	Production			
		1969	1970	1971	1972
Reed-beds	1	NS	NS	24.0	61.0
Inshore zone	10	7.0	1.7	6.0	12.5
Offshore zone	89	1.2	0.3	1.0	2.1
Lake	100	1.8	0.4	1.7	3.7

production of about 80 g dry wt m^{-2} yr^{-1} reported by Berczik (1967) for the shallow but sheltered Lake Velence (Hungary) belongs to the highest ever recorded for a natural lake and is within the range of those reported for fish ponds (Zieba 1971). Low production values as in Tjeukemeer were also recorded for the large and shallow Neusiedlersee in Austria (Schiemer 1979). Chironomid production in the large, shallow and wind-exposed Lake Arresø was in fact a magnitude lower than in Tjeukemeer and Neusiedlersee (Jónasson & Lindegaard 1979). Both in these lakes and in Tjeukemeer, severe wave action causing instability in the upper layers of the substrate in their large central areas may explain the low production rates (see also discussion on density and biomass gradients, page 292).

The diets of the two most important benthic predators in the lake, the bream and the eel, were studied by Hazelzet (1972), Lammens (1982) and de Nie (1982), but only the last two authors determined the share of the different chironomid spp in terms of biomass. According to Hazelzet and de Nie, the diet of bream and eel contain respectively about 50 and 60% chironomid biomass. During the study period the bream production was estimated to be 34 kg and the eel production about 6 kg fr wt ha^{-1} yr^{-1} (Goldspink & Banks 1975; Goldspink 1978). Using these production estimates and assuming a growth efficiency of 0.10 (Humphreys 1979), an assimilation efficiency of 0.8 (Winberg 1956) and chironomid dry weight to be 20% of fresh weight (Winberg *et al.* 1971), the annual benthic food requirements for these two fish populations were estimated. The bream consumed annually approximately

$$\frac{34}{10} \times \frac{100}{10} \times \frac{10}{8} \times \frac{50}{100} \times \frac{20}{100} = 4.3 \text{ g dry wt m}^{-2}.$$

This was for the eel approximately:

$$\frac{6}{10} \times \frac{100}{10} \times \frac{10}{8} \times \frac{60}{100} \times \frac{20}{100} = 0.9 \text{ g dry wt m}^{-2}.$$

These figures on the mean annual consumption, together with data on the share of the different chironomid spp. found in the fish stomachs, were used to partition their food requirements according to food species (Table 8). The total estimated production is lower than the estimated total consumed production, this also being the case for the individual taxa with the exception of *Chironomus*. This

Table 8. A comparison of the mean annual production of chironomid taxa (g dry wt m^{-2}) and their utilization by benthivorous fish in Tjeukemeer.

	Production	Production utilized	
	(1969–72)	Bream	Eel
C. anthracinus *C. plumosus*	1.01	0.0	0.7
G. pallens	0.27	0.8	0.2
P. scalaenum	0.17	0.6	0.0
C. mancus	0.36	2.7	0.0
S. histrio	0.06	0.2	0.0
P. nubeculosum	0.06	0.0	0.0
Total	1.93	4.3	0.9

points to a high utilization of production. High values were also reported by MacKay (1977b) for roach and bleak feeding on chironomids in the River Thames. Gerking (1962) observed approximately 50% utilization of the chironomid production during summer by the bluegill sunfish, the dominant benthivorous fish in Wyland Lake (Indinana, U.S.A.).

The observed discrepancies between the production and the food requirements of the fish may have several reasons. First, with the method employed here to assess the chironomid production, the production value depends mainly on three different quantities: 1) the number of generations per year; 2) mean standing crop biomass per larval generation; 3) the productivity per cohort (P/B). The estimate of the first is subject to greater error than the other two. For most species generally three generations per year were estimated, but in the small species considerable overlap of instars during summer was found. In laboratory cultures of *Cladotanytarsus* at 15 °C, MacKay (1977a) found a generation time of less than 2 weeks. Since we sampled at 2 week intervals we may have missed several summer generations of *C. mancus* and *P. scalaenum*. Consequently, their real production could be a factor of 2 higher than shown in Tables 5 and 8. Secondly, although the larval population densities in the offshore zone were much lower than in the reed-beds and the inshore zone, the offshore zone, because of its much larger surface area, contributed more to the larval production in the lake than the other two zones together. However, sampling in the offshore zone was much less intensive than in the inshore zone. Consequently, the production estimates for this zone, and thus for the whole lake area, are subject to inaccuracies. Thirdly, since the calculation of the food requirements of the fish, are based upon assumptions from the literature, and the observations on production and fish diets are from different years, our estimates are crude. Finally, there is some evidence that larger food particles such as chironomid larvae are digested more slowly, and consequently are present in the fish stomach for longer than smaller food particles such as zooplankton (Berrie 1972; Fänge & Grove 1979). Since we did not correct the diet figures to allow for relative digestion rates, our food consumption estimates will be too high.

Notwithstanding all this, the conclusions that the mean annual chironomid production in Tjeukemeer is low compared with most other lakes and that a large part of the produced biomass is utilized by fish, are still valid. Assuming that both the standing crop biomass in the offshore zone and the annual productivities of *C. mancus* and *P. scalaenum* were a factor of 2 higher than estimated, the mean annual production for the whole lake area thus calculated would still range from 1.2 to 5.7 g dry wt m^{-2} yr^{-1}. Such low production values may be a characteristic of all large, shallow, wind exposed lakes.

Summary

The density and biomass distribution of larval chironomids was measured in a moderately large, shallow, wind exposed lake during 1969–72. The production was calculated, using the measured mean standing crop biomass per cohort, the estimated number of generations per year and a cohort P/B ratio of 5 taken from the literature. Seven species contributed together more than 95% of all the individuals collected: *Chironomus anthracinus, Chironomus plumosus, Cladotanytarsus mancus, Glyptotendipes pallens, Polypedilum nubeculosum, Polypedilum scalaenum* and *Stictochironomus histrio*. Numerically, the community is dominated by two small species: *C. mancus* and *P. scalaenum;* but in terms of larval biomass the larger species, *Chironomus* spp. and *G. pallens*, usually dominated.

The populations are characterized by a low production, due to a low standing crop biomass and by their instability. Density and biomass were mainly distributed according to substrate stability and substrate type. The standing crop biomass in the mud was about 5 times higher than in the sand. The substrates in the reed-beds showed the highest production: 24.0–61.0 g dry wt m^{-2} yr^{-1}. In the lake itself production is highest in a narrow inshore zone: 1.7–12.1 g dry wt m^{-2} yr^{-1}. It is suggested that wind induced wave action, causing substrate instability, is the main factor responsible for the low production in this zone. Because of its large surface area the offshore zone contributes more to the larval production of the lake than the other two zones combined. The mean annual larval production in the lake was: 1.8 g (1969), 0.4 g (1970), 1.7 g (1971) and 3.7 g dry wt m^{-2} (1972). A large part of the produced biomass is utilized by benthivorous fish.

Acknowledgements

I am indebted to all the members of the Tjeukemeer Foodchains and Production Research Group of the Limnological Institute, Tjeukemeer Laboratory, Oosterzee. I would especially like to thank Miss E. van Biemen, Mr. L. Lemsma and Mr. P. J. MacGillavry for their practical assistance and Dr. J. Vijverberg for useful advice and discussions. I am also grateful to Mr. B. P. M. Krebs of the Delta Institute for Hydrobiological Research, Yerseke, for kindly checking the classifications of the adult chironomids and to Prof. K. Bakker, Drs. R. D. Gulati and H. L. Golterman for their interest and highly valued criticism during the preparation of the manuscript. Finally I would like to thank Miss M. E. McLaughlin who improved the English.

References

Anderson, R. O., 1959. A modified flotation technique for sorting bottom fauna samples. Limnol. Oceanogr. 4: 223–225.

Beattie, D. M., 1979. A modification of the Ekman-Birge bottom sampler for heavy duty. Freshwat. Biol. 9: 181–182.

Beattie, D. M., 1981. Investigations into the occurrence of midge plagues in the vicinity of Harderwijk (Netherlands). Hydrobiologia 80: 147–159.

Beattie, D. M., Bromley, H. J., Chambers, M., Goldspink, C. R., Vijverberg, J., Zalinge, N. P. van & Golterman, H. L., 1972. Limnological studies on Tjeukemeer, a typical Dutch 'polder reservoir'. In: Kajak, Z. & Hillbricht-Illkowska, A. (Eds.) Productivity Problems of Freshwaters, pp. 421–447. Proceedings of an IBP-UNESCO Symposium on Productivity Problems of Freshwaters. Kazimierz Dolny, Poland.

Berczik, A., 1967. Zur Populationsdynamik des Macrobenthos in Velencer See. Opusc. Zool. Budapest 6: 247–265.

Berrie, A. D., 1972. Productivity of the River Thames at Reading. Symp. Zool. Soc. Lond. 29: 69–86.

Bonomi, G., 1962. La dinamica productiva della principali popolazioni macrobentoniche del Lago di Varese. Mem. Ist. Ital. Idrobiol. 15: 207–254.

Bryce, D., 1960. Studies on the larvae of the British Chironomidae (Diptera), with keys to the Chironominae and Tanypodinae. Trans. Soc. Br. Entomol. 14: 19–60.

Chambers, M. R., 1977a. The population ecology of Gammarus tigrinus (Sexton) in the reed-beds of the Tjeukemeer. Hydrobiologia 53: 155–164.

Chambers, M. R., 1977b. A comparison of the population ecology of Asellus aquaticus (L.) and Asellus meridianus Rac. in the reed-beds of the Tjeukemeer. Hydrobiologia 53: 147–154.

Charles, W. N., East, K., Brown, D., Gray, M. C. & Murray, T. D., 1974. The production of larval Chironomidae in the mud at Loch Leven, Kinross. Proc. Roy. Soc. Edinb. 74: 241–258.

Chernovskii, A. A., 1949. Identification of larvae of the midge family Tendipedidae. USSR Academy of Sciences Press, Moscow. Eng. Trans. Lees, E., 1961. National Lending Library, Boston Spa, England. 300 pp.

Fänge, R. & Grove, D., 1979. Digestion. In: Hoar, W. S., Randall, D. J. & Brett, J. R., (Eds.) Bioenergetics and Growth. Fish Physiology 8: 161–260. Academic Press, New York. 786 pp.

Fisher, J., 1974. Experimentelle Beiträge zur Ökologie von Chironomus (Diptera). I. Dormanz bei Chironomus nuditarsus und Ch. plumosus. Oecologia 16: 73–95.

Gerking, S. D., 1962. Production and food utilization in a population of bluegill sunfish. Ecol. Monogr. 32: 31–78.

Gizinski, A. & Wisniewski, R., 1971. An attempt to determine the dynamics of number, biomass and production of main components of the profundal fauna in the southern part of Lake Jeziorak. Zesźyty Nankowe Limnological Papers 6: 115–132.

Goldspink, C. R., 1978. The population density, growth rate and production of bream, Abramis brama, in Tjeukemeer, The Netherlands. J. Fish Biol. 13: 499–517.

Goldspink, C. R. & Banks, J. W., 1975. A description of the Tjeukemeer Fishery together with a note upon the yield statistics between 1964 and 1970. J. Fish Biol. 7: 687–708.

Hazelzet, J. C. A. M., 1972. The bream population of the Tjeukemeer. Student Report 1972–3, Limnological Institute, Nieuwersluis/Oosterzee, The Netherlands. 96 pp.

Hilsenhoff, W. L., 1966. The biology of C. plumosus (Diptera, Chironomidae) in Lake Winnebago, Wisc. Ann. ent. Soc. Am. 59: 465–473.

Hruska, V., 1974. The changes of benthos in Slapy reservoir in the years 1960–61. Hydrobiol. Stud. Prague 2: 213–247.

Humphreys, W. F., 1979. Production and respiration in animal populations. J. Anim. Ecol. 48: 427–453.

Jancovič, M., 1971. Anzahl der Generationen der Art Chironomus plumosus in den Karpfenteichen Serbiens. Limnologica 8: 203–210.

Johannsen, O. A., 1937a. Aquatic Diptera. Part 3. Chironomidae: sub-families Tanypodinae, Diamesinae and Orthocladiinae. Mem. Cornell Univ. Agric exp. Stn. 205: 1–84.

Johannsen, O. A., 1937b. Aquatic Diptera. Part 4. Chironomidae: sub-family Chironominae. Mem. Cornell Univ. Agric. exp. Stn. 210: 1–56.

Jónasson, P. M., 1965. Factors determining population size of C. anthracinus in Lake Esrom. Mitt. int. Verein. Limnol. 13: 139–162.

Jónasson, P. M., 1972. Ecology and production of the profundal benthos. Oikos Suppl. 14: 1–148.

Jónasson, P. M. & Lindegaard, C., 1979. Zoobenthos and its contribution to the metabolism of shallow lakes. Arch. Hydrobiol. 13: 162–180.

Judd, W. W., 1953. A study of the production of insects emerging as adults from the Dundas Marsh, Hamilton, Ontario, during 1948. Am. Midl. Nat. 49: 801–824.

Kajak, Z., Bretschko, G., Schiemer, F. & Lévêque, C., 1980. Zoobenthos. In: Le Cren, E. D. & Lowe-McConnel, R. H., (Eds.) The functioning of Freshwater Ecosystems, pp. 285–307. International Biological Programme 22, Cambridge University Press, Cambridge. 588 pp.

Klötzli, A. M., Romer, F. & Rosen, S., 1971. Jahreszyklus grossenvariation bei Chironomus plumosus L. Rev. Suisse Zool. 78: 587–603.

Konstantinov, A. S., 1958. The effect of temperature on growth rates and development of chironomid larvae. Dokl. Akad. Nanuk. SSSR 120: 1362–1365.

Lammens, E. H. R. R., 1982. Growth, condition and gonad development of bream (Abramis brama L.) in relation to its feeding conditions in Tjeukemeer. Hydrobiologia 95: 311–320.

Lindner, E., 1962. Die Fliegen der Palaearktischen Region. Schweizerbart'sche Verlagbuchhandlung, Stuttgart.

Mackay, A. P., 1977a. The growth and development of larval Chironomidae (Diptera). Oikos 28: 270–275.

Mackay, A. P., 1977b. Quantitative studies on the Chironomidae (Diptera) of the Rivers Thames and Kennet. Part 4. Production. Arch. Hydrobiol. 80: 327–348.

Maitland, P. S., Charles, N. W., Morgan, N. C., East, K. & Gray, M. C., 1972. Preliminary research on the production of Chironomidae in Loch Leven, Scotland. In: Kajak, Z. & Hillbricht-Illkowska, A., (Eds.) Productivity Problems of Freshwaters. Proceedings of an IBP-UNESCO Symposium on Productivity Problems of Freshwaters, pp. 795–812. Kazimierz Dolny, Poland.

Maitland, P. S. & Hudspith, P. M. G., 1974. The zoobenthos of Loch Leven, Kinross, and estimates of its production in the sandy littoral area during 1970 and 1971. Proc. Roy. Soc. Edinb. 74: 219–239.

Miller, R. B., 1941. A contribution to the ecology of the Chironomidae of Costello Lake, Algonquin Park, Ontario. Univ. Toronto Studs. Biol. 49: 1–63.

Moller Pillot, H. K. M., 1978–79. De larven der Nederlandse Chironomidae (Diptera). Nederl. Faun. Meded. 1: 1–276.

Nie, H. W. de, 1982. A note on the significance of larger bivalve molluscs (Anodonta spp. and Dreissena spp.) in the food of the eel (Anguilla anguilla) in the Tjeukemeer. Hydrobiologia 95: 307–310.

Parma, S., 1971. Chaoborus flavicans (Meigen) (Diptera, Chaoboridae): an autecological study. Ph.D. thesis, University of Groningen. 128 pp.

Pinder, L. C. V., 1978. A key to the adult males of the British Chironomidae (Diptera), the non-biting midges. Freshwater Biological Association, Scientific Publ. no. 37, vols. 1 and 2. 281 pp.

Potter, D. W. B. & Learner, M. A., 1974. A study of the benthic macro-invertebrates of a shallow eutrophic reservoir in South Wales with emphasis on the Chironomidae (Diptera); their life-histories and production. Arch. Hydrobiol. 74: 186–226.

Ringe, F., 1974. Chironomiden-Emergenz 1970. In: Breitenbach und Rohrweisenbach Schlitzer Produktionsbiologische Studien. Arch. Hydrobiol. Suppl. 45: 212–304.

Schiemer, F., 1979. The benthic community of the open lake. In: Löffler, H. (Ed.) Neusiedlersee, the Limnology of a Shallow Lake in Central Europe. Monographiae Biologicae, Vol. 37. Junk, The Hague. 543 pp.

Sokolova, N. J., 1961. Die Entwicklung der Bodenfauna des Utscha-Wasserbeckens. Verh. int. Verein. Limnol. 14: 640–642.

Tubb, R. A. & Dorris, T. C., 1965. Herbivorous insect populations in oil refinery effluent holding pond series. Limnol. Oceanogr. 10: 121–134.

Waters, T. F., 1977. Secondary production in inland waters. Adv. Ecol. Res. 10: 91–163.

Waters, T. F., 1979. Influence of benthos life history upon the estimation of secondary production. J. Fish. Res. Bd Can. 36: 1425–1430.

Winberg, G. G., 1956. Rate of metabolism and food requirements of fish. Fish. Res. Bd Can., Transl. Ser., 194.

Winberg, G. G. (Ed.) & collaborators, 1971. Symbols, units and conversion factors in studies of freshwater productivity. IBP Central Office, London. 23 pp.

Zbarakh, T. I., 1971. On the biology of Chironomus behningi in lake Biili-Kul (River Talas Basin). Limnologica 8: 107–109.

Zieba, J., 1971. Production of macrobenthos in fingerling ponds. Pol. Arch. Hydrobiol. 18: 235–246.

A note on the significance of larger bivalve molluscs (*Anodonta* spp. and *Dreissena* sp.) in the food of the eel *(Anguilla anguilla)* in Tjeukemeer

H. W. de Nie
Limnological Institute, Tjeukemeer Laboratory, De Akkers 47, 8536 VD Oosterzee, The Netherlands

Keywords: feeding relations, *Anodonta, Dreissena, Anguilla anguilla,* eutrophic lake

Abstract

The stomach contents of 517 eels were analysed during 1979. In that particular year the diet consisted predominantly of chironomids, but large bivalve molluscs were of secondary importance. The meaning of this observation is discussed in relation to food conditions and the food preference of the eel.

Introduction

Within the framework of a foodchain and production study, the growth, feeding relations and population dynamics of the eel *(Anguilla anguilla)* are being investigated. A part of the study on the feeding ecology of the eel is presented here, with emphasis on the importance of larger bivalve molluscs in the diet during September and October 1979.

Methods

The eel population was sampled at night after sunset by means of a small fine meshed trawl net at one location (site 4) in the open water zone of Tjeukemeer (De Haan 1982). A total of 1381 eels were caught. Catches are considered to be representative of the eel population. After deep freezing the catches were measured and weighed. Subsamples were taken for the analysis of the stomach contents per 5 cm class: 10.0–14.9, 15.0–19.9, 20.0–24.9, 25.0–29.9, 30.0–34.9, > 35 cm. The eels between 20 and 30 cm length constituted from 43 to 70% of the catches, 5% (or less) were larger than 35 cm or smaller than 15 cm. Five stomachs per catch were taken at random from each of the four length classes between 15 and 35 cm; eels larger than 35 cm and smaller than 15 cm were not subsampled.

From each sampling date, 19 to 127 stomachs were analysed. From the total of 517, 95 stomachs, were empty.

The food items in the stomach contents were determined to the lowest possible taxonomic level, mainly species and genus. The food items were counted and measured. The total stomach content was reconstructed to the weight from the length-weight relationships of the food items. The occurrence is the percentage of stomachs filled with that particular (group of) food item(s), related to the total number of stomachs, empty ones included.

The percentages were corrected with the aid of the length frequency distribution of the population so that they were representative for the whole catch on the date of sampling.

Results

The differences in frequency of occurrence of the food items within the length classes smaller than 25 cm was small. The corrected percentages of filled stomachs and the frequency of occurrence of chironomids, bivalve molluscs, gastropod molluscs and fish in the stomachs are given in Table 1. The pupae

Hydrobiologia 95, 307–310 (1982). 0018-8158/82/0953-0307/$00.80.

Table 1. Seasonal variations in the frequency of occurrence of food items in eel stomachs in 1979 in Tjeukemeer. N = number of examined stomachs.

	Filled stomachs	Chironomids	Molluscs		Fish	N
			Bivalves	Gastropods		
	(%)	(%)	(%)	(%)	(%)	
22 May	88	85	1.6	0.92	0	60
12 June	84	82	2.1	6.0	0	53
24 July	73	71	0	0	5.0	36
7 Aug.	79	64	4.7	0	24	19
21 Aug.	81	71	3.7	0.50	0.94	62
4 Sept.	78	66	7.1	5.4	0.36	127
25 Sept.	95	81	26	11	0	114
2 Oct.	67	38	20	17	1.2	46

of *Chironomus plumosus* and *Glyptotendipes pallens* were the dominant food items found. In Table 2 the corrected mean stomach contents, expressed as biomass, and the share of the different food items in the total stomach content are shown; the latter is given separately for the total eel population and for eels longer than 25 cm. During summer the eel fed mainly on chironomids. In mid-summer 0+ fish, mostly ruffe *(Gymnocephalus cernua)* and roach *(Rutilus rutilus)*, were also frequently found in the stomachs. Bivalve molluscs (mussels) were particularly important during September and October. The gastropod molluscs were quantitatively unimportant. Their highest frequency of occurrence was 17% but this represented only 4.1% of the biomass of the total stomach content (Tables 1 and 2).

Only parts of the mussels, generally pieces of the gills or the foot, were found in the stomach. In large eels especially, mussel particles (no shells) up to 2.5 cm long, corresponding to a weight of 1 to 3 g fr wt, were found. I frequently observed eggs or juvenile instars of the watermite *Unionicola aculeata* in the gills of the mussel. This mite is dependent upon the mussel *Anodonta anatina* to deposit its eggs and uses it as a shelter during the early state of its life cycle (Davids 1973). The proportion of *Dreissena polymorpha* in the total bivalve mollusc biomass in the eel stomachs was about 20% on 4 September, 10–25% on 25 September and about 60% on 2 October. On the last date it was the most important food item. Sometimes pieces of the shells of *Dreissena* were found in the stomach. The gastropod molluscs were quantitatively unimportant.

Table 2. Seasonal variations in the stomach content of the eel in 1979 in Tjeukemeer. Contribution of the different food items as a percentage of the total stomach content.

	All length classes					Longer than 25 cm		
	Mean content (mg fr wt stom^{-1})	Chir. (%)	Bivalve (%)	Gastr. (%)	Fish (%)	Bivalve (%)	Gastr. (%)	Fish (%)
22 May	220	90	9.4	0.025	0	15	0	0
12 June	91	89	9.5	0.65	0	25	0	0
24 July	79	66	0	0	33	0	0	61
7 Aug.	210	20	2.6	0	77	3.3	0	96
21 Aug.	64	74	11	0.039	9.4	8.8	0.051	12
4 Sept.	63	50	28	0.40	6.9	45	0.026	20
25 Sept.	210	34	53	0.79	0	63	0.93	0
2 Oct.	110	10	62	4.1	21	66	3.9	23

Discussion

The chironomids are the main food item for eels in Tjeukemeer although the chironomid production in the lake is exceptionally low: 2–12 g dry wt m^{-2} yr^{-1} (Beattie 1982). The bream *(Abramis brama)*, which dominates the fish biomass in the lake, also predates intensively on chironomids (Lammens 1982). Larger eels have at least two other options in their food choice: young fish and molluscs. Of the eel stomachs filled with bivalve molluscs, 75% also contained chironomids. During 1979 young fish were extremely scarce while in 1980 young perch *(Perca fluviatilis)* and young smelt *(Osmerus eperlanus)* were very abundant (van Densen & Vijverberg 1982). In 1980 bivalves were seldom found in eel stomachs; the larger eel ate mainly fish (de Nie, unpubl.). Only 20% of the stomachs filled with fish contained other food items. The eel probably has a stronger preference for fish than for molluscs.

The presence of freshwater mussels in eel stomachs was previously reported by Moriarty (1972) for Lough Corrib (Ireland) and Biró (1974) for Lake Balaton (Hungary). *Anodonta* and other larger bivalves were frequently observed in stomachs from Lough Corrib. In the open water zone of Lake Balaton, however, *Anodontas* were found only sporadically, while *Dreissena polymorpha* was encountered there more often.

Relatively more is known about the significance of gastropod and small bivalve molluscs in the diet of longer eels in fresh water lakes. Frost (1946) found 86% occurrence of molluscs in eel stomachs from Windermere (England); Moriarty (1972) found many gastropods in eels longer than 40 cm; Biró (1974) mentioned the gastropod *Lithoglyphus* as being of secondary importance next to chironomids in the eel diet; Opuszyński & Leszczyński (1967, cited in Tesch 1977) called *Bithynia tentaculata* an important food organism for eels longer than 30 cm. Gastropods, mainly *Bithynia tentaculata* and *Valvata piscinalis* do occur in Tjeukemeer (Chambers 1971). They are probably numerous, at least in the inshore zone where they form the main food source for the abundant 1+ and older roach *(Rutilus rutilus)*. The roach grows very well on this diet (Lammens, pers. comm.). It is not clear why the eel did not utilize this alternative option. It is unlikely that the mite infested *Anodontas* were easier to catch. The juvenile instars of the watermite *Unionicola* do not damage the mussel (Davids, pers. comm.). Uninfested mussels were eaten too. To prey upon mussels the eel requires a strong jaw apparatus. Eel is able to attack larger crustaceans by biting and thereafter making jerky movements, until manageable pieces have been torn off (Tesch 1977). Dröscher (1879, cited in Tesch 1977) found great differences in the diets of eel from neighbouring stretches of the Baltic coast, with different types of bottom substrate. In one area he found the bivalve *Macoma baltica* to dominate the diets, and in another area he found fish predominant in the stomachs.

From our results we conclude that, although bivalves may contribute significantly to the eel's diet, fish is a more preferred food.

Summary

The stomach contents of 517 eels *(Anguilla anguilla)* were analysed during 1979. The diet consisted predominantly of chironomids, mainly pupae of *Chironomus plumosus* and *Glyptotendipes pallens*, but larger bivalve molluscs were of secondary importance. Although gastropods were probably quite numerous (the main food source of the roach in Tjeukemeer), they were not important as a food item for the eel. In another year when 0+ fish were abundant, eel preferred 0+ fish above larger bivalves.

Acknowledgements

I thank Mr. H. Hermans, Mr. A. Bijlard and Mr. P. J. MacGillavry for help in the analysis of the eel stomach contents, Dr. C. Davids, for the identification of the watermites, Dr. J. Vijverberg, Dr. A. E. Jansen and Mr. W. L. T. van Densen, for commenting on the manuscript, and Miss M. E. McLaughlin for improving the English text.

References

Beattie, D. M., 1982. Distribution and production of the larval chironomid populations in Tjeukemeer. Hydrobiologia, this volume.

Biró, P., 1974. Observations on food of the eel (Anguilla anguilla L.) in Lake Balaton. Ann. Biol. Tihany 41: 133–153.

Chambers, M. R., 1971. Studies on the littoral fauna of Tjeukemeer. Ph.D. thesis, University of Liverpool.

Davids, C., 1973. The relations between mites of the genus Unionicola and the mussels Anodonta and Unio. Hydrobiologia 41: 37–44.

De Haan, H., 1982. Physico-chemical environment in Tjeukemeer (The Netherlands) with special reference to speciation of algal nutrients. Hydrobiologia 95: 205–221.

Densen, W. L. T. van & Vijverberg, J., 1982. The relations between 0+ fish density, zooplankton size and the vulnerability of pikeperch, Stizostedion lucioperca, to angling in the Frisian lakes. Hydrobiologia 95: 321–336.

Frost, W. E., 1946. Observations on the food of eels (Anguilla anguilla) from the Windermere catchment area. J. Anim. Ecol. 15: 43–53.

Lammens, E. H. R. R., 1982. Growth, condition and gonad development of bream (Abramis brama L.) in relation to its feeding conditions in Tjeukemeer. Hydrobiologia 95: 311–320.

Moriarty, C., 1972. Studies of the eel Anguilla anguilla in Ireland I. In the lakes of the Corrib System. Irish Fish. Invest. Ser. A, 10: 39 pp.

Tesch, F. W., 1977. The Eel. Chapman & Hall, London. 434 pp.

Growth, condition and gonad development of bream (*Abramis brama* L.) in relation to its feeding conditions in Tjeukemeer

E. H. R. R. Lammens
Limnological Institute, Tjeukemeer Laboratory, De Akkers 47, 8536 VD Oosterzee, The Netherlands

Keywords: bream, growth, condition, fecundity, zooplankton, chironomids

Abstract

The bream (*Abramis brama* L.) population of Tjeukemeer was studied for three successive years. The growth, condition, gonad development, diet and feeding conditions of the fish are described. Food competition for two years with the then abundant young planktivorous smelt *(Osmerus eperlanus)* and perch *(Perca fluviatilis)* resulted in a decreased growth rate, condition and gonad development of the larger bream. The biomass of chironomids was too small in relation to that of the zooplankton to be important for the bream population, although they were eaten more efficiently.

Introduction

This study is part of a larger study on the population structure and feeding ecology of bream (*Abramis brama* L.), roach (*Rutilus rutilus* L.) and white bream (*Blicca björkna* L.), which are the most important cyprinids in Tjeukemeer. Of these three species, bream is dominant in biomass and occurs both in the open water and in the littoral zone, while the other two species are restricted to the littoral zone. Also, in relation to the other species, the bream is dominant in Tjeukemeer.

The growth of the bream in Tjeukemeer was first studied by Goldspink (1978). Compared with other waters (Backiel & Zawisza 1968; Cazemier 1975), the growth rates of the bream in this lake were medium to poor (e.g. 400–500 g after 9 years). The food of the bream in this lake was first examined by Hazelzet (1972) who found that bream in the open water depended largely on zooplankton and in the inshore zone on chironomids. The latter finding is in accordance with the distribution of the chironomids, which are mainly restricted to the 100 m inshore zone (Beattie 1982). Other studies on bream (Laskar 1948; Cazemier 1975) showed the importance of chironomid larvae for good growth. In Tjeukemeer this food source is scarce.

This paper describes the growth, condition and gonad development of bream in relation to its feeding conditions and diet in three successive years (1979 to 1981).

Study area

Tjeukemeer is a shallow (mean depth 1.5 m), eutrophic, freshwater lake with a surface area of 21.0 km^2. The littoral zone is poorly developed, the index of development (D_l) (Hutchinson 1957) being 1.25. The lake is part of an interconnected system of water bodies, the Friesian 'boezem', which act as water reservoirs for the surrounding polders (Beattie *et al.* 1978). The hydrology is described in this volume by Leenen (1982).

Materials and methods

Fish were caught at least once a month at five sites in the lake, with a 5 m small meshed (5.5 mm cod end) trawl. In this way the three variables of time, site and size could be considered properly.

Hydrobiologia 95, 311–320 (1982). 0018-8158/82/0953-0311/$02.00.

The total catch per date was less than 0.05% of the total population. The length of all individuals was measured in forklength to the nearest 0.5 cm and length-frequency distributions were constructed. On each sampling occasion the forklength (cm), weight (g), sex and gonad weight of 100–200 fishes were determined and the relationships between length and weight, and weight and gonad weight were determined. Both relationships were regarded as power functions:

$$W = aL^b \qquad (1)$$

$$W_{gon} = cW^d \qquad (2)$$

in which W = weight, L = length, W_{gon} = gonad weight and a, b, c and d are constants, calculated using the geometrical mean functional regression after Ricker (1973).

The age was determined using both scales and finrays. Scales were taken between the pelvic and dorsal fin, just above the lateral line, and finrays were taken from the dorsal fin. The average length was calculated in spring and autumn for each year class; thus the seasonal growth could be calculated.

The guts of 25 fish of the most abundant size class, namely 25–30 cm, were collected from the five sites on each sampling date. Gut contents were lumped per site, preserved in 4% formalin, and the qualitative food composition determined. From each site and date three subsamples of the lumped gut contents were analysed: the food organisms were identified, counted and measured. The calorific values of the food organisms were calculated by using the relationships given by Vijverberg & Frank (1976) and Vijverberg (unpubl.) for zooplankton and chironomid larvae, respectively. For chironomids, 1 g fresh weight was assumed to represent 1 kcal (Winberg *et al.* 1971).

In spring 1980 and 1981 the ovaries from three length classes were collected and preserved in modified Gilson fluid (Bagenal 1968) and subsampled with the whirling apparatus of Kott (1953).

Results

Length and age composition

The length composition of the bream population changed most in 1979 (Fig. 1); the whole length-frequency distribution shifted to the right and the average size of bream increased a few centimetres. In 1980 only the left part of the length-frequency distribution shifted to the right, while the right part did not change. The frequency distribution data are based on catches in April and May (969 fishes) and September and October 1979 (996 fishes), and September and October 1980 (773 fishes). These dates were chosen because no growth occurs during the winter months.

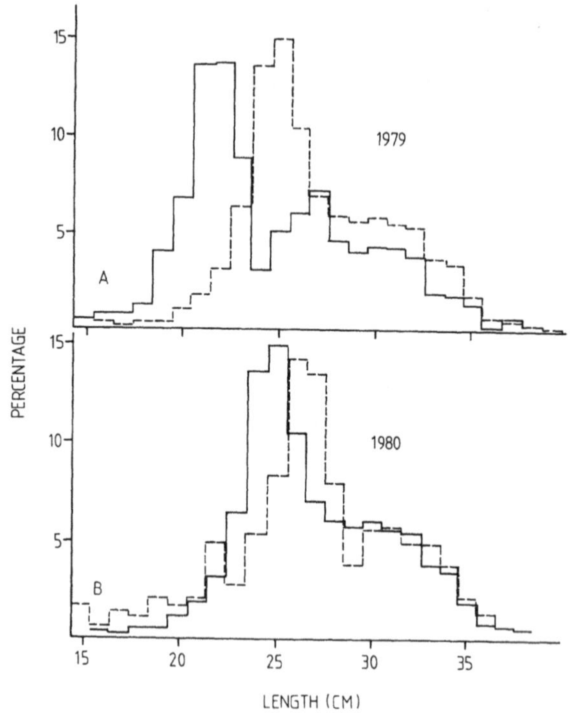

Fig. 1. A: Length-frequency distributions in the Tjeukemeer during 1979 of bream larger than 15 cm in spring (straight line) and autumn (broken line). B: Length-frequency distributions in the Tjeukemeer of bream larger than 15 cm in autumn 1979 (straight line) and autumn 1980 (broken line).

The population was dominated by the year-classes of 1975, 1973 and 1970, which contributed about 75% to the total population (Table 1). The first strong year-class after 1975 was formed in 1979 (not shown in figures and tables). As a consequence there is a large gap in the left part of the length-frequency distribution. The 1975 year-class is easy to follow in the L-F distribution of 1979 and 1980, but the 1973 and 1970 year-classes were indistinguishable in 1980.

Table 1. Average length (cm) of the most abundant year-classes in spring 1979 and autumn of 1979 and 1980, and their relative part in the population in 1980.

	1976	1975	1974	1973	1972	1971	1970	1969 and older
Spring 1979		21.1	24.1	26.7	28.8		31.1	33.4
Autumn 1979	18.1	24.4	26.7	28.6	30.4		32.3	34.2
Autumn 1980	22.1	26.3	28.0	29.4			32.2	34.4
Percentage 1980	3	52	6	15	5	2	10	7

Condition

The condition of bream, expressed as the total weight at a certain length class, was better in 1979, for both males and females (Fig. 2) than in the subsequent two years. This applies particularly to the larger bream, whose condition had deteriorated since spring 1980. The 25–30 cm bream had a similar condition only in the early summers of the three years.

Since condition is related to both sex and size (LeCren 1951), it was considered for each separately. The weights were calculated using the length-weight relationship (Eq. 1). The females almost always weighed more than the males, except just after spawning when the females weighed less, and in the summer of 1981 when there was hardly any difference. The weights of the immature males and females (≤25 cm) were similar (Fig. 2). Both sexes recovered very fast after spawning in 1979, but did not do so in 1980 and 1981.

The deteriorating condition of the larger bream is reflected in the coefficient 'b' of Eq. 1, which became steadily smaller in the course of the three years (Table 2). After spawning it was always higher than 3 in 1979, lower than 3 in 1981 and had values both below and above 3 in 1980.

Growth

In 1979 all year-classes increased in length (Fig. 3A) while in 1980 this was true only for those with

Table 2. The average values and standard deviations of the constants 'b' and 'ln a' in Eq. (1) for males and females in the three successive years.

Year	b	s_b	ln a	$s_{\ln a}$
July 1979–May 1980	♀ 3.22	0.09	-10.10	0.54
	♂ 3.09	0.10	- 9.41	0.58
July 1980–May 1981	♀ 3.05	0.08	- 9.07	0.47
	♂ 2.98	0.07	- 8.65	0.36
June 1981–Oct. 1981	♀ 2.80	0.04	- 7.58	0.23
	♂ 2.86	0.06	- 7.92	0.35

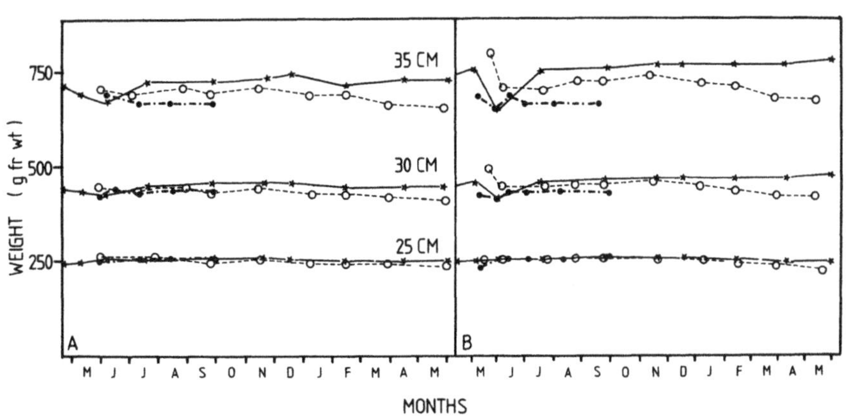

Fig. 2. The total individual weight in g fr wt of three length classes of bream in the course on 1979 (straight line), 1980 (dashed line) and 1981 (dash-dotted line). A = males, B = females.

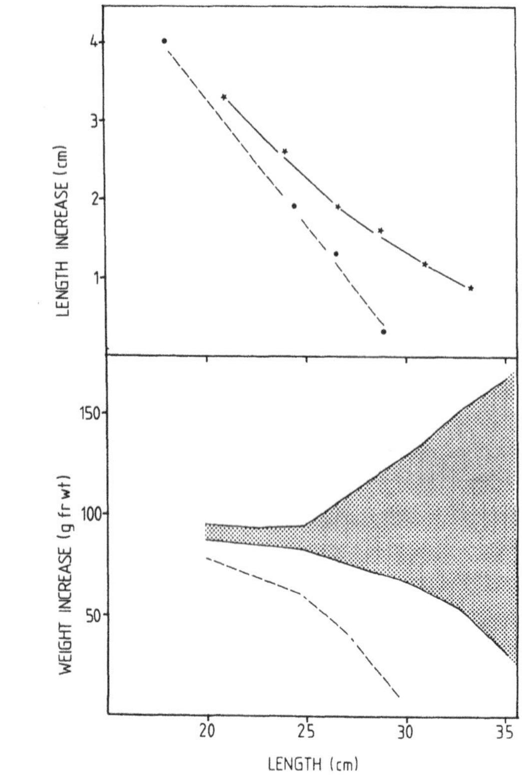

Fig. 3. A: The length increase (cm) of different year-classes of bream per year in 1979 and 1980. The length increase of each year-class is plotted against its length at the beginning of the growing season in 1979 (straight line) and 1980 (broken line). B: The weight increase (g fr wt) of different year-classes of bream per year in 1979 (straight line) and 1980 (broken line). The weight increase of each year-class is plotted against its length at the beginning of the growing season. For 1979 this weight increase is differentiated into weight increase due to condition recovery and that due to length increase.

an average length smaller than 30 cm. This seasonal growth is also reflected in the change of the length-frequency distribution (Fig. 1).

So, the growth of the younger year-classes in 1980 was less affected than that of the older ones. If the seasonal length increase for all length classes had been reduced to the same extent in 1980, the quotient of the length increase of comparable length classes in 1979 and 1980 would have been the same for all length classes. This quotient is slightly higher than 1 for 20 cm bream but very much higher for 30 cm bream (Fig. 3A). These quotients become greater when weight increase is considered (Fig. 3B) instead of length increase, because then condition recovery is also taken into account. It is clear that even when length increase is small, the weight increase of a mature bream due to condition recovery is considerable.

No distinction was made between the sexes since the growth rates did not differ.

Gonad development

The length at which 50% of the fishes matured in that length class was for the males 24.7 cm in 1979, 25.7 cm in 1980 and 25.0 cm in 1981 (Fig. 4A); for females these figures were 26.8, 27.5 and 26.0 cm (Fig. 4B) respectively. This implies a poor recruitment to the mature part of the population in 1980 and a strong recruitment in 1981.

The gonad development of both sexes in these years started in late summer and was interrupted in the winter months (Fig. 5). The gonad weights were higher in 1979 than in 1980. This difference continued until spawning time.

The relationship between the gonad weight and the total weight of the fish was described by a power function (Eq. 2). The constant d in this function is nearly 2 in both years (Table 3), implying that the gonads increase relatively more in weight than the total weight with increasing size. Another power

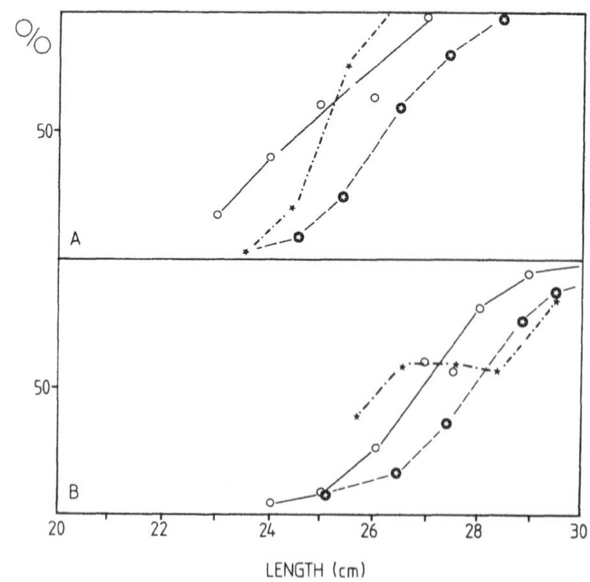

Fig. 4. The maturity ogives of the bream in 1979 (straight line), 1980 (dashed line) and 1981 (dash-dotted line). The percentage mature bream per cm length class are indicated. A = males, B = females.

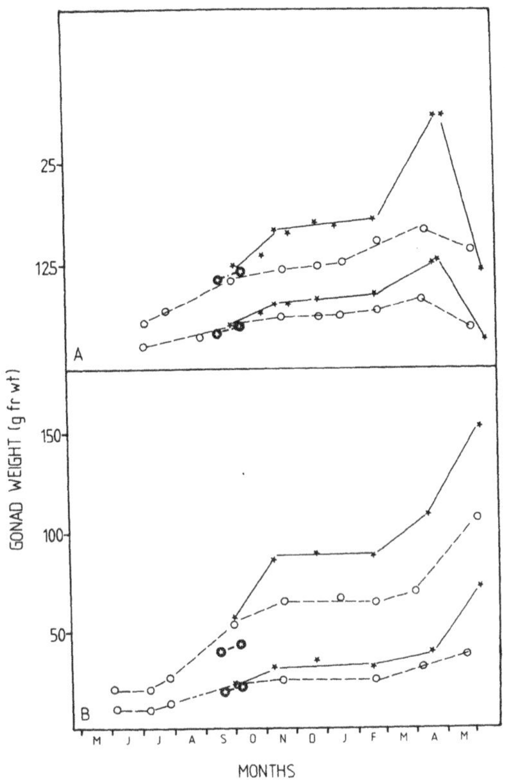

Fig. 5. The gonad weights (g fr wt) of two length classes (30 and 35 cm) of bream in the course of 1979 (straight line), 1980 (dashed line) and 1981 (dash-dotted line). A = males, B = females.

function arises relating gonad weight to length:

$$W_{gon} = ca^d L^{bd} \qquad (3)$$

with the same symbols as used for Eqs. (1) and (2). In 1980 the power bd was lower than in 1979. This means that the gonad weight of the larger individuals was more affected than that of the smaller ones.

Table 3. The average values and standard deviations of the constants 'd' and 'ln c' in Eq. (2) for males and females in two successive years.

Year	d	s_d	ln c	$s_{\ln c}$
Sept. 1979–May 1980	♀ 1.95	0.15	-8.37	-1.23
	♂ 1.59	0.16	-6.63	-1.36
Sept. 1980–May 1981	♀ 1.92	0.11	-8.14	-1.05
	♂ 1.46	0.08	-5.73	-0.71

Table 4. Absolute fecundity and number of eggs per gram roe of bream for three size classes in spring 1980 and 1981. Absolute fecundity is expressed as total number of eggs per fish.

	Absolute fecundity		Number of eggs per gram roe	
Size (cm)	15-4-80	1-4-81	15-4-80	1-4-81
30	54 210	63 863	1 390	2 021
32.5	96 447	95 273	1 323	1 989
35	158 037	141 360	1 226	1 950

Although the gonad weights clearly differed, the absolute fecundity per length class seemed to be less affected (Table 4). The total number of eggs of the smaller individuals was slightly (although not significantly) higher in spring 1981 than in 1980, while the reverse applied to the larger ones. The lower weight per individual egg in 1981 together with an almost unchanged fecundity (Table 4) explains the lower gonad weight.

The development of the gonads was accompanied by a decreasing weight of the somatic tissue (Fig. 6). The somatic weight of the males is almost

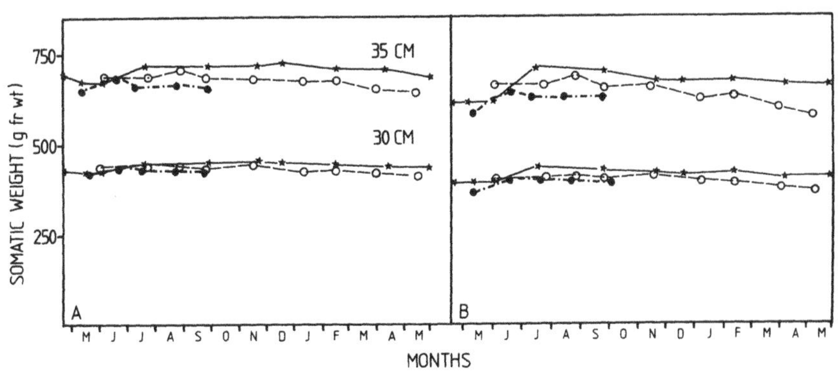

Fig. 6. The somatic weights (g fr wt) of two length classes (30 and 35 cm) of bream in the course of 1979 (straight line), 1980 (dashed line) and 1981 (dash-dotted line). A = males, B = females.

Table 5. Food composition (%) based on energy content of food organisms for the 25–30 cm length class during the growing seasons of 1979 and 1980.

	1979				1980			
	June	July	Aug.	Sept.	June	July	Aug.	Sept.
Daphnia hyalina	89	2	20	38	62	10	10	3
Copepoda	9	36	22	45	1	7	10	28
Bosmina coregoni		1	1	6	2	3	4	3
Alona sp.			3	3			2	5
Cladotanytarsus sp.	2	27	8	3	35	42	43	29
Cryptochironomus sp. / *Einfeldia* sp. / *Procladius* sp.		34	46	5		38	31	32

always higher than that of the females, which is probably related to their relatively low gonad weight. Even though the mature individuals did not increase in weight in 1980, they were still able to form gonads at the expense of somatic tissue formed the year before.

Food

The most important food organisms for the 25–30 cm length class were small crustaceans and chironomids (Table 5). Their relative significance, however, differed from 1979 to 1980. In 1979 zooplankton made up the largest part of the diet (65%), while in 1980 chironomids did so (62%).

Comparing the average biomass composition of the food organisms of the standing stock (Table 6) with that found in the guts of bream (Table 5), among the zooplankton larger *Daphnia* were preferred and among the chironomids the smaller *Cladotanytarsis.*

Because these figures apply to the 25–30 cm size class from sites in the open water and the influence of size and site is not considered here (Lammens, in prep.), these figures are biased. This length class is, however, the most abundant and the open water represents approximately 90% of the total surface area.

The average biomass of zooplankton and chironomids differed greatly in both 1979 and 1980. In both years the biomass of zooplankton was from one to two orders of magnitude higher than that of chironomids. The difference between the biomasses of zooplankton in both years was less, by a factor 3. So although the biomass of chironomids is much lower, its share in the diet of this size class of bream is relatively great, in 1980 even greater than that of zooplankton.

As the consumption must have been higher in 1979 than in 1980 because of the better growth in that year (Fig. 3), the diets in these years cannot be compared directly. If consumption in 1979 were twice as great as in 1980, the amount of consumed chironomids in both years would be similar, while

Table 6. Average standing stock biomass (kcal m^{-2}) of the relevant food organisms of the bream during the growing seasons of 1979 and 1980. For *Daphnia hyalina* only the average lengths are also included.

	1979				1980			
	June	July	Aug.	Sept.	June	July	Aug.	Sept.
Daphnia hyalina	21.6	2.1	2.9	5.8	8.6	0.3	0.6	0.3
Average lengths (mm)	1.10	1.22	1.07	1.04	1.09	0.74	0.74	0.64
Copepoda	1.5	2.8	6.2	4.8	2.0	1.4	2.7	1.1
Bosmina coregoni	0.2	0.2	0.3	0.4	0.5	0.9	0.9	0.3
Cladotanytarsus sp.	0.05	0.09	0.06	0.03	0.05	0.08	0.06	
Cryptochironomus sp.		0.09	0.03	0.01	0.15	0.04	0.13	0.02
Einfeldia sp.			0.28	2.59	0.51	0.01	0.14	0.69

the amount of zooplankton consumed in 1979 would be four times that consumed in 1980. If weight increase is taken as a measure of consumption above maintenance level, then Fig. 3B is a good illustration of this relative consumption in 1979 and 1980.

Discussion

In 1979 and 1980 the feeding conditions for the bream differed both qualitatively and quantitatively, especially with respect to the availability of zooplankton. In 1979 the numbers of large *Daphnia* and copepods were high throughout the year, while in 1980 the larger zooplankton species reached high densities only in June (Table 6). The standing stock biomass of the chironomids was very low in both years.

The high standing stock of zooplankton in 1979 largely determined condition, growth and gonad development in that year, while in 1980 consumption of zooplankton was much less, which could not be compensated for by chironomids. As a result, condition, growth and gonad development of the larger bream decreased.

The number and size of the zooplankton species in these years were largely influenced by the amount of 0+ fish: in 1980 a large amount of 0+ fish recruited and depressed by size selective feeding (Van Densen, in prep.) the size and numbers of *Daphnia*. Only the smaller fish were able to handle the small zooplankton and eat sufficient quantities to restore condition and grow in length. In 1979 there was hardly any recruitment of 0+ fish, and the large daphnids particularly were present in high numbers. They were consumed by the larger bream, who could recover quickly after spawning and maintain a stable condition. In 1980 and 1981 they were not able to do so.

The dependence of fish on zooplankton in certain periods and places was also found for another large cyprinid, the carp (*Cyprinus carpio* L.) (Fanget 1972). Especially in early and late summer when the zooplankton was abundant it was eaten in large amounts; the small particles seem to be mechanically sieved by the gill-rakers (Uribe-Zamora 1975).

The influence of the feeding situation on condition, growth and gonad development was differentiated into size classes. The larger bream, in particular, deteriorated in condition, growth was retarded, and eggs were smaller. This was reflected in the values of b and bd (Eqs. 1 and 3), both of which decreased constantly after spawning in 1980.

Because of this relation of size and sex with condition, a distinction was made between the sexes as well as between the size classes, making use of Eq. (1), calculated by the geometrical mean functional regression. LeCren (1951) pointed out the limitations of the conventional condition factors, while Ricker (1973) showed the limitations of the generally used predictive regression.

The condition recovery and growth in length proved to be related: only after condition recovery was realized did length growth occur. Therefore it is likely that growth in length starts at the moment the condition reaches a certain level, which level was not reached by bream >30 cm in 1981. Geyer (1939) studied the growth and condition of cyprinids in 5 Ostholsteinian lakes. Although the growth rates of fish in those lakes differed, he could not find differences in condition. Only one lake deviated from the general picture with a deteriorated bream condition. Therefore, Geyer stated that the condition is independent of growth rate; in the one deviating lake the food conditions were deteriorated as a result of a very successful reproduction and consequent overcrowding. A similar situation prevailed in Tjeukemeer in both 1980 and 1981.

This phenomenon has consequences for the length–weight relationship: only the growing individuals satisfy this relationship. If fishes with a deteriorated condition are also included, the log-transformed values of length and weight cease to be linear. If the larger individuals have a deteriorated condition, their weight will be overestimated. The curve will be linear for the fishes with a stable condition and will bow at the size where condition deteriorates. A calculated line will overestimate the extremes while the middle part will be underestimated. In Fig. 7 three curves are drawn which show these effects. In this study only fishes larger than 20 cm are considered. Although this relationship is not linear, the straight line used fits very well because most data are in the short interval 25–35 cm. If small fishes are also considered the deviation is much greater (Fig. 7).

Not only were the condition and growth influenced by the feeding conditions, but also the gonad development. The gonad development starts at the

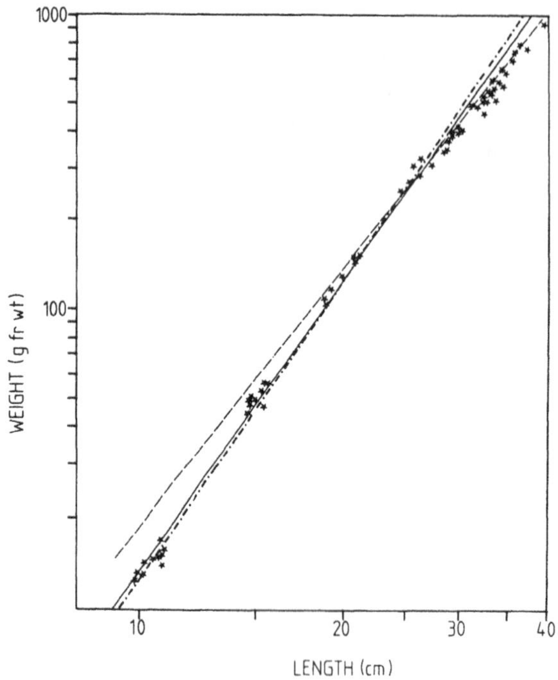

Fig. 7. The total weights of female bream caught on 30-9-1981 are plotted against their lengths on a double log scale. Three lines were calculated: for 10–40 cm (straight line), for 10–25 cm (dash-dotted line) and for 25–40 cm (dashed line).

end of July and is interrupted during the winter months. During this process the somatic weight decreases. In 1979 somatic tissue was formed in June and July and partly transformed to gonad tissue thereafter. In 1980, however, there was hardly any condition recovery, so the development of gonads took place at the expense of somatic tissue formed in the preceding year and consequently broke into the reserves.

Volodin (1979) showed that foraging conditions influenced the gonad development. In ponds where bream competed with yearling carp, condition decreased severely and yolk formation in most fish failed and an increased mortality was noticed, while non-competing bream in a different pond increased in condition and matured normally. The deterioration of the condition in Tjeukemeer resulted in a worse gonad development in 1980 and 1981 than in 1979. Yolk formation, however, still occurred. Ivlev (1961) found that yearling bream did not die until they lost 30% of their weight. The larger bream in Tjeukemeer had lost 15% of their original weight in late summer of 1981. If for this size class the same value of 30% holds, increased mortality is not to be expected until the end of 1982, if the competition for zooplankton continues.

In the literature several cases have been reported on the relation between fish fecundity and feeding conditions. Fecundity has received much attention from Russian workers who regard it as one of the most important responses to feeding conditions. Absolute fecundity and egg weight of bream in relation to size and for different years are given by Kuznetsov (1977), Tsyplakov (1969) and Vasilchenko (1970). These studies show that when growth was retarded, bream matured at a smaller length and the egg weight decreased. Absolute fecundity increased for the smaller bream, while that of the larger bream decreased. In Tjeukemeer the egg weight and absolute fecundity changed in the same way, although the latter changed only slightly. The size at which the bream matured was larger in 1980 than in 1979. This is in contradiction with the above-mentioned studies. The extensive study of Alm (1959), in which he relates maturity to size and age, also contradicts these findings. In 1981, however, the bream matured at a smaller size than in 1979. The reason for this difference is not clear.

A few studies have been made concerning the relation between the food supply and the growth of bream. The most extensive comparative study was made in the Ostholsteinian lakes in Germany (Lundbeck 1926; Geyer 1939; Laskar 1948). Another comparative study (Cazemier 1975) related growth to food supply. Both studies show that if the chironomid standing stock exceeds 20 g fr wt m^{-2} the growth of bream is very good. In those waters bream attained a weight of 1 kg after nine years. In waters with a medium growth, e.g. 0.5 kg after nine years, the chironomid biomass amounted to 5–20 g fr wt m^{-2}. Poor growth, e.g. 0.2 kg after nine years, is associated with a standing stock of chironomids of less than 5 g fr wt m^{-2}. In this respect Tjeukemeer, with less than 2 g fr wt m^{-2} (Vijverberg, unpubl.), had extremely poor feeding conditions. The standing stock of zooplankton was, however, much higher in both years. In 1979 it amounted to 5–20 g fr wt m^{-2}; in 1980 it was less than 5 g fr wt m^{-2}.

So, in Tjeukemeer, zooplankton was the most important food for the bream when its standing stock was more than 5 g fr wt m^{-2}. Planktivorous young fish such as smelt and perch were important competitors when their densities were high. When the standing stock of zooplankton was less than 5 g fr wt m^{-2}, chironomids became important although their biomass was even less than 2 g fr wt m^{-2}. For bream, benthic feeding appears to be more efficient than planktivorous feeding. Because of its low efficiency of planktivorous feeding, bream cannot compete with other planktivorous fish, but due to its high efficiency in feeding on chironomids, it can compete strongly with other benthic feeders. It is not unlikely that the low chironomid biomass in Tjeukemeer is caused by the bream population, because every year almost the whole production of chironomids is consumed (Lammens, in prep.; Vijverberg, unpubl.), except for the large *Chironomus plumosus*. This species is almost only eaten by the eel *(Anguilla anguilla)* (de Nie 1982).

In Tjeukemeer the standing stock of the larger zooplankton species is influenced largely by the size selective predation of the planktivorous 0+ fish (van Densen, in prep.); the amount of 0+ fish in its turn varies greatly from year to year depending upon the success of larval recruitment and upon the predation by older pikeperch *(Stizostedion lucioperca)*. Adverse weather conditions can hamper reproduction in the lake, while heavy precipitation can depress the passive immigration of larvae (van Densen & Vijverberg 1982). As a consequence, condition, growth and gonad development of bream in Tjeukemeer are mainly dependent on the weather conditions in early summer on the one hand and on the standing stock of pikeperch on the other.

Summary

The bream population in Tjeukemeer was sampled from 1979 to 1981 to describe growth, condition, gonad development and diet. Simultaneously, the pelagial zooplankton and the benthic chironimid larvae were sampled.

The condition, growth and gonad development of the larger bream decreased constantly during the course of the study. This was associated with a deterioration in feeding conditions as a consequence of two successive strong recruitments of young fish in 1980 and 1981. Due to the size selective predation of the young fish, the zooplankton food organisms decreased in size and were too small to be efficiently eaten by the larger bream. The chironomid population was, with a mean standing crop of less than 2 g fr wt m^{-2}, poorly developed and its relative contribution to the food of the bream was only important when the zooplankton biomass was very low.

Acknowledgements

I would like to express my gratitude to all the members of the Tjeukemeer Foodchains and Production Research Group of the Limnological Institute in Oosterzee, especially Dr. J. Vijverberg and Mr. W. L. T. van Densen. I am also indebted to Prof. J. W. M. Osse from the Agricultural University of Wageningen and the students A. Bijlard, G. Hampsink and J. Stöpetie. Finally I would like to thank Dr. R. Gulati for valuable criticism, Miss M. E. McLaughlin for the improvement of the English text, and Miss A. Schoon for typing.

References

Alm, G., 1959. Connection between maturity, size and age in fishes. Inst. Freshwater Res., Drottingholm Rep. No. 40.

Backiel, T. & Zawisza, J., 1968. Synopsis of biological data on the bream, Abramis brama L. FAO Fisheries Synopsis No. 36.

Bagenal, T. B., 1968. Eggs and early life history. In: Ricker, W. E. (Ed.) I.B.P. Handbook No. 3. Methods of Assessment of Fish Production in Fresh Waters, Chap. 7. Blackwell, Oxford. 159 pp.

Beattie, D. M., Golterman, H. L. & Vijverberg, J., 1978. An introduction to the limnology of the Friesian Lakes. Hydrobiologia 58: 49–64.

Beattie, D. M., 1982. Distribution and production of the larval chironomid populations in Tjeukemeer. Hydrobiologia, this volume.

Cazemier, W. G., 1975. Onderzoek naar de oorzaken van groeiverschillen bij brasem (Abramis brama L.) (A study concerning the causes of differences in growth rate of bream). Visserij 28: 197–207.

Densen, W. L. T. van (in prep.). The ecology of 0+ fish in Tjeukemeer, The Netherlands. II. Feeding behaviour.

Densen, W. L. T. van & Vijverberg, J., 1982. The relations between 0+ fish density, zooplankton size and the vulnerability of pikeperch, Stizostedion lucioperca, to angling in the Frisian lakes. Hydrobiologia 95: 321–336.

Fanget, R., 1972. Contribution à l'ecologie des étangs piscicoles de la Dombes: sur le régime alimentaire de la carpe à miroir, Cyprinus carpio 1. Thèse Univ. Sci. Lyon. pp. 62, annexes.

Geyer, F., 1939. Alter und Wachstum der wichtigsten Cypriniden Ostholsteinische Seen. Arch. Hydrobiol. 34: 543–644.

Goldspink, C. R., 1978. The population density, growth rate and production of bream, Abramis brama L., in Tjeukemeer, The Netherlands. J. Fish. Biol. 13: 499–517.

Hazelzet, J. C. A. M., 1972. The bream population of the Tjeukemeer. Internal Report (1972-3), Limnological Institute, Oosterzee, The Netherlands. 96 pp.

Hutchinson, G. E., 1957. A Treatise on Limnology, Vol. I. Wiley, New York. 1015 pp.

Ivlev, V. S., 1961. Experimental Ecology of the Feeding of Fishes. Yale University Press, New Haven. 302 pp.

Kott, P., 1953. Modified whirling apparatus for sub-sampling of plankton. Aust. J. mar. freshwat. Res. 4: 387-393.

Kuznetsov, V. A., 1973. The fecundity and quality of eggs of bream (Abramis brama L.). J. Ichthyol. 13: 669-677.

Lammens, E. H. R. R. (in prep.). The food and feeding of the bream (Abramis brama) in Tjeukemeer in relation to size, site, season and time of the day.

Laskar, K., 1948. Die Ernahrung des Brassens (Abramis brama L.) im eutrophen See. Arch. Hydrobiol. 42: 1-165.

LeCren, E. D., 1951. The length-weight relationship and seasonal cycle in gonad weight and condition in perch (Perca fluviatilis L.). J. Anim. Ecol. 20: 201-219.

Leenen, J. D., 1982. Hydrology of Tjeukemeer. Hydrobiologia 95: 199-203.

Lundbeck, J., 1926. Die bodentierwelt Norddeutscher Seen. Arch. Hydrobiol. Suppl. 7: 1-473.

Nie, H. W. de, 1982. A note on the significance of larger bivalve molluscs (Anodonta spp. and Dreissena sp.) in the food of the eel (Anguilla anguilla) in Tjeukemeer. Hydrobiologia 95: 307-310.

Ricker, W. E., 1973. Linear regressions in fishery research. J. Fish. Res. Bd Can. 30: 409-434.

Tsyplakov, E. P., 1969. Variation in reproduction of the bream (Abramis brama L.) population in the Kuybyshev reservoir. J. Ichthyol. 9: 66-75.

Uribe-Zamora, M., 1975. Selection des proies par le filtre branchial de la carpe à miroir (Cyprinus carpio L.). Thèse Univ. Sci. Lyon. 127 pp.

Vasilchenko, O. N., 1970. The fecundity and the state of the gonads of common carp and bream used for breeding in the Volga Delta. J. Ichthyol. 10: 67-76.

Vijverberg, J. & Frank, Th. H., 1976. The chemical composition and the energy contents of copepods and cladocerans in relation to their size. Freshwat. Biol. 6: 333-345.

Volodin, V. M., 1979. The influence of feeding and overwintering conditions on the duration of egg resorption and the development of the next generation of oocytes in the bream, Abramis brama, of Rybink reservoir. J. Ichthyol. 19: 147-151.

Winberg, G. G. (Ed.) & collaborators, 1971. Symbols, units and conversion factors in studies of freshwater productivity. I.B.P. Central Office, London. 23 pp.

The relations between 0+ fish density, zooplankton size and the vulnerability of pikeperch, *Stizostedion lucioperca*, to angling in the Frisian lakes

W. L. T. van Densen[1, 2] & J. Vijverberg[1]
[1]*Limnological Institute, Tjeukemeer Laboratory, De Akkers 47, 8536 VD Oosterzee, The Netherlands*
[2]*Department of Fish Culture and Inland Fisheries, Agricultural University, Wageningen, The Netherlands*

Keywords: *Daphnia*, 0+ fish, catchability, angling, pikeperch

Abstract

During a five year study, the size and abundance of zooplankton and 0+ fish in September were registered in a number of Frisian lakes. The mean size of *Daphnia* was related to the abundance of planktivorous 0+ fish in September. Immigration of fish larvae into the Frisian Lake District from the IJsselmeer in June was observed in 1979 and 1980. The abundance of 0+ fish was negatively related and the size of *Daphnia* was positively related to the vulnerability to angling of predatory pikeperch.

Introduction

This study deals with the consequences of varying amounts of 0+ fish on the numbers and size composition of zooplankton and on the condition and vulnerability to angling of pikeperch, *Stizostedion lucioperca*, in the Frisian Lake District.

The size composition of the zooplankton in Tjeukemeer, one of the Frisian lakes, is influenced by size-selective predation by 0+ fish, as was concluded from studies on the populations of *Daphnia hyalina* and *D. cucullata* (Vijverberg & Richter 1982) and from the feeding behaviour of 0+ fish (van Densen, in prep.). In this lake the growth of 0+ smelt, *Osmerus eperlanus*, and perch, *Perca fluviatilis*, is depressed when they are abundant and thus exhaust the stock of large zooplankters (van Densen, in prep.).

The first aim of this study was to investigate whether the fluctuations in abundance of 0+ fish and the related size of preferred zooplankton species act synchronously in every lake of the district. One reason for a synchronous fluctuation in abundance of 0+ fish could be the distribution of fish larvae by the water currents flowing through the lake district. This is possible because the Frisian Lake District is an open water body supplied with water from the IJsselmeer in the S. during summer. There is a continuous outflow of water to the Wadden Sea in the N. and to the province of Groningen to the E.

The second aim of this study was to examine to what extent the immigration of fish larvae with the water currents influences the numbers and distribution of 0+ fish in the lake district at the beginning of the growing season.

The third aim of this study was to link the vulnerability to angling of pikeperch, via the abundance of 0+ fish, to the mean size of *Daphnia*. Mean *Daphnia* size in Tjeukemeer was depressed in the presence of high densities of planktivorous 0+ fish. High densities of 0+ yellow perch, *Perca flavescens*, result in low vulnerability to angling of walleye *Stizostedion vitreum vitreum*, in Oneida Lake, New York (Forney 1980), so a small mean size of *Daphnia* might be related to a low vulnerability to angling for the pikeperch in the Frisian lakes.

Study area

The Frisian lakes form a 14 000 ha waterbody of

Hydrobiologia 95, 321–336 (1982). 0018-8158/82/0953-0321/$03.20.

interconnected lakes (Fig. 1). Seventeen of these were studied during one or more sampling trips. All the lakes are shallow and have a similar mean depth of about 1.5 m (range: 1.0–2.5 m). Fourteen of them have been described by Beattie et al. (1978). Among the other three, Oude Gaaster Brekken (lake no. 12) and Gaastmeer (lake no. 13) are small lakes located N. of the Fluessen. They have a surface area of 293 and 205 ha, respectively. The much larger (about 2 000 ha) Lauwersmeer (lake no. 1) was created by separating the former Lauwers Sea from the Wadden Sea in May 1969. Though the salinity of the lake has decreased sharply in recent years, the chloride concentration is still much higher here than in the rest of the Frisian lakes (Table 1). The chloride

Table 1. Chloride concentration ($mg.l^{-1}$) of the water in the Frisian lakes in 1980.

Lake	June	September
Lauwersmeer	551 (398–806)	584 (474–804)
Bergumermeer	139	125
Pikmeer	143	–
Snekermeer	161	156
Langweerder Wielen	133	161
Heegermeer	134	161
Fluessen	234	185
Morra	131	181
Gaastmeer	152	153
Slotermeer	146	157
Grote Brekken	171	172
Tjeukemeer	135	125

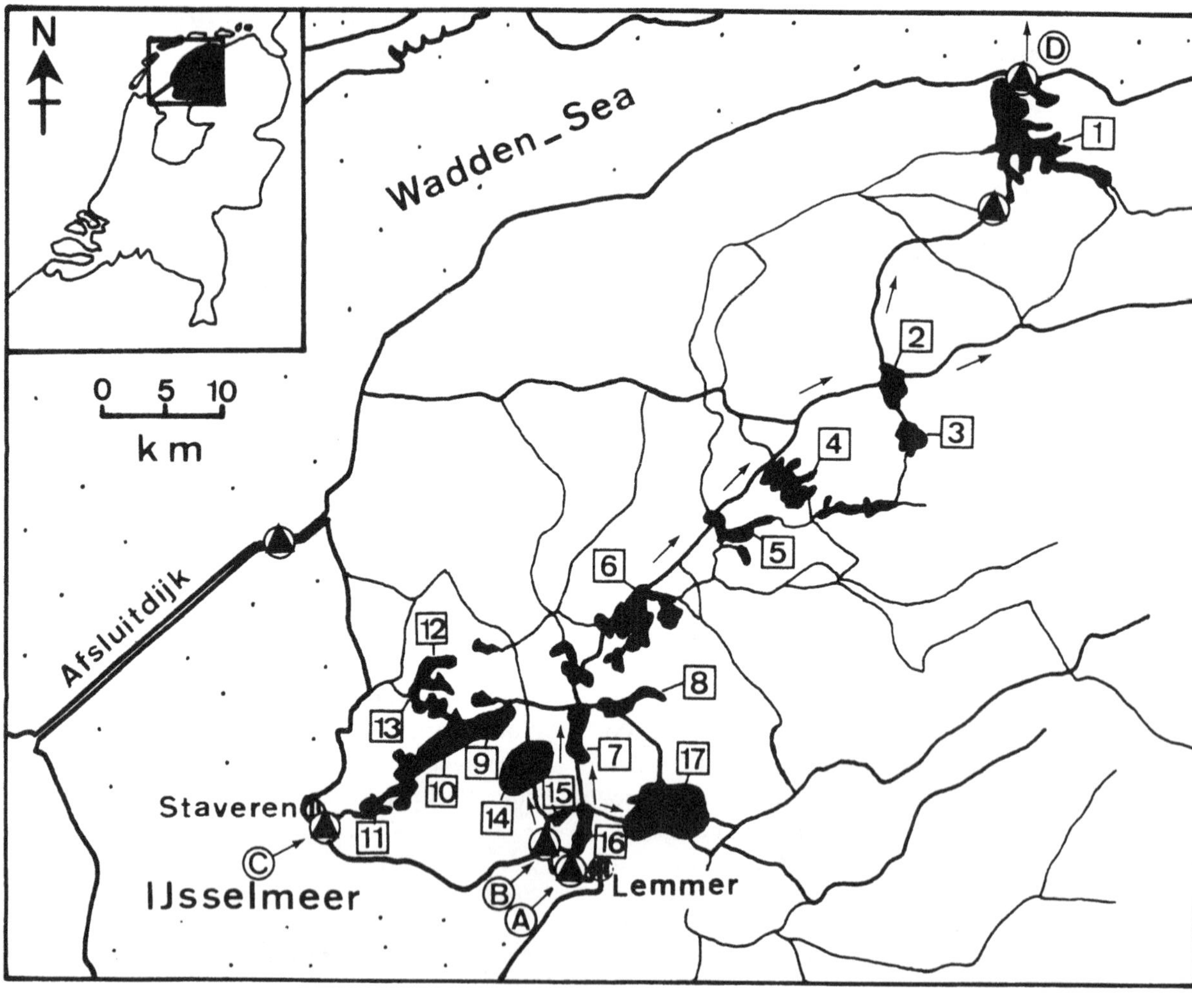

Fig. 1. The Frisian Lake District. The lakes are indicated by numbers: 1. Lauwersmeer; 2. Bergumermeer; 3. Leijen; 4. Princenhof; 5. Pikmeer; 6. Sneekermeer; 7. Koevordermeer; 8. Langweerder Wielen; 9. Heegermeer; 10. Fluessen; 11. Morra; 12. Oude Gaaster Brekken; 13. Gaastmeer; 14. Slotermeer; 15. Brandemeer; 16. Grote Brekken; 17. Tjeukemeer. The sluices are indicated: A. Lemmer; B. Tacozijl; C..Staveren; D. Lauwersoog. The main water flow during summer is indicated by arrows.

concentration in the Lauwersmeer has a distinct N.-S. gradient. The depth of its main waterbody ranges from 2.5 to 9 m, the various bays being much shallower (0.5–1.5 m). The lake receives water via Dokkumerdiep in the S.W. and Zoutkamperril in the S.E. The outflow of water is regulated by sluices near Lauwersoog.

During summer the Frisian Lake District is supplied with water from the IJsselmeer via the sluices near Lemmer (A) and Tacozijl (B). The main water flow is via the Prinses Margriet Canal to the Wadden Sea and the province of Groningen in the E. During winter the sluices near Lemmer and Tacozijl are closed and the water surplus, which is a consequence of precipitation, is let out to the N. as during summer. The volumes of water entering the Frisian lakes during early summer 1974–1980 are given in Fig. 2. The total water intake for April, May and June 1979 was extremely low. In the other years the water intake during these months was between 113 and 200 $\times 10^6$ m^3. In this way approximately 50% of the total water volume in the lake district was replaced by that from the IJsselmeer.

Observations on the chemistry, zooplankton and zoobenthos of the Frisian lakes from 1970 to 1972 were reported by Beattie *et al.* (1978). The main fish species are: smelt, *Osmerus eperlanus;* perch, *Perca fluviatilis;* pikeperch, *Stizostedion lucioperca;* ruffe, *Gymnocephalus cernua;* roach, *Rutilus rutilus;* bream, *Abramis brama;* white bream, *Blicca björkna;* and eel, *Anguilla anguilla.*

Material and methods

Sampling of zooplankton

Because the summed seasonal effect of fish predation is manifested most clearly at the end of the growing season, the densities and size of zooplankton were registered in September 1974, 1975, 1976, 1979 and 1980. Zooplankton was sampled just below the surface with a 5 l Friedinger sampler (column height, 60 cm). In most cases 5–7 samples per lake were taken on each date from sampling sites distributed regularly over the open water zone. Samples were filtered through 120 μm mesh plankton gauze and preserved in 4–5% formalin. The samples collected on a particular date were combined for each lake and sub-sampled following Kott (1953). Length was measured under the microscope with an eyepiece micrometer. Cladocerans were measured from the top of the head or, if a 'helmet' was present, from the base of the helmet to the base of the tail spine, and copepods from the anterior to the posterior end of the cephalothorax. The subsampled individuals were generally sorted and individuals per species counted, only cyclopoid copep-

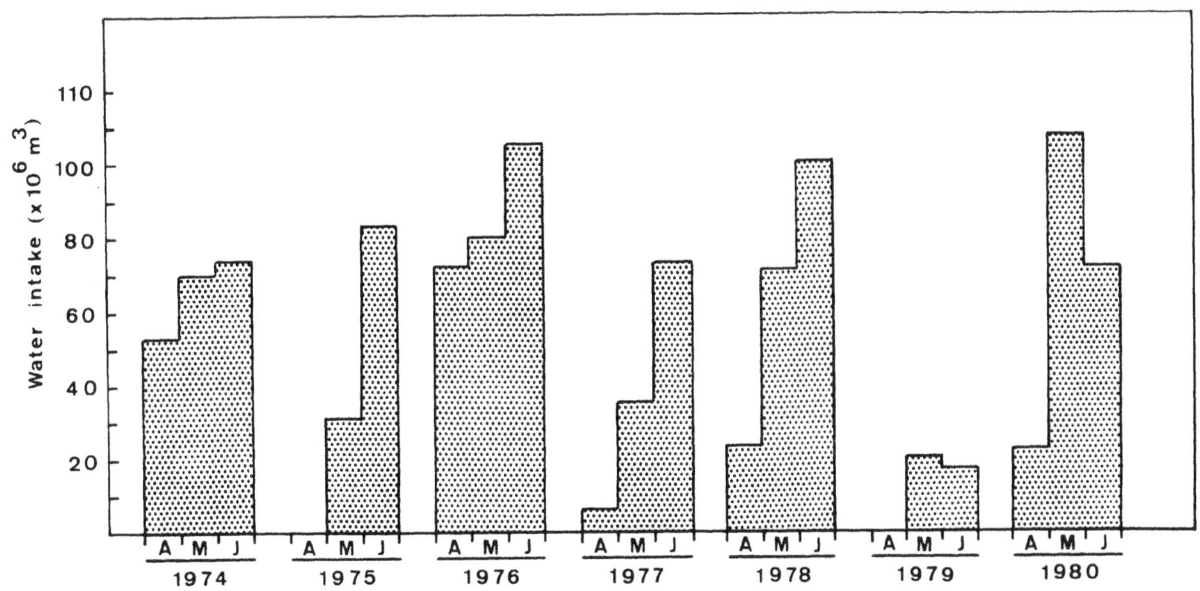

Fig. 2. Water intake (10^6 m^3) of the Frisian Lake District during April, May and June 1974 to 1980.

ods being grouped together. *Daphnia hyalina* and *D. cucullata* were counted separately in 1974–1976, but grouped together in 1979 and 1980. The size of grouped *Daphnia* spp. was measured.

Sampling of fish

The 0+ fish were sampled with 40 cm larvae nets, a 3 m trawl and a 5 m trawl, all described in van Densen (in prep.). For studying the transport of larvae with the water movements, 30 s hauls with larvae nets were made at 14 sampling sites between IJsselmeer and Tjeukemeer and in Tjeukemeer itself in May and June 1979 and 1980 (Fig. 3). Also, in May 1980 the larvae populations in IJsselmeer in the vicinity of Lemmer were sampled at one depth (40 cm), at 29 sampling sites covering an area of 2 000 ha. The distribution of 0+ fish in the lake district in June 1979 and 1980 was registered with the 3 m trawl, taking six to ten 5 min hauls in each lake. At the end of the growing season 0+ fish were sampled with the 5 m trawl, taking one 20 min haul in each lake in 1974, 1975 and 1976, and six to ten 5 min hauls in each lake in 1979 and 1980. For comparison, all trawl catches were converted to catches per 5 min haul.

The 1+ and older pikeperch were sampled in autumn with the 5 m trawl and with gill nets of tiptolon 210/3 with meshes of 30 to 55 mm half mesh size.

Both catches of larvae in May and June and of 0+ fish in June were preserved in 5% formalin. The

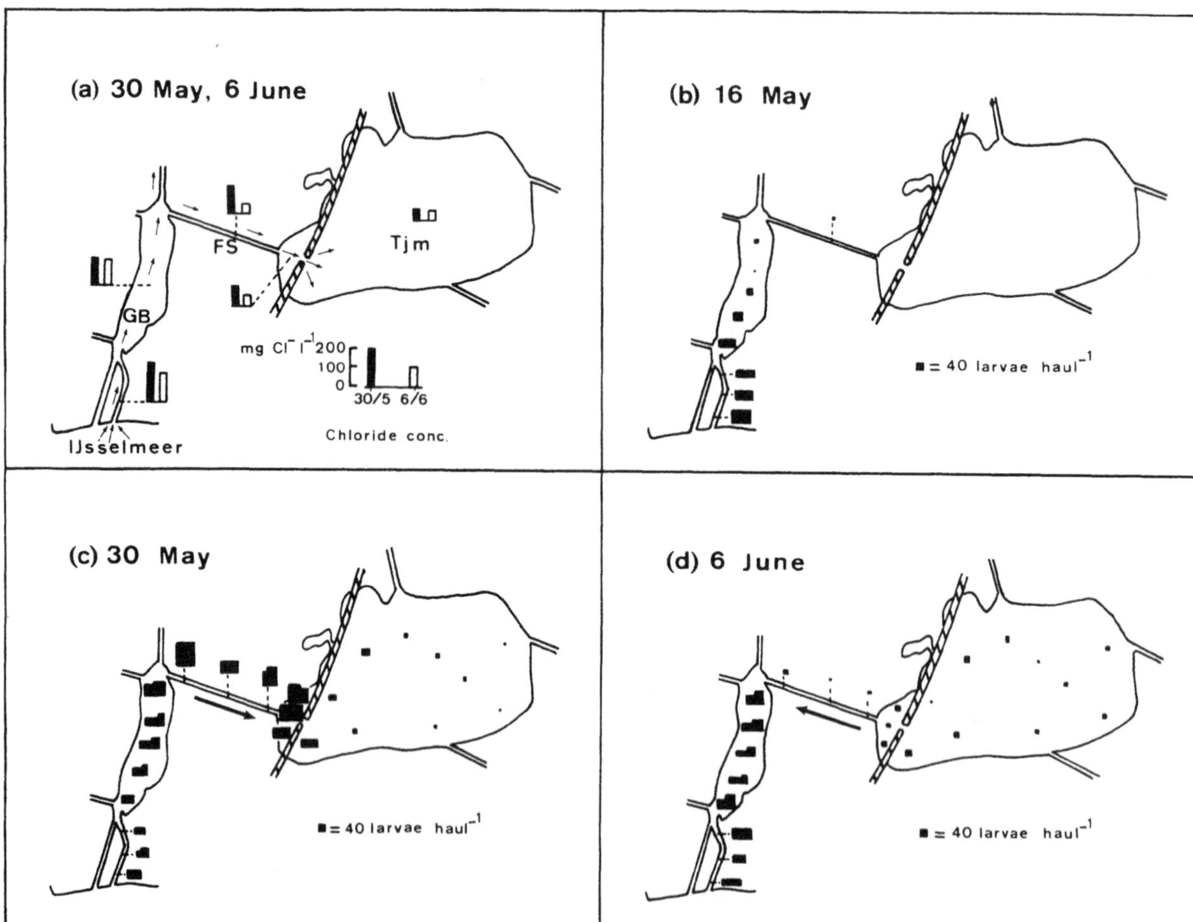

Fig. 3. Chloride concentration ($mg\ Cl^- l^{-1}$) and density of smelt larvae (per 30 s haul with larvae net) on a traject between IJsselmeer and Tjeukemeer in the early summer of 1979 (GB = Grote Brekken; FS = Follegasloot; Tjm = Tjeukemeer). The arrows in (c) and (d) indicate the direction of water flow.

species identification of percid larvae was made according to the number of myomeres. No distinction was made between 0+ bream and 0+ white bream. Bream dominated white bream in the catches of older fish.

The 0+ pikeperch at the end of the growing season is separated into three size categories: i) small summer generation pikeperch, ii) slow-growing, mainly planktivorous, spring generation pikeperch, and iii) fast growing, mainly piscivorous, spring generation pikeperch.

The fork length of 0+ fish was measured to the nearest mm, and of older fish to the nearest 5 mm. For the length–weight relationship of older pikeperch, the length was measured to the nearest millimetre and the weight to the nearest 0.1 g fresh weight. Functional geometric mean regressions (Ricker 1973) were used in the calculation of the length–somatic weight relationship.

The stomachs of older pikeperch were dissected after measuring and weighing and the contents determined while fresh.

Results

Density and size of zooplankton in September

Eight of the 13 zooplankton species found were cladocerans and five were copepods. The cladoceran species are: *Bosmina coregoni, B. longirostris, Ceriodaphnia pulchella, Chydorus sphaericus, Daphnia hyalina, D. cucullata, Diaphanosoma brachyurum* and *Leptodora kindtii*. The copepod species are: *Acanthocyclops robustus, Cyclops vicinus vicinus, Diacyclops bicuspidatus, Eurytemora affinis* and *Mesocyclops leuckarti*.

Densities of grouped *Daphnia* spp. varied strongly both from lake to lake in the same year and from year to year in the same lake (Table 2). Low densities occurred in 1974 and 1980. The minor exceptions were the high densities for the Leyen in 1974 and for Lauwersmeer and Grote Brekken in 1980.

The further separation of the larger *Daphnia hyalina* from the smaller *D. cucullata* in 1974, 1975 and 1976 showed the former to be dominant in 1976 in almost every lake (Table 3). The numerical dominance of the larger species in 1976 was also expressed by the larger mean size of grouped *Daphnia* spp. per lake in 1976 than in 1974 and 1975 (Fig. 4). *Daphnia* spp. were not separated in 1979 and 1980 but the mean size differed greatly in these years: high densities and a large size in 1979, low densities and a small size in 1980. Only in the Grote Brekken were higher densities and a larger size observed in 1980 than in 1979.

It is concluded that at low densities of grouped *Daphnia* spp. their mean size is small (1974, 1980) but at higher densities this relationship between numbers and size does not necessarily hold, because

Table 2. Numbers of grouped *Daphnia* spp.(ind.l^{-1}) in the Frisian lakes in September 1974, 1975, 1976, 1979 and 1980.

Lake	1974	1975	1976	1979	1980
Lauwersmeer	–	–	–	127	72
Bergumermeer	1	52	19	270	14
Leyen	67	597	104	–	–
Princenhof	–	3	–	–	–
Pikmeer	–	20	–	117	9
Snekermeer	–	63	–	101	13
Koevordermeer	–	29	124	–	–
Langweerder Wielen	–	57	–	97	17
Heegermeer	0	269	53	98	17
Fluessen	0	280	46	162	32
Morra	0	252	33	–	22
Oude Gaater Brekken	0	118	78	–	–
Gaastmeer	0	200	48	–	–
Slotermeer	0	62	65	72	5
Brandemeer	–	13	–	–	–
Grote Brekken	2	22	76	18	110
Tjeukemeer	0	15	122	86	16

Table 3. Numbers (ind.l^{-1}) of *Daphnia hyalina* (D.h.) and *D. cucullata* (D.c.) in the Frisian lakes in September 1974, 1975 and 1976.

Lakes	16–19 Sept. 1974		1–4 Sept. 1975		13–14 Sept. 1976	
	D. h.	D. c.	D. h.	D. c.	D. h.	D. c.
Bergumermeer	0	1	4	48	5	14
Leyen	1	66	6	591	50	54
Princenhof	–	–	1	2	–	–
Pikmeer	–	–	1	19	–	–
Snekermeer	–	–	6	57	–	–
Koevordermeer	–	–	3	26	122	2
Langweerder Wielen	–	–	1	56	–	–
Heegermeer	0	0	4	265	44	9
Fluessen	0	0	4	276	37	9
Morra	0	0	4	248	24	9
Oude Gaaster Brekken	0	0	4	114	74	4
Gaastmeer	0	0	2	198	40	8
Slotermeer	0	0	5	57	62	3
Brandemeer	–	0	0	13	–	–
Grote Brekken	0	2	9	13	74	2
Tjeukemeer	0	0	1	14	90	32

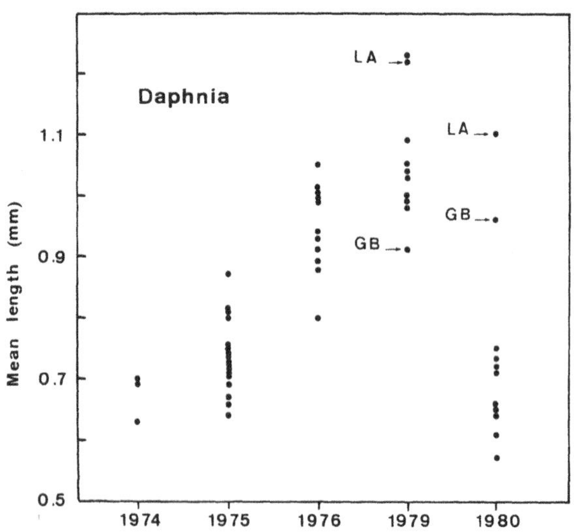

Fig. 4. Mean length (mm) of grouped *Daphnia* spp. in the Frisian lakes in September 1974, 1975, 1976, 1979 and 1980 (LA = Lauwersmeer, GB = Grote Brekken).

of the possible dominance of *D. cucullata.* In any case, the mean size of grouped *Daphnia* spp. fluctuates synchronously from year to year in almost every Frisian lake and a very sharp fluctuation was observed between September 1979 and 1980.

The fluctuations in density (Table 4) and size (Table 5) of the much smaller cladoceran, *Chydorus sphaericus,* showed resemblances with that for *Daphnia* spp.; low densities occurred in 1974 and 1980, the density in the Leyen in 1974 was relatively high, densities were higher and mean size larger in 1979 than in 1980, and the densities in Grote Brekken were in opposition to this trend in 1979 and 1980.

A larger size in 1979 than in 1980 was also found for both *Bosmina coregoni* and *B. longirostris* (Table 5). As with *Daphnia* spp. and *C. sphaericus,* the lowest densities of *B. coregoni* were observed in 1974.

At the extremely low densities of cladocerans in the Frisian lakes in 1974, cyclopoid copepods formed the dominant taxon. Compared with *Daphnia, Chydorus* and *Bosmina* the densities of cyclopoid copepods in 1974 to 1980 varied less, both from lake to lake and from year to year (Table 6). Six out of ten lakes had a higher density of cyclopoid copepods in 1979 than in 1980. Seven out of the ten lakes had a larger mean size in 1979.

Immigration of fish larvae in 1979 and 1980

The short-living smelt in the Frisian lakes matures at the end of the first growing season. A restricted sampling trip for 0+ fish in Slotermeer, Heegermeer and Tjeukemeer in the autumn of 1978 showed low densities of the year-class 1978. Therefore the recruitment of smelt larvae in 1979 in these lakes was anyway very dependent on the possible immigration of smelt larvae from the IJsselmeer.

Table 4. Numbers (ind. l^{-1}) of *Chydorus sphaericus* in the Frisian lakes in September 1974, 1975, 1979 and 1980.

Lake	1974	1975	1976	1979	1980
Lauwersmeer	–	–	–	61	3
Bergumermeer	35	94	229	98	38
Leyen	1039	1032	205	–	–
Princenhof	–	17	–	–	–
Pikmeer	–	39	–	155	29
Snekermeer	–	270	–	162	75
Koevordermeer	–	597	624	–	–
Langweerder Wielen	–	805	–	221	99
Heegermeer	1	334	200	179	83
Fluessen	0	406	248	229	99
Morra	0	362	231	–	84
Oude Gaaster Brekken	0	895	261	–	–
Gaastmeer	0	114	125	–	–
Slotermeer	0	402	579	130	69
Brandemeer	–	402	–	–	–
Grote Brekken	14	613	550	78	210
Tjeukemeer	5	758	993	196	37

Table 5. Mean size (mm) of cyclopoid copepods, *Bosmina coregoni*, *B. longirostris*, and *Chydorus sphaericus* in the Frisian lakes in September 1979 and 1980.

Lake	Copepods		*B. coregoni*		*B. longirostris*		*C. sphaericus*	
	1979	1980	1979	1980	1979	1980	1979	1980
Lauwersmeer	0.35	0.44	0.47	0.37	0.32	0.29	0.28	0.29
Bergumermeer	0.40	0.35	0.41	0.35	0.33	0.29	0.29	0.25
Pikmeer	0.37	0.34	0.43	0.32	0.33	0.28	0.28	0.25
Snekermeer	0.37	0.36	0.42	0.34	0.31	0.28	0.28	0.24
Langweerder Wielen	0.36	0.36	0.43	0.35	0.32	0.28	0.27	0.24
Heegermeer	0.37	0.35	0.44	0.34	–	0.25	0.27	0.25
Fluessen	0.36	0.38	0.45	0.34	–	0.27	0.28	0.24
Morra	–	0.37	–	0.33	–	0.29	–	0.25
Slotermeer	0.33	0.32	0.49	0.35	–	0.27	0.27	0.25
Grote Brekken	0.37	0.25	0.46	0.41	–	0.32	0.27	0.27
Tjeukemeer	0.41	0.39	0.42	0.35	0.32	0.29	0.28	0.27
Mean	0.37	0.36	0.44	0.35	0.32	0.28	0.28	0.25

Because of heavy precipitation in the spring of 1979 the water inlet from IJsselmeer started relatively late, on 15 May. On the next day high densities of smelt larvae were observed on the trajectory between IJsselmeer and Tjeukemeer (Fig. 3).

During the weeks thereafter the passive movement of smelt larvae with the water current to the Tjeukemeer was followed.

The water masses from IJsselmeer were traced by their much higher chloride concentration than the Frisian lake water. On 30 May water from the IJsselmeer with a high chloride concentration and a high larval density entered Tjeukemeer. Another period of heavy precipitation at the beginning of June caused a water surplus in Tjeukemeer. Then a water flow in the opposite direction, out of the lake, through Follegasloot to the W. occurred. This resulted in low chloride concentrations in Follagasloot and a 'loss' of large numbers of larvae. In this wet period the sluices near Lemmer even had to be closed at the beginning of June. The water surplus was let out to the N. and into IJsselmeer in the S.W. via the sluices near Stavoren (Fig. 1C). The consequent poor immigration of smelt larvae into Tjeukemeer as well as the low reproduction success of other fish species resulted in the lowest stock of 0+

Table 6. Numers (ind. l^{-1}) of cyclopoid copepods in the Frisian lakes in September 1974, 1975, 1976, 1979 and 1980.

Lake	1974	1975	1976	1979	1980
Lauwersmeer	–	–	–	34	246
Bergumermeer	17	55	67	31	34
Leyen	64	312	105	–	–
Princenhof	–	113	–	–	–
Pikmeer	–	75	–	76	49
Snekermeer	–	81	–	104	43
Koevordermeer	–	108	72	–	–
Langweerder Wielen	–	156	–	138	81
Heegermeer	34	189	76	90	19
Fluessen	80	171	92	169	40
Morra	81	144	51	–	60
Oude Gaaster Brekken	67	339	251	–	–
Gaastmeer	45	137	107	–	–
Slotermeer	93	225	78	105	186
Brandemeer	–	84	–	–	–
Grote Brekken	8	53	76	18	58
Tjeukemeer	21	120	193	156	37

fish in this lake during 1976 to 1980 (van Densen, in prep.). The exceptional position of Tjeukemeer in 1979 within the Frisian lakes could be seen from the results of the sampling with the 3 m trawl in June (Fig. 5). Because of water movements northwards and westwards the other lakes received more young smelt.

Immigration was of no importance for the recruitment of juvenile perch and pikeperch in 1979. Both species occurred in low numbers in the larvae samples on the trajectory between IJsselmeer and Tjeukemeer in early summer. The densities of perch in June showed low recruitment levels in every lake, most of them originating from local reproduction (Fig. 5). Recruitment of pikeperch appeared to be more succesful, except in Tjeukemeer, Grote Brekken and Langweerder Wielen.

The low densities of 0+ smelt, perch and pikeperch together, in June 1979, resulted in a high depensatory mortality rate (Forney 1976) for all three species. As a consequence the abundance of 1+ smelt in the spring of 1980 was low in almost every Frisian lake. So, recruitment of larval smelt in the Frisian lakes was dependent on immigration from the IJsselmeer as in 1979.

The continuous inflow of water from IJsselmeer in the dry early summer of 1980 (Fig. 2) brought large quantities of smelt and perch larvae with it. Sampling of the larvae population at 29 sites in IJsselmeer in the vicinity of the sluices near Lemmer on 28 May 1980 showed comparable quantities of smelt and perch larvae. The mean catch with 95% confidence limits was 194 ± 56 smelt and 207 ± 95 percid larvae per 30 s haul. These figures correspond with about 50 larvae m^{-3} smelt and percid larvae combined.

The intake of a large volume of water with similar concentrations of smelt and percid larvae and the high densities of 0+ perch found in the southern lakes in June (Fig. 5) suggest that the 0+ perch population in the Frisian lakes in 1980 were immigrated individuals from IJsselmeer. In 1980 also, no immigration of larval pikeperch was observed. Their densities and distribution pattern in June were similar to densities and distribution in June 1979. Again, low densities occurred in Tjeukemeer, Grote Brekken and Langweerder Wielen.

The high densities of 0+ smelt, perch and pikeperch together in June 1980 resulted in a low depensatory mortality rate and many times higher 0+ densities of all species in September 1980 than in September 1979.

It is concluded that immigration was important in 1979 and 1980 for stocking the Frisian lakes with 0+ prey fish. Not only reproduction success in IJsselmeer, as was clear from the difference with perch in both years, but also a continuous water inlet into the Frisian lakes during early summer, as in 1980, is of relevance for an extensive immigration of smelt and perch larvae. A possible immigration of pikeperch larvae could not be proved.

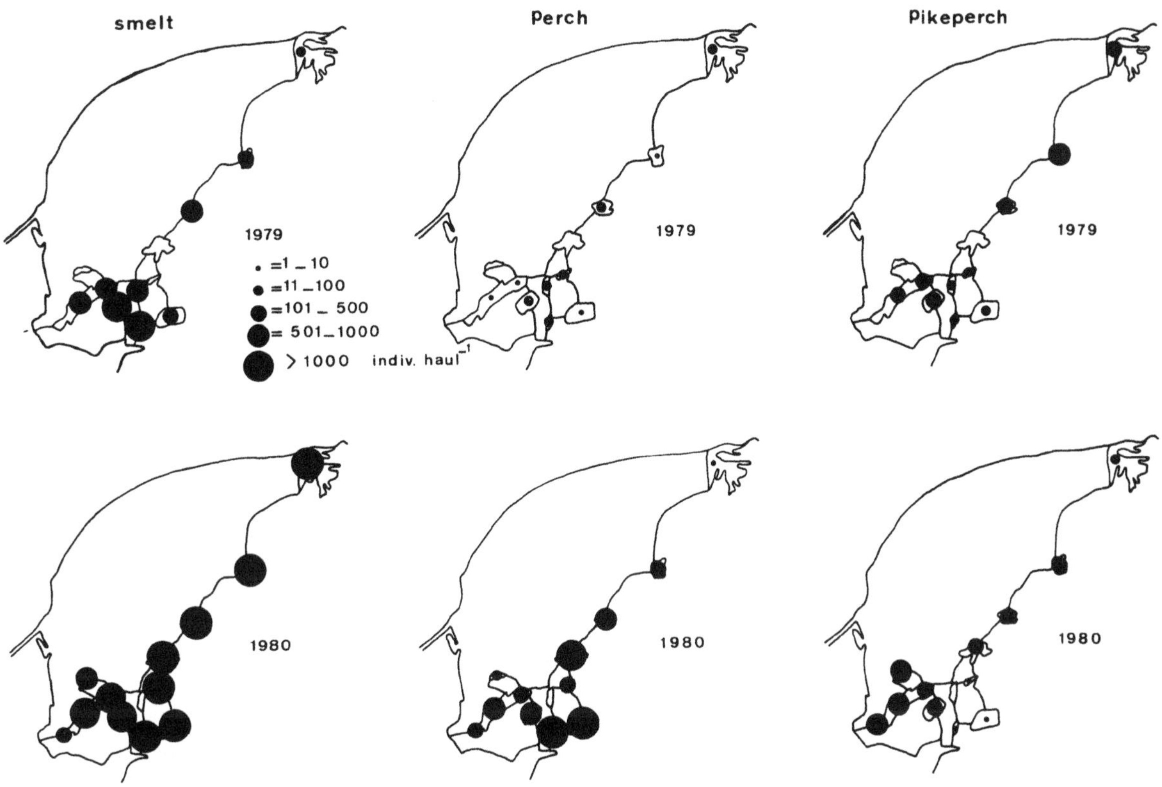

Fig. 5. Density (Nos. per 5 min 3 m trawl haul) of 0+ smelt, perch and pikeperch in the Frisian lakes in June 1979 and 1980.

Numbers and size of 0+ fish at the end of the growing season

Smelt. Numbers of 0+ smelt were high, dominating the catch in almost every lake, in 1974, 1975 and 1980 (Fig. 6). In November 1976 and September 1979 much lower smelt densities were observed. Because more trawl hauls per lake were taken in 1979 and 1980, the observed densities in these years are more reliable. The high depensatory mortality rate in 1979 will have resulted in similarly low densities in November 1979 as in November 1976, both being almost three orders of magnitude lower than in the other three years. At higher 0+ densities in June the depensatory mortality rate is low and therefore the higher densities in September 1980 than in September 1979 are already explained by the small and the extensive immigration in the early summer of 1979 and 1980 respectively. Generally there is a synchronous fluctuation in the densities of 0+ smelt in the Frisian lakes.

The size of 0+ smelt in the lakes is clearly inverse-

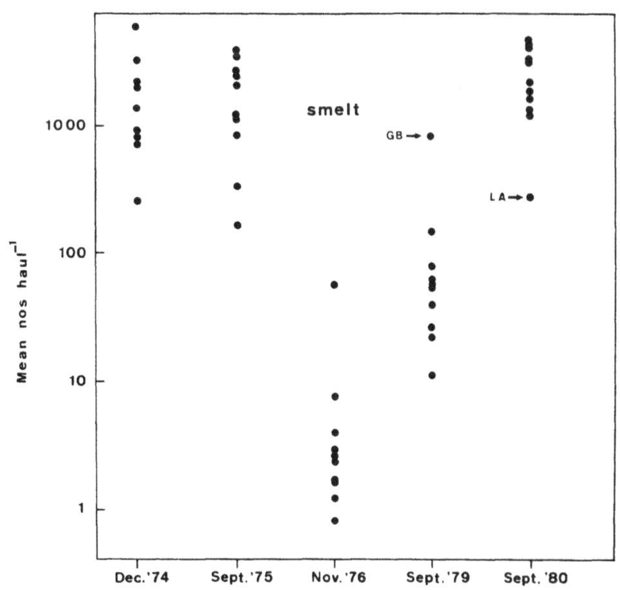

Fig. 6. Density (Nos. per 5 min 5 m trawl haul) of 0+ smelt in the Frisian lakes in Dec. 1974, Sept. 1975, Nov. 1976, Sept. 1979 and Sept. 1980 (GB = Grote Brekken; LA = Lauwersmeer).

ly related to their density (Fig. 7). The largest sizes of 0+ smelt were observed in 1976 and 1979. Food conditions in the form of abundant large *Daphnia* in the second part of the growing season were optimal in those years. Leyen and Lauwersmeer are lakes with consistently large sizes of 0+ smelt.

Perch, ruffe, roach and bream. At the end of 1974 and 1976 the catches of 0+ fish other than that of smelt were extremely small in all the lakes. Quite different from smelt, this is related to their actual density as well as to a lower catchability in winter time. The comparison of abundance and size for species other than smelt is restricted to the catches in September 1975, 1979 and 1980 (Table 7, Fig. 8).

The densities of 0+ perch were generally high in 1975 and 1980, and low in 1979. The difference between the last two years has already been explained by the more extensive immigration of smelt and perch larvae in 1980. As with smelt the size of this mainly planktivorous species was better in 1979 than in 1980. The mean size in Bergumermeer was consistently larger than in other lakes. This was related to the temperature effect of the Bergum Power Station (van Densen & Hadderingh 1982) rather than to food supply.

The catches of 0+ ruffe varied greatly between lakes. The highest mean density for the district was recorded in 1980 when, particularly in Bergumermeer and Morra, extremely high densities were observed. Because the ruffe is benthivorous its growth does not depend on the availability of zooplankton.

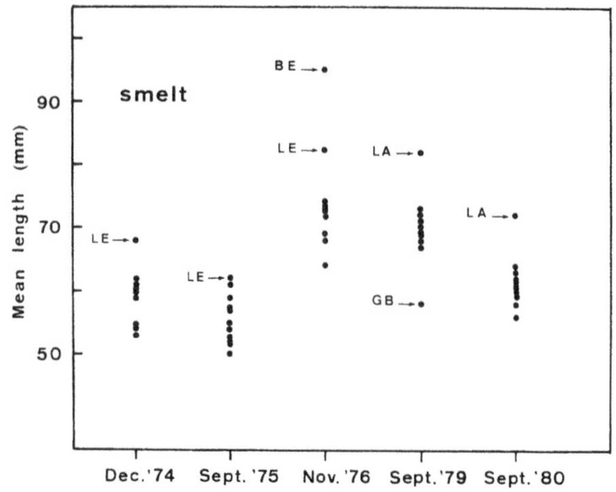

Fig. 7. Mean length (mm) of 0+ smelt in the Frisian lakes in Dec. 1974, Sept. 1975, Nov. 1976, Sept. 1979 and Sept. 1980 (LE = Leijen; BE = Bergumermeer; LA = Lauwersmeer; GB = Grote Brekken).

The mean density of roach and bream for the whole lake district was high in 1980. The decisive influence of temperature on the growth of 0+ cyprinids is seen from their larger mean length in the thermally affected Bergumermeer. The larger mean length at the end of a warm summer, e.g. 1975, than after the colder summers 1979 and 1980 is in line with this observation.

Pikeperch. The separation between spring and summer generation pikeperch and the further dis-

Table 7. Catches (Nos. per 5 min 5 m trawl haul) of 0+ perch, ruffe, roach and bream in the Frisian lakes in September 1975, 1979 and 1980.

Lake	Perch			Ruffe			Roach			Bream		
	1975	1979	1980	1975	1979	1980	1975	1979	1980	1975	1979	1980
Lauwersmeer	–	0	0	–	0.3	0.5	–	0.4	3	–	14	88
Bergumermeer	21	2	45	300	210	712	19	12	152	80	556	559
Leyen	148	–	–	78	–	–	25	–	–	956	–	–
Pikmeer	19	0	7	10	30	180	169	4	90	898	380	359
Snekermeer	4	0	68	4	0.7	34	28	20	89	73	103	434
Langweerder Wielen	8	0	152	4	0.8	5	64	19	404	551	124	955
Heegermeer	126	0.8	186	8	11	67	4	5	329	17	13	399
Fluessen	1070	1	756	25	30	55	5	27	194	20	28	114
Morra	2	–	347	134	–	1781	2	–	259	358	–	1227
Slotermeer	2138	0	540	91	2	49	0	10	28	36	27	95
Grote Brekken	80	0.4	1871	2	0	11	69	7002	4448	187	744	756
Tjeukemeer	199	0	1454	17	0.6	3	100	6	446	154	37	541

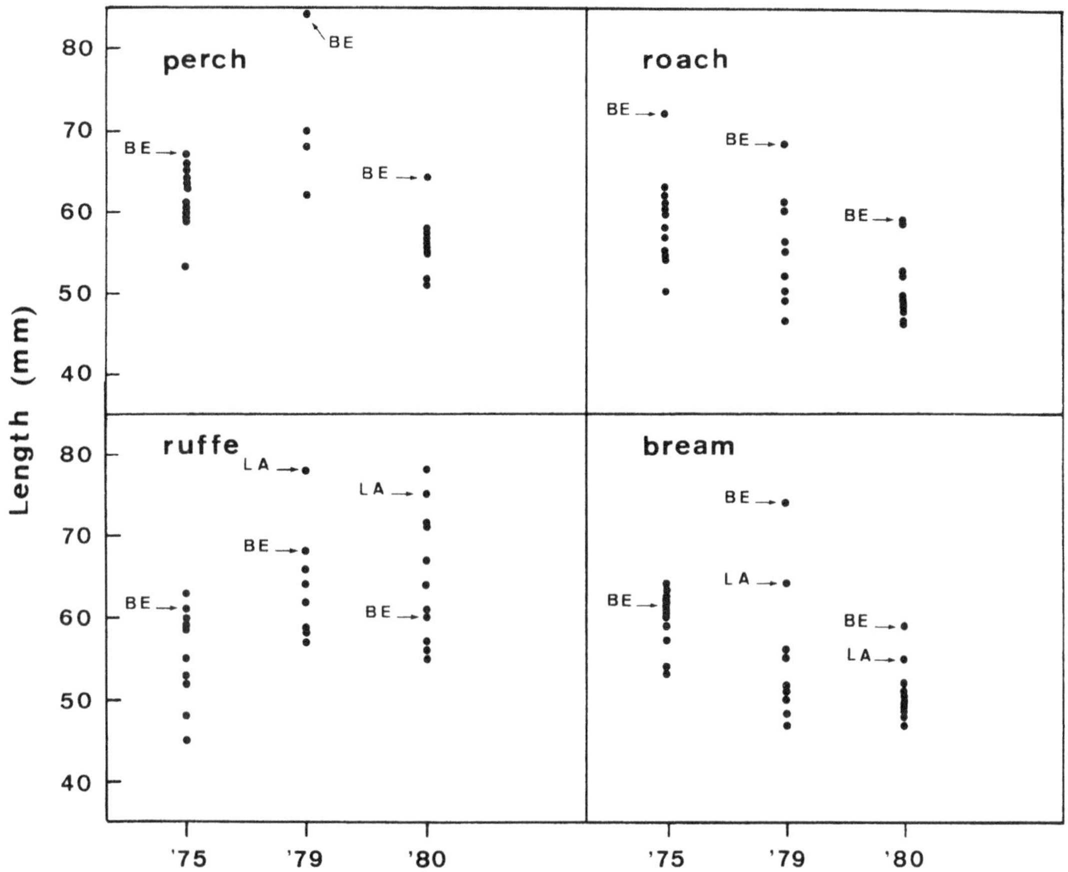

Fig. 8. Mean length (mm) of 0+ perch, ruffe, roach and bream in the Frisian lakes in September 1975, 1979 and 1980 (BE =Bergumermeer, LA = Lauwersmeer).

tinction in the spring generation between slow growing, mainly planktivorous, and fast growing, mainly piscivorous, individuals was based on the length–frequency distributions.

Individuals of summer generation pikeperch were only observed in 1979 and 1980, but occurred in very low densities (Table 8).

The total numbers of pikeperch, all of which belonged to the spring generation, were low in September 1975. Large-sized piscivorous individuals, which were dominant in some of the lakes, thrived on the vast amount of smelt present in that year. Within the spring generation in 1979 and 1980, small-sized individuals dominated in every lake, except for Grote Brekken and Tjeukemeer in 1980. The number of large-sized individuals in most lakes, Slotermeer and Tjeukemeer in particular, were higher in 1980 than in 1979. High densities of 0+ smelt in June 1980 both facilitated the onset of piscivority and buffered (Forney 1976) the 0+ pikeperch against predation by adult pikeperch during the rest of the season. The size to which 0+ pikeperch grow determines their survival rate, and hence the year-class strength since larger fish can escape predation better (Forney 1976; Willemsen 1977).

In addition to the denstity of 0+ smelt, the onset of piscivority also depends on the size of the 0+ pikeperch relative to the size of the smelt. This mean relative size of pikeperch was 0.78 ± 0.14 in June 1979 and 0.74 ± 0.043 in June 1980. The mean absolute size of 0+ smelt at the same moment was remarkably constant: 25.4 ± 0.70 mm in June 1979 and 26.4 ± 1.6 mm in June 1980. Thus, the large variation in the relative size of pikeperch in June 1979 was due to its mean absolute size, which varied from 14.8 mm in Slotermeer to 24.7 mm in Bergumermeer. Despite the large relative size in the

Table 8. Catches (Nos. per 5 min 5 m trawl haul) of 0+ pikeperch in the Frisian lakes in September 1975, 1979 and 1980 (ss = small-sized spring generation; ls = large-sized spring generation; sg = summer generation).

Lake	1975			1979			1980		
	ss	ls	sg	ss	ls	sg	ss	ls	sg
Lauwersmeer	–	–	–	114.7	2.0	0.1	1.5	0.4	1.0
Bergumermeer	1	2	0	1.4	0.7	0.6	12.5	6.1	0.2
Leyen	2	0	0	–	–	–	–	–	–
Pikmeer	0	0	0	19.0	0.5	0.7	10.0	3.3	1.0
Snekermeer	0	0	0	0	0	0.5	13.0	0.3	1.0
Langweerder Wielen	2	0	0	1.0	0.2	0.3	6.5	0.3	1.3
Heegermeer	5	5	0	39.2	0	0.5	42.2	2.8	0
Fluessen	0	5	0	56.7	0.3	0.5	25.3	1.7	0
Morra	4	0	0	–	–	–	66.3	11.0	0
Slotermeer	0	11	0	32.8	0.3	0.3	127.0	20.5	0
Grote Brekken	0	4	0	5.4	5.4	5.4	1.2	3.0	0.8
Tjeukemeer	0	3	0	2.5	1.4	1.4	3.0	19.4	0

northern lakes in June 1979, the numbers of large-sized piscivorous individuals in September were not high, since the densities of 0+ smelt during the summer were low and sharply decreasing. This made the pikeperch weakly buffered at the end of the season. However, the relative size in 1980 was not large, the density dependent slower growth of 0+ smelt in 1980 making the latter increasingly more available to the piscivorous 0+ pikeperch.

Sprat. The sprat, *Sprattus sprattus,* was encountered in Lauwersmeer in September 1979, confined to the area with a higher chloride content. The sprat reproduces in the sea. The youngest stages probably succeeded in entering Lauwersmeer when ships were locked from the Wadden Sea to Lauwersmeer. In the western Dutch Wadden Sea the size of young sprat is 3–6 cm in June and August, increasing in size to a maximum (6–8 cm) in October–November (Fonds 1978). The sprat caught in Lauwersmeer was established to be mainly 0+ fish, using the length-frequency distribution.

Condition and vulnerability to angling of pikeperch

The condition of the pikeperch in Slotermeer and Grote Brekken was compared with that in Tjeukemeer in September 1979 and in April and May 1980 respectively. For this we assessed length–weight relationships for somatic weight of both males and females and calculated the weight of a 450 mm pikeperch (Table 9). The lower numbers of 0+ prey fish during the summer of 1979 in Tjeukemeer than in Slotermeer and Grote Brekken is reflected in the

Table 9. Length–weight relationships ($W = aL^b$) for pikeperch in Tjeukemeer, Slotermeer and Grote Brekken and the calculated somatic fresh weight of 450 mm pikeperch.

Lake	Period	Sex	b	ln a	r	n	W 450	Catching method
Tjeukemeer	Sept. 1979	♂	3.301	-13.462	0.973	55	816	Trawl
		♀	3.071	-12.050	0.991	37	823	Trawl
Slotermeer	14-9-1979	♂	3.102	-12.140	0.995	27	908	Gill nets
		♀	2.942	-11.176	0.992	23	891	Gill nets
Tjeukemeer	April/May 1980	♂	3.505	-14.885	0.929	67	850	Gill nets
		♀	3.227	-13.057	0.929	131	777	Gill nets
Grote Brekken	April/May 1980	♂	2.911	-10.985	0.922	33	896	Gill nets
		♀	3.233	-13.037	0.965	89	822	Gill nets
Tjeukemeer	Sept. 1980	♂	3.305	-13.257	0.936	90	1030	Gill nets
		♀	3.080	-11.920	0.982	52	989	Gill nets

lower somatic weight of the 450 mm pikeperch in Tjeukemeer at the end of the summer of 1979 and in the spring of 1980.

After the immigration of larvae in the early summer of 1980, and thus the building-up of a large stock of 0+ fish, the condition in Tjeukemeer increased constantly to a much higher level than in all three lakes at the end of 1979 (Table 9). A similar increase in condition must have occurred in all other lakes of the district during the summer of 1980 because the densities of prey fish were much higher.

The condition of the pikeperch in Slotermeer was not registered in 1978. However, stomach analysis of the 1+ and older pikeperch caught with a trawl on 25 October 1978 indicated a poor food situation: 19 of the 47 fish caught had empty stomachs, 8 contained the remains of 0+ fish and 23 contained, sometimes besides fish, about 15 individuals of *Neomysis integer* per stomach. The low numbers of 0+ fish caught (13 per 5 min 5 m trawl haul) confirmed the poor food situation.

Until 15 March 1977 there was a commercial gill net fishery for pikeperch and perch in the Frisian lakes. Since this date only angling for these species has been allowed. Angling is forbidden from 15 March to 1 June. The fishing mortality of the pikeperch caused by the sporting fishermen since 1977 is substantially less than that caused by the commercial fishermen before 1977. In 1978 the strong 1975 year-class became available to the sport fisheries by exceeding the legal minimum size of 42 cm total length.

Both seasonal and annual variations in angling success appeared to be more related to the condition of the adults than to its population density. Optimal catches were always realized in June, when adult pikeperch condition is poor due to spawning. Angling success was high in 1979 and in June 1980, when adult condition was low, but about five times as low during the second half of 1980 and in 1981 when adult condition was high (pers. comm., P. Kooy, Secretary of the Organization of Frisian Angling Clubs). A particularly sharp decrease in angling success was reported during July 1980. Simultaneously we observed a marked increase in adult condition but no sharp decrease in the number of pikeperch caught per gill net set in Tjeukemeer.

Discussion

The similarities between the Frisian lakes at the end of the growing season are clear from the synchronous fluctuations in: densities of *Daphnia hyalina*, mean size of most zooplankton taxa, numbers and size of 0+ smelt and the size of the 0+ fish of most other fish species. The varying density of 0+ smelt and 0+ perch is considered to be the main causative agent for the synchronous fluctuations in the abundance of *Daphnia hyalina* and the mean size of *Daphnia*. These two fish species generally dominate the open water zone in mid-summer. The 0+ cyprinids, particularly 0+ bream, invade the open water zone at a much later date than smelt and perch.

The synchronous fluctuation in abundance of 0+ smelt in the Frisian lakes at the end of the growing season can be explained both by a similarity in reproduction succes in the lakes and by an equal spread of immigrated larvae over the lakes. For 1979 and 1980 the latter explanation holds. In 1975 and 1976 the parent stock of smelt was large enough to make high recruitment by local reproduction possible. From the data, however, nothing can be concluded about the possible importance of immigration in those years. Notable exceptions in the similarity at the beginning of the growing season 1979 and 1980 can be explained from a reflux of water (Tjeukemeer–1979) or by the distance from the water inlet near Lemmer (Morra–1980).

The parent stock of perch in the Frisian lakes at the end of the seventies was much lower than that of pikeperch (van der Eyk 1981). Slotermeer and Hegermeer had relatively higher densities and the large numbers of 0+ perch in September 1975 in these lakes is in line with this observation. From the sampling of 0+ fish in June and September 1980 it is concluded that the immigration of perch larvae was less far-reaching than that of 0+ smelt in the same year.

A notable exception in the relationship between 0+ fish density and *Daphnia* size was the relatively large size of *Daphnia* in the Grote Brekken in September 1980 with high densities of 0+ fish. This was possibly caused by an influx of water containing large *Daphnia* from the IJsselmeer.

In the late seventies the population of older pikeperch in Tjeukemeer was dominated by the strong 1975 year-class and to a lesser extent by the 1976 year-class (van Densen, in prep). Willemsen (1979), after reading scales of pikeperch from Slotermeer, Langweerder Wielen and Koevordermeer, found the 1975 year-class dominant in all three lakes in 1979 (Fig. 9). The dominance of the 1975 year-class was also found in trawl samples taken in 1977 and 1978 in a number of Frisian lakes (van der Eijk 1981). Our limited sampling of 0+ fish in September 1975, with one haul per lake, revealed small numbers of 0+ pikeperch in the lakes. These, however, were strongly buffered against predation by adult pikeperch because of the large numbers of 0+ smelt. In Tjeukemeer and Slotermeer the 0+ pikeperch were all large-sized piscivorous individuals which are known to have a higher survival rate because of their larger size (Forney 1976; Willemsen 1977).

That the food situation for the adult pikeperch in Slotermeer stayed poor from the moment of our observation in October 1978 until the beginning of the following summer is confirmed by the observations of Willemsen (1979). He found that of 6 June 1979 only 23% of the pikeperch contained food, mostly *Neomysis integer* and chironomid larvae. He also found the relative condition of the pikeperch in Slotermeer on that date to be 80% of that of a reference sample caught mainly in IJsselmeer. The immigration of smelt larvae improved the food situation during the summer of 1979 and the condition of the adult pikeperch in September was 96% of the earlier mentioned reference sample. The resultant food situation at the end of the summer was poor, because the limited stock of 0+ fish was intensively exploited. The condition of the pikeperch in Slotermeer, however, must have improved further during the summer of 1980 when an extensive amount of larvae immigrated into the lake. The resultant stock of prey fish in September 1980 was large.

The success of fishing methods using bait fish depends on the abundance of prey fish, being depressed with large amounts of 0+ fish. Forney (1980) showed the importance of food abundance in judging possible angling success for walleye in Oneida Lake, New York. With about the same number of walleye in 1958 and 1959 the angling success in terms of man hours to catch walleye was five times better in 1959 than in 1958. The hook-and-line fishery of the commercial fishermen in Friesland is directed at catching eel with bait fish, which fish they collect from the fykes set for eel and tench, *Tinca tinca.* Pikeperch was repeatedly caught by the hook-and-line method in late summer 1979 at the low densities of 0+ prey fish which occurred (pers. comm. J. Poepjes, commercial fisherman, Oosterzee). This was not the case in 1980 when 0+ prey fish was abundant and the vulnerability of pikeperch to angling was greatly depressed.

The inverse relation between the density of 0+ fish and the mean size of *Daphnia* can be applied also to the inverse relation between abundance of 0+ prey fish and the vulnerability of pikeperch to angling. From the data gathered in the Frisian lakes in 1979 and 1980 it is concluded that a mean *Daphnia* size of over 0.9 mm in September guarantees good fishing quality in terms of vulnerability of pikeperch to angling in the succeeding period of the angling season.

The numbers of large *Daphnia* were used as an index for fishing quality by Galbraith (1975) for rainbow trout, *Salmo gairdneri,* in small lakes in Michigan, USA. Johnson & Hasler (1954) showed that the growth of rainbow trout depends on the amount of zooplankton present, from which the rainbow trout consumes *Daphnia* larger than 1.3 mm (Galbraith 1967). Galbraith (1975) first indexed the fishing quality of a number of small lakes with the aid of questionnaires in subjective terms and in terms of the number of rainbow trout caught per gill net set overnight. He found a possitive correlation between this index and the numbers of large *Daphnia* present in the lake. This application of the numbers of large *Daphnia* as an index of fishing quality is fundamentally different from ours. We used mean *Daphnia* size, being related to the numbers of large *Daphnia,* as an index for the vulnerability to angling of a piscivorous predator irrespective of the stock size of this predator. Anyway, no sharp decrease in the numbers of pikeperch caught per gill net set in Tjeukemeer was observed from 1979 to 1980. Galbraith (1975) used the numbers of large *Daphnia* as an index of the amount of food available to a planktivorous fish and so for its sustainable stock, irrespective of its vulnerability to angling.

Lammens (1982) observed that, since the standing stock of zoobenthos in Tjeukemeer is low, older bream has a dominant planktivorous feeding behaviour, when zooplankton is largely available.

The condition of the bream improved in 1979, when the availability of large *Daphnia* was good but the condition worsened in 1980 when availability was poor. For adult pikeperch quite the opposite trend was observed in 1979 and 1980. The causative agent for these opposite trends was the density of 0+ fish. This density was much higher in 1980 as a consequence of the succesful immigration of smelt and perch larvae from the IJsselmeer.

Acknowledgements

The investigations were partially financed by grant no. 14-85.09 of the Foundation for Fundamental Biological Research (BION), which is subsidized by the Netherlands Organization for the Advancement of Pure Research (ZWO).

We wish to thank Mr. J. Willemsen for his critical comment on the manuscript and the permission to use his data for constructing Fig. 9. We are also indebted to skipper L. Lemsma, and to Mrs. A. Frank-Landman and P.McGillavry for practical assistance. Miss A. Schoon, J. Meischke and Miss A. Kroon participated in the research during their study at the Agricultural University of Wageningen. Finally we wish to thank Miss M. McLaughlin who helped considerably with the English.

References

Anonymous, 1975. Annual Report, Ministery of Agriculture and Fishery. Directorate of Fisheries, The Netherlands. Visserij 29: 244–368.

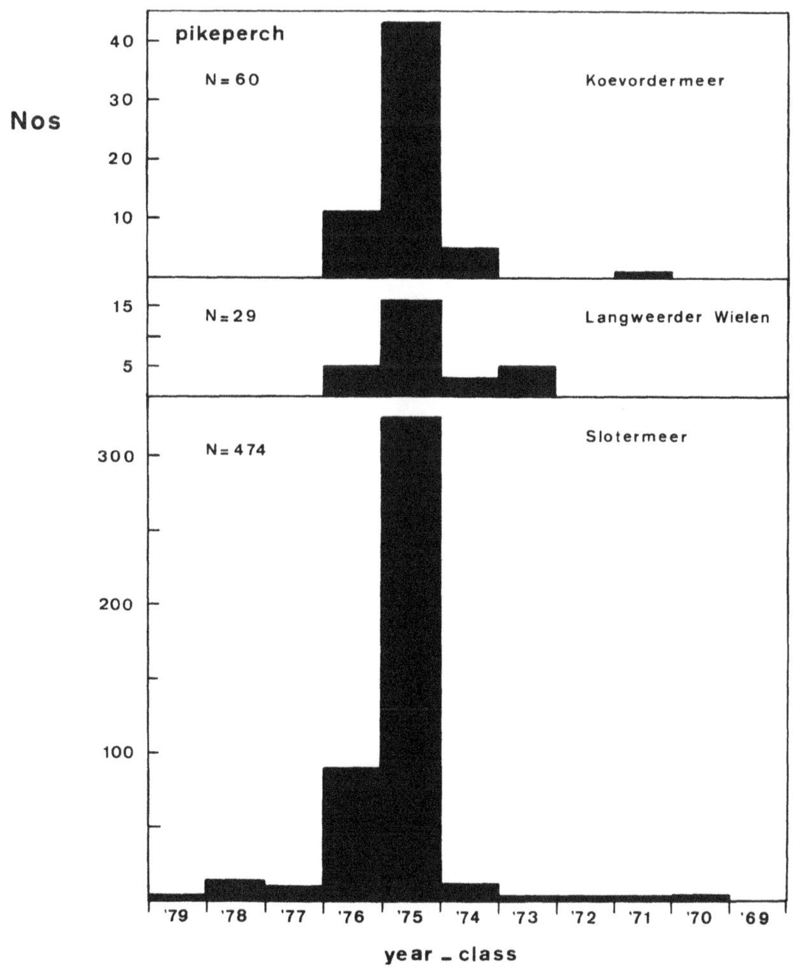

Fig. 9. Relative year-class strength (Nos. in sample) of pikeperch in Slotermeer in September 1979 (after data of Willemsen (1979)).

Anonymous, 1979. Annual Report, Ministery of Agriculture and Fishery. Directorate of Fisheries, The Netherlands. Visserij 33: 184–322.

Beattie, D. M., Golterman, H. L. & Vijverberg, J., 1978. An introduction to the limnology of the Friesian Lakes. Hydrobiologia 58: 49–64.

Densen, W. L. T. van & Hadderingh, R. H., 1982. Effects of entrapment and cooling water discharge by the Bergum Power Station on 0+ fish in the Bergumermeer. Hydrobiologia 95: 351–368.

Eijk, M. van der, 1981. Visserijkundige waarnemingen in de friese boezemwateren in de jaren 1977–1979. Visserij 34: 21–32.

Fonds, M., 1978. The seasonal distribution of some fish species in the western Dutch Wadden Sea. In: Dankers, N., Wolf, W. J. & Zijlstra, J. J. (Eds.). Fish and Fisheries of the Wadden Sea, pp. 5/42–5/77. Report no. 5 of the Wadden Sea Working Group.

Forney, J. L., 1976. Year-class formation in the walleye, Stizostedion vitreum vitreum, population of Oneida Lake, New York, 1966–1973. J. Fish. Res. Bd Can. 33: 783–792.

Forney, J. L., 1980. Evolution of a management strategy for the walleye in Oneida Lake, New York. N.Y. Fish Game J. 27: 105–141.

Galbraith, M. G., Jr., 1967. Size-selective predation on Daphnia by rainbow trout and yellow perch. Trans. Am. Fish. Soc. 96: 1–10.

Galbraith, M. G., Jr., 1975. The use of large Daphnia as indices of fishing quality for rainbow trout in small lakes. Verh. int. Verein Limnol. 19: 2485–2492.

Johnson, W. E. & Hasler, A. D., 1954. Rainbow trout production in dystrophic lakes. J. Wildl. Mgmt. 18: 113–135.

Kott, P., 1953. Modified whirling apparatus for subsampling of plankton. Aust. J. mar. Freshwat. Res. 4: 387–393.

Lammens, E. H. R. R., 1982. Growth, condition and gonad development of bream (Abramis brama L.) in relation to its feeding conditions in Tjeukemeer. Hydrobiologia 95: 311–320.

Ricker, W. E., 1973. Linear regressions in fishery research. J. Fish. Res. Bd Can. 30: 409–434

Ricker, W. E., 1973. Linear regressions in fishery research. J. Fish. Res. Bd. Can. 30: 409–434.

Vijverberg, J., 1977. Population structure, life histories and abundance of copepods in Tjeukemeer, The Netherlands. Freshwat. Biol. 7: 579–597.

Vijverberg, J. & Richter, A. F., 1982. Population dynamics and production of Daphnia hyalina Leydig and Daphnia cucullata Sars in Tjeukemeer. Hydrobiologia 95: 235–259.

Willemsen, J., 1977. Population dynamics of percids in Lake IJssel and some smaller lakes in The Netherlands. J. Fish. Res. Bd Can. 34: 1710–1719.

Willemsen, J., 1979. Voedsel, groei en jaarklassterkte van snoekbaars in enkele friese wateren. Rapport ZS 79–03. Rijksinstituut voor Visserijonderzoek, IJmuiden. 9 pp.

Effects of thermal effluents from the Bergum Power Station on the zooplankton in the Bergumermeer

H. W. de Nie
Limnological Institute, Tjeukemeer Laboratory, De Akkers 47, 8536 VD Oosterzee, The Netherlands

Keywords: thermal pollution, eutrophic lake, copepods, cladocerans

Abstract

The influence of thermal discharge from a 640 MW steam electric power station on the Bergumermeer (shallow, eutrophic lake) was investigated. The temperature of the thermal effluent did not exceed 30 °C. The lake was sampled extensively from 1975 to 1978. Zooplankton densities at different depths in the heated and non-heated parts of the lake were compared. A vital staining technique was used to assess entrainment mortality of cladocerans. The entrainment mortalities for the most abundant cladocerans were only 2 to 3% but for *Leptodora kindtii* 19% and about 25% for *Diaphanosoma brachyurum, Ceriodaphnia pulchella* and the cyclopoid nauplii. The ecological consequences for the lake ecosystem are discussed.

Introduction

In 1974 a 640 MW electric power station near Bergum started production. Water from the shallow eutrophic Bergumermeer was used for cooling purposes. The effects of the thermal discharge on the Bergumermeer were extensively investigated. The physical and chemical environment was studied by Allersma (1969), Koops (1981) and Van Wakeren (1979), the rotifers by Donze (1978), the macrofauna by Hadderingh (1975) and Beattie & de Kruijff (1978), and the fish by Schaap & Willemsen (1978), Hadderingh (1978, 1979) and van Densen & Hadderingh (1982). This paper deals with the crustacean zooplankton and is part of a study on plankton (de Nie 1979; Wanders 1979). Major findings have already been reported as an Internal Report (Wanders *et al.* 1980).

Study area

The lake's morphometry and location are summarized in Table 1. The Bergumermeer is connected with the other Frisian lakes by a main canal (Prinses Margriet Kanaal) and a subsidiary canal (De Lits) (Fig. 1). The chemistry of the lake water is, therefore, similar to that of the other Frisian lakes (Beattie *et al.* 1978).

During winter, up to 100×10^6 m^3 month^{-1} flows into the lake via the western part of the main canal. It flows out mainly via the power station to the discharge canal to the north. The retention time in the area north of the main canal is less than 1 day but in the southern part is about 10 days.

During spring and summer, 10 to 40×10^6 m^3

Table 1. Some morphological characteristics of the Bergumermeer.

Geographical position	53°12′N; 6°02′W
Elevation	0.5 m below sea level
Area	4.64 km^2
Volume	6.0×10^6 m^3
Max. length	3.5 km
Max width	1.9 km
Mean depth	1.3 m
Max. depth	3.5 m
Shore length	16.4 km (dam included)

Hydrobiologia 95, 337–349 (1982). 0018-8158/82/0953-0337/$02.60.

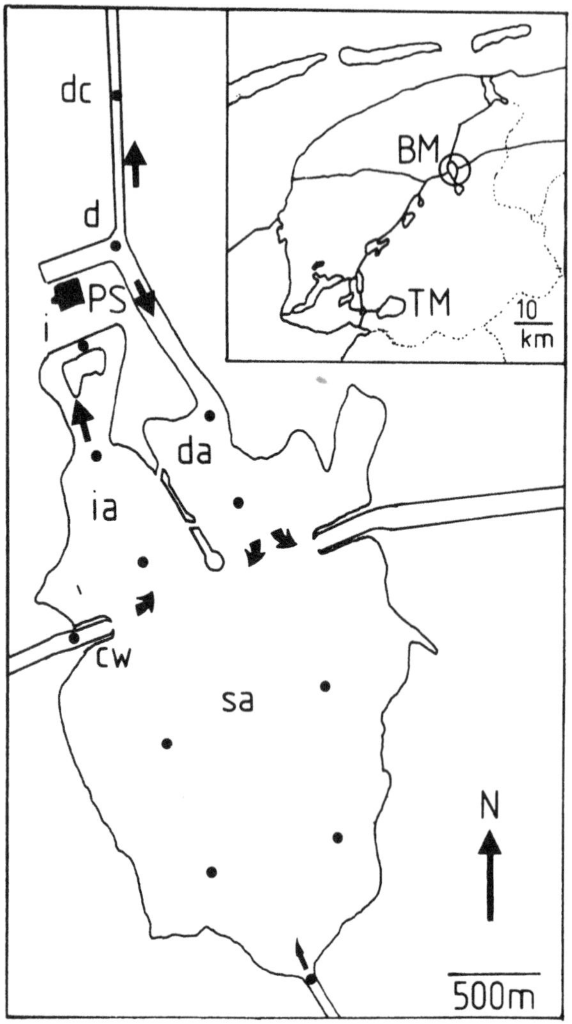

Fig. 1. Map of the Bergumermeer, showing the position of the power station (PS) and the sampling sites (●). The direction of the main water flows are indicated by arrows. dc = discharge canal; d = discharge; da = discharge area; i = intake; ia = intake area; cw = Princes Margriet Canal at West side; and sa = southern part of the lake with subsidiary canal (De Lits). Inset, top right: Map of Friesland. TM = Tjeukemeer and BM = Bergumermeer.

month^{-1} flows out of the lake via the eastern side of the main canal. About 40 to 90% of the water recirculates around the dam between the intake and discharge point. Complete recirculation happens occasionally, during severe frost periods. Incomplete recirculation is most frequent. Then, the retention time in the northern part is 2 to 3 days and in the southern part 3 to 6 weeks.

The lake's hydrology was modified in 1971–1972 due to the construction of a dam and rechanneling of the discharge canal in the north to make the water follow a longer route between intake and discharge points. Water enters the power station in the N.W. side of the lake (Fig. 1). The maximum water intake capacity is 27.8 $m^3 s^{-1}$ (73×10^6 m^3 month^{-1}). The water is filtered through 5 mm meshed rotating drum screens before entering the cooling water pumps and is heated for 4 s in the condenser. The total retention time in the power station is 5 min. The water cascades into the discharge canal over a concrete wall, 1.5 m above the water level.

The mean difference in water temperature between the intake and discharge (ΔT) is 4.5–5.3 °C. During incomplete recirculation 2% of the lake surface is heated more than 4 °C, and 4, 6.5 and 23% is heated more than 3, 2 and 1 °C, respectively. The lake water which passes through the power station cools down to 50% of the ΔT in 4–5 h; after 10–15 h the ΔT is less than 1 °C. Reheating starts after 20–24 h. In the discharge canal in the north, the water hardly cools at all; 15 km downstream from the power station, we measured 4 °C (ΔT) when the main flow direction was to the north.

Material and methods

The influence of the cooling water was studied for the most abundant species (Table 2).

Field work was carried out between September 1974 and September 1978. We sampled at weekly intervals during spring and summer, but at fortnightly intervals from November to March. There were 13 sampling stations (Fig. 1). After May 1976 three points in the southern part of the lake were omitted. Additional samples were taken at different depths near the discharge and during winter in the discharge canal between the power station and the sluices 18 km downstream to the north.

Samples were taken with a Friedinger-sampler (volume 5 l; height 60 cm). The zooplankton samples were concentrated by filtration through a 80 μm filter and preserved in 4–5% formaline. These samples were subsampled with the whirling apparatus of Kott (1953). In 1978 *Leptodora* samples were taken using a 335 μm plankton net, which was horizontally hauled for 15, 30 or 60 s at two different depths separately near intake and discharge. In 1976 we sampled in the foam layers near

Table 2. List of the most abundant crustacean zooplankton species observed in the Bergumermeer (1974–1978).

Cladocera	Copepoda
Bosmina coregoni Baird	*Cyclops vicinus* (Uljanin)
Bosmina longirostris (O. F. Müller)	*Acanthocyclops robustus* (Sars)
Chydorus sphaericus (O. F. Müller)	*Mesocyclops leuckarti* (Claus)
Daphnia hyalina Leydig	*Eurytemora affinis* (Poppe)
Daphnia cucullata Sars	
Ceriodaphnia pulchella Sars	
Diaphanosoma brachyurum (Lieven)	
Leptodora kindtii (Focke)	

the discharge of the power station and in the wind induced foam far from the point of discharge. An area of 180 cm^2 was sampled by means of a funnel. Results are expressed as proportional cover of the water surface:

$$\frac{N_i \times s_i}{180} \times 100\%,$$

in which N_i = total number i^{th} species and s = its mean individual area in cm^2. At moderate wind speed a rough and perhaps too high estimate was made of the total quantity of foam produced by the power station (3.3×10^5 cm^2 day^{-1}). Thus the proportion of zooplankton caught in the foam layer after passing through the power station could be estimated.

The vital staining experiments were conducted on six different dates between 18 May and 22 June 1977. Two 5 l samples were taken near the intake and one from the discharge cascade. They were stored in 2 l bottles. A pair containing a sample each from discharge and intake was hung in the lake near the intake for 3 to 3.5 h. Similarly, another pair was hung near the discharge cascade. After this a carmine solution (1:100 000 after dilution) was added. The samples were filtered after 17.5 to 22.5 min and concentrated in 4–5% (acetate buffered) formaline. The ΔT during the six experiments ranged between 4 and 6 °C; the water temperature varied between 15.5 °C on 1 June at the intake station and 24.9 °C on 15 June in the discharge cascade.

The samples were counted as soon as possible. Cladocerans having no carmine in their guts were considered as dead. The results varied greatly; therefore, a large number of individuals had to be counted to assess mortality percentages.

By the computation of mean proportional near-surface density all the non-zero counts, expressed as number of individuals per litre, were log (n + 1) transformed. The mean proportional near-surface density for a certain species is defined as:

$$\exp \bar{y}_j \times 100\%,$$

in which:

$$\bar{y}_j = \frac{\sum_{t=1}^{N} y_{jt}}{N},$$

$$y_{jt} = \ln(n_{jt} + 1) - \ln(n_{It} + 1),$$

N = number of sampling dates,

n_{jt} = individuals per litre on date t, sampling station j,

n_{It} = individuals per litre on date t, near intake.

The variable y was considered as a random normally distributed stochastic variable; 95% confidence limits for the mean value were calculated.

Results from additional sampling at different depths were not log-transformed. Then the proportional vertical density means that the near-surface density at that station is 100%, and the proportional near-surface density of a certain station means that the near-surface density near intake is 100%.

Although sometimes slightly heated, the intake area represented a good reference station, being situated midway along the cooling water trajectory. By assuming the intake zooplankton density to be 100%, the effects immediately after entrainment through the power station were shown more clearly.

Results

The mean proportional near-surface densities are given in Fig. 2. The most significant reduction, 20–60% was found at the discharge point. The proportional near-surface densities were compared with results from additional observations on the vertical migration. Where the proportional near-surface density was <100%, the vertical distribution was less homogeneous (Fig. 3). The high reduction in numbers near the surface at the discharge point can be explained by a downward movement of the cladocerans.

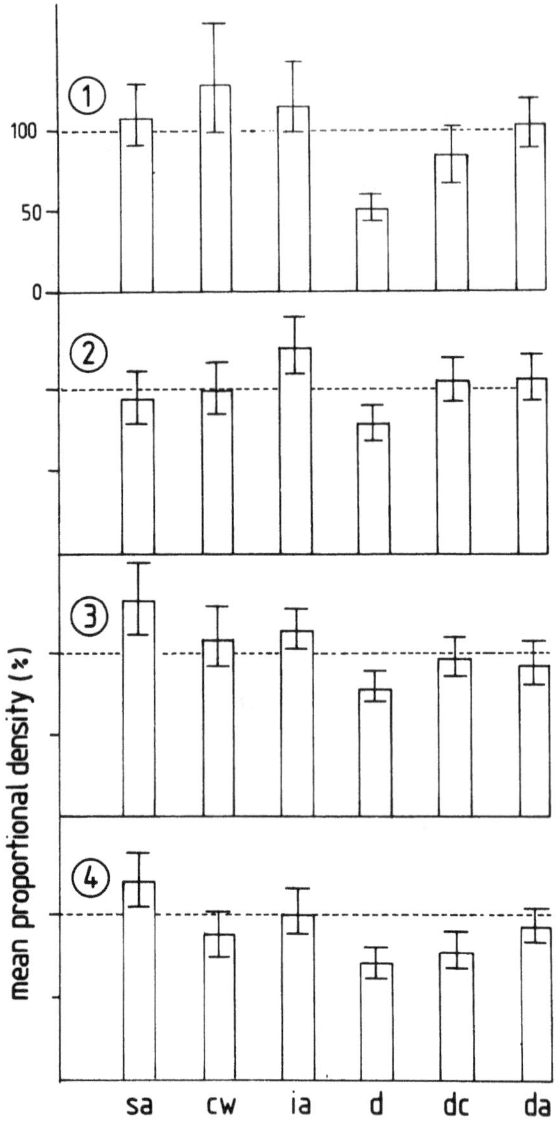

Fig. 2. Mean proportional cladoceran densities (intake = 100%) in different areas of the Begumermeer (see Fig. 1); 95% confidence intervals are indicated. 1. *Bosmina coregoni;* 2. *B. longirostris;* 3. *Chydorus sphaericus;* 4. *Daphnia* spp.

Mortality assessed directly by vital staining was low. An increase of 0.3 to 1.7% due to heating resulted in 2.0 to 3.4% mortality (Table 3). Entrainment plus thermal shock followed by a rapid cooling down resulted in 3.0–4.3%. Entrainment and thermal shock not followed by cooling resulted in 4.5–4.9% mortality; a 1.8 *(D. hyalina)* to 2.9 *(C. sphaericus)* factor increase of its control mortality. The number of crustacean zooplankton caught in the foam layers near the discharge cascade was also a means of direct mortality assessment. The cover percentages ranged from 4 to 160% (more than one layer). The mean percentage during summer was 35%; the maxima per species were 40% for *Daphnia hyalina* and *Bosmina* spp. and 90% for *C. sphaericus.* Cover percentages of 0.03 to 2% for cladocerans were found in wind induced foam. The estimated proportion of cladocerans killed in foam in relation to the total numbers passing through the power station did not exceed 1%. No copepods were found in the foam layers.

Between 24 May and 5 June 1977 the thermal effluent recirculated almost completely. The proportion of adult *Daphnia hyalina* females and the egg number per female were measured in both the heated and the non-heated areas (Table 4). Before recirculation the proportion of adult females hardly differed in the heated and non-heated parts. During recirculation, however, their proportion in the heated area decreased from 52 to 35.8%. On 9 June, after recirculation, this decrease continued due to the inflow of water from the main canal. During recirculation there was little transport of water through the main canal. The number of eggs per female decreased in both the heated and the non-heated parts of the lake.

The available data for *C. pulchella* and *D. brachyurum* were sufficient for calculation of the mean near-surface densities (Fig. 4) but too few for calculations of the proportional near-bottom densities. At 700 m from the power station the near-surface numbers of *C. pulchella* in the discharge canal were reduced by 41%, but nearer to the discharge point the reduction was not significant. For *D. brachyurum* the reduction was 25% in the whole discharge area, while all the other cladocerans in this area had densities comparable to those in the intake. This reduction of *C. pulchella* and *D. brachyurum* was probably caused by mortality.

The seasonal variations in densities of *Leptodora kindtii* in 1978 near the power station are shown in Fig. 5. Near the bottom, at the discharge sampling point, the proportional density ranged from 19 to 170% (mean 63%). Near the water surface the proportional densities ranged from 7.6 to 175% (mean 101%). According to the Wilcoxon paired-sample test the near-bottom samples at the discharge point differed significantly ($P = 0.025$) from those at the intake. The near-surface samples did not differ ($P > 0.25$). At both sampling points the densities near the

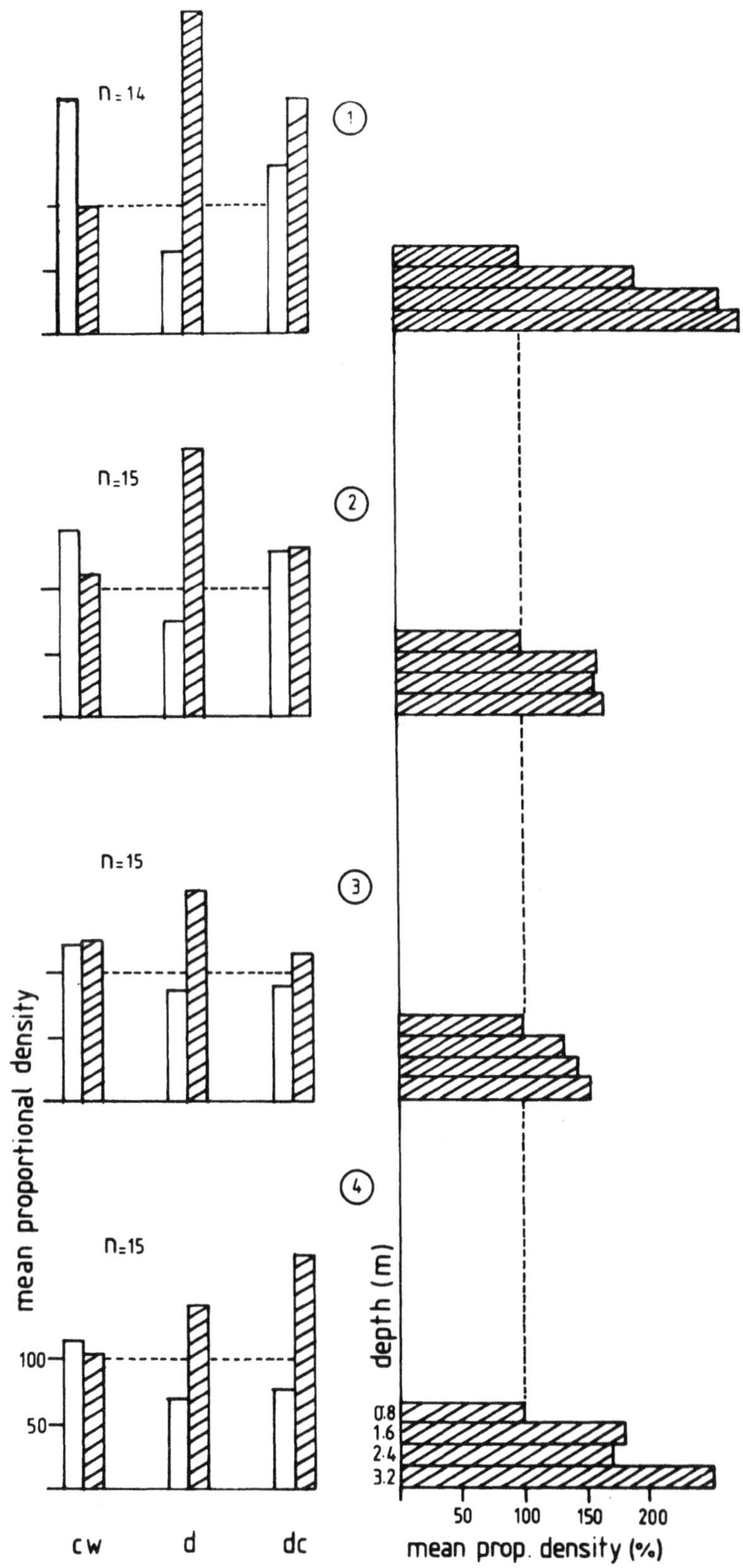

Fig. 3. Proportional distribution of some cladocerans. Right, the proportional vertical densities near the discharge (surface = 100%). Left, shaded, the proportional near bottom density (surface = 100%) and non-shaded, the proportional surface density (intake 100%) at three different sampling stations (see Fig. 1). 1. *Bosmina coregoni;* 2. *B. longirostris;* 3. *Chydorus sphaericus;* 4. *Daphnia* spp.

Table 3. Mortality percentages (in 3.5 h) as a result of vital staining. Number examined in parentheses.

Experiment	Species			
	Bosmina coregoni	*Bosmina longirostris*	*Chydorus sphaericus*	*Daphnia hyalina*
1. Control, inlet temp.	1.7 (425)	2.7 (1 889)	1.7 (1 797)	2.5 (1 244)
2. Inlet water, heated to discharge temp.	2.0 (353)	3.7 (1 443)	3.4 (1 291)	2.8 (1 010)
3. Discharge water, cooled down to inlet temp.	3.0 (336)	4.0 (1 224)	3.6 (1 122)	4.3 (862)
4. Discharge water, not cooled down.	4.5 (313)	4.9 (1 335)	4.9 (1 105)	4.5 (995)
P (χ^2 test)	<0.10	<0.005	<0.001	<0.01

Table 4. Percentage of adult females and number of eggs per adult female of *Daphnia hyalina* before, during and after recirculation of the heated effluent. Number examined in parentheses.

		% adult females		No. of eggs per adult female	
		North (heated)	South	North (heated)	South
Before recirculation	23 May	24.3 (206)	22.6 (115)	5.0 (34)	5.2 (9)
During recirculation	26 May	52.0 (250)	49.7 (175)	5.2 (50)	4.7 (30)
	1 June	35.8 (227)	50.0 (158)	3.4 (45)	3.4 (40)
After recirculation	9 June	25.1 (255)	46.0 (124)	2.3 (31)	2.7 (27)

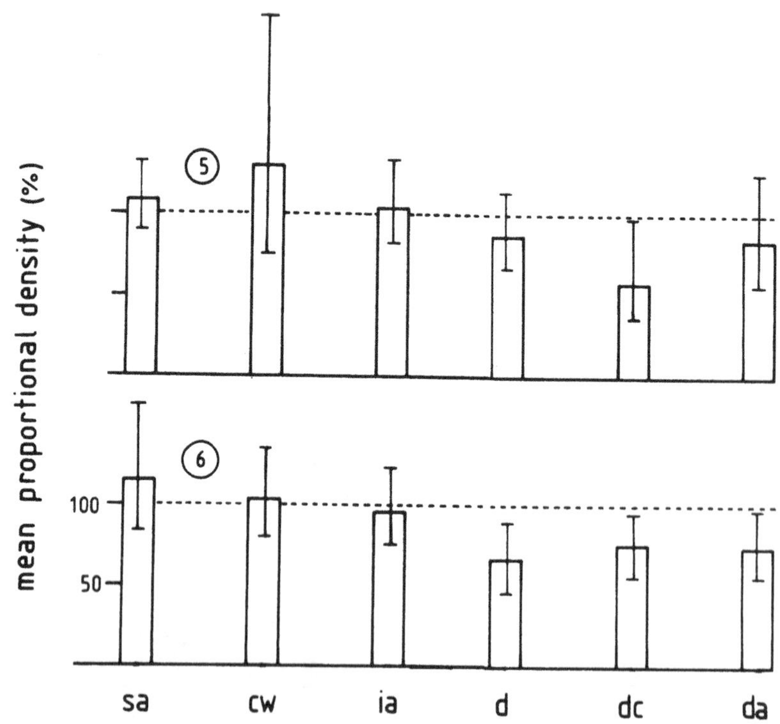

Fig. 4. Mean proportional cladoceran densities (intake = 100%) at different areas in the Bergumermeer (see Fig. 1); 95% confidence intervals are indicated. 5. *Ceriodaphnia pulchella;* 6. *Diaphanosoma brachyurum.*

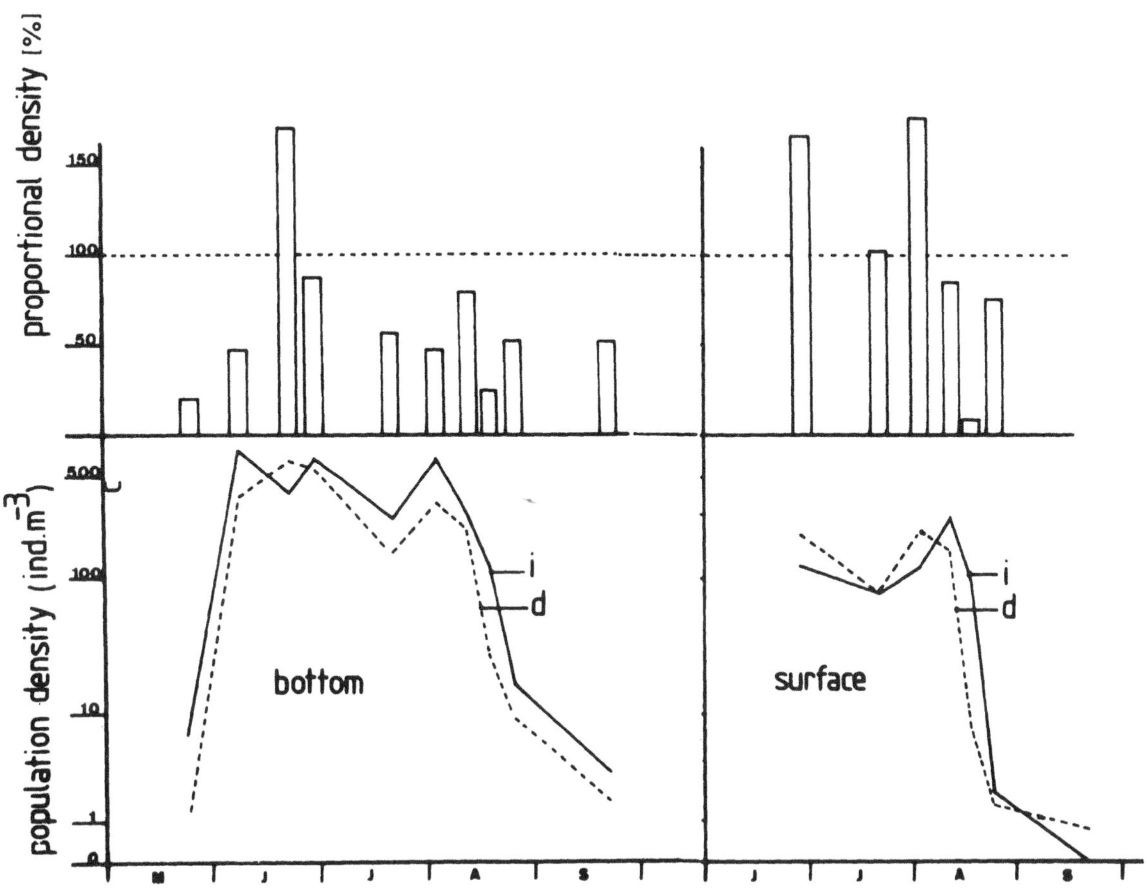

Fig. 5. Population density of *Leptodora kindtii;* results from near-bottom (left) and near-surface sampling (right). Below: numbers near intake (—) and discharge (---). Top: proportional densities (intake = 100%).

bottom were higher. We calculated the proportional difference in density for the whole water column and found a reduction of 19%, caused undoubtedly by entrainment mortality.

In Fig. 6 the mean proportional near-surface densities are given for nauplii and the copepodite instars of *E. affinis* and cyclopoids (mainly *A. robustus* and *C. vicinus*). The reduction in numbers was significant for the cyclopoids. At 700 m from the power station the mean proportional near-surface density of the copepodites was <100%, but densities were higher near the bottom. The vertical distribution of the nauplii was homogeneous near the power station (Fig. 7). The data for the nauplii have been split into two sets: 16 samples at relatively high ΔT and 16 at low ΔT (Fig. 8). The reduction in the mean proportional near-surface density was less at lower ΔT. In the discharge canal, far up to the north, the proportional near-surface density decreased from 78 to 56%; the proportional vertical density remained about 100% (Fig. 9). The results were tested applying the Friedman rank sum test for ordered alternatives (Hollander & Wolfe 1973). The proportional vertical density did not differ ($P > 0.10$) but there was strong evidence for a decrease of 22 to 44% in numbers with increasing distance from the power station ($P < 0.05$). For *E. affinis* the near-surface distribution did not differ significantly at the discharge point. In the discharge area, the mean proportional near-surface density was about 100% (Fig. 6).

The reduction in near-surface density of the copepodite instars can be explained by downward movement. No evidence for such a movement was found for nauplii. Thus a reduction of 25% should be due to mortality in the discharge area. This mortality increased in winter to 44% in the discharge canal.

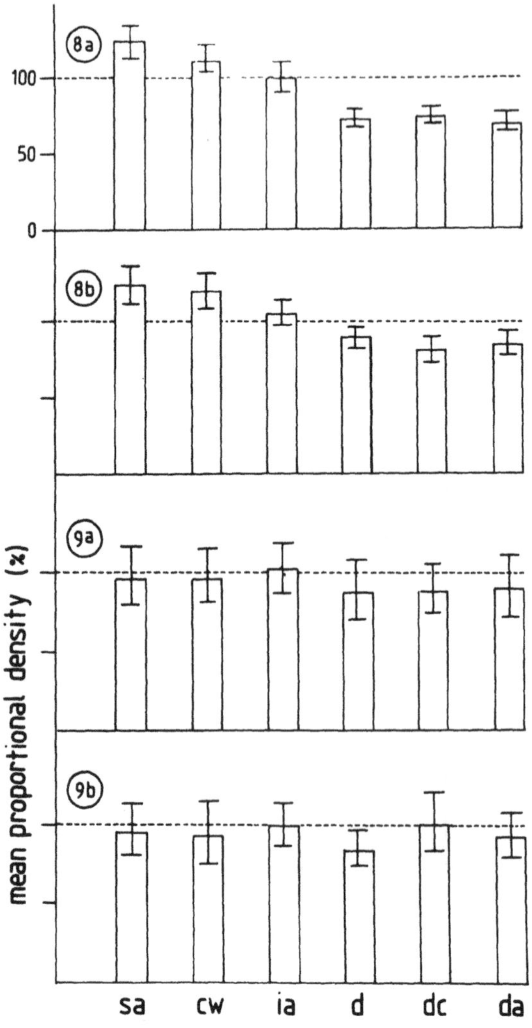

Fig. 6. Mean proportional copepod densities (intake = 100%) at different areas in the Bergumermeer (see Fig. 1); 95% confidence intervals are indicated. 8. cyclopoids; 9. *Eurytemora affinis* (a: naupliar instars, b: copepodite instars).

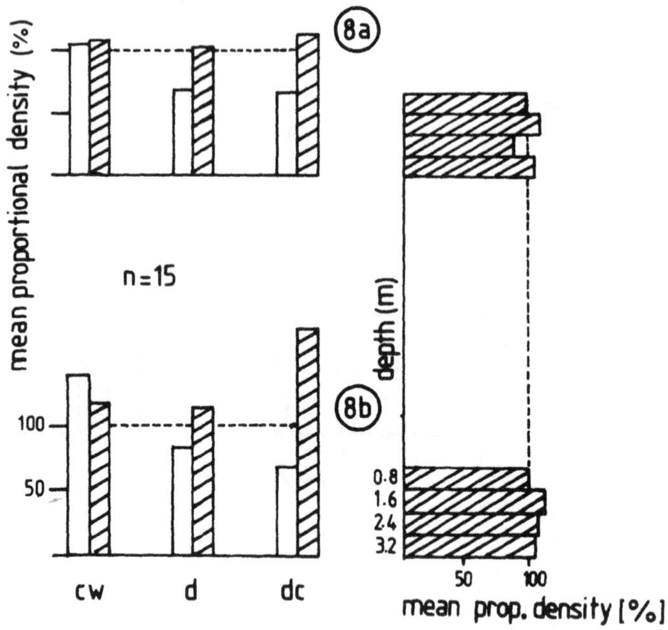

Fig. 7. Proportional distribution of cyclopoid copepods. Right, the proportional vertical densities near the discharge (surface = 100%). Left, shaded, the proportional near bottom densities (surface = 100%) and non-shaded, the proportional near surface densities (intake = 100%) at three sampling stations (see Fig. 1). 8a: naupliar instars; b. copepodite instars.

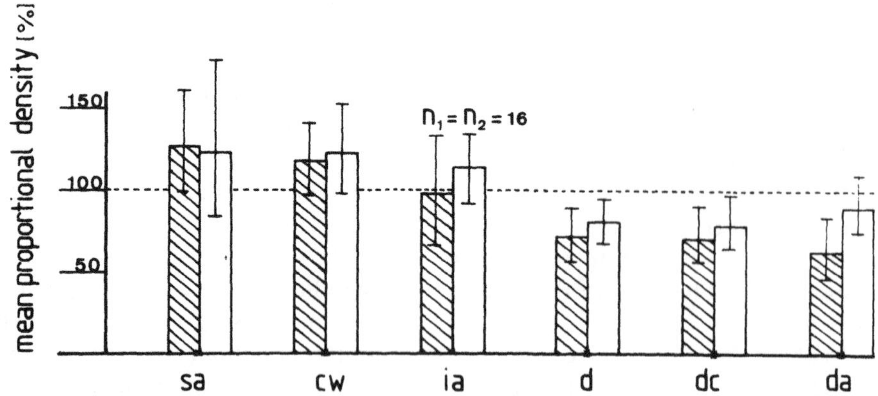

Fig. 8. Mean proportional densities (intake = 100%) of naupliar instars of cyclopoids at different areas in the Bergumermeer (see Fig. 1). Shaded, for high ΔT and non-shaded, for low ΔT.

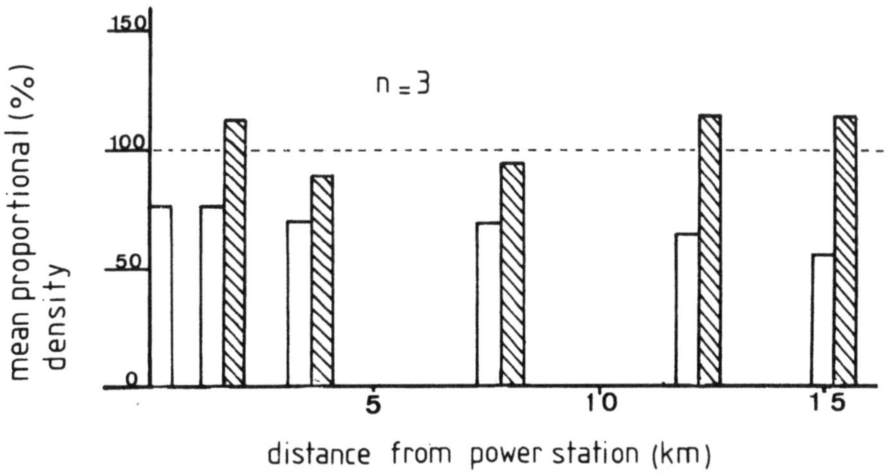

Fig. 9. Mean proportional densities of naupliar instars during winter (1976–1977) in the discharge canal, in relation to the distance downstream from the power station. Non-shaded, proportional near-surface density (intake = 100%) and shaded, proportional near-bottom density (near-surface = 100%).

Discussion

Effects near the power station

The use of lake water for cooling purposes has two direct effects on the zooplankton in the lake near the discharge: a disturbance of the vertical distribution and entrainment mortality. *B. coregoni, B. longirostris, Daphnia* spp. and *C. sphaericus* move to deeper water near the discharge of the power station. About 700 to 1 000 m from the discharge point their vertical distribution is homogeneous, as in Tjeukemeer (de Nie *et al.* 1980). Brauer *et al.* (1974) and Carpenter *et al.* (1974) also found changes in the vertical distribution near the discharge of power stations. Gehrs (1974) proved experimentally that a slight increase in temperature stimulated the downward movement of *Diaptomus sanguineus* and *Daphnia parvula.* Thus these changes in the vertical distribution are probably caused by a thermal shock. On the other hand, for *L. kindtii* the difference between near surface and near bottom density was smaller near the discharge point. In this case the population was probably mixed due to water turbulence, because near the intake *L. kindtii* was concentrated near the bottom.

A high entrainment mortality of 19% was found for *L. kindtii.* Bunting (1974) reported 90% for *L. kindtii* near a similar power station. Bunting also mentioned that *Diaphanosoma leuchtenbergianum* is very vulnerable to entrainment. Herzig (1974) found a high correlation between death rate and wind speed (water turbulence) for this species. Thus it is likely that the decrease in *D. brachyurum* in the whole discharge area of the Bergumermeer was due to mortality. In a study similar to ours, Davies & Jensen (1974) mentioned 6% entrainment mortality for *Bosmina* and *Daphnia.* This is of the same order of magnitude as our findings (about 3%) for these genera. Higher mortalities are reported near power stations with high ΔT and temperatures of 32–35 °C in the discharge area (Heinle 1969; Suchanik & Grossman 1971 cited in Carpenter *et al.* 1974; Brauer *et al.* 1974; Carlson 1974; Davies & Jensen 1974; Goss & Bunting 1976; Donze 1978).

According to Davies *et al.* (1976) the copepods were less vulnerable to entrainment. However, Benda & Gulvas (1976) and Storr (1974) reported copepod mortalities ranging from 7 to 18% and Carpenter *et al.* (1974) up to 70%, five days after entrainment. We found 25% mortality for the naupliar instars but found no significant mortality for the copepodite instars.

The observed entrainment mortality for *L. kindtii, C. pulchella, D. brachyurum* and the cyclopoid nauplii were the highest. The death rate for natural populations of *Daphnia,* reviewed by Hebert (1978), ranged from 10 to more than 25% per day during the growing season. Assuming the same order of magnitude for the other crustaceans, the highest entrainment mortalities occur around the upper limit of the reported natural mortality. The en-

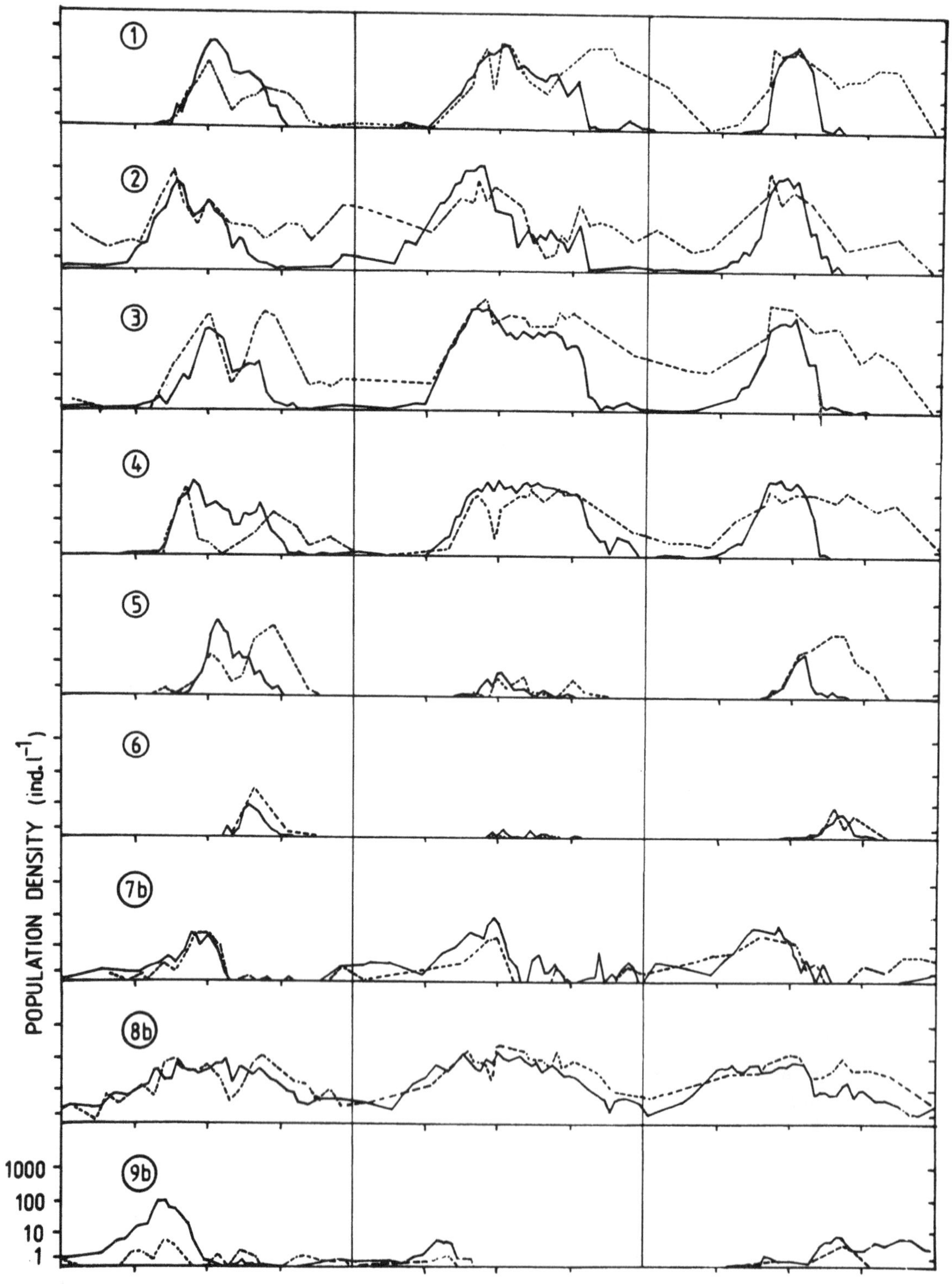

Fig. 10. Composition of the seasonal variation in population densities of cladocerans and copepods in Bergumermeer (—) and Tjeukemeer (---). 1. *Bosmina coregoni;* 2. *B. longirostris;* 3. *Chydorus sphaericus;* 4. *Daphnia* spp.; 5. *Ceriodaphnia pulchella;* 6. *Diaphanosoma brachyurum;* 7b. copepodite instars of *Cyclops vicinus;* 8b. copepodite instars of cyclopoids (lumped together); 9b. copepodite instars of *Eurytemora affinis.*

trainment mortalities for *B. coregoni, B. longirostris, Daphnia* spp. and *C. sphaericus* (about 3%) are much lower than the reported minimum rate.

Effects on the zooplankton abundance in the lake

We evaluated the indirect effects of the power station on the abundances of the different zooplankton species by comparing data from Bergumermeer with Tjeukemeer both before and after the power station was built (i.e. 1973). Between 1970 and 1972, Beattie *et al.* (1978) observed that *Mesocyclops leuckarti* and all the cladocerans attained lower densities in Bergumermeer than in Tjeukemeer, while *E. affinis* was more abundant in Bergumermeer. Between 1975 and 1977 the crustacean zooplankton in Bergumermeer did not differ from that in Tjeukemeer either in species composition, the period of increase, or the time of maximum abundance. *E. affinis* remained more abundant in Bergumermeer (Fig. 10). In autumn lower densities than in Tjeukemeer were found in the Bergumermeer, also in parts not affected by the thermal discharge. This is in agreement with the findings of Whitehouse (1971). He studied the crustacean zooplankton in the partly heated Lake Trawsfynydd in Wales and found no effect on the wax and wane of the most abundant species. Only authors working on rotifers have been able to report changes in the open water zooplankton community in several types of waterbodies (Lanner & Pejler 1973; Horkan *et al.* 1977; Lair 1980; Leentvaar 1980).

Possible consequences of the zooplankton entrainment mortality for the 0+ fish could not be detected. *L. kindtii* is a preferred food item for 0+ smelt *(Osmerus eperlanus)*, perch *(Perca fluviatilis)* and pikeperch *(Stizostedion lucioperca)* in Tjeukemeer (van Densen, in prep.). However, these fish species did not show a limited growth in Bergumermeer (van Densen & Hadderingh 1982). Cyclopoid nauplii are essential food items for fish larvae which have just hatched, but there was no indication of lower recruitment of fish larvae in comparison with Tjeukemeer (van Densen & Hadderingh 1982).

The abundance of zooplankton in the open water of Bergumermeer is influenced by three factors: the water movements, the fish predation and heating by the power station. These factors are interdependent. The power station causes heating and water movements. According to Richter's mathematical simulation model (Wanders *et al.* 1980), heating alone increases the productivity of the *Daphnia* population, especially in spring. Heating and water movements directly influence the distribution of 0+ fish, especially smelt (Hadderingh 1978, 1979; van Densen & Hadderingh 1982), while fish predation is an important factor on the population dynamics of larger zooplankton species, especially *Daphnia* spp. (van Densen & Vijverberg 1982).

Generally we can conclude that although the zooplankton in the discharge area was clearly affected by the Bergum power station, the influence on the abundance of crustacean zooplankton in the open water zone of the lake was small and difficult to distinguish from other factors acting on the abundance.

Summary

The influence of thermal discharge from a 640 MW electric power station on the Bergumermeer, a eutrophic lake with a surface area of 4.64 km^2 and mean depth of 1.3 m, was investigated. The lake has a complex hydrology. The temperature increase ΔT averaged from 4.5 to 5.5 °C. Only 6.5% of the surface area of the lake showed an increase of >2 °C due to thermal discharge.

Near the discharge *Bosmina longirostris, B. coregoni, Chydorus sphaericus, Daphnia hyalina* and *D. cucullata* and the copepodite instars of cyclopoids, mainly *Acanthocyclops robustus*, showed higher densities near the bottom. At 0.7 to 1.0 km from the power station the number of organisms found near the water surface and the bottom were similar.

A vital staining technique was used to assess entrainment mortality. For *B. longirostris, B. coregoni, Daphnia hyalina* and *C. sphaericus* 2–3% mortality was found. The differences in density of *Leptodora kindtii* near the discharge are attributed to mortality (about 19%). This is also true for differences of about 25% found for *Ceriodaphnia pulchella, Diaphanosoma brachyurum* and the cyclopoid nauplii. The nauplii decreased in numbers by up to 44% during winter in the discharge canal 18 km downstream.

No evidence was found to suggest a negative

influence on the abundance of zooplankton in the lake.

Acknowledgements

The financial support of the Ministry of Health and the Environment is gratefully acknowledged. I thank Dr. H. L. Golterman for guidance. Drs. R. D. Gulati and J. Vijverberg evinced interest during the whole research period and commented upon the manuscript. Mr. J. Swart counted the large number of zooplankton samples and took care of transport and other facilities. Dr. A. E. Jansen critically read the first draft of the manuscript and Ms. M. E. McLaughlin corrected the English and typed the script.

References

Allersma, E., 1969. Centrale Bergumermeer, capaciteit uitwendig koelcircuit. Waterloopkundig Lab., Delft. R 562.

Beattie, D. M., Golterman, H. L. & Vijverberg, J., 1978. An introduction to the limnology of the Friesian Lakes. Hydrobiologia 58: 49–64.

Beattie, D. M. & De Kruijff, H., 1978. Population dynamics and biomass production of Neomysis integer (Leach) in the Bergumermeer. Verh. int. Verein. Limnol. 20: 2566–2571.

Benda, R. S. & Gulvas, J., 1976. Effects of the Palisades Nuclear Power Plant on Lake Michigan. In: Esch, G. W. & McFarlane, R. W. (Eds.) Thermal Ecology II, pp. 243–250. ERDA Symposium Series (Conf. 750425), Augusta, Ga.

Brauer, G. A., Neill, W. H. & Magnuson, J. J., 1974. Effects of a power plant on zooplankton distribution and abundance near plant's effluent. Wat. Res. 8: 485–489.

Bunting, D. L., 1974. Zooplankton: thermal regulation and stress. In: Gallagher, B. J. (Ed.) Proc. Symp. Energy Production and Thermal Effects, pp. 50–55. Milwaukee, Wis.

Carlson, D. M., 1974. Responses of planktonic cladocerans to heated waters. In: Gibbons, J. W. & Sharitz, R. R. (Eds.) Thermal Ecology, pp. 186–206. AEC Symposium Series (Conf. 730505), Augusta, Ga.

Carpenter, E. J., Peck, B. B. & Anderson, S. J., 1974. Survival of copepods passing through a nuclear power station on northeastern Long Island Sound. U.S.A. Mar. Biol. 24: 49–55.

Davies, R. M. & Jensen, L. D., 1974. Effects of entrainment of zooplankton at three Mid-Atlantic power plants. Electric Power Research. Institute Publication No. 74-049-11-1. Paulo Alto, Calif. 75 pp.

Davies, R. M., Hanson, C. H. & Jensen, L. D., 1976. Entrainment of estuarine zooplankton into a Mid-Atlantic Power Plant: delayed effects. In: Esch, G. W. & McFarlane, R. W. (Eds.) Thermal Ecology II, pp. 349–357. ERDA Symposium Series (Conf. 750425) Augusta, Ga.

Densen, W. L. T. van (in prep.) The ecology of 0+ fish in Tjeukemeer (The Netherlands) I. Feeding behaviour.

Densen, W. L. T. van & Hadderingh, R. H., 1982. Effects of entrapment and cooling water discharge by the Bergum Power Station on 0+ fish in the Bergumermeer. Hydrobiologia 95: 351–368.

Densen, W. L. T. van & Vijverberg, J., 1982. The relations between 0+ fish density, zooplankton size and the vulnerability of pikeperch, Stizostedion lucioperca, to angling in the Frisian lakes. Hydrobiologia 95: 321–336.

Donze, M., 1978. Measurements of the effect of heating on survival and growth of natural plankton populations. Verh. int. Verein. Limnol. 20: 1822–1826.

Gehrs, C. W., 1974. Vertical movement of zooplankton in response to heated water. In: Gibbons, J. W. & Sharitz, R. R. (Eds.) Thermal Ecology, pp. 285–290. AEC Symposium Series (Conf. 730505) Augusta, Ga.

Goss, L. B. & Bunting, D. L., 1976. Thermal tolerance of zooplankton Wat. Res. 10: 387–398.

Hadderingh, R. H., 1975. Littoral macroinvertebrates in Lake Bergum in 1972, 1973 and 1974, prior to the commissioning of the Bergum power station. KEMA, report 6902–75.

Hadderingh, R. H., 1978. Mortality of young fish in the cooling water system of the Bergum Power Station. Verh. int. Verein. Limnol. 20: 1827–1832.

Hadderingh, R. H., 1979. Fish intake mortality at power stations. The problem and its remedy. Hydrobiol. Bull. 13: 83–93.

Hebert, P. D. N., 1978. The population biology of Daphnia. Biol. Rev. 53: 387–426.

Heinle, D. R., 1969. Temperature and zooplankton. Chesapeake Sci. 10: 186–209.

Herzig, A., 1974. Some population characteristics of planktonic crustaceans in Neusiedler See. Oecologia 15: 127–141.

Hollander, M. & Wolfe, D. A., 1973. Non Parametric Statistical Methods. John Wiley, New York. 503 pp.

Horkan, J. P. K., Gallup, D. N. & Nursall, J. R., 1977. Effects of thermal effluent upon the planktonic Rotifera survival and egg production. Arch. Hydrobiol. Beih. 8: 84–87.

Koops, F. B. J., 1981. De invloed van de koelwaterlozing door de centrale Bergum op chemie en plankton in het Bergumermeer. I, II and III. KEMA report 6632-81.

Kott, P., 1953. Modified whirling apparatus for subsampling of plankton. Aust. J. mar. Freshwat. Res. 4: 387–393.

Lair, N., 1980. The rotifer fauna of the River Loire (France) at the level of the nuclear plants. Hydrobiologia 73: 153–160.

Lanner, M. & Pejler, B., 1973. The effect of cooling water discharges on zooplankton in a bay of Lake Mälaren. Inst. Freshwat. Res. Drottningholm Rept 53: 31–34.

Leentvaar, P., 1980. Note on some Brachionidae (rotifers) from the Netherlands. Hydrobiologia 73: 253–262.

Nie, H. W. de, 1979. Effects of thermal effluents from the Bergum power station on the zooplankton in Lake Bergum. (Summary). Hydrobiol. Bull. 13: 100.

Nie, H. W. de, Bromley, H. J. & Vijverberg, J., 1980. Distribution patterns of zooplankton in Tjeukemeer, The Netherlands. Plankton Res. 2: 317–334.

Schaap, A. & Willemsen, J., 1978. Verslag over bemonstering van eieren en larven van diverse vissoorten in verband met koelwatergebruik door electriciteitsgebruik. RIVO report z.s. 78-2.

Storr, J. F., 1974. Plankton entrainment by the condenser system of nuclear power stations on Lake Ontario. In: Gibbons, J. W. & Sharitz, R. R. (Eds.) Thermal Ecology, pp. 291–295. AEC Symposium Series (Conf. 730505), Augusta, Ga.

Wakeren, J. H. A. van, 1979. Verspreiding van koelwater rondom centrale Bergum. KEMA report 4783-79.

Wanders, J. B. W., 1979. Effects of thermal effluents from the Bergum Power Station on the phytoplankton in the Bergumermeer (Summary). Hydrobiol. Bull. 13: 98–99.

Wanders, J. B. W., Nie, H. W. de, Richter, A. F., Frank-Landman, A. G. & Swart, J. S., 1980. Gevolgen van koelwaterlozing op groei en voorkomen van plankton in het Bergumermeer. Intern verslag 1980-2. Limnological Institute, Tjeukemeer Laboratory, The Netherlands. 144 pp.

Whitehouse, J. W., 1971. Some aspects of the biology of Lake Trawsfynydd; a power station cooling pond. Hydrobiologia 38: 254–288.

Effects of entrapment and cooling water discharge by the Bergum Power Station on 0+ fish in the Bergumermeer

W. L. T. van Densen[1] & R. H. Hadderingh[2]

[1] *Limnological Institute, Tjeukemeer Laboratory, De Akkers 47, 8536 VD Oosterzee, The Netherlands, and Department of Fish Culture and Inland Fisheries, Agricultural University, Wageningen, The Netherlands*
[2] *N.V. KEMA, 6800 ET Arnhem, The Netherlands*

Keywords: thermal pollution, fish ecology, 0+ fish, entrapment mortality, cooling water

Abstract

The numbers, distribution and size of 0+ fish in Bergumermeer and in the cooling water system of the Bergum Power Station were registered from 1978 to 1981. An exponentially decreased vulnerability to entrapment by the power station was found with increasing size of fish. Because of the large water intake in relation to the lake volume, entrainment mortality of fish larvae in May and June is large and probably of the same order of magnitude as the natural mortality. In the same period the Bergumermeer can be supplied with large amounts of fish larvae, immigrating passively with the water current. Impingement mortality later on in the season is insignificant compared with natural mortality. The heated effluents from the power station influence the distribution and the size of 0+ fish in the lake. Year-class strength of pikeperch may be improved due to the larger size of 0+ pikeperch. Less water intake as a possibility to minimize entrainment is discussed.

Introduction

This paper deals with the effects of the Bergum Power Station on the numbers, spatial distribution and size of 0+ fish in Bergumermeer. The direct influence on the numbers in the lake acts via the entrapment of 0+ fish in the cooling water system of the power station (Hadderingh 1978). Theoretical modelling on the possible effects of entrapment on fish populations is well developed (Swartzman *et al.* 1977). However, very few studies until now have given data on the consequences of entrapment mortality for the size of the population (Hess *et al.* 1975; Teleki 1976). When entrapment mortality is large it will have an influence on the distribution pattern of 0+ fish in the lake by depleting the stock in the intake area. The heated discharge of a power station can also influence the distribution pattern by attracting or repelling fish (Horoszewicz & Morawska 1979). The size of 0+ fish in a heated lake can increase due to an increase in growth rate and to a lengthening of the growing season. The latter is caused by an advancement of the spawning time (Boytsov 1971; Zawisza & Backiel 1972; Hadderingh 1975) and by a later termination of the growth in the autumn (Boytsov 1971; Brown 1973; Wilkońska & Zuromska 1977).

Observations from 1975 to 1978 showed a slightly decreased primary production rate (Wanders 1979) and higher mortality rates for cyclopoid nauplii, *Diaphanosoma brachyurum* and *Leptodora kindtii* (de Nie 1982) in the cooling water of the Bergum Power Station.

For this study observations were made on numbers, spatial distribution and the size of 0+ fish in the lake from 1978 to 1981. These observations were compared with the numbers and size of 0+ fish sampled simultaneously in the cooling water system of the power station. The size of 0+ fish in the lake is also compared with the size in unheated reference lakes in the Frisian Lake District. A possible long-term effect of the functioning of the power station on the fish population in the lake will be discussed.

The main fish species in the lake are smelt, *Osme-*

Hydrobiologia 95, 351–368 (1982). 0018-8158/82/0953-0351/$03.60.

rus eperlanus; perch, *Perca fluviatilis;* pikeperch, *Stizostedion lucioperca;* ruffe, *Gymnocephalus cernua;* bream, *Abramis brama;* white bream, *Blicca björkna;* roach, *Rutilus rutilus;* three-spined stickleback, *Gasterosteus aculeatus;* and eel, *Anguilla anguilla.*

Study area

The Bergumermeer

The Bergumermeer is one of the northern lakes in the Frisian Lake District; morphological and topographical details are given in Table 1. The shore of the lake is covered with a vegetation dominated by *Typha latifolia, Phragmites australis* and *Scirpus lacustris* spp. *lacustris.* Aquatic plants with floating leaves *Nymphoides peltata, Nymphaea alba* and *Nuphar luteum* are present in the bays.

Table 1. Physical features of the Bergumermeer.

Elevation	0.5 m below sealevel
Area	4.64 km^2
Volume	6.0×10^6 m^3
Maximum length	3.5 km
Maximum breadth	1.9 km
Mean depth	1.3 m
Maximum depth	3.5 m (in canals)
Shore length	16.4 km (dam included)
Geographical position	53° 12′ N 6° 02′ W

The lake is connected with the other Frisian lakes by a main canal, the Prinses Margriet canal, and a subsidiary one, the Lits (see Fig. 1). Therefore the

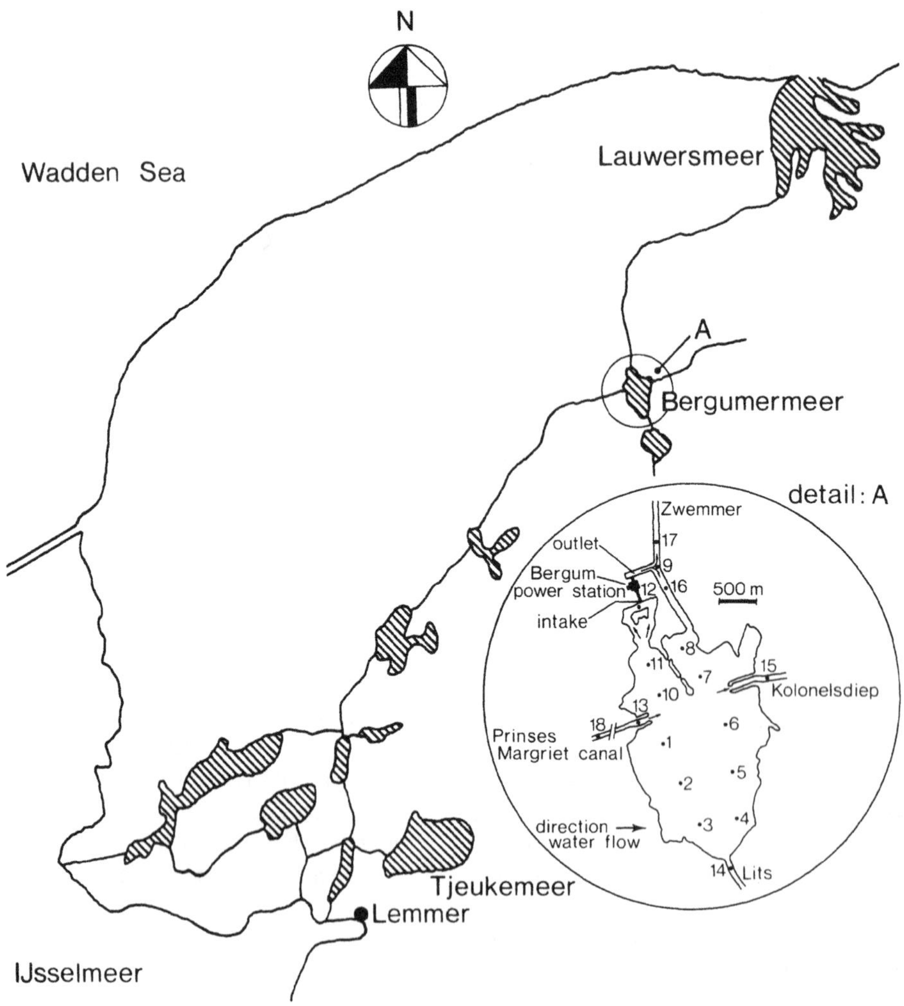

Fig. 1. The location of the Bergumermeer in the Frisian Lake District and the position of the sampling sites.

chemical properties of the lake are similar to those of the other Frisian lakes (Beattie *et al.* 1978).

Bergum Power Station

Bergum Power Station is situated on the N.W. shore of the Bergumermeer (Fig. 1 and Plate 1) and has two 320 MW units, both having a once-through cooling water system. The power station has been in operation since September 1974. Cooling water is withdrawn from the lake via the intake, situated on the N.W. shoreline. The maximum cooling water

Plate 1. Bergum Power Station on the northern shore of the Bergumermeer (Aerophoto Teuge).

flow is 2.4 × 10^6 m^3 day^{-1} (27.8 m^3 s^{-1}) which is about 40% of the whole lake volume. The cooling water enters the intake through a coarse steel grid with openings of 3.3 cm and is sieved afterwards by four rotating screens with 0.5 cm openings. After a temperature increase of about 5 °C in the condenser, the cooling water is discharged via a weir into the outlet canal. The water velocity ranges between 30 cm s^{-1} just outside the steel grid and 200 cm s^{-1} inside the cooling water system. Fish larvae and juveniles smaller than about 4 cm in length pass through the cooling water system and are discharged in the outlet canal (entrainment). Larger fish are retained by the rotating screens (impingement) and are discharged via a pipeline into the Zwemmer (see Fig. 1). The main causes of fish mortality in the cooling water system are mechanical injury and pressure fluctuations (Marcy 1976; Hadderingh 1979).

Water movements and water temperature

The hydrology of the Bergumermeer is affected by both the in- and the outflow of water via the interconnecting canals and by the pumping of cooling water of Bergum Power Station. The inflow of water into the lake is mainly via the Prinses Margriet canal in the W. (Fig. 1) and the outflow via the Zwemmer in the N. to the Wadden Sea; during summer outflow occurs also via the Kolonelsdiep to the eastern province of Groningen. From late spring to early autumn the lake received water from the IJsselmeer, which enters the lake district via the inlet near Lemmer. During winter the lake receives surplus water from the Frisian 'polders' (Beattie *et al.* 1978).

The distribution pattern of the heated cooling water depends on the ratio of the volume of cooling water to the volume of water through the Zwemmer and the Kolonelsdiep. When the volume of cooling water exceeds that of the outflow via the Zwemmer, part of the heated effluent finds its way to the lake.

The volumes of water flowing weekly through the canals and the power station for the period May–September 1979 and 1980 are presented in Fig. 2. Greater water movements in early summer 1979 and in July 1980 were caused by a water surplus in the lake district. The volume of cooling water from the power station depends on whether two or four pumps are in operation. During most weeks in the period May to September 1979 and 1980, recirculation of a part of the heated cooling water to the Bergumermeer occurred.

For representative examples of water temperature distribution in the lake, data from 28 May and 31 July 1980 are given in Fig. 3. On these dates the

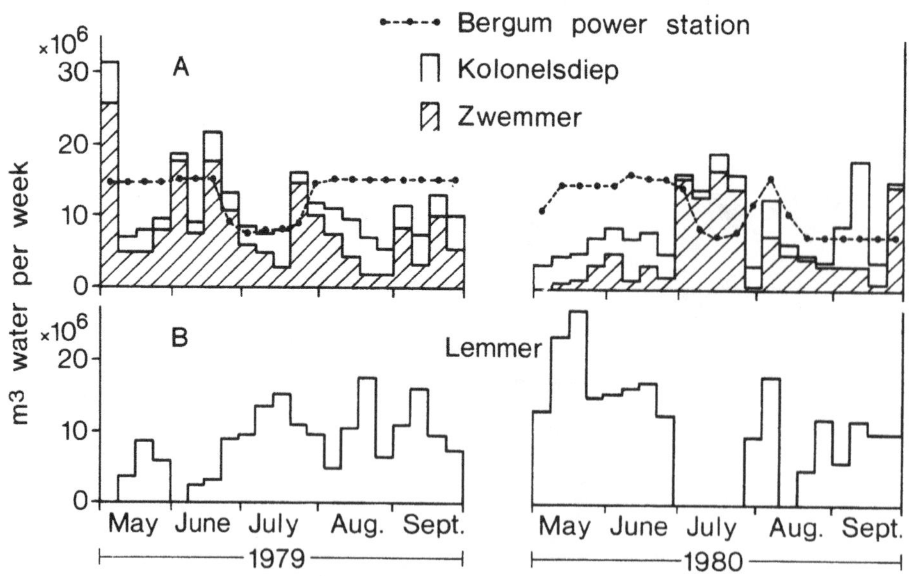

Fig. 2. (A) Water flow through the Bergum Power Station, the Kolonelsdiep and the Zwemmer. (B) Water intake of the Frisian Lake District from the IJsselmeer via the sluices near Lemmer.

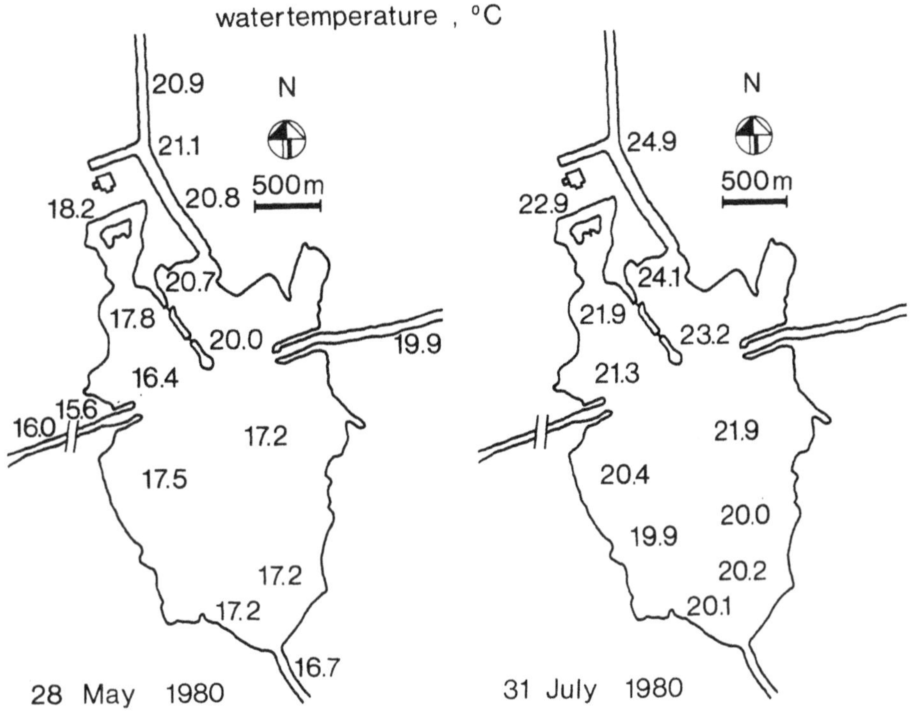

Fig. 3. Water temperature in The Bergumermeer on 28 May and 31 July 1980.

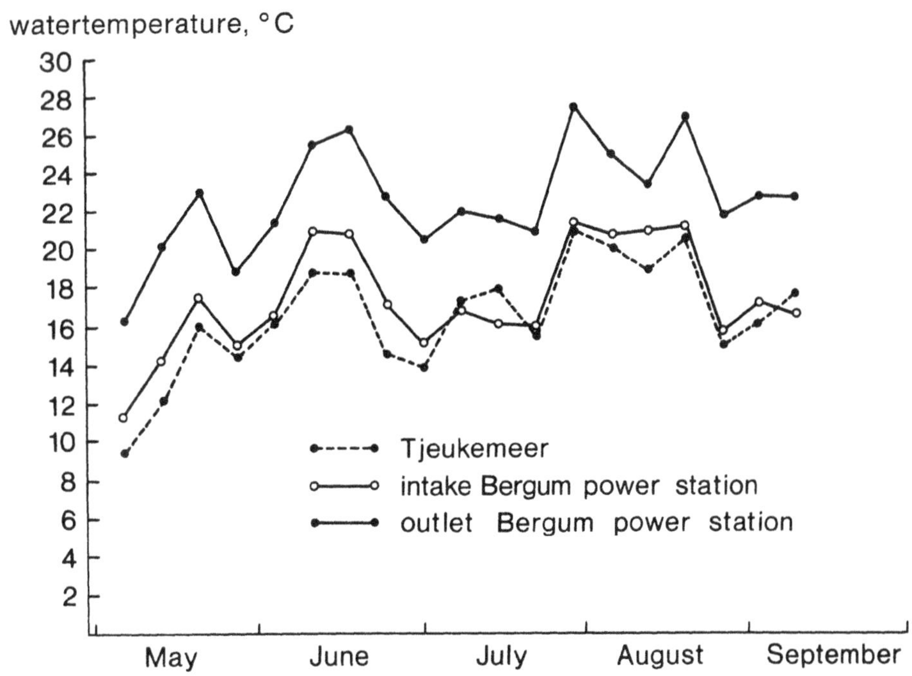

Fig. 4. Water temperature during the summer of 1980 near the Bergum Power Station and in the unheated Tjeukemeer.

highest water temperatures were recorded in the Zwemmer, the N.E. part of the lake, the Kolonelsdiep and near the intake of the power station. The data clearly show incomplete recirculation of heated cooling water.

The weekly temperature data near the intake and the outlet recorded for the period 6 May–10 September 1980 show an average increase in cooling water temperature of 5 °C (Fig. 4). Relatively low water temperatures occurred in July with an average intake temperature of 16 °C. This can be attributed to the cold weather in this month when both the average air temperature (15.5 °C) and the hours of sunshine (131) were well under normal (data Royal Netherlands Meteorological Institute). The water temperature at the intake of the power station appears to be up to 2 °C higher than in the unheated Tjeukemeer, a reference lake in the S. of the Frisian Lake District.

Table 2. Sampling dates, type of fishing gear, sampling time and number of samples per site.

Date	Type of gear	Sampling time (s)	No. of samples per site
28-08-78	5 m trawl	300	1
03-05-79	50 cm larvae net	60	3–4
18-06-79	3 m trawl	300	1
12-09-79	5 m trawl	300	1
06-05-80	50 cm larvae net	60	3–4
28-05-80	30 cm larvae net	30	3–4
16-06-80	3 m trawl	300	1
31-07-80	5 m trawl	300	1
10-09-80	5 m trawl	300	1
24-06-81	3 m trawl	300	1
29-07-81	5 m trawl	300	1
09-09-81	5 m trawl	300	1

Material and methods

Sampling of the fish population in the lake

The sampling sites for 0+ fish were evenly distributed over the lake and in the canals (Fig. 1). According to their length the 0+ fish can be distinguished as larvae (May) and juveniles (June–September). Sampling for larvae was carried out with plankton nets at sites 1 to 18, excluding sites 2 and 5. Additional sampling was done at eight sites within 2 m from the littoral vegetation of the lake and canals. Trawling for juveniles was conducted at sites 1 to 13. An overview of the sampling scheme is presented in Table 2.

At the beginning of May 1979 and 1980 larvae were sampled using a conical plankton net with a mouth diameter of 50 cm and a length of 190 cm. This net was kept beside the boat which sailed at a speed of approx. 1.2 m s^{-1}. On 28 May 1980, smelt larvae showed net avoidance at this speed. As net avoidance decreases with increasing towing speed, sampling was then carried out with a smaller conical plankton net (mouth diameter 30 cm, length 150 cm) which could be towed at the higher speed of 1.8 m s^{-1}. Both nets had a mesh size of 0.5 mm and an open area ratio, defined as the ratio between total open mesh area and the opening (Tranter & Smith 1968), of about 5. The towing speed and the sampled water volume were determined using a Hydrobios flow meter, model 438110, mounted in the mouth of the nets. On 3 May 1979 and 6 May 1980 hauls were made at 0.25 and 1.0 m depths and on 28 May 1980, at 0.30, 0.65 and 1.0 m depths.

The lake population of juvenile fish in June was sampled with a 3 m trawl making 5 min hauls at a towing speed of about 0.5 m s^{-1}. The population in July to September was sampled with a 5 m trawl making 5 min hauls at a towing speed of about 1 m s^{-1}. A description of the trawls together with the fishing efficiency is given by van Densen (in prep.).

Sampling fish entrapped in the power station

Entrapment of 0+ fish was established, if possible, on the same dates as chosen for the sampling of the population. The number of entrapped fish was always estimated for 24 hour periods because the number of fish entrapped during day and night differed substantially (Hadderingh 1978). The two fractions of entrapped fish, the impinged and the entrained individuals respectively, were sampled simultaneously.

The impinged fraction was determined by collecting all individuals retained in 24 hours by the screens at the intake.

The entrained fraction, passing through the screens, was estimated by sampling with conical plankton nets placed in the cooling water stream just before the weir at the outlet. In May and June, when high numbers of larvae and juveniles were caught, six samples were taken by keeping the net in the water flow for 30 min at four hourly intervals for 24 hours. The nets used in these months had 0.5 mm meshes, a 50 cm mouth diameter and a length of 190 cm. From July to September it was necessary to sample for longer periods because the number of entrained juveniles decreased very strongly. Therefore the net was kept in the water flow for 24 hours. This net had 1.5 mm meshes, a 50 cm mouth diameter and a length of 160 cm. The water volume sampled in the cooling water flow was also estimated with the Hydrobios flowmeter. The number of larvae and juveniles passing through the cooling water system in 24 hours was calculated by multiplying the fish concentration in the sample with the actual volume of cooling water flowing through the power station in 24 hours.

Sampling analysis

The fish in the samples were separated by species, enumerated, and the fork length was measured to the nearest millimetre. Some material was preserved in 5% formalin for further identification in the laboratory. The percid larvae were identified according to the number of myomeres. Larvae of roach and bream are presented as cyprinids, since identification proved problematical. No distinction was made between 0+ bream amd 0+ white bream. With older fish in the catch, bream dominated white bream to a large extent. Individuals of 0+ pikeperch were separated in three categories based on their length-frequency distributions. In August and September the small individuals of summer generation pikeperch are easily separated from spring generation pikeperch. The spring generation comprised small, mainly planktivorous, fish and large, mainly piscivorous, fish.

Results

Numbers, size and distribution of 0+ fish in the lake

Numbers. The mean numbers of fish in the lake were based on the samples taken at the open water sites 1–8, 10 and 11. Large differences in the number of 0+ fish in the lake were observed from year to year (Table 3). Smelt was dominant in the summers of 1980 and 1981. The numbers of smelt were highest in June 1980, being 6 times higher than in June 1979, while this ratio increased to 20 when comparing the numbers in September 1980 with 1979. Perch was most abundant in 1980. Numbers of pikeperch in June 1979, 1980 and 1981 varied only to a small extent but differed greatly at the end of the growing season. The lowest number, as for smelt, was found in September 1979. Numbers of ruffe and bream in September were stable from one year to another and were relatively high compared with those in the growing season, but were never of large relative importance. At the end of the growing season it became more difficult to distinguish between 0+ and 1+ bream from their length.

Table 3. Mean catch of 0+ fish per unit volume or area in the Bergumermeer.

Date	Unit volume or area	Smelt	Perch	Pikeperch			Ruffe	Roach	Bream	Three-spined stickleback
				pl	pi	sg				
28-08-78	360 m²	186	5.0	26	–	80	150	10	434	28
03-05-79	10 m³	5.4	273					0.06		0
18-06-79	50 m²	327	7.0	542	0	0	31	0.8	2.5	0
12-09-79	360 m²	22	2.2	1.4	0.9	0.6	210	12	556	6.1
06-05-80	10 m³	0.80	1.9	35	0	0	2.4	0.28		0
28-05-80	10 m³	114	3.9	11	0	0	6.3	3.1		0
16-06-80	50 m²	1 859	234	342	0	0	101	98	82	0
31-07-80	360 m²	10 723	111	98	26	0	146	205	445	323
10-09-80	360 m²	4 219	45	12	6.1	0.2	712	152	559	56
24-06-81	50 m²	591	0.5	184	0	0	26.6	0.1	1.8	2.6
29-07-81	360 m²	2 335	0.4	85.0	35.6	0	129.7	1.0	304	67.5
09-09-81	360 m²	1 465	1.7	10.0	5.2	1.6	53.2	47.3	479	5.2

pl = planktivorous; pi = piscivorous; sg = summer generation.

Table 4. Length (mm) of 0+ fish entrapped in the power station (P) and of 0+ fish in the Bergumermeer (L).

Date	Smelt		Perch		Pikeperch						Ruffe		Roach		Bream		Three-spined stickleback	
					pl		pi		sg									
	P	L	P	L	P	L	P	L	P	L	P	L	P	L	P	L	P	L
28-08-78	61	65	66	77	87	91	x	x	37	37	62	68	54	55	43	52	34	-
03-05-79	7	6	6[1]	6[1]	6[1]	6[1]	x	x	x	x	6[1]	6[1]	6[2]	7[2]	6[2]	7[2]	x	x
18-06-79	23	26	25	25	21	25	x	x	x	x	22	22	21	22	18	25	24	x
12-09-79	70	69	87	84	77	82	137	152	39	47	65	67	66	68	57	74	38	38
06-05-80	7	7	7	7	7	6	x	x	x	x	5	5	7	7	7	7	x	x
28-05-80	15	15	10	10	9	8	x	x	x	x	7	6	7	7	7	7	x	x
16-06-80	26	26	21	23	17	19	x	x	x	x	17	22	20	23	16	21	23	x
28/31-07-80	50	48	46	46	44	47	93	70	x	x	41	48	38	44	31	37	29	29
09/10-09-80	60	60	57	64	76	79	148	138	39	44	53	61	51	58	44	58	36	36
25-06-81	35	-	38	28	39	34	x	x	-	x	29	28	-	19	-	31	27	26
29-07-81	55	55	56	48	48	50	x	93	-	x	36	46	31	50	27	45	29	27
09-09-81	62	67	71	82	67	76	153	155	39	45	56	64	58	64	53	68	32	-

pl = planktivorous; pi = piscivorous; sg = summer generation; - = not measured; x = not observed.
[1] Percidae.
[2] Cyprinidae.

Large changes in the number of 0+ fish in the catches can also be observed within one growing season. Increase in numbers, occurring mainly during May and June, can be found with smelt (1979, 1980), perch (1980), roach (1979, 1980, 1981) and bream (1979, 1980, 1981). Only three-spined stickleback showed an increase at the end of the growing season (1979, 1980, 1981). These increases in the catches can be explained by inflow of 0+ fish into the Bergumermeer via the Prinses Margriet canal and/or by migration from the shore area to the open water, where the sampling was carried out. The sharp increase in density of smelt in May 1980 from 0.8 (6 May) to 114 (28 May) individuals per 10 m^3 due to the inflow from the Prinses Margriet canal is illustrated by the distribution pattern (see later). This is in agreement with the large intake of water from the IJsselmeer during this period (Fig. 2) and with observations in the whole lake district (van Densen & Vijverberg 1982). The number of ruffe was fairly stable among the catches throughout the seasons 1979, 1980 and 1981.

Mean length. The mean length of all fish species showed a distinct increase during the growing seasons of 1979, 1980 and 1981 (Table 4). All species except pikeperch attained greatest length in 1979.

Horizontal distribution pattern of 0+ fish in the lake. Smelt larvae, sampled with the plankton nets in May 1979, were uniformly distributed both in the lake and the Prinses Margriet canal. The densities in the outlet area and in the canals to the E. and to the S. were low. The distribution of smelt larvae differed considerably on the two sampling dates in May 1980. On 6 May larvae were absent in the W. part of the lake and in the Prinses Margriet canal. In contrast, on 28 May high numbers occurred in these areas (Fig. 5). Both the distribution pattern and the high densities on 28 May demonstrated a considerable import of smelt larvae into the Bergumermeer via the inflow from the Prinses Margriet canal. From June onwards, juvenile smelt always occurred in low numbers in the comparatively warmer N.E. part of the lake (Fig. 6). This decrease in density of smelt may be due to two factors: firstly, the continuous effect of entrapment mortality in the power station; secondly, the avoidance by smelt of the heated N.E. area since it prefers relatively low water temperatures (Belyanina 1969; Effer & Bryce 1975). No size related distribution pattern was observed for 0+ smelt.

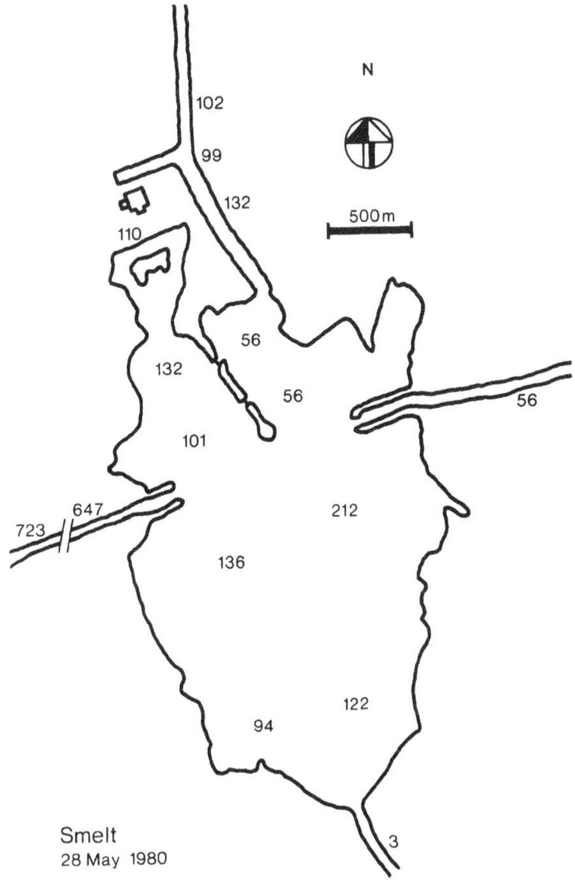

Fig. 5. Distribution (Nos. per 10 m^3) of smelt larvae on 28 May 1980.

In 1980, when 0+ perch was abundant, no distinct distribution pattern was observed for the larvae in May, but from June to September the highest numbers of juveniles were always caught at the warmer site 7 (Fig. 7). In June 1980 the largest mean size was also recorded at this site.

With 0+ pikeperch neither larvae nor juveniles showed a distinct distribution pattern. The largest mean length in June was always observed in the warmer N. part of the lake.

Both ruffe larvae and juveniles were evenly distributed within the lake. The largest mean size of juveniles, however, was observed in the heated N.E. part of the lake on most sampling dates from 1978 to 1981.

At the beginning of May 1979 and 1980 cyprinid

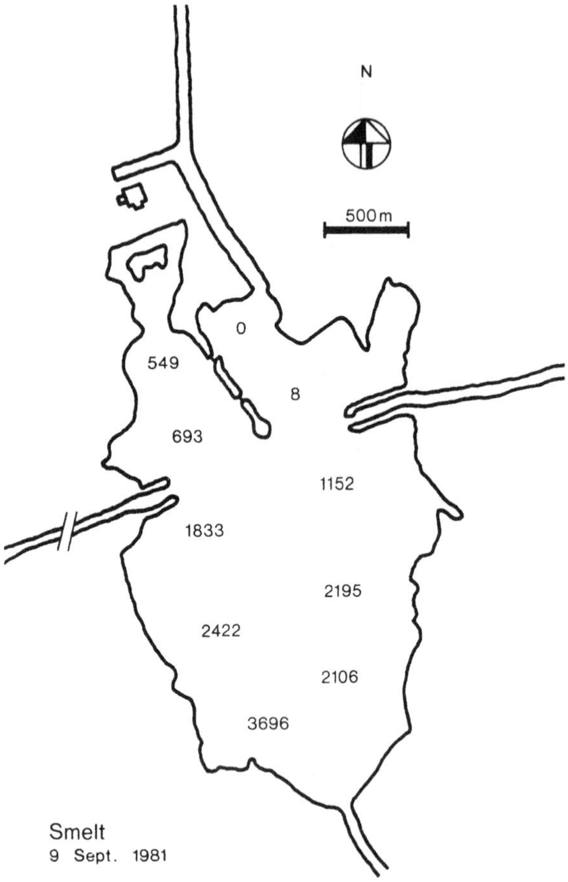

Fig. 6. Distribution (Nos. per 5 min 5 m trawl haul) of 0+ smelt on 9 September 1981.

(bream and roach) larvae, though low in numbers, occurred almost exclusively in areas with relatively high water temperatures, namely in the N.E. part of the lake and in the outlet area.

On 28 May 1980 the average density of larvae in the shore zone was seven times higher than in the open water zone. On the same date several larvae between 10 and 18 mm were caught in the shore zone of the heated N.E. area of the lake and near the outlet of the power station, but such large larvae were rarely found in the unheated areas. This larger size suggests an advancement in spawning time and/ or faster growth of the fish in this heated area. In June 1980 relatively high numbers (Fig. 8) and sizes of juvenile bream were recorded in the warm N.E. area. In each year these fish must have spread over the lake after June because they were not encountered in definitely higher numbers in this area during the rest of the season. Considerable numbers of juvenile roach occurred only in 1980, with highest densities in the heated area of the lake in June, July and September (Fig. 8). Only in July were size differences observed within the lake, i.e. the largest fish occurred in the heated area. The distribution pattern of both the numbers and the size of larvae and juveniles indicates that the heated part of Bergumermeer may function as an important spawning and nursery area for bream and roach.

The three-spined stickleback was caught in high numbers only in 1980. A relation between number or size and the water temperature in the lake was lacking.

Vertical distribution of larvae. In 1979 the larval densities at 0.25 and 1.0 m water depth did not differ. In contrast, in 1980 the densities of all species

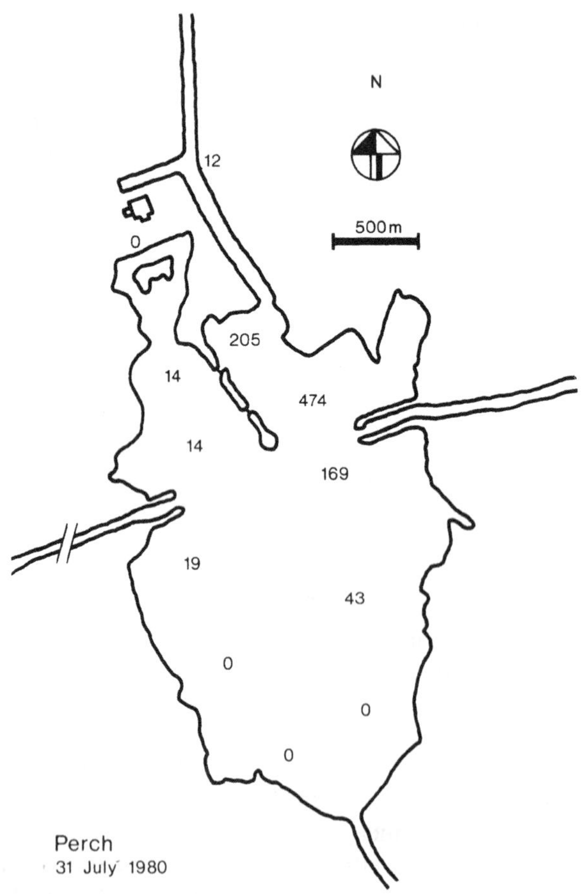

Fig. 7. Distribution (Nos. per 5 min 5 m trawl haul) of 0+ perch on 31 July 1980.

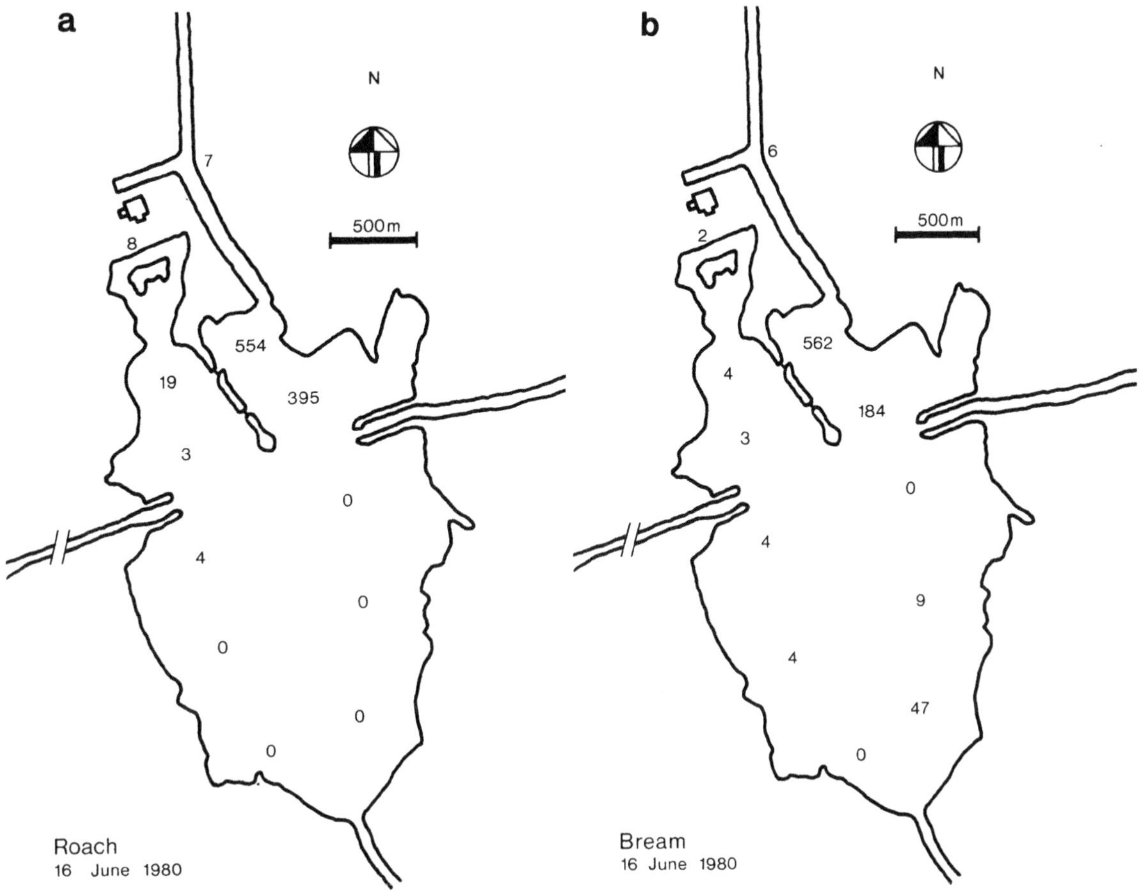

Fig. 8. (A) Distribution (Nos. per 5 min 3 m trawl haul) of 0+ roach on 16 June 1980. (B) Distribution (Nos. per 5 min 3 m trawl haul) of 0+ bream on 16 June 1980.

in the upper water layers, except for smelt on 6 May, were higher (Table 5).

Table 5. Vertical distribution of fish larvae in 1979 and 1980 in the Bergumermeer.

Depth (m)	No. of samples	% distribution of larvae in the watercolumn				
		Smelt	Pikeperch	Perch	Ruffe	Cyprinids
			3 May 1979			
0.25	4	58	50	–	–	–
1.00	4	42	50	–	–	–
			6 May 1980			
0.25	4	47	69	83	71	65
1.00	4	53	31	17	29	35
			28 May 1980			
0.30	8	59	52	46	43	61
0.65	8	17	41	44	33	24
1.00	8	24	7	10	24	15

Number and size of entrapped 0+ fish

The numbers and composition of 0+ fish entrapped in 24 hours in the cooling water system of Bergum Power Station were determined for 12 dates from 1978 to 1981. The numbers of 0+ fish entrapped differed from year to year and from month to month within one year (Table 6).

Among the entrapped fish, smelt and bream were most abundant in August 1978.

In 1979 the numbers of fish entrapped were the highest in May and June. Smelt and pikeperch were the most numerous species.

In 1980 smelt and pikeperch were again the most numerous. The amount of fish entrapped in this year was the highest in the 1978–1981 period.

Table 6. Numbers of 0+ fish entrapped in the power station per 24 hours.

Date	Smelt	Perch	Pikeperch			Ruffe	Roach	Bream	Three-spined stickleback
			pl	pi	sg				
28-08-78	2 466	16	48	7	182	104	150	3 474	663
03-05-79	305 355	—	—	15 160 002	—	—	— 61 071 —		0
18-06-79	961 011	11 379	198 892	0	0	43 079	12 366	12 442	3 412
12-09-79	1 957	20	64	5	1	100	58	1 873	5 817
06-05-80	29 705	101 847	1 659 253	0	0	241 886	— 16 974 —		0
28-05-80	21 510 426	886 170	1 251 489	0	0	915 106	— 567 872 —		0
16-06-80	3 265 071	439 089	302 931	0	0	223 885	68 079	208 807	12 793
28-07-80	523 671	8 139	19 270	13	0	4 835	3 292	44 530	53 251
09-09-80	18 885	202	24	3	4	239	485	4 797	11 790
25-06-81	1 122 462	2 276	19 430	0	-	10 387	-	-	8 797
29-07-81	28 174	57	3 155	0	-	1 134	850	23 917	22 690
09-09-81	32 686	4	45	6	6	87	74	2 456	6 143

pl = planktivorous; pi = piscivorous; sg = summer generation; - = not sampled.

In 1981 most 0+ fish were entrapped in June with a clear dominance of smelt.

The general pattern of fish entrapment during the growing season shows a continuous decrease in numbers. This pattern can be attributed to both a decline in the number of fish in the Bergumermeer (Table 3) and to an increased swimming ability as the fish grows larger, thus reducing passive movements. The sharp increase in the number of entrapped smelt larvae from 6 to 28 May 1981 corresponds with the observations in the lake as described earlier.

The mean length of all entrapped fish species showed a distinct increase during the 1979 to 1981 growing seasons (Table 4). Smelt and three-spined stickleback from the power station and the lake are similar in size. The length of the other entrapped fish species is slightly smaller than that of the fish in the lake. This applies mainly to the juvenile individuals from June onwards. The size of the larvae in May is very similar for both groups.

Entrapment mortality relative to the numbers in the lake

The numbers of 0+ fish in the lake are subject to both natural mortality and entrapment mortality.

$$\frac{dN}{dt} = -(M + E).N \qquad (1)$$

where N = numbers in the lake,
M = instantaneous rate of natural mortality,
E = instantaneous rate of entrapment mortality.

In this equation it is supposed that individuals surviving their presence in the cooling water system of the power station return to the lake. The instantaneous rate of entrapment mortality depends on the cooling water flow (I) relative to the lake volume (V), on the concentration of 0+ fish in the cooling water [P] relative to the concentration in the lake [L] and on the mortality in the cooling water system (C).

$$E.day^{-1} = \frac{I}{V}\frac{[P]}{[L]}.C. \qquad (2)$$

During this study period the cooling water flow amounted from 18 to 37% of the total lake volume per day.

With its increasing size towards the end of the growing season, the 0+ fish can withstand the suction force of the power station better. Therefore the concentration of 0+ fish in the cooling water relative to that in the lake will diminish with the increasing size of the fish. To quantify this relationship all catches in the cooling water and in the open water zone of the lake were converted to numbers per unit water volume. The ratio of both concentrations was related to the mean length of the fish in the lake according to the formula:

$$\ln\frac{[P]}{[L]} = a + b.L. \qquad (3)$$

The calculated regression lines for six fish species are drawn in Fig. 9. The parameters of these lines are given in Table 7. At larval fish size the ratio is

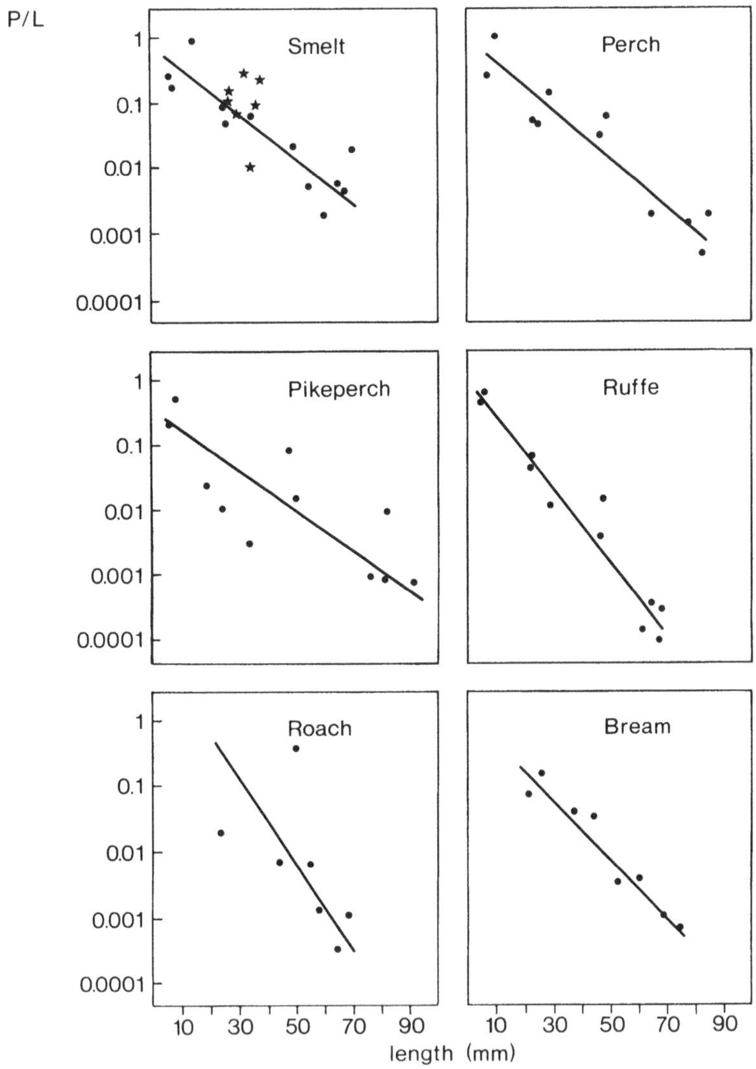

Fig. 9. Regression lines for concentration ratio on the mean length (mm) of the 0+ fish in the Bergumermeer. The stars indicate the observations for three-spined stickleback.

close to unity; thus, at that time the concentration of 0+ fish in the cooling water is almost equivalent to the one in the lake. At larger fish size a definite decrease of ratio was observed for every species. The most sharp and significant decrease was observed for ruffe. The ratio for 60–70 mm ruffe for instance is at least ten times lower than for the same-sized smelt. This may be attributed to the more benthic life of the ruffe. The length range for three-spined stickleback was too small to calculate a regression line but the mean concentration ratio was similar to the same-sized smelt.

Table 7. Parameters for the regression of the ln-transformed concentration ratio on the mean length of 0+ fish in the lake (mm), together with the 95% confidence limits of the regression coefficient (RC), the correlation coefficient (r) and the number of observations (n).

Species	RC ± 95% CL	Intercept	r	n
Osmerus eperlanus	-0.0785 ± 0.0256	-0.310	-0.886	12
Perca fluviatilis	-0.0862 ± 0.0221	0.0267	-0.940	11
Stizostedion lucioperca	-0.0743 ± 0.0339	-0.854	-0.796	11
Gymnocephalus cernua	-0.131 ± 0.0244	0.155	-0.969	11
Rutilus rutilus	-0.152 ± 0.105	2.686	-0.722	8
Abramis brama	-0.104 ± 0.0277	0.417	-0.964	8

Table 8. Species composition of 0+ fish in the catch on the lake (L) and in the power station (P) in percentages.

Date	Location	Smelt	Perch	Pikeperch	Ruffe	Roach	Bream	3-spined stickleback
28-08-78	L	20	0.6	11	16.3	1.1	47.2	3.1
	P	35	0.2	3.2	1.5	2.1	48.7	9.3
03-05-79	L	1.9		98.0		0.022		0
	P	2.0		97.6		0.40		0
18-06-79	L	35.9	0.77	59.5	3.4	0.09	0.27	0
	P	77.3	0.92	16.0	3.5	0.99	1.0	0.28
12-09-79	L	2.7	0.3	0.3	25.9	1.5	68.6	0.8
	P	19.8	0.2	0.7	1.0	0.6	18.9	58.8
06-05-80	L	2.0	4.7	86.7	5.9	0.69		0
	P	1.5	5.0	80.9	11.8	0.83		0
28-05-80	L	82.4	2.8	8.0	4.6	2.2		0
	P	85.6	3.5	5.0	3.6	2.3		0
16-06-80	L	68.4	8.6	12.6	3.7	3.6	3.0	0
	P	72.2	9.7	6.7	4.9	1.5	4.6	0.28
31-07-80	L	88.0	0.92	1.0	1.2	1.7	3.7	2.7
	P	79.6	1.2	3.0	0.74	0.51	6.8	8.1
10-09-80	L	73.2	0.78	0.33	12.4	2.6	9.7	0.96
	P	51.1	0.57	0.084	0.66	1.4	13.2	33.0
24-06-81	L	73.3	0.062	22.8	3.3	0.012	0.22	0.32
	P	96.5	0.19	1.7	0.89	–	–	0.76
29-07-81	L	78.9	0.014	4.1	4.4	0.034	10.3	2.3
	P	35.2	0.071	3.9	1.4	1.1	29.9	28.4
09-09-81	L	70.8	0.082	0.81	2.6	2.3	23.2	0.25
	P	78.8	0.0096	0.14	0.21	0.18	5.9	14.8

Generally, the representation of the species in the lake and in the cooling water agrees well (Table 8). However, for ruffe and three-spined stickleback the differences are larger, especially at the end of the growing season when the former is under-represented and the latter over-represented in the cooling water.

The mortality rates in the cooling water system are different per species and type of entrapment (Hadderingh 1978). Total entrainment mortality, including the mortality registered after passage, is 87% for small smelt and 37% for small Percidae. For impingement mortality these figures were 95 and 67% respectively.

In Table 9 the data on the cooling water flow, the concentration ratio and the mortality in the cooling water system are combined for the calculation of entrapment mortality of 0+ smelt and pikeperch. Generally, a decrease in mortality was observed with increasing size of fish. The percentages for these species are given because they occurred in considerable numbers in the open water zone of the lake throughout the whole of the 1978 to 1981 season. From the course of the concentration ratios in Fig. 9 it is clear that there is a general sharp decrease of mortality rate in a comparable order of magnitude for the other species.

The high mortality rate of 0+ smelt on 28 May 1980 is distorted and can be explained by the exten-

Table 9. Percentage entrapment mortality per day relative to the population of 0+ smelt and pikeperch in the Bergumermeer. Mean length of entrapped fish also given.

Date	Smelt		Pikeperch	
	%	L (mm)	%	L (mm)
03-05-79	8.1	6	3.5*	6
06-05-80	5.4	7	2.9	6
28-05-80	27.5	15	7.2	8
16-06-80	1.7	26	0.36	19
18-06-79	2.8	26	0.15	25
25-06-81	1.8	35	0.070	34
29-07-81	0.089	55	0.19	50
31-07-80	0.36	48	0.99	47
28-08-78	0.10	65	0.010	91
09-09-80	0.17	67	0.023	76
10-09-81	0.033	60	0.010	81
12-09-79	0.67	69	0.25	82

* Percidae.

sive passive immigration of smelt larvae into the lake at that time. A considerable part of the immigrating larvae will have streamed northwards to the power station. At the same time the mean concentration of smelt in the lake was still relatively low and this together resulted in a high concentration ratio and so in a high instantaneous rate of entrapment mortality.

Size of 0+ fish in Bergumermeer and in unheated reference lakes

The size of 0+ fish in the Frisian lakes is known for September 1975, 1979 and 1980 (van Densen & Vijverberg 1982). The mean length of 0+ perch and roach in Bergumermeer was always the largest when compared with other lakes in the district. Also, the largest size of 0+ bream was recorded in Bergumermeer in September 1979 and 1980, but not in September 1975. For 0+ smelt and ruffe the size in Bergumermeer in September was not distinctly larger.

Because of the low numbers of pikeperch caught in September in most lakes and because of the three separate categories, size comparison for this species is difficult. However, Bergumermeer and Tjeukemeer were extensively compared in the years 1978 to 1981. The length-frequency distribution data of 0+ pikeperch in these lakes show the length in the former lake to be always larger, from the beginning of the growing season onwards (Fig. 10). A larger size of 0+ pikeperch in June promotes the onset of piscivority and therefore the development of bimodality in the length-frequency distribution of the spring generation. Of course, a larger absolute size of fish is only beneficial when the size of its prey, mainly smelt, is not larger either. When 0+ smelt originates from outside Bergumermeer, as in 1980, its size in June will not be enlarged by the functioning of the power station. The faster growth of other 0+ prey species in Bergumermeer such as 0+ pikeperch, roach and bream makes them less available for 0+ pikeperch.

Figure 10 shows that summer generation pike-

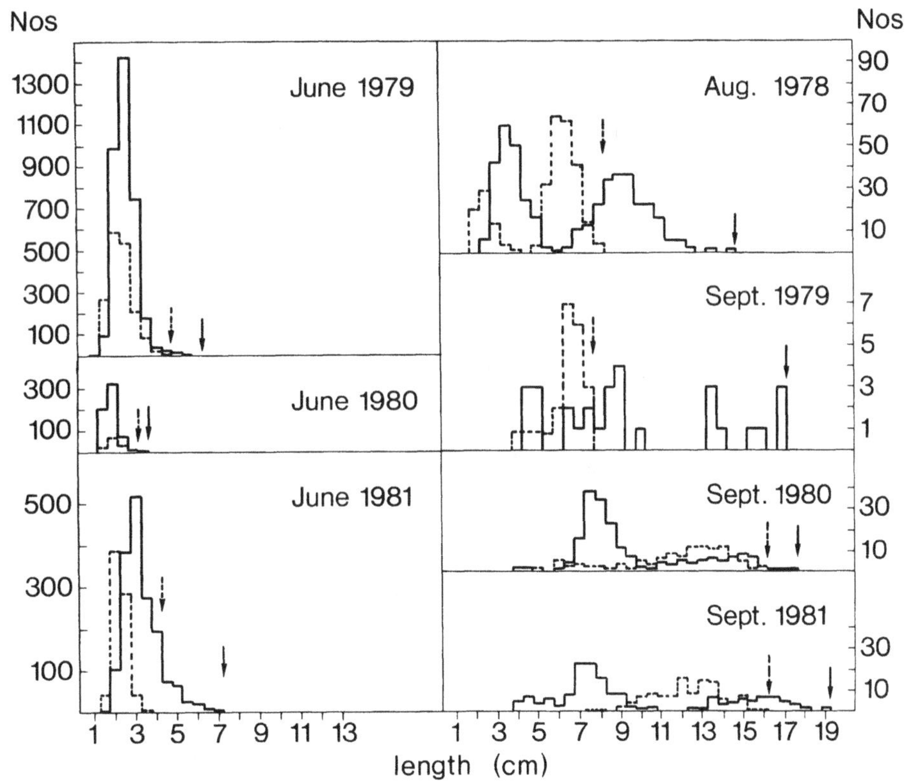

Fig. 10 Length-frequency distributions of 0+ pikeperch in Bergumermeer (———) and in Tjeukemeer (--------) in August 1978 and in June and September 1979 to 1981. The observed maximum lengths are indicated by arrows.

perch was abundant in 1978 only and had a mean length at the end of August of 37.9 mm in Bergumermeer and 31.5 mm in Tjeukemeer. In September 1979, 1980 and 1981 their numbers were minimal. Bimodality of the spring generation was clearly seen in 1980 and 1981 in both lakes, but was absent in Tjeukemeer in 1979, being hampered by the low numbers of 0+ smelt (van Densen, in prep.). At higher densities in Bergumermeer in June 1979 piscivority was observed in the case of 0+ pikeperch measuring 27 mm and more. In August 1978 bimodality was not observed in the length-frequency distribution of the spring generation. The apparent bimodality in the total length-frequency distributions was due to the presence of large numbers of summer generation pikeperch.

Discussion

The circumscription of the population of 0+ fish in the Bergumermeer is hampered by the openness of the lake. As in most Frisian Lakes (van Densen & Vijverberg 1982), in some years an extensive immigration of smelt and perch larvae can be observed during early summer. The immigration of smelt larvae into the Bergumermeer in 1980 was clearly proved during this study. In this situation the entrapment mortality of 0+ smelt originating from the population in the lake was largely overestimated. When passive immigration into the lake became insignificant, from the end of June onwards, the estimates of relative entrapment mortality were more realistic.

The greater capability of larger fish to withstand entrapment by the power station is manifested in two different ways; firstly, from the decreasing concentration ratio towards the end of the season (Fig. 9), and secondly, from the generally smaller mean size of the fish in the cooling water system than those in the lake on the same date (Table 4).

The impingement mortalities for smelt, Percidae and bream are all above 60%, while roach and three-spined stickleback have much lower mortalities of 25% and 3% respectively (Hadderingh 1978). The latter figure shows that the greater survival of this species compensates for its size-related higher vulnerability to entrapment.

In the 1976 to 1980 period the natural mortality rate from July to September in Tjeukemeer varied between 1.4 and 6.7% per day for 0+ smelt and between 3.0 and 13.3% per day for 0+ pikeperch (van Densen, in prep.). Thus, the entrapment mortality of both species in this period (Table 9) is much lower than their natural mortality; therefore, the effects of entrapment will be difficult to measure. This is hard to imagine when sometimes more than 400 kg of 50 mm smelt is entrapped per 24 hours from a 400 ha lake, as on 28 July 1980. No data are available on natural mortality in May and June. Dahlberg (1979) listed mortality rates for yolk-sac larvae and post-larvae of several freshwater and an anadromous fish species. Those for yolk-sac larvae ranged from 6.9 to 27.4% per day, those for post-larvae from 3.0 to 9.9% per day. The entrapment mortality observed in the Bergumermeer in May was of the same order of size (Table 9). Therefore, it is believed that entrapment mortality can have a considerable influence on the larval density of smelt and percid larvae at the beginning of growing season. However, densities in Bergumermeer observed during June were not markedly lower than in other lakes. The density of 0+ pikeperch in June 1979 and 1980 even ranked among the highest in the lake district (van Densen & Vijverberg 1982). Furthermore, long-term effects are masked by the sometimes extensive immigration of larvae. In contrast to the situation in 1980 the relative amount of immigrant fish larvae was small in 1975 and 1976, when Schaap & Willemsen (1978) made larvae surveys in Bergumermeer. They found high concentrations of pikeperch larvae in the area near the inlet of the power station. Our observations on the vertical distribution of larvae are very similar to those of Schaap & Willemsen. Their nightly observations revealed a more homogeneous distribution of perch and pikeperch larvae over the water column.

This study illustrates clearly that the passive migration of smelt larvae out of the Prinses Margriet canal via Bergumermeer to the Lauwersmeer in the N. is obstructed. For cyprinid and ruffe larvae the entrapment mortality might initially be overestimated because they recruit to the open water zone at a larger size. Some bias in the mortality rates is introduced by the uncertainty in the efficiency of the 3 m trawl. This may be twice as high (van Densen, in prep.), but the application of this higher efficiency did not change our conclusions.

It was estimated that entrainment mortality of young fish, mainly rainbow smelt, *Osmerus mor-*

Table 10. Wet weight (g) of 0+ fish in heated and in unheated lakes in September.

Fish species	Poland (Zawisza & Backiel 1972)		The Netherlands (van Densen & Vijverberg 1982)	
	Lichenskie lake (heated)	Šlesinskie lake	Bergumermeer (heated)	Other Frisian lakes
Perca fluviatilis	15.0 (1969)	5.0 (1969)	3.4 (1980)	2.2 (1980)
Gymnocephalus cernua	3.7 (1968)	1.2 (1967)	3.0 (1980)	3.0 (1980)
Rutilus rutilus	9.7 (1968)	3.4 (1968)	2.6 (1980)	1.6 (1980)
Abramis brama	27.0 (1968)	3.2 (1968)	2.6 (1980)	1.6 (1980)

dax, in Nanticoke Generating Station, Ontario, in 1974 resulted in a decrease of 12% in the mean annual catch of one trawl fisherman operating in the eastern basin of Lake Erie (Teleki 1976). This author applied a general 1% survival rate, from larvae to harvestable recruit.

The relative importance of an earlier spawning time and of temperature induced faster growth on the mean size of 0+ fish is difficult to indicate. In Lake Licheńskie, Poland, with a rise in the lake water temperature from 8 to 11 °C due to the effluent from a power station, the spawning time of several fish species, e.g. perch, bream and roach, began 3 to 8 weeks earlier than that in the neighbouring lakes (Zawisza & Backiel 1972). The rise in the mean lake temperature due to the effluent of the Bergum Power Station is much less (de Nie 1982). In 1976 spawning of perch, roach and bream in Bergumermeer began only 1 to 2 weeks earlier than in the unheated Leyen, a lake S. of Bergumermeer (Anonymous 1976, 1977; Schaap & Willemsen 1978).

The differences in net weights of 0+ perch, bream and roach in Licheńskie lake and its reference lake during September were much greater than between 0+ fish in Bergumermeer and in the other Frisian Lakes in the same month (Table 10). For 0+ ruffe no difference at all was observed between Bergumermeer and the other Frisian lakes. Most interesting is the larger size of 0+ pikeperch in Bergumermeer at the end of the growing season. A large size ensures better escapement from predation and is one of the conditions for the formation of strong year-classes (Forney 1976; Willemsen 1977). This would mean that in the presence of enough small sized prey fish, especially smelt, the power station indirectly propagates the formation of strong year-classes of pikeperch. Entrainment mortality of pikeperch larvae early in the season depressed larval recruitment but a larger size at the end of the growing season is more important for the year-class formation.

To minimize the mortality caused by the power station it is better to focus on the period May–June. As long as the consequent higher effluent temperature is not detrimental to the organisms in the lake, less water intake during these months would be beneficial to the larval population in the lake. The natural lake temperature in May and June is low enough to prevent lethal effluent temperatures. Officially in the Netherlands, the discharge temperature may not exceed 30 °C and the difference in temperature between intake and outlet during the summer may not be greater than 7 °C. With a lower water intake by the power station, the water flow from the Prinses Margriet canal via the Bergumermeer to the N. would need to be constantly high. In this way the high water temperature of the discharge water in the Zwemmer would be cooled.

Acknowledgements

This work was partially supported by the Foundation for Fundamental Biological Research (BION), which is subsidized by the Netherlands Organization for the Advancement of Pure Research (ZWO).

We wish to thank Prof. E. A. Huisman, Dr. J. Vijverberg and Mr. J. Willemsen for their critical comments on the manuscript. We are grateful to G. H. F. M. van Aerssen and J. W. van der Stoep for their assistance in the field and for the calculation of the data of the fish at the power station.

References

Anonymous, 1976, 1977. Annual report 1975, 1976. Directorate of Fisheries, Ministry of Agriculture and Fisheries.

Beattie, D. M., Golterman, H. L. & Vijverberg, J., 1979. An introduction to the limnology of the Friesian Lakes. Hydrobiologia 58: 49–64.

Belyanina, T. N., 1969. Synopsis of biological data on smelt Osmerus eperlanus (Linnaeus) 1758. FAO Fisheries Synopsis No. 78.

Boytsov, M. P., 1971. The effect of warm water discharged by the Konakovo power station on the distribution and growth of young fishes of Ivan'kovo reservoir. J. Ichthyol. 11: 257–262.

Brown, D. J. A., 1973. The effect of power station cooling water discharges on the growth of coarse fish fry. Proc. 6th br. Coarse Fish Conf.: 191–202.

Dahlberg, M. D., 1979. A review of survival rates of fish eggs and larvae in relation to impact assessments. Mar. Fish. Rev. 41: 1–12.

Densen, W. L. T. van & Vijverberg, J., 1982. The relations between 0+ fish density, zooplankton size and the vulnerability of pikeperch, Stizostedion lucioperca, to angling in the Frisian lakes. Hydrobiologia 95: 321–336.

Effer, W. R. & Bryce, J. B., 1975. Thermal discharge studies of the Great Lakes – the Canadian experience. In: Environmental Effects of Cooling Systems at Nuclear Power Plants. Proc. Symp. Oslo, 26–30 August, 1974, pp. 371–388. I.A.E.C., Vienna.

Forney, J. L., 1976. Year-class formation in the walleye, Stizostedion vitreum vitreum, population of Oneida Lake, New York, 1966–1973. J. Fish. Res. Bd Can. 33: 783–792.

Hadderingh, R. H., 1975. Effects of the cooling water discharge on the macro-invertebrates and fish populations around Flevo Power Station. Verh. int. Verein Limnol. 19: 2214–2218.

Hadderingh, R. H., 1978. Mortality of young fish in the cooling water system of Bergum Power Station. Verh. int. Verein. Limnol. 20: 1827–1832.

Hadderingh, R. H., 1979. Fish intake mortality at power stations, the problem and its remedy. Hydrobiol. Bull. 13: 83–93.

Hadderingh, R. H. Experimental reduction of fish impingement by artificial illumination at Bergum Power Station. (in prep.).

Hess, K. W., Sissenwine, M. P. & Saila, S. B., 1975. Simulating the impact of the entrainment of winter flounder larvae. In: Saila, S. B. (Ed.) Fisheries and Energy Production: A Symposium, pp. 1–29. Lexington Books, Lexington, Mass.

Horoszewicz, L. & Morawska, B., 1979. Behaviour of fish in a thermally differentiated environment. In: Luczak, J. (Ed.) Pol. ecol. Stud. 5: 33–42.

Marcy, B. C., 1976. Planktonic fish eggs and larvae of the lower Connecticut river Yankee plant including entrainment. In: Merriman, D. & Thorpe, L. M. (Eds.) The Connecticut River Ecological Study – The Impact of a Nuclear Power Plant. Am. Fish. Soc. Monograph. 1: 115–139.

Nie, H. W. de, 1982. Effects of thermal effluents from the Bergum Power Station on the zooplankton in the Bergumermeer. Hydrobiologia 95: 337–349.

Schaap, L. A. & Willemsen, J., 1978. Verslag over bemonstering van eieren en larven van diverse vissoorten, in verband met koelwatergebruik door elektriciteitscentrales. 1974–1977. Report ZS 78-02 (in Dutch), Rijksinstituut voor Visserijonderzoek. IJmuiden. 34 pp.

Swartzman, G., Deriso, R. & Cowan, C., 1977. Comparison of simulation models used in assessing the effects of power plant-induced mortality on fish populations. In: Van Winkle, W. (Ed.) Assessing the Effects of Power-Plant-Induced Mortality on Fish Populations, pp. 333–361. Pergamon Press, New York.

Teleki, G. C., 1976. The incidence and effect of one-through cooling on young-of-the-year fishes at Long Point Bay, Lake Erie. A preliminary assessment. In: Esch, G. W. & McFarlane, R. W. (Eds.) Thermal Ecology II, pp. 387–393. United States Energy Res. Developm. Adm.

Tranter, D. J. & Smith, P. E., 1968. Filtration performance. In: Tranter, D. J. (Ed.) Zooplankton Sampling. UNESCO Monographs on Oceanographic Methodology. No. 2, pp. 27–56.

Wanders, J. B. W., 1979. Effects of thermal effluents from the Bergum Power Station on phytoplankton in the Bergumermeer. Hydrobiol. Bull. 13: 98–99.

Wilkonska, H. & Zuromska, H., 1977. Growth of fry in the heated Konin lakes complex. Roczn. Nauk. roln. 97: 91–111.

Willemsen, J., 1977. Population dynamics of percids in Lake IJssel and some smaller lakes in The Netherlands. J. Fish. Res. Bd Can. 34: 1710–1719.

Zawisza, J. & Backiel, T., 1972. Some results of fishery biological investigation of heated lakes. Verh. int. Verein. Limnol. 18: 1190–1197.

Temporal and spatial variations of dissolved oxygen concentrations in some Dutch polder ditches

R. Veeningen
Limnological Institute, 'Vijverhof' Laboratory, Rijksstraatweg 6, 3631 AC Nieuwersluis, The Netherlands

Keywords: dissolved oxygen, polder ditch, diurnal and spatial variations, macrophytes, filamentous algae, water quality

Abstract

This study deals with the temporal and spatial variations of dissolved oxygen concentration, temperature and occasionally pH in three polder ditches in the Netherlands. A routine programme including different environmental parameters served to characterize the ditch environment. Conclusions concerning water quality criteria are presented briefly.

Introduction

Until recently research studies with respect to pollution control and water quality management in the Netherlands was confined mainly to large water bodies, i.e. lakes, rivers, canals and streams. Therefore, most of the existing guidelines and standards have been designed for and tried out in these types of water. For many centuries artificial waterways, namely ditches and small canals, have been constructed for quantitative water management and especially for land reclamation in the western part of the country. These relatively small water bodies differ in many respects from the other surface waters. In all probability the water quality guidelines and standards, e.g. dissolved oxygen (DO) concentration, primarily designed as quality criteria for large water bodies, are not valid for ditches. Now the questions arise what criteria should be used to judge water quality in the ditches and how should they be applied. In the Netherlands water pollution is controlled legally by the Pollution of Surface Water Act of 1970. A large number of sewage treatment plants have since been built all over the country. Thus, generally speaking the pollution of surface waters has decreased. The policy in the past was aimed at preventing gross pollution and for special uses standards were set. The recent so called Water Action Program (1980–1984) of the Government is intended to give the water quality management an ecological base, in that the guidelines or standards should be relevant to the functioning of the aquatic ecosystems.

The present study is aimed at understanding the dynamics of the DO concentration and to apply it as a water quality standard in polder ditches. The oxygenbalance underlying these dynamics is an important measure of the functioning of the whole aquatic ecosystem, i.e. the biological, chemical and physical processes. The project is financed by the Ministry of Public Health and the Environment, the Netherlands.

Study site description

Polder ditches are artificial, more or less permanent, oblong bodies of water not exceeding 8 m in breadth and generally 0.5–1.5 m deep. The waterflow is artificial and temporary or periodic and in such a case it may constitute an ecological factor (see also de Lange 1972). The total length of the

Hydrobiologia 95, 369–383 (1982). 0018-8158/82/0953-0369/$03.00.

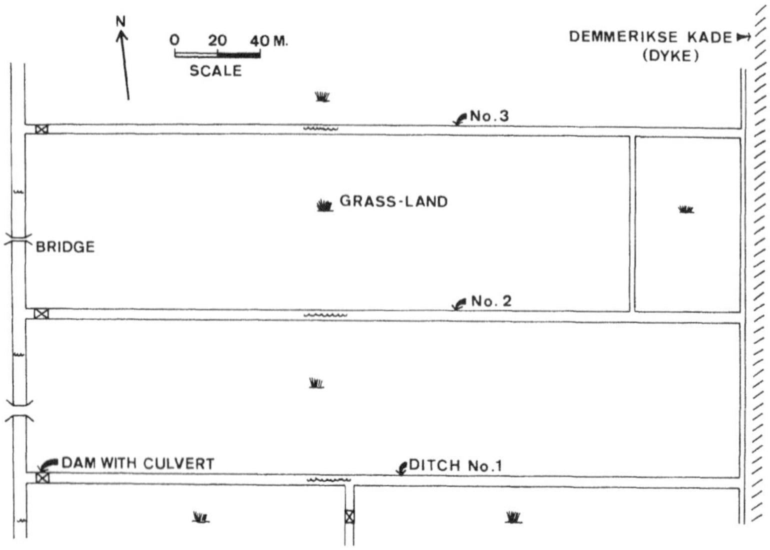

Fig. 1. Topography of the area of the three ditches (Nos. 1, 2 and 3).

ditches in the Netherlands is estimated at 3–4.10^5 km. Under normal management the ditches are cleaned chemically or dredged manually or mechanically at least once a year to facilitate normal inflow and discharge.

The three polder ditches chosen to study the temporal and spatial variation of the DO concentration are situated centrally in the Netherlands in the polder Groot-Wilnis-Vinkeveen (52° 13′N, 4° 55′E). The main morphometric data are: length, 320 m; width, 4–7 m; and depth 0.3–0.7 m (mean depth 0.5 m). They lie in grasslands and are interconnected by a canal at the west and a small ditch at the east (Figs. 1 and 2). The grasslands serve as cattle grazing grounds or for harvest of hay. Within a radius of about 1 km there are no agricultural

Fig. 2. The ditches 1, 2 and 3 from left to right. Photos taken in October 1981 from the dam.

Table 1 Dominant macrophyte species in ditches 1, 2 and 3 in 1980 and 1981.

Emergent	Floating leaved	Submerged
Sparganium erectum L.	*Nyphaea alba* L.	*Potamogeton pectinatus* L.
Scirpus maritimus L.	*Nuphar luteum* Sm.	*Ranunculus circinatus* Sibth.
Typha angustifolia L.		*Myriophyllum spicatum* L.
Alisma plantago-aquatica L.		*Elodea* spp.
		Chara spp.

farms nor any discharge of sewage. The average water level in the ditches is -2.12 m N.A.P. (Netherlands Ordnance Datum); it is artificially regulated. Generally the polder has a water deficit in summer and Rhine-water is supplied via the Amsterdam–Rhine canal and the polder reservoir. In autumn and winter there is a surplus of water which is pumped out in the polders reservoir. The chloride concentration and the conductivity were measured to get an insight into the water budget. They show high levels in the inflow. During the summers of 1980 and 1981 the water column was dominated by macrophytes and filamentous algae, the latter forming temporarily a thick mat at the surface at some locations; at other sites a thick algal mat occurred at the bottom. Dominant species of macrophytes in 1980 and 1981 are listed in Table 1. Except for *Chara* spp., all other species listed are very common in polder ditches. The ditches were cleaned mechanically in autumn 1979 and the macrophytes, filamentous algae and the upper sediment layer were removed and left on the banks. The cleaning in autumn 1980 was restricted to ditch 1 and the west part of ditch no. 2. The species composition and abundance of the macrophytes in 1981 was similar to that in 1980 except for *Chara* spp. which were absent in 1981. The location of the mats of filamentous algae in ditch 2 and 3 differed during the two years.

The variations in the presence, structure, biomass and physiological state of the macrophytes and filamentous algae imply a great spatial heterogeneity in the ditch environment. Characterization of the ditch as a single habitat is impossible (Caspers & Heckman 1981). The ditch ecosystem can in some respects be well compared with the littoral zone of lakes and with marshes.

Methods

Ditches 1 and 2 were sampled monthly at four locations in each ditch. The phosphates (3 fractions), ammonia, nitrate, chloride and alkalinity were determined according to Golterman *et al.* (1978). The chlorophyll-*a* concentration was determined by extraction with 80% ethanol and corrected for degradation products (Moed & Hallegraeff 1979). Dissolved organic matter was measured as absorbance at 250 nm of the 0.45 μm filtered water samples (de Haan *et al.*, in press), and compared with measurements using a TOC analyzer. Conductivity and pH were measured *in situ* with submersible electrodes (WTW, Federal Republic of Germany).

The DO concentration and temperature were measured *in situ* with an oxygen meter (Yellow Springs model 57) equipped with a submersible stirrer. The oxygen electrode was calibrated in air-saturated water according to the instructions given by the manufacturers. Oxygen concentrations referring to 100% saturation were derived from Standard Methods (1971). The DO concentrations was measured up to five times a day starting at just after sunrise up to just before sunset at approximately 40 locations in the three ditches from August 1980 to August 1981. The oxygen electrode was lowered to half way down the water column. Oxygen production and consumption in the free water were measured in light plexiglas and dark plastic syringes (100 ml volume). The DO concentration in the syringes was measured both at the start and after 4–5 h incubation by emptying the syringes in a flow-through-stirring cell (6 ml volume) in which the oxygen electrode is mounted (Veeningen 1980).

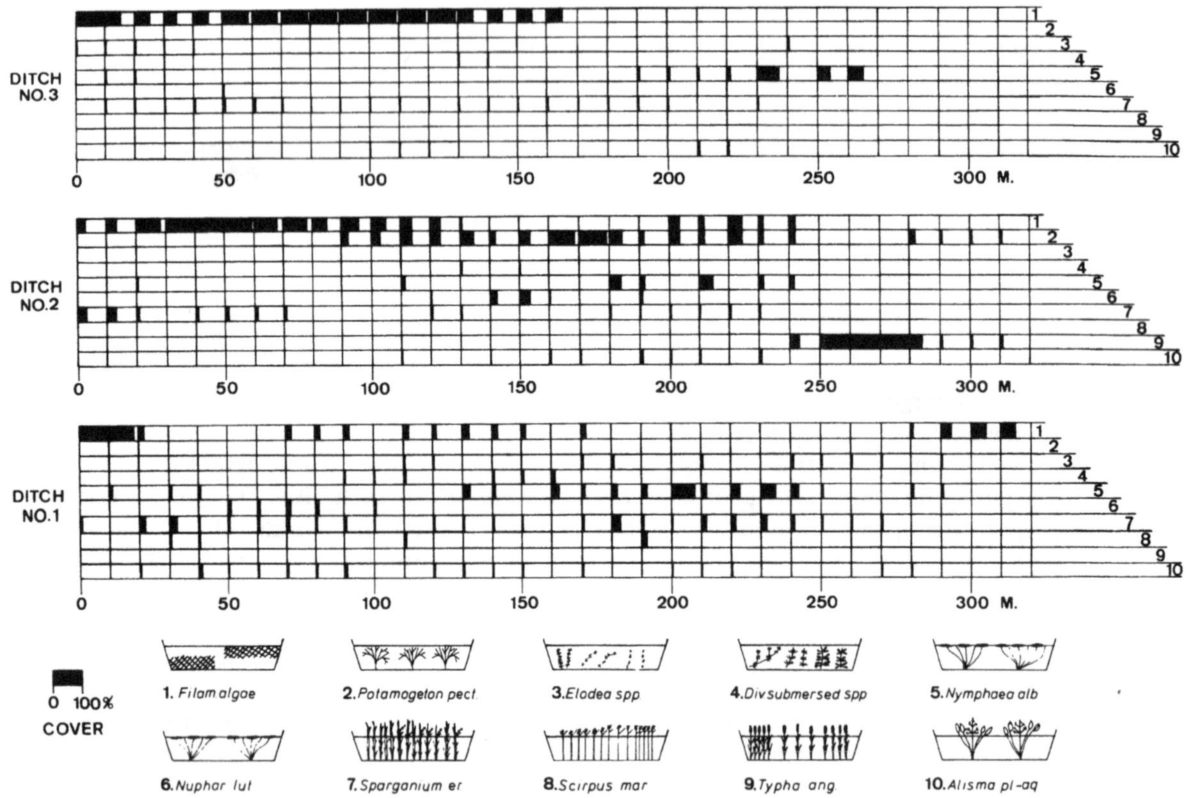

Fig. 3. The percent coverage by macrophytes and filamentous algae in the ditches; ditches 1 and 2 were surveyed in summer 1980 and ditch 3 in summer 1981. Each box represents a 10 m stretch in the length of the ditch. The cover percentage is determined on the basis of vertical projection of the species in the box which when fully shaded has a 100% cover. The physical structure of the vegetation is sketched below.

Results

Chemical and biological factors

The results of the monthly sampling summarized in this paper pertain to the period February 1980–August 1981.

Macrophytic biomass, though it forms an important part of the autotrophs, was not quantified. Figure 3 gives an illustration of the denseness and patchiness of the macrophyte vegetation in the summer period. Chlorophyll-*a* concentrations in ditch water indicate the abundance of phytoplankton (Fig. 4). In the spring period of both 1980 and 1981 there was a bloom of phytoplankton giving

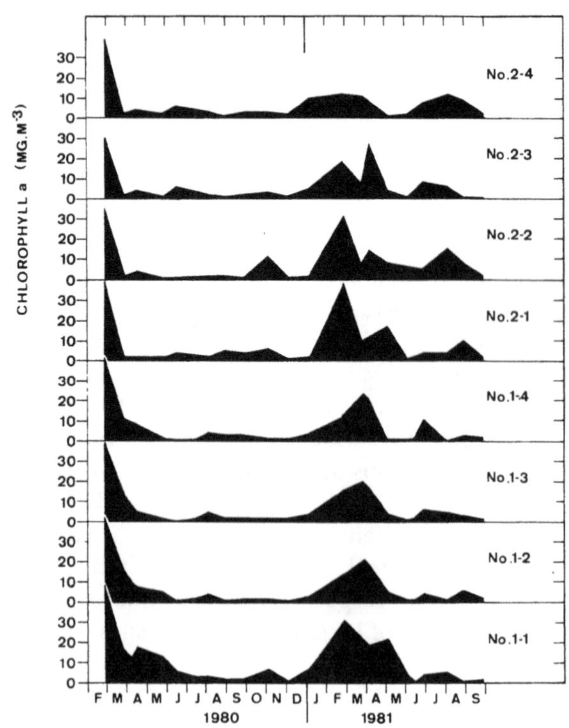

Fig. 4. Chlorophyll-*a* concentration ($\mu g \cdot l^{-1}$) in the free water in ditches 1 and 2 from February 1980 till October 1981. The numbers on the right concern the four points in each ditch, e.g. 1-1 is ditch 1, station 1 and so forth. The distance between each sampling point is about 90 m.

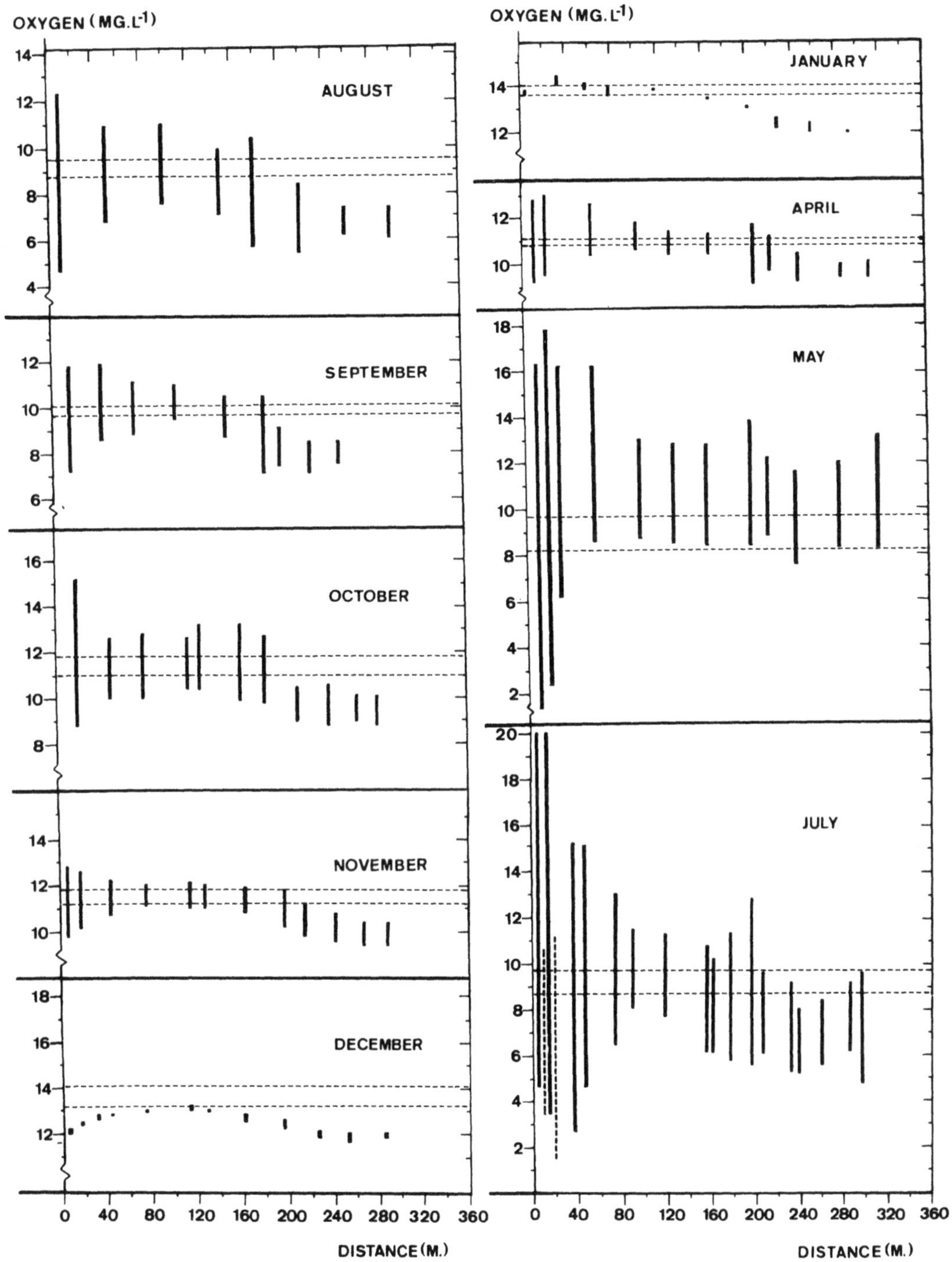

Fig. 5. Diurnal and spatial variation of DO concentration in ditch 1 from August 1980 to August 1981. The abscissae give the distances of the successive locations measured from the dam. The vertical bars represent the diurnal variation. The broken bars indicate the diurnal variation below a mat of filamentous algae. The two broken horizontal lines are the theoretical DO concentrations at 100% saturation for the average maximum and minimum temperature.

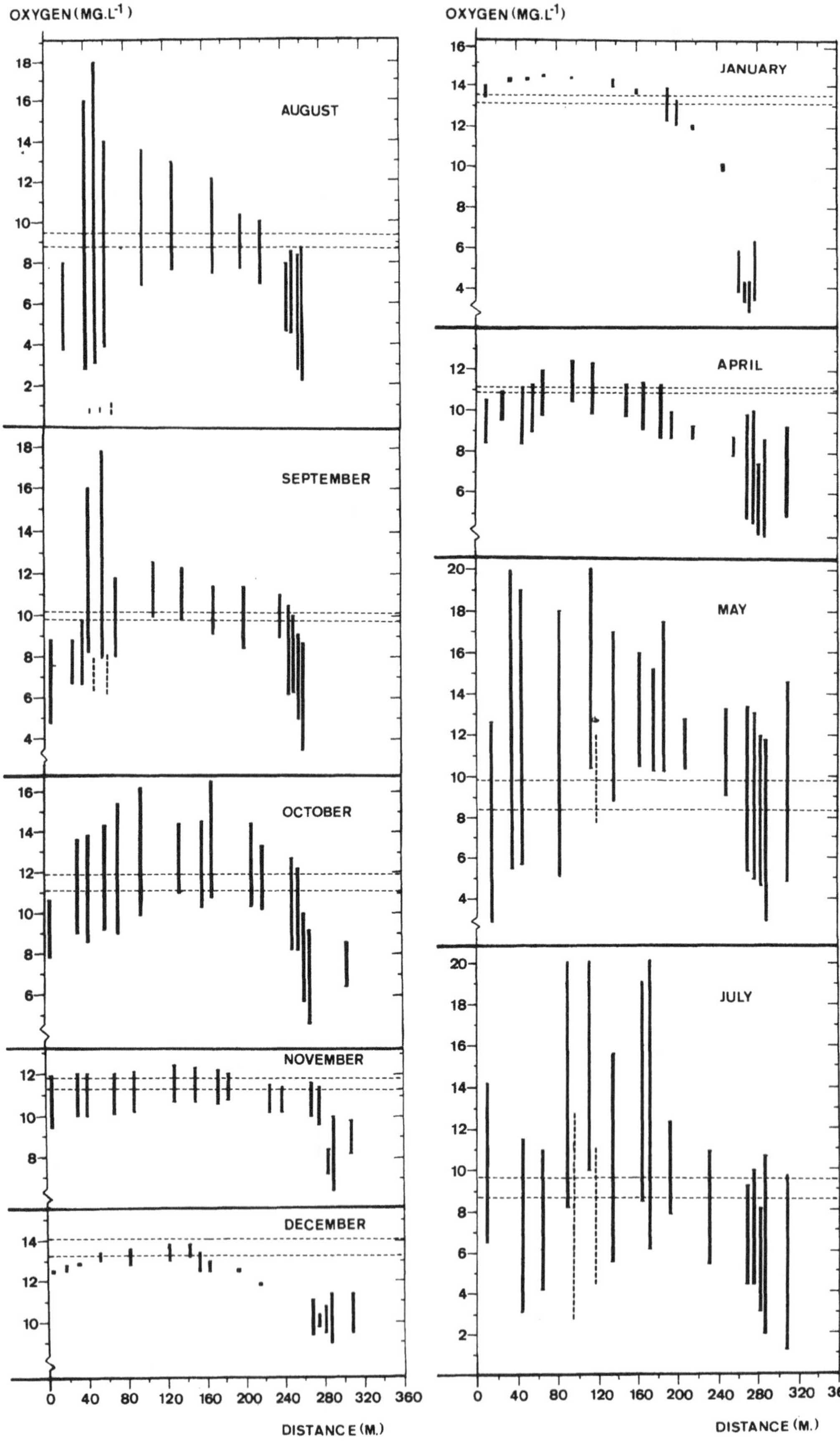

Fig. 6. Diurnal and spatial variations of DO concentration in ditch 2 from August 1980 to August 1981. For details see Fig. 5.

rise to a chlorophyll.*a* concentration of up to 50 $\mu g \cdot l^{-1}$. In the summer and winter periods the chlorophyll-*a* concentrations did not exceed 10 $\mu g \cdot l^{-1}$ and were even below 5 $\mu g \cdot l^{-1}$ during a prolonged period. Maximal chlorophyll-*a* concentrations of the sediment, measured occasionally, was about 50 mg $\cdot$ m^{-2}. The average concentrations of PO_4-P and Tot-P were 14 (range, 5–110) and 53 μg P. $\cdot$ l^{-1} (range, 20–160), respectively, and did not show a seasonal periodicity. The concentrations of both NO_3-N and NH_4-N were below 30 μg N $\cdot$ l^{-1} in the summer but increased tol 000 μg N $\cdot$ l^{-1} in the winter. Bicarbonate alkalinity varied from 1.5 to 3.0 meq $\cdot$ l^{-1}. The water in the ditches is intensively coloured due to dissolved humic compounds. The concentration of these compounds measured as ab-

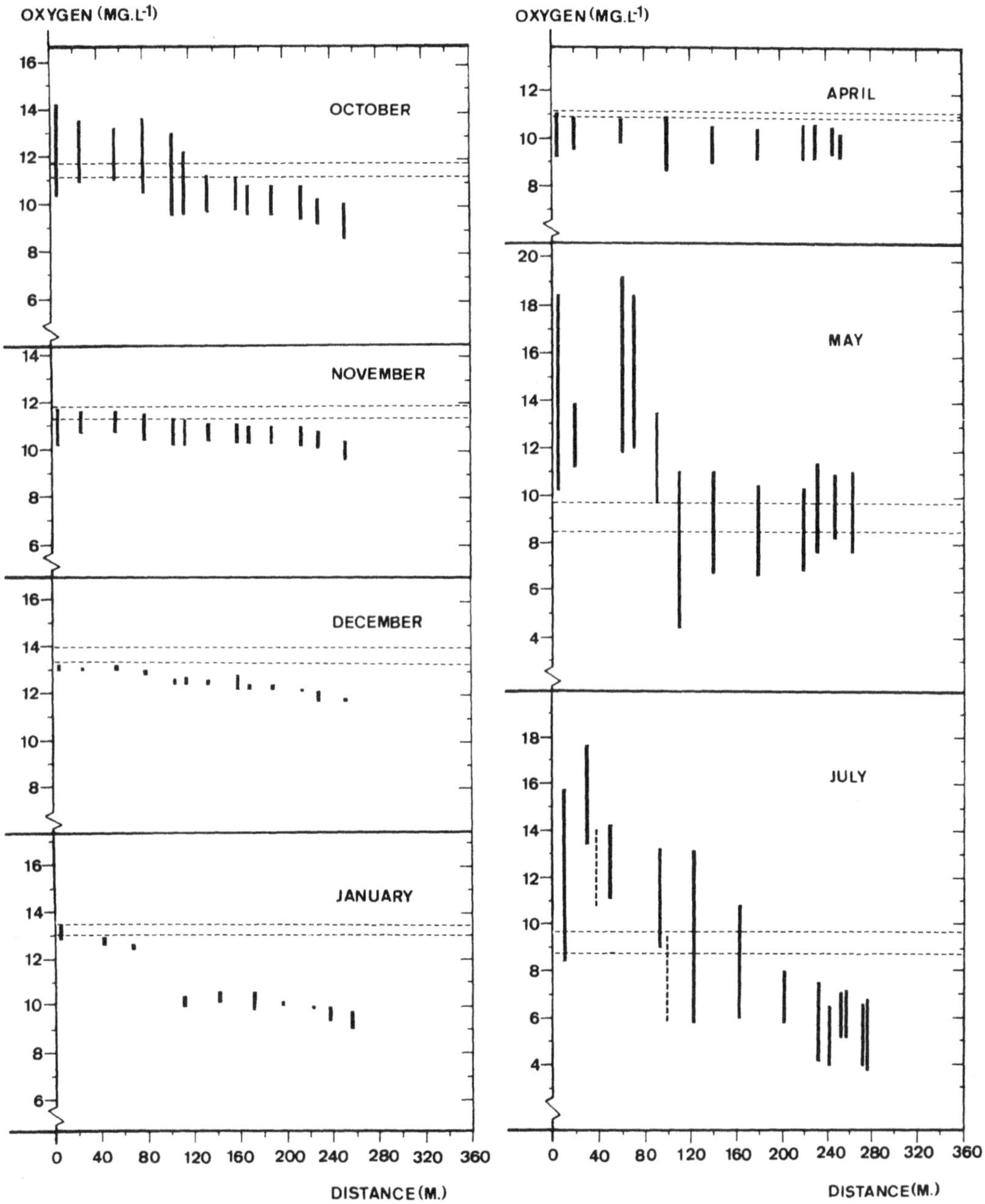

Fig. 7. Diurnal and spatial variations of DO concentration in ditch 3 from October 1980 to August 1981. For details see see Fig. 5.

sorbance at 250 nm (1 cm cuvette) ranged seasonally between 0.6 and 1.2 (extinction units). The DOC values varied from 10 to 30 mg C · l^{-1}, and showed good correlation with the absorbance values ($R^2 = 0.95$; n = 16). The light absorption in a water column without macrophytes or filamentous algae, measured occasionally, was 50% of the subsurface intensity at the bottom (depth, about 0.6 m).

In spring the chloride concentration was 80 mg Cl · l^{-1} but increased gradually to 240 mg Cl · l^{-1} in the course of the early summer. In autumn and winter the chloride concentration decreased to the level in the spring. The conductivity was related directly to the chloride concentration. In the early summer period (May–June) strong gradients of chloride and conductivity were observed in the longitudinal axis of the three ditches (0.2 mg Cl · l^{-1} · m^{-1} and 1 μS · cm^{-1} · m^{-1}; decrease to the east).

Dissolved oxygen

The diurnal and spatial variations in the DO concentration in the three ditches from August 1980 to August 1981 are shown in Figs. 5–7. The variations are the greatest in summer, but generally decrease in autumn and are reduced to less than 1 mg · l^{-1} in the winter when the water temperature is below 5 °C, except in ditch 2 during winter (Fig. 6). In most locations and on most dates there was an alternation of supersaturation (around sunset) and undersaturation (around sunrise). Undersaturation was observed throughout the 24 hours during the

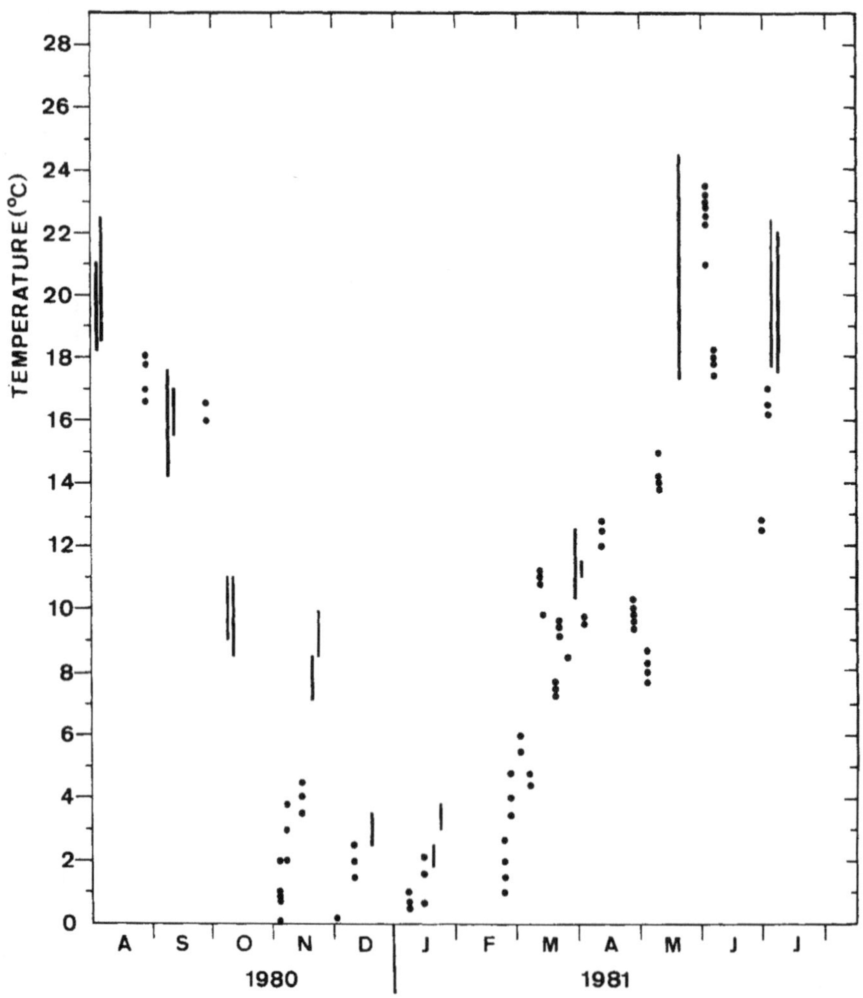

Fig. 8. Temporal and spatial variations of temperature in ditches 1, 2 and 3 from August 1980 to August 1981. The vertical bars refer to the diurnal variations during semi-continuous measurements presented in Figs. 5, 6 and 7. The dots refer to incidental measurements generally between 2 and 4 hours after sunrise, indicating the spatial variation on the days of sampling.

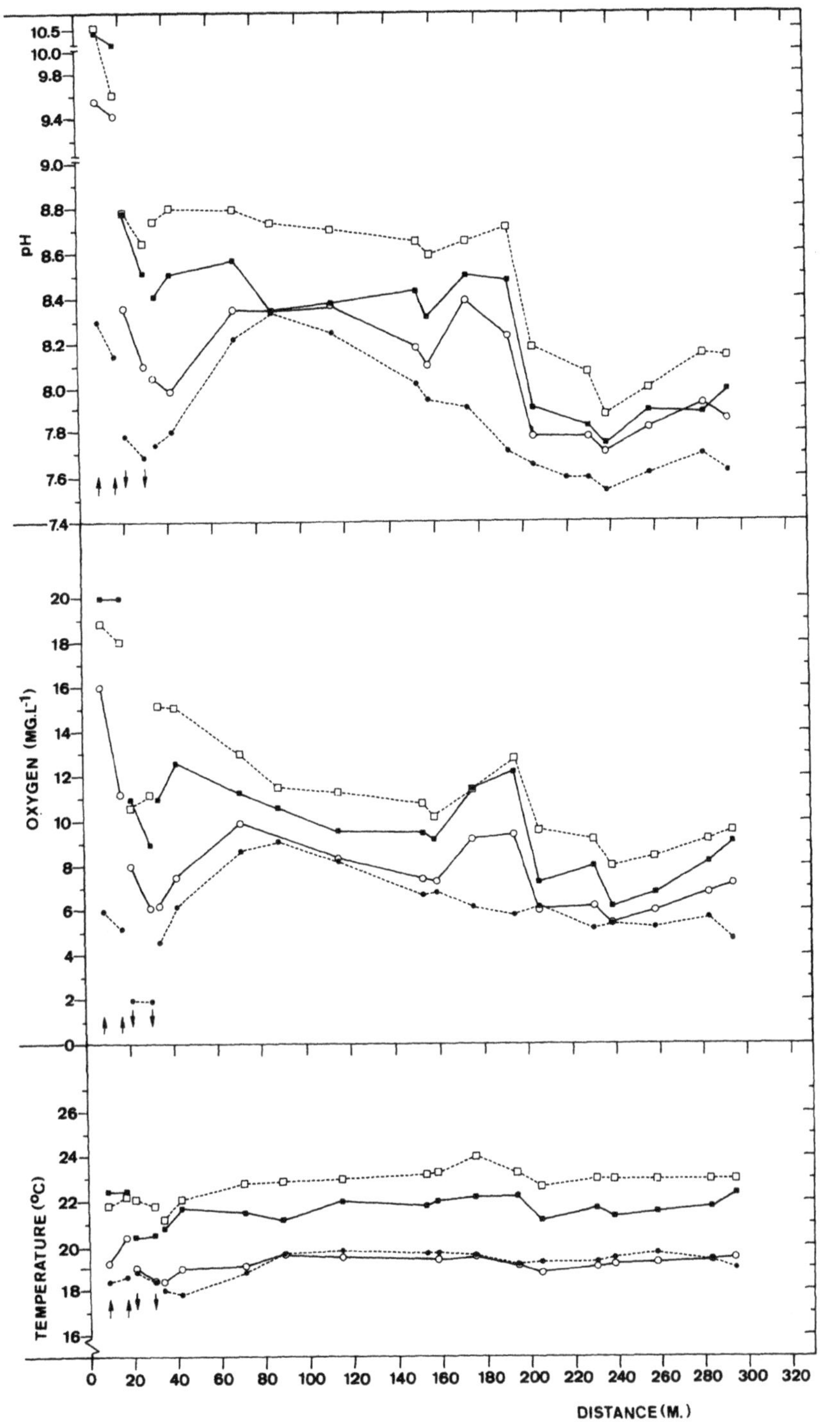

Fig. 9. The DO concentration, pH and temperature in ditch 1 (July 1981). The abscissae give the successive stations in the ditch. All parameters were measured simultaneously on four successive moments:
11.30–12.45 a.m. (7 July), open circles;
15.30–16.40 p.m. (7 July), closed squares;
20.55–22.15 p.m. (7 July), open squares and
06.45–08.00 a.m. (8 July), closed circles. (Sunset on 7 July: 22.01 p.m.; sunrise on 8 July: 05.29 a.m.). The arrows indicate the locations with a mat of filamentous algae (upwards directed: at the top of the mat; downwards: below the mat).

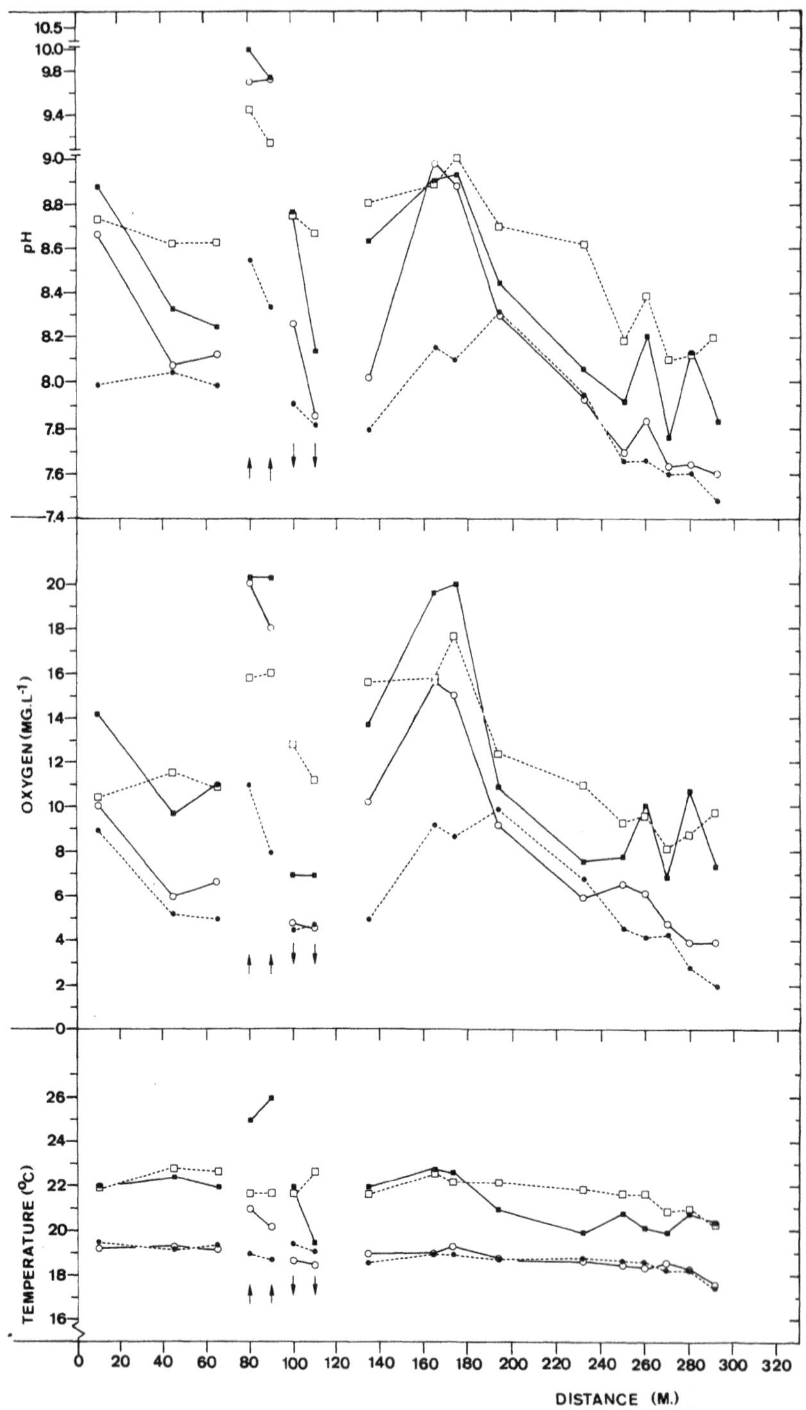

Fig. 10. The DO concentration, pH and temperature in ditch 2 (July 1981). For details see Fig. 9.

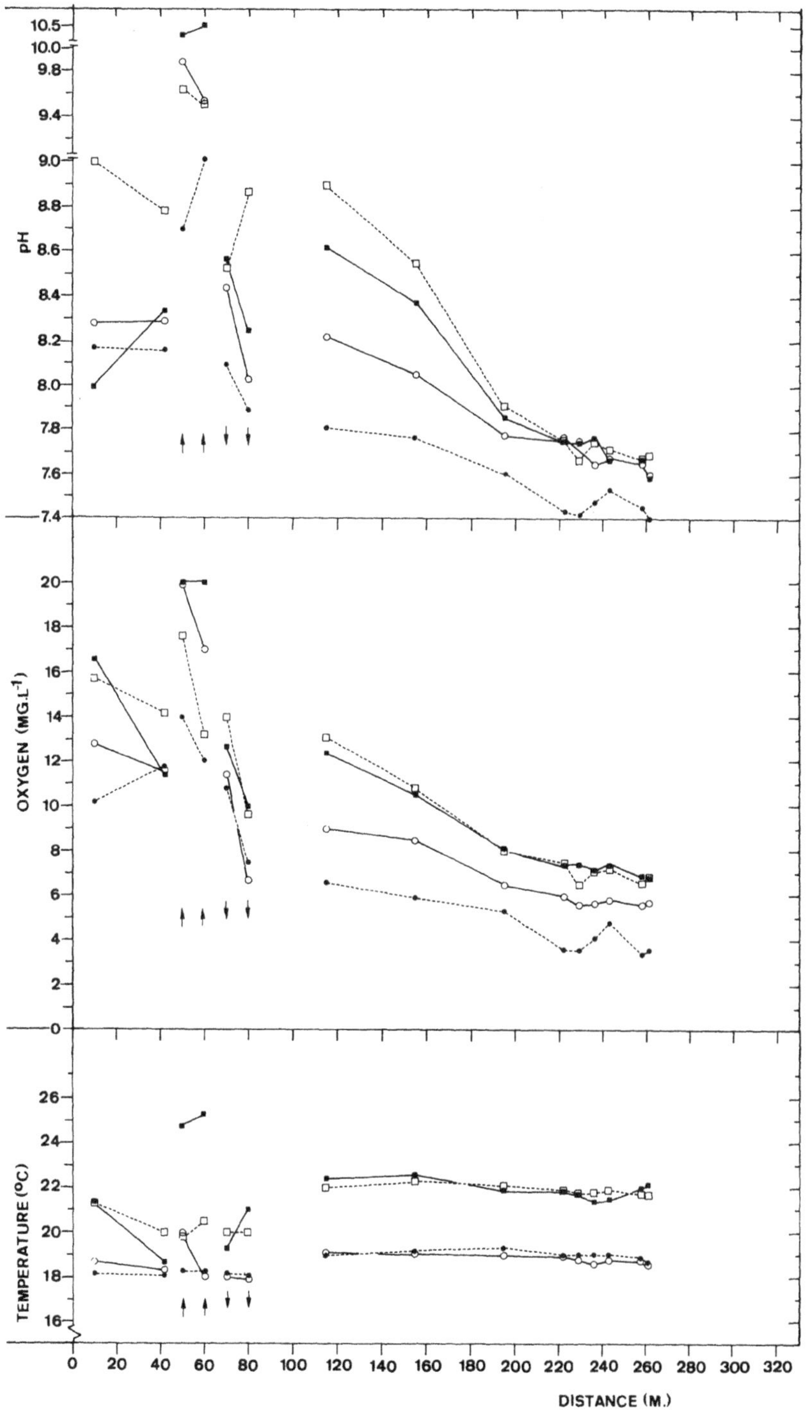

Fig. 11. The DO concentration, pH and temperature in ditch 3 (July 1981). For details see Fig. 9.

whole year in the east part of ditches 1 and 3 stretching from 200 tot 320 m (Figs. 5 and 7) except in May 1981.

Supersaturation over the whole period of measurement occurred only in ditch 2 in January 1981 (Fig. 6). In summer, strong vertical gradients were observed at sites with thick mats of filamentous algae at the water surface. At stations without filamentous algae vertical gradients were rare and occurred only in the late afternoon when wind was calm and the water surface screened by emerged vegetation.

The temperature, like the DO concentration, showed the greatest diurnal variations in summer (Fig. 8) but the spatial variations were less pronounced, except at locations with mats of filamentous algae. There was no long period with ice-cover in the winter of 1980–1981.

The results of the simultaneous measurements of DO, pH and temperature in July 1981 are shown in Figs. 9–11. On most locations the diurnal and spatial variations of DO were simultaneous with those of pH.

The rates of oxygen production and consumption in the free water were hardly detectable with the DO electrode, indicating that these rates were below 0.025 mg · l^{-1} · h^{-1}. Only in the spring when chlorophyll-*a* concentration reached high levels are the data reliable; net production and respiration rates were, respectively, 0.2–0.3 and 0.05–0.10 mg · l^{-1} · h^{-1} (chlorophyll-*a* concentration = 80 μg · l^{-1}).

Discussion

General

A polder ditch can be considered as an open ecosystem, with inlet and discharge of water and the relatively large drainage area being important for the water balance. Because of the small depth of ditches and thus the large surface–volume ratio the bottom acts as an important sink and source for seston mass and nutrients. The small depth of the water column implies that situations with complete and continuous darkness are rare. In most ditches there is an intensive growth of macrophytes which form a substrate for other organisms. Because of shallowness, the water mass is exposed intensively to the diurnal and seasonal changes in temperature and other climatic conditions and sensitive to drainage and run-off water from the surrounding grasslands – a 10% increase of the water volume after a heavy rainfall is not uncommon. All these factors contribute to the complexity and dynamics of a ditch ecosystem. Therefore, in studies of such an ecosystem a good insight into the climatic and hydrological changes in the environment as well as frequent and detailed measurement of the parameters in the ditch itself are indispensable.

Chemical and biological factors

The macrophyte and filamentous algae were mapped continuously. They appear to predominate the plant biomass in the three ditches. However, it is practically impossible to quantify macrophyte biomass on an areal basis, viz. dry weight or kg · C · m^{-2}, because of the both vertical and horizontal heterogeneity. The phytoplankton could be quantified easily, and was very low but for the bloom in spring. The macrophytes, especially the *Potamogeton pectinatus* L., showed a rich population of epiphytic algae. Intact chlorophyll-*a* was detected also at the sediment surface. Thus, the photosynthetic microorganisms in the ditches are highly underestimated if only the chlorophyll-*a* content in the free water is used as a measure. It appears as though the phytoplankton is 'filtered out' by the macrophytes. The phytoplankton settle on the macrophytes and at the sediment where, light and nutrition conditions, for example, might be favourable compared with those in the free water.

The nutrients in the water that can be taken up directly by the autotrophs are low in concentration in summer when they may limit primary production in the water phase. The high concentrations of ammonia and nitrate in winter may be due to their mineralization in the ditches or to input via the run-off and drainage water. The phenomenon was also observed in ditches by Beltman & Vink (1980) and Caspers & Heckman (1981). For phosphorus there are no indications of mineralization in the ditch or input from allochthonous sources. However, the nutrients were only analysed in the water phase; the sediment can contribute significantly to the supply of nutrients, especially for the macrophytes, most of which derive them from the sediment.

From the close relationship between the absorbance of the filtered water at 250 nm and DOC concentrations it is likely that a major part of the latter is of humic origin. It is known that humic compounds do not break down easily. Their origin is from the decay of plant materials, the leaching of the peaty soil or both. Because of their intensive brown colour the humic compounds contribute greatly to the absorption of light in the ditches.

The water economy of the ditches is under strong influence of polder management. During dry periods, when there is a water deficit, water is pumped into the polder. Rhine water is generally used for this in the area where the ditches are located and is marked by a relative high chloride concentration. The inlet which is started in May can be followed easily by following the courses of the chloride concentrations and conductivity. There are no indications that dissolved organic matter or mineral nutrients are introduced with the inlet water. This is explained by the long distance covered by the inlet water before it reaches the ditches, its velocity and the retention capacity of the macrophytes, phytoplankton and sediments. The sharp longitudinal gradients of the chloride concentration and the conductivity when the inflow of Rhine water is started indicate a very weak current in the ditches.

Dissolved oxygen

The most striking aspect of my research in the polder ditches is the enormous spatial and temporal variations of the DO concentration. Spatial variations occur along the length and in some cases along the breadth and vertically. Temporal variations are observed especially during day and night but also seasonally.

The variations in DO, both spatial and temporal, can be explained by the metabolic activity of the autotrophic and heterotrophic populations and by the changes in the climatic conditions as solar radiation and temperature. The biomass of both the macrophytes and the filamentous algae and the climatic factors all have their maxima in summer and minima in winter. The variations of the DO concentration confirm this pattern. Because of the enhanced solubility of oxygen with decrease in water temperatures, the DO concentration is generally higher in the winter.

The diurnal variation of temperature, like that of DO, is the greatest in summer and reduced in winter. The spatial variation may be explained by the water depth, shading by trees or emergent macrophytes, and stratification at the locations with mats of filamentous algae.

The spatial and temporal variations of DO concentration coincide with those of the pH. The changes in the pH are caused by a shift in the carbonate–bicarbonate equilibrium which in turn is affected by the uptake and release of inorganic carbon. It seems the buffering capacity is so low that it causes the apparent pH changes. Similar relationships between DO and pH were observed in orchard drainage ditches (Caspers & Heckman 1981), in a drainage canal and small ponds (Reddy 1981), in mangrove waterways (Boto & Bunt 1981) and in a littoral community (Kairesalo 1980).

The spatial and temporal variations of DO concentration, pH and to a lesser extent of the temperature reflect the heterogeneity in the ditches resulting from the heterogeneous growth of macrophytes and filamentous algae. The changes in the physiological state and physical structure of the biomass must also be taken into account. The biomass of phytoplankton was low and explains partially the low oxygen metabolism in the free water. The dissolved organic matter that consists mainly of humic material which does not break down easily contributes to the low oxygen consumption rates. Thus on most locations the contribution of both the phytoplankton and the bacteria in the free water to the oxygen balance is negligible.

Studies on the contribution of the sediment to the oxygen balance are in progress. Being rich in photosynthetic and heterotrophic organisms the upper layers of the sediment will be a major factor, especially where the water column is small. The sediment metabolism may explain the diurnal variations of DO at locations without macrophytes and filamentous algae. However, the metabolism of the macrophytes and filamentous algae cause the variations in the DO concentration in most sections of the ditches.

The exchange of oxygen between water and atmosphere depends on several factors and is not easy to quantify (Wilson & Macleod 1974). It can be assumed that but for the actual DO concentrations the factors determining the rate of exchange are generally the same at all locations in the ditches. Therefore quantification of the exchange cannot help much in explaining the spatial and temporal variations.

As remarked in the introduction, the ditches can be best compared with shallow waters which are more or less dominated by macrophytes. The magnitude of the temporal and spatial variations of DO concentration was explained by the amount of macrophytes and filamentous algae in Caspers & Heckman (1981), Reddy (1981) and Kairesalo (1980). Neuhuber & Hammer (1979) also studied the variation in a littoral community and Simonsen & Harremoës (1978) in a small river. However, the magnitude of spatial and diurnal variations of DO over relatively short stretches as reported here has not been reported earlier.

Water quality criteria

With regard to the standards for water quality management, it must be remarked that the spatial and temporal variations of DO concentration reflect clearly the functioning of the ditch ecosystem. However, the results presented here are quite different from those that have led to the DO standards for large surface waters. In large surface waters the dynamics of the DO concentration reflects the metabolism of the free water only, as against the ditches where it is a reflection of total metabolism, i.e. of the activity of both the macrophyte and sediment community and of the free water. Although spatial and temporal variations of DO can also be observed in great surface waters, it is obvious that incidental measurements of DO concentration make no sense at all in the ditches because of the enormous variations. Likewise, incidental measurements cannot be used to discriminate between the less or more polluted ditches. The results and conclusions based on the study of the three polder ditches can be extrapolated to the other polder ditches which are similar in their composition and abundance of macrophytes and filamentous algae. Whether the results of this study can be applied to the ditches in general needs to be examined.

Summary

Ditches in general but especially those in the polders represent a large area and volume of surface water in the Netherlands. However, when water quality criteria are set, little or no account is presently taken of this type of surface water. In order to explore useful criteria, a study was made of the temporal and spatial variations of dissolved oxygen concentration together with those of water temperature and occasionally pH in three polder ditches (dimensions: length $\pm$ 320 m; breadth 4–7 m and depth 0.3–0.7 m). A routine analysis of the different environmental parameters served to characterize the polder ditch environment. The diurnal and spatial variations of dissolved oxygen are large in summer (up to 10 mg $\cdot$ l^{-1}) but much smaller in winter (up to 1 mg $\cdot$ l^{-1}). The variations are the highest (up to 15 mg $\cdot$ l^{-1}) at locations with mats of filamentous algae; this is true also to a lesser extent for the variations of temperature. The temporal and spatial variations of the dissolved oxygen concentration are coupled with those of pH, and reflect the heterogeneity and dynamics of the ditch ecosystem in both quantitative and qualitative ways. The causal factors are: the changes in metabolism, in biomass and in species of macrophytes and filamentous algae. Conclusions concerning the water quality criteria are presented briefly.

Acknowledgements

Thanks are due to Dr. R. D. Gulati for reading the English text and for valuable comments, to Mr. E. M. Mariën for photographic work, and to Mr. R. H. Kwist and Mr. G. L. S. Tripels for assistance in the field and at the laboratory.

References

American Public Health Association, 1971. Standard Methods for the Examination of Water and Waste Water, 13th edn. APHA, New York, 874 pp.

Beltman, B. & Vink, J., 1980. Onderzoek in sloten, een schakel in het effectenonderzoek in het landschap. Meded. Werkgem. landschapsekol. Onderz. (in Dutch) 7: 41–45.

Boto, K. G. & Bunt, J. S., 1981. Dissolved oxygen and pH relationships in northern Australian mangrove waterways. Limnol. Oceanogr. 26: 1176–1178.

Caspers, H. & Heckman, C. W., 1981. Ecology of orchard drainage ditches along the freshwater section of the Elbe Estuary, Arch. Hydrobiol./Suppl. 43 (Untersuch. Elbe-Aestuar 4) 4: 347–486.

Golterman, H. L., Clymo, R. S. & Ohnstad, M. A. M., 1978. Methods for Physical and Chemical Analysis of Fresh Waters. IBP Handbook no. 8, 2nd edn. Blackwell Scientific, Oxford. 214 pp.

Haan, H. de, Boer, T. de, Krammer, H. A. & Voerman, J., (in press) Applicability of light absorbance to estimate organic carbon in Dutch lake water. Wat. Res.

Kairesalo, T., 1980. Diurnal fluctuations within a littoral plankton community in oligotrophic Lake Pääjärvi, southern Finland. Freshwat. Biol. 10: 533–537.

Lange, L. de, 1972. An ecological study of ditch vegetation in the Netherlands. Ph. D. thesis, University of Amsterdam. 112 pp.

Moed, J. R. & Hallegraeff, G. M., 1978. Some problems in the estimation of chlorophyll-a and phaeopigments from pre- and post-acidification spectrophotometric measurements. Int. Revue ges. Hydrobiol. 63: 787–800.

Neuhuber, F. & Hammer, L., 1979. Oxygen conditions. In: Löffler, H. (Ed.). Neusiedlersee: Limnology of a Shallow Lake in Central Europe, pp. 121–130. Dr. W. Junk, The Hague.

Reddy, K. R., 1981. Diel variations of certain physico-chemical parameters of water in selected aquatic systems. Hydrobiologia 85: 201–207.

Simonsen, J. F. & Harremoës, P., 1978. Oxygen and pH fluctuations in rivers. Wat. Res. 12: 477–489.

Veeningen, R., 1981. Project polder research. In: Parma S. & Gulati, R. D. (Eds.) Progress Report Limnological Institute, Verh. kon. ned. Akad. Wet. Afd. Natuurk. Rks 2, 77: 36–38.

Wilson, G. T. & Macleod, N., 1974. A critical appraisal of empirical equations and models of the coëfficient of reaeration of deoxygenated water. Wat. Res. 8: 341–366.

GPSR Compliance
The European Union's (EU) General Product Safety Regulation (GPSR) is a set of rules that requires consumer products to be safe and our obligations to ensure this.

If you have any concerns about our products, you can contact us on

ProductSafety@springernature.com

In case Publisher is established outside the EU, the EU authorized representative is:

Springer Nature Customer Service Center GmbH
Europaplatz 3
69115 Heidelberg, Germany

www.ingramcontent.com/pod-product-compliance
Lightning Source LLC
Chambersburg PA
CBHW080934240926
51849CB00003BA/454
9789061937623